Handbook of
Organic Farming and Biofertilizers

NIPA® GENX ELECTRONIC RESOURCES & SOLUTIONS P. LTD.
New Delhi-110 034

Handbook of Organic Farming and Biofertilizers

A.C. Gaur

NIPA® GENX ELECTRONIC RESOURCES & SOLUTIONS P. LTD.
New Delhi-110 034

NIPA® GENX ELECTRONIC
RESOURCES & SOLUTIONS P. LTD.
101,103, Vikas Surya Plaza, CU Block
L.S.C. Market, Pitam Pura, New Delhi-110 034
Ph : +91-11-43860225, Mob.: +91 9717133558, 9540816132
E-mail: newindiapublishingagency@gmail.com
Website: www.nipaersources.com

Print ISBN: 978-93-58877-29-8
ebook ISBN: 978-93-58874-70-9

Composed and Designed by NIPA®.

M S Swaminathan Research Foundation
3rd Cross Street, Institutional Area, Taramani, Chennai 600 113, India

Foreword

This book by Dr A.C Gaur on Organic Farming and Biofertilizers is a very timely one, since there is growing interest in organic farming both in India and globally. Organic farming based on the use of techniques of soil and plant health management without the application of mineral fertilizers and chemical pesticides, requires a considerable amount of scientific underpinning. We must standardize methods of maintaining and enhancing soil productivity in perpetuity without associated ecological harm. This is a pathway to an Evergreen Revolution which is essential to enabling farm families with small holdings to produce adequate marketable surplus for a reasonable income. Dr. Gaur has given extensive information on methods of improving soil health based on the application of biofertilizers. green manures and cereal-legume rotation. Organic Farming needs more science and not less. Scientific Organic Farming is not the same as "traditional farming" characterized by low productivity. We must remember that when no mineral fertilizers and pesticides were applied, our country was experiencing frequent famines. Between 1870 and 1900 nearly 30 Million children, women and men died out of hunger. Therefore we need methods of enhancing productivity without adverse long term environmental consequences. This is why the present book by Dr. A.C. Gaur is a timely one. I hope it will be widely read and used.

M.S. Swaminathan
Chairman
M.S. Swaminathan Research
Foundation Chennai

Preface

There is increasing awareness all over the world regarding the need for establishing of organic and sustainable system of agriculture. The humus and organic fraction of tropical and subtropical soils are under constant threat and depletion due to environmental and inadequate replenishment. The maintenance of a good soil fertility and the regular use of organic manures and bioiertilizers are of great importance to keep up the farm in good state of productivity involving natural processes in its place.

This book highlights various sources of organic and biological agents which can be used for organic farming and as well as for maintenance of soil fertility over years'. Organic manures are byproducts of processes of decomposition of organic materials available both in rural and urban areas. Organic manures have been used by the farmers from the beginning of agriculture. But it was neglected in India during the past two to three decades.

The production of bioiertilizers is a new development and they are produced by isolation, selection and culturing on mass scale of specific microorganisms such as bacteria or fungi. Nitrogen fixing and phosphate solubilizing microorganisms are produced as biofertilizers arid made available to farmers to supplement nitrogen and phosphorus needs of the crops with either organic or chemical fertilizers. Moreover, these inputs are inexpensive and eco-friendly.

The potential, technology and uses of different types of organic sources such as animal wastes, biogas slurry, poultry and sea food wastes; agricultural, tree and urban wastes; wastes from agro-industries such as oil cakes, sugar industry byproducts, canning and food processing, fermentation and tannery wastes have been highlighted. The characteristic, potential and use of rock phosphate, basic slag and flyash have also been covered.

The concepts of organic Agriculture, scientific basis and technologies for composting oi organic wastes and the underground biodiversity as basis for biofertilizers have been dealt in this book. Nitrogen fixing biofertilizers such as Rhizobia, Frankia, Azolla, blue-green algae, Azotobacter and Azospirillum and P-solubilizing bacteria and Mycorrhizae involved in nitrogen and phosphorus nutrition of crops have been selectively discussed.

The production technology of biofertilizers and methods and media for analysis of organic manures have been subsequently outlined and discussed. The topics have been convered in 21 chapters and would be useful to farmers, students, extension workers and scientists. I believe that the book in its present form would convey the readers a clear and useful information on the topics covered.

The author wishes to express deep sense of gratitude to Dr. M.S. Swaminathan encouraging me to bring out this book and for writing a Foreword by sparing his valueable time from his busy schedule. My special thanks are due to the Secretary, Department of Science and Technology, Ministry of Science and Technology, New Delhi for supporting this project. I am thankful to Dr. A. K. Singh, Energy Research Applications, New Delhi for his cooperation in this endeavour. The author is grateful to his family members in particular to Shri Amitabh Gaur and Dr. Anupam Gaur for encouraging me to complete this work.

Delhi **A.C. Gaur**

Contents

1

Introduction and Basics

I. Introduction

India has been deeply concerned about the food security on the one hand and health and welfare of its people on the other. Agriculture has been traditionally the backbone of Indian economic progress and development. The role of agriculture including animal husbandry, dairying and fisheries accounts for some 33 per cent of the country's gross domestic product (GDP) and about 65 per cent of the work force depends directly or indirectly on agriculture for its livelihood and support. Along with the food grain, the production of vegetables and fruits has also been increased to meet the nutritional security of the people.

India is the second most populated nation with a geographical land area of about 329 million hectares (Table 1). The gross cropped area is 185.5 million hectares with a net sown area of about 142.5 million hectares. Currently area under assured irrigation is estimated to be 50.10 million hectares which is 35.15 percent of the cultivated area which has progressively risen to this level due to the expansion of major and medium irrigation projects including harnessing of ground water reserves excluding tanks (Shantaram, 1998). A substantial part of the net cultivated area which works out to be about 92.3 million hectares is rainfed depending on the vagaries of rains. The forest area (68.07 million hectares) only 22.3 per cent which is much less than the desired level because the national forest policy demands it at least 66 per cent in hills and 33 per cent in plains. However, it is not feasible to divert land for forest plantation due to obvious population pressure on agricultural land. However, the advantages of forest can be harnessed by growing trees on marginal lands alongwith crops through agroforestory practices for restoring ecological balance.

Table 1: Distribution of Land Use in India

	1950-51	1970-71	1990-91	1992-93
1. Geographical Area	328.73			
II. Reported Area for Land Utilisation Statistics (1 to 5)	284.32	303.76	305.02	305.01
1. Forest	40.48	63.91	67.98	68.07
2. Not Available for Cultivation (A+B) A. Area under Non-Agricultural use B. Barren and unculturable land	47.52 9.36 38.16	44.64 16.48 28.16	40.84 21.24 19.60	41.26 21.88 19.38
3. Other uncultivable land excluding fallow land (A+B+C) A. Permanent pasture and other grazing lands B. Land under not included in net area sown C. Culturable waste	49.45 6.68 19.83 22.94	35.06 13.26 4.30 17.50	30.56 11.80 3.96 15.07	29.61 11.30 367 14.64
4. Fallow Lands (A+B) A. Fallow land other than current fallows B. Current fallow	28.12 17.44 10.65	19.88 8.76 11.12	23.39 9.58 13.81	23.56 9.65 13.91
5. Net Area Sown (A-B) A. Total cropped area (gross cropped area) B. Area sown more than once	118.75 131.89 13.14	14027 16579 25.52	142.25 185.91 43.66	142.51 185.49 42.98
6. Cropping intensity	111.10	118.20	130.69	130.16
III. Net Irrigated Area	20.85	31.10	47.78	50.10
IV. Gross Irrigated Area	22.56	38.11	62.47	66.14

Source: Agricultural Statistics at a Glance, Directorate of Economics and statistics, Department of Agriculture and Co-operation, Ministry of Agriculture, Government of India, March, 1996.

Instability of Indian Agriculture

India agriculture has made commendable progress in the last four decades, particularly after mid sixties, leading to more than three time increase in irrigation potential, about 200 times increase in fertilisers consumption, enormous increase in the area under high yielding cultivars and a record food grain production of about 200 million tonnes. Despite these impressive achievements not only the yield levels of several crops are lower than those in many other countries (Table 2) but also marked fluctuations are witnessed in different years and regions largely due to weather aberrations, poverty ridden high population growth and land degradation of various kinds. Significant positive shifts in productivity of rice and wheat are largely responsible for a quantum jump in food grain production but these crops *are* also showing a declining trend in productivity in rice-wheat cropping system inspite of application of recommended levels of chemical fertilisers indicating the deterioration in soil productivity (Gaur, 1998).

Table 2: Area, production, and yield of cereals and pulses in some major production countries.

Country	Cereals			Pulses		
	Areas (million ha)	Prodn. (million t)	Yield (kg/ha)	Areas (million ha)	Prodn. (million t)	Yield (kg/ha)
India	101.4	199.6	1969	22.1	12.9	582
Egypt	2.5	14.7	5869	0.2	0.5	2308
Canada	20.1	49.1	2450	0.6	0.9	1612
Mexico	10.1	24.7	2438	1.4	1.0	766
USA	65.9	352.3	5360	0.7	1.2	1668
Argentina	8.2	24.8	3026	0.2	0.3	1272
Brazil	20.7	44.2	2138	5.3	2.8	539
Bangladesh	10.8	28.5	2640	0.7	0.5	719
China	91.1	400.4	4397	4.4	6-5	1478
Indonesia	14.3	55.8	3908	0.4	0.5	1302
Pakistan	11.5	22.1	1916	1.4	0.7	481
France	9.3	60.4	6516	0.7	3.4	4623
Germany	6.5	34.7	5334	0.1	0.2	2102
Netherlands	0.2	1.3	7454	<1.1	0.1	3643
Spain	7.4	14.4	1949	0.3	0.2	688
UK	3.5	22.2	6365	0.2	0.7	3177
Australia	14.1	24.9	1764	1.6	1.7	1059
World	699.6	195.2	2791	67.5	57.5	852

1. Area of pulses is 0.014 millions ha.
Source: FAO (1992).

Approximately 194 million ha (about 60 percent of total land area) are degraded to a greater or lesser degree because of water and wind erosion, waterlogging, floods, soil salinization etc. in India. Land degradation of physical, chemical and biological nature is affecting the welfare of millions of people. The problem of land degradation started with human cut the first tree to practice arable farming. Anonymous (1927) has narrated the history of severe erosion in various parts of India. For example, extensive destruction of natural teak forests of west coast of Bombay Presidency (now Maharashtra state) prior to 1927 resulted in denudation of Western Ghats and Deccan Plateau, excessive erosion and gradual silting up the rivers. Another such example is the erosion in the hills of the state of Punjab and Uttar Pradesh.

II. What is Sustainable Agriculture?

Soil and water are basic resources essential for sustainable agriculture. But unfortunately very few people realise the importance of conserving and judiciously utilising the soil as a basic resource by successfully managing the soil both internal, external, renewable or non renewable inputs. Sustainable agriculture is defined as the successful management of resources for agriculture production to satisfy changing human needs while maintaining or enhancing the quality of the environment and conserving natural resources (CGIAR, 1989). The Key words found in the above definition are (a) resource management (b) changing human needs (c) environmental quality and natural resources. While keyword (b) is involved in the concept of agricultural productivity keyword (c) is involved in the protection of the environment concept. Both concepts are combined into keyword (a) which refers to resource management. Therefore, the basic concept of sustainable is to define the combined concept of productivity (or profitability) and environmental soundness in farm management or in agriculture at the national level.

World commission on Environment and Development (1987) defined sustainability as "a system which can be considered sustainable if it ensures that today's economic development is not at the expense of tomorrow's development prospects".

Sustainability is not a fixed phenomenon, it changes with time, requirement, places and so on. A system, which is sustainable today, may not be sustainable tomorrow and vice versa. The Rio declaration of 1992, while enunciating many principles for sustainability of environment and development, emphasised the following principles most relevant to sustainable agriculture.

- Human beings are at the centre of concern for sustainable development and environmental issues are best handled with the participative co-operation of all concerned.
- No development can be considered complete unless it meets the criteria of productivity, equity and environmental safety for the present and future generations.
- To ensure development and sustainability, it is necessary to remove all the negative factors leading to unsustainability.

Environmental policies must form an integral part of development policies and strategies.

Environment

The definition of "sustainable land use" is in consonance with "safe environment" or "environmental protection", because harmonization of all the land uses in a given area guarantees safe environment conditions in the sense of a safe protected habitat for human and other living organism". Therefore, sustainable land use creates and guarantees a safe environment by definition (Blum, 1994).

The conventional method of farming helped the developing countries to raise the agricultural production but it has shown some undesirable effects such as discouragement of mixed cropping, minimal tillage practice, dominance of mineral fertilizers and chemical pesticides, lesser use of organic, crop residue, green manures and biological inputs/approaches, neglects, of rainfed and semi-arid areas and crops.

Recently; the areas of sustainable agriculture has been receiving attention of agricultural scientists and policy makers. This will, hopefully, lead to the development, testing and refinement of technologies for sustainable farming and development. Such data and specific models are very much needed because present information is based on scattered reports, scientific analysis and possibilities.

Sustainable Agriculture has several positive points such as lesser use of pesticides and chemical fertilizers because when these are used in excess not only pollute the environment but cause health hazards contaminating food grains, vegetables, fruits, fodder and milk. However, sustainable agriculture should not be equated with subsistence farming and perpetual low yields.

III. Area of Research and Development in Sustainable Agriculture

The foremost emphasis should be on the improved management of biological system, based on a better understanding of their feedback and balancing process. Integrated plant nutrient supply (IPNS) system and integrated pest management (IPM) are appropriate examples to substitute for heavy reliance on chemical-based applications. The other important emphasis should be on better information management, implying the need for sound data on natural resources, land use and farming systems, etc. to improve environmental monitoring capability and make better use of natural resource potential. Finally the emphasis should be on improved farm-household system management, in order to obtain a better understanding of the differing systems while considering in an integrated way household farm and off-farm activities.

Moving to a more operational level, two central thrusts can be identified, each having a set of priorities. One thrust is aimed at promoting sustainable increase in productivity in the higher potential areas. The second should target on marginal and fragile environments where soil degradation must be reversed and production stabilized or raised. These must be supplemented by two cross-cutting and highly complementary approaches, that of rehabilitation and restoration of ecology, and that of exploiting the synergism of indigenous technical knowledge and modern science. All four actions must be supported by specific efforts to strengthen the national agricultural research system, both institutionally and financially, since they will have to undertake much of the adaptive and applied research.

The priorities for making progress toward sustainable growth in productivity include: expanding on-farm/biological production and recycling of farm inputs and lowering the needs and use of mineral fertilizer and pesticides; raising crop yield targets; improving irrigation, ameliorating soil acidification; using more efficiency and promoting renewable energy sources; and efficient use of labour in some multiple cropping systems.

On farm biomass production or recycling of inputs and use of microbial inoculants can serve three main purposes, (Gaur, 1998b), First, they can provide small-scale farmers with a profitable alternative to high cost external input system, which though effective in technical terms, incur financial and environmental risks, Second, they can help to prevent soil nutrient mining and the excessive build up of mineral fertilizer and pesticides residues in soil and ground and surface waters. Third, through the greater use of cereal and leguminous mulches, green manures and other organic residues, they can be improve soil organic matter, microbial biomass, soil structure, maintain soil fertility and enhance the soil's role as sink for carbon dioxide. This requires a much deeper knowledge of agro-ecosystem function, intercropping and so on. It is difficult to imagine, for example, how some of the complex intercropping and relay planting system currently used in India and China to achieve three crops a year, which involve a lot of hard work, can survive in a higher labour cost environment. Research priorities include nutrient recycling process and techniques, natural resources management at village level, and integrated crop/livestock management system.

Raising yield ceiling in difficult environments has been an issue for a number of years. There is now a better understanding of the issues involved, and progress is being made with some crops like millets and legumes. Similarly, there is renewed emphasis on raising yield ceiling of staples food crops in the high potential areas, which have been the centre of the successes of the "green revolution" (the irrigated rice and wheat areas of Asia) and where experimental yields seems to have reached a plateau and have bee virtually static for the past ten years or more (Pingali *et al.*, 1990). Although average farm yields in these areas are still well below experimental yields, and so yield levels could continue to rise up to 2010, thereafter there would be rapid declines in yield growth unless research managers breakthrough the plateau and bring about another shift in the production frontier (Hayani and Otsuka, 1991). The challenge has become a major priority for IRRI and other research institutions, but their efforts must be backed up by national action. Recent widespread introduction of hybrid rice in some countries of Asia, first for temperate regions but subsequently for tropical conditions, offer great promise of significantly raising the yield ceiling of rice. Priority research include: (a) production of crops varieties with enhanced tolerance or resistance to moisture stress and soil constraints; (b) studies to overcome micro-nutrients deficiencies; and (c) investigating soil conditions under continuous intensive crop production as well as under low inputs conditions.

A common outcome of more intensive land use, whether it be through high or low external input systems, is increasing competition between crops and weeds for the available soil nutrients and water, and for light. The conventional responses are either more hand weeding or more herbicides. Labour for the former is becoming increasingly scarce and expensive, and the increasing quantities of pesticides are growing threat to the environment and human health. The key priorities are (a) greater emphasis on weed management; (b) increased research on biological control methods and biodegradable herbicides; and (c) researching innovative ways to reduce herbicide applications.

Fulfilling the energy needs of agriculture and/or rural services is at the core of improving productivity, income-generation and overall well-being of farmers. Land preparation, harvesting, irrigation and processing require different types and levels of energy inputs, both in a direct (mechanical, thermal and electrical energy) and indirect forms (fertilizer and manures). Without these energy inputs, agricultural productivity will remain low and probably well below its full potential. At the same time, unsustainable energy system based on renewable energy- agricultural linkages, and to promote sustainable energy system based on renewable energy sources, mainly biomass, solar and wind energies, the potential of agriculture itself as an energy producer requires further studies and research into the use of biomass residues, energy plantations and combined energy and food production schemes. Key priorities are: (a) evaluation of the energy- agriculture interrelationship for different ecosystem, and in relation to high and low potential areas; (b) better understanding of the integrated management of energy and other inputs (water, fertilizer, pesticides, mechanization); and (c) assessment of potential of biofuels for different environmental and land-use policy situations.

It is increasingly acknowledged that there are mainly indigenous techniques for on-farm or local water conservation that can be used now or quickly adapted to complement the above actions. Maintaining progress after 2010 will, however, requires more basic and applied actions research in the next two decades. Priorities include: (a) development of minimal tillage system for low-income farmers in the drylands of developing countries; and (b) methods to improve pastures in the extensive range areas, both under tropical and temperate conditions.

While there are many opportunities for developing sustainable lower inputs, systems, complete independence from external inputs is rather difficult except in a few special circumstance. More research is therefore required on lowering the unit cost of external inputs, or achieving the same objective by raising input use efficiency or reducing needs through innovative ways of overcoming the factors that currently impose marginality. Some of the opportunities are well illustrated by the widespred problem of phosphate deficient soils. Phosphate is an essential chemical for plant growth but many soils are highly deficient in it, or the present phosphate is not available to plant because of other soil factors: e.g. aluminum or iron toxicity. Organic manures are seldom the long-term solution if high yields are to be achieved while manure made from biomass grown on phosphorous- deficient soils will itself be deficient in phosphorus. Therefore, biomass low in phosphate can be composted with rock phosphates to produce P-enriched manure (Gaur, 1998).

The conventional productions of phosphate fertilizers is expensive and when long distance transport costs are added they tend to become even more uneconomic for farmers in marginal areas. Many countries, however have low-grade phosphate rock or other phosphate bearing materials that could be used if cheap methods could be found for treating them so that phosphate becomes readily available to plants. Alternatively, and in the longer term, there seems to be possibilities for genetically transferring to other crops the properties of pigeon peas to release bound phosphate in the soil and made them available for plant growth. Without such technological breakthroughs, however, sustainable development in many marginal areas will be impossible. Research priorities, therefore include: (a) development of inexpensive techniques for improving the effectiveness of low-grade phosphate bearing rock such as incorporating organic matter and promoting P-solubilising bacterial/fungal activity in the soil and the role of mycorrhizae, (b) determination of the pigeon peas mechanism of releasing bound soil phosphate, and possible transfer of the mechanism to other crops, and (c) promote phosphate solubilising microorganisms as biological inputs.

Increasing competition between crop and livestock production for land in both high potential and marginal areas, and the threat to crop yields from declining soil fertility, will be a positive force for integrating the two system. This will, however, intensify the demand on research to come up with more satisfactory solutions to a number of problems and opportunities, notably : (a) reduction in the labour needs and the other constraints to the adoption of livestock oriented alley cropping and other stylo-pastoral system taking into account competition for water, light and plant nutrients (b) development of legumes based relay, intercropping, soil conservation, and other practices to raise the supply of high protein feeds and forages: and (c) introduction and refinement of ley farming systems for arid and other low fertility soil based on legume-grass pastures.

The foregoing indicates about the needs and opportunities for laying the foundation for sustainable agriculture in the 21st century. Unless technological progress is accompanied by institutional changes and the other improvements in the incentives for agricultural development, many of the research finding will not leave the laboratory or experimental stations or be developed on farm themselves.

Policy on Agricultural Development

The achievements on the agricultural front in India have been possible through implementation of a systematic five-year planned development having successfully completed nine five-year plans, the country embarked on the tenth five-year plan in 1996. On the basis of the review of the performance of the eight five-plan, the Planning Commission has already finalized the priority areas for development, the various target sectors, and the resources allocation for the tenth five-year plan in consultation with the concerned ministries of the government of India and state governments. Based on country's requirements and realizing the urgency of increased agricultural production towards economic prosperity of India, the government has committed itself to invest fifty percent of its expenditure on the agricultural and rural sectors.

In order to attain faster and balanced regional growth of agricultural and to overcome interregional disparities and inequitable development, agro-climatic zonal planning has been recently introduced. This strategy involves uniform development of all regions in the country utilizing natural resources and covering all sector of the farming community. Based on homogeneity of agro-climatic zones and attributes such as rainfall, temperature, soil topography, and cropping system of the zones, the country has been divided into 15 agro-climatic regions.

Within each zone, sub zones have been demarcated and detailed exercise have been made for balanced agricultural development in a holistic way including farmers' participatory approach. Constraints and the type of developmental program needed in future have been identified for sustainable agricultural development of the country. Having considered the requirement of food grain and other agricultural commodities for the projected population of

1,000 million by the year 2000, and beyond some important projections are mentioned in Table 3.

Table 3: Important Targets ad Projections by the year 2000

	Unit	**2000**
Population	Millions	1,000
Total geographical area	Million hectares	329
Net sown area	Million hectares	150
Gross irrigated area	Million hectares	84
Gross cropped area	Million hectares	200
Food grains production	Million tons	240
Oilseeds production	Millions tons	25
Cropping intensity	Percent	133
Gross area under H^{g}V	Million hectares	61
Area under food grain crops	Million hectares	123
Per capita availability of land	Million hectares	0.17
Fertilizer consumption (N + P + K)	Million tons	20.0
Gross area under food grain		
(a) Rice	Million hectares	32.00
(b) Wheat	Million hectares	17.55
(c) Other cereals	Million hectares	48.50
(d) Pulses	Million hectares	25.00
Total		**123.05**
Yield		
(a) Rice	Tons/hectare	2.5
(b) Wheat	Tons/hectare	2.9
(c) Pulses	Tons/hectare	1.4
(d) Oilseeds	Tons/hectare	1.0

Source: S. K. Saxena, 1993

Increased irrigation facilities, adoption of high-yield varieties (HYV), and better management practices have also contributed to this growth (Table 3). Productivity of practically all crops has increased.

The success attained in food production in India following the Green Revolution has not been without costs. Most of these costs relate to varied responses which the Green Revolution evoked in different crop and region. Area under paddy and wheat has continuously increased in many states at the cost of coarse cereals, millets, pulses and in some area, cotton due to the technological supports and price supports that were provided which made paddy and wheat production profitable. The important cost of this development are serious imbalances in the cropping pattern, widened regional disparities and increased instability in production.

IV. Plant Nutrients

Plants require at least 13 mineral elements for their growth and development besides carbon, hydrogen and oxygen. All these essential nutrients are present in all soil but one or more of these invariably is not present in optimum amounts of plant-usable form which make external additions necessary (Table 4). Since each element has certain specific functions to perform, one mineral can not substitute for another. If a soil is deficient in even one nutrient, it can not support a good crop unless deficiency of this element is made up. This is essentially where mineral fertilizers and other sources of nutrients such as organic manure and microbial inoculants are recommended for application.

Table 4: Essential mineral nutrients their forms absorbed by plants, concentration in dry matter and major functions.

Nutrients (Symbol)	Forms absorbed	Contents in dry matter	Major functions in the plant
Macronutrients			
Nitrogen (N)	NH^{+}_{4}, NO^{-}_{3}	1-5%	Growth, protein production
Phosphorus (P)	$H_2PO^{-}_{4}$, $HPO^{=}_{4}$	0.1-0.4%	Energy compound, root development, ripening.
Potassium (K)	K^{+}	1-5%	Crop quality, water regulation, stress tolerance.
Sulphur (S)	$SO^{=}_{4}$	0.1-0.4%	Protein and oil production.
Calcium (Ca)	Ca^{++}	0.2-1%	Cell structure, cell division.
Magnesium (Mg)	Mg^{++}	0.1-0.4%	Part of Chlorophyll, energy transfer.
Micronutrients			
Boron (B)	$H_3BO_3^{-}$	6-60 ppm	Cell wall development, seed setting
Chlorine (Cl)	Cl^{-}	0.2-2%	Tissue hydration, photosynthesis
Copper (Cu)	Cu^{++}	5-20%	Enzyme activity, chlorophyll formation.
Iron (Fe)	Fe^{++}	50-200 ppm	Respiration, chlorophyll formation.
Manganese (Mn)	Mn^{++}	20-500 ppm	Enzyme activation, photosynthesis.
Molybdenum (Mo)	MoO_4	<1 ppm	Nitrate reduction, N-fixation.
Zinc (Zn)	Zn^{++}	25- 150 ppm	Enzyme activity, protein synthesis

Diverse source of nutrients can be gainfully used as long as they are able to furnish the nutrient in readily or potentially usable forms. If a plant root is about to absorb N as ammonium or nitrate, to the best of our knowledge it neither knows nor distinguishes between it's source be it a fertilizer, FYM, green manure or microbial fertilizer. Such differences are taken care of in nutrient cycle which are working in all soils and inorganic sources enter the same cycle (fig. 1). In soil, organic and inorganic forms are inter-convertible through mineralization and immobilization process mediated by soil microorganisms.

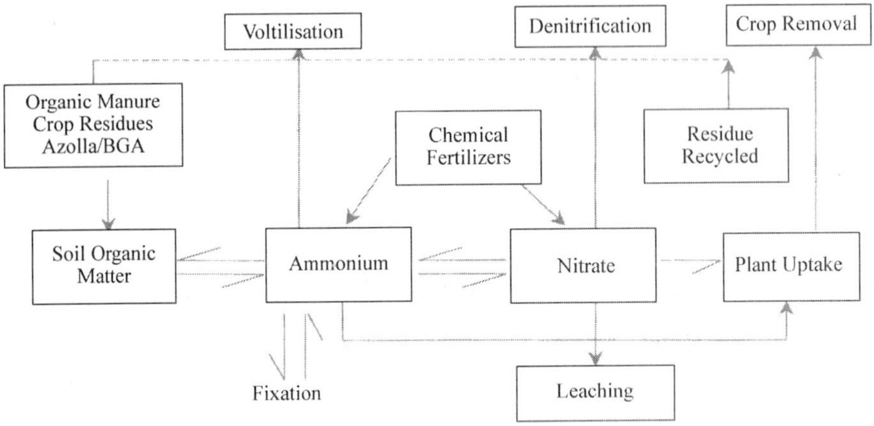

Fig. 1: A simplified version of the nitrogen dynamics through organic and mineral inputs in crop nutrition.

Fertility Status of Soil

The fertility status of soil indicates their nutrient supplying capability. When seen alongwith crop demand, this is the major determinant whether and to what extent / nutrient input is required. Only a fraction of the total nutrients present in soil exist in readily or potentially usable form on a short medium term basis and therefore knowledge of available nutrient status of soils is very valuable in planning nutrient needs of agriculture. It serves as an indicator of soil fertility and to a fair degree can be estimated by getting soil sample tested in a good laboratory.

The overall fertility status of a soil has to be assessed for each nutrient separately. Thus a composite fertility profile of a field is made up of sub-profiles, one of each nutrient. At present, deficiencies of five nutrients are of large scale practical importance in India and therefore, much of the focus is on N, P, K, S and Zn. An overall picture of fertility status of soil is provided in Table 5. The fertility status of soil which are categorized as low is the poorest and they can not support even moderate crop yields without nutrient addition.

Table 5: Nutrient status of Indian soils and the deficiencies

Nitrogen	Low in 228 districts, medium in 118, higher in 18 districts
Phosphorous	Low in 170 districts, medium in 1 84, higher in 17 districts
Potassium	Low in 47 districts, medium in 192, higher in 122 districts
Sulphur	Deficiencies scattered in 100-120 districts
Magnesium	Kerala and other southern states, very acid deficient, soil
Zinc	50% of 150,000 soils analyzed found deficient
Iron	On upland calcareous soils for rice, groundnut, sugarcane
Boron	Parts of Bihar, Karanataka, West Bengal

In reality, available nutrient status of a soil does not have alone a meaning but is demand-dependent. A soil, which has medium fertility for raising a 3 tonnes crop will be low for growing 6t/ha grain. A soil, which is medium for local varieties may turn out to be low (deficient) for high yielding varieties (HYV). It is for this reason that soil, test-crop response correlations assume importance and soil, crop and production potentials are to be taken into account while developing soil fertility classes.

Field experiments can be conducted to know the crop yields a soil can support on its own nutrients reserve. The rate at which soil fertility declines can be estimated from long-term experiments. Results from such experiments show that virtually no soil can support medium-high yields indefinitely without external nutrient additions (Nambiar and Ghosh 1984) unless this is done, available nutrient status can decline from sufficiency to deficiency levels. Even if this happens for one nutrient, the crop is not spared from its adverse effect.

All available data show that the nutrient reserves of Indian soils are being depleted (Tandon 1987, 1991 a,b Tandon and Sekhon 1988) and nutrient status of soil is declining. This is primarily due to negative balances caused by nutrient removals which persistently exceed nutrient additions. Besides, chemical fertilizers, plant nutrients, demand is met from soil reserves, organic manures/compost and biofertilizers as the main inputs for fertilization.

Nutrient Uptake and Removal by Crops

To produce each tonne of grain, cereal crop absorbs 50-65 Kg of N + P_2O_5 + K_2O plus lesser amounts of other plant nutrients. Absorbed nutrients are differentially distributed in different plants parts. While 70-80% of absorbed N and P and 40-60% of S end up in the grains, 70-80% of potash remains in the straw/stover. The distribution of absorbed nutrients varies largely with the nature of nutrients in questions and crop species and to lesser extent with growth conditions. Such distribution has important implications from organic recycling for nutrient supplement which would be maximum in case of K and least for N and P.

Some estimates of nutrients absorbed by crop in relation to yields are given in table 6. Nutrient uptakes by intensive crop rotations can be very high, approaching 600-900kg/ha/yr of N + P2O + K2O (Tandon and Sekhon 1988). The extent of nutrient removal from a field depends on the relative composition of main and bi-produce, extent of leaf fall, residue retained at harvest and those recycled later. Actual studies on these aspects in quantitative terms are not many.

Table 6: Average nutrient removal by diverse crops under fields conditions.

S.No.	Crop	Produce	Kg/tonne Produce		
			N	PA	K_2O
1.	Rice	Paddy	20.1	11.2	30.0
2.	Wheat	Grain	24.5	8.6	32.8
3.	Sorghum	Grain	22.4	13.3	34.0
4.	Chickpea	Grain	46.3	8.4	49.3
5.	Pigeonpea	Grain	63.8	17.7	42.3
6.	Groundnut	Grain	58.1	19.6	30.1
7.	Mustard	Grain	32.8	16.4	41.8
8.	Cotton	Seed	44.5	28.3	74.7
9.	Jute	Fibre	23.5	13.0	41.7
10.	Sugarcane	Cane	1.7	0.2	2.0
11.	Potato	Tuber	3.9	1.4	4.9
12.	Cassava	Tuber	7.8	1.2	5.1
13.	Berseem	Fodder	8.6	1.4	3.8
14.	Fodder grasses	Dry fodder	9.4	1.5	14.2
15.	Grapes	Fruit	4.9	1.5	5.9
16.	Banana	Fruit	8.2	3.0	32.3
17.	Tomato	Fruit	4.1	1.5	5.9
18.	Tea	Made tea	110.0	38.0	44.0
19.	Coffee	Clean beans	37.0	6.0	47.0
20.	Tobacco	Leaf (dm)	17.0	5.0	26.0
21.	Coconut	1000 nuts	7.1	3.5	10.7
22.	Java citronella	Herb	0.8	0.3	1.0

V. Mineral Fertilizers

Fertilizers are mined or manufactured commercial products which contain one or more essential plants nutrients. For a mineral to qualify as a fertilizer, it should contain nutrients in appreciable amount and in reality or potentially usable form. In addition, a fertilizer (plant-food carrier), just like any food item should not contain or produce any substance, which is toxic to the soils, plant or humans above permissible limits. Many countries including India have laws as to what can or can not be labeled as a fertilizer.

Fertilizers are used with the sole purpose of improving soil fertility so that it can support larger yields. Fertilizer represents the conventional source used by farmers to replenish plant nutrients into soils to ensure that adequate nutrients are available to the crop. Plant roots do not absorb fertilizer granules as they come out of the bag or as they are in manure or compost heaps. Plant absorb nutrients in specific ionic forms which either a fertilizer furnishes when it dissolves in soil water or various chemical and biological agents in the soil convert fertilizer form into ionic form acceptable to roots. Fertlizer's use is determined primarily by soil fertility status and nutrient requirement of crops, or the cropping system.

Fertilizers and Food Production

Agricultural production target are usually set for a given size of population and to meet the need of the societies and export market. To give the expected yields, each crop must absorb a certain amount of each essential mineral nutrient from the soil. Where the nutrient supply rates of the soil are low, as is the case with a majority of Indian soils, medium to high crop yields can only be harvested if the nutrient supplies of the soils are augmented through the addition of fertilizers, manures and bio-fertilizers. Since each crop yield results in removal of nutrients from the soil, nutrients applications have to be repeated, if the yield targets are to be attained. In the ultimate analysis, nutrients balances and efficient nutrient cycling hold the key to a soil ability to sustain yields at a satisfactory level (Gigou, 1986, von, Uexkul, 1988).

It has been estimated that fertilizer account for 50% of the total food grain output in India (Randhawa and Tandon, 1982). This means that the food needs of two out of five persons are being met because fertilizer are being used.

This estimated removal of three major plant nutrients (N, P, K) by grain and straw residues of crops show the demand which are being put on the soil system (Table 7). The magnitude of demand cannot be met by our soils themselves, either on a short term or a long-term basis. Even though India is the fourth largest user of fertilizers in the world, its soils are still being depleted of their nutrient reserves as a result of the wide gap between additions and removals (Tandon and Sekhon, 1988). It is not fully recognized that nutrients depleted soil also mean the degradation of the environment but such soils must be fertilized to reduce this depletion and enable them to sustain good harvests.

Table 7: Estimated removal of NPK by Crops in India

Nutrient	**Removal, Million Tonnes**		
	1985	**2000 (a)**	**2015 (b)**
Nitrogen (N)	5.9	8.2	9.6
Phosphorus (P_2O_5)	2.6	3.6	4.2
Potassium (K.O)	10.3	14.4	16.9
Total	18.8	26.2	30.7

(a) Associates with 255 mt food grain production (b) associates with 300mt food grain production. *Source:* Government of India (1976)

Historical Development and Fertilizer Use

In India use of artificial fertilizer was first initiated in 1876 with import of Chilean nitrate. By 1905 calcium nitrate, calcium cyanamide, ammonium sulphate, super-phosphate and potassium sulphate were also imported.

Parry and Company (presently known as E.I.D. Parry India Ltd.) were the pioneers in the fertilizers scene appearing in 1906 when their plant in Tamil Nadu produced single super-phosphate by treating powdered animal bones with sulphuric acid. The first nitrogenous fertilizer, i.e. ammonium sulphate, was produced by Tata Iron and Steel Company limited (TISCO) during 1930 as a by-product of the steel industry. However, ammonium sulphate was produced through a synthetic process by Mysore Chemical and Fertilizers in 1941 in Mysore, which was followed by setting of the factory at AlwayN in 1947. For production of ammonium sulphate the Government of India setup a fertilizer factory at Sindri in 1951, which was expanded in 1959 to produce limited quantity of ammonium sulphate, nitrate and urea. For production of calcium ammonium nitrate, Nangal fertilizers factory was set up in 1961 and Rourkela in 1962. More factories for the manufacture of fertlizer were set up at a number of places *viz.* Trombay, Neyvelli etc. Upto 1960, ammonium sulphate and single superphosphate were the main fertilizers produced in India.

During 1951, the beginning of the First Five-Year Plan, the entire nitrogenous and phosphatic production in the country was in the form of ammonium sulphate and single superphosphate. Urea a high-grade fertilizer was introduced during the 1960s and Diammonium Phosphate (DAP), another high grade phosphate was introduced during the 1970's revolutionized the product pattern of fertilizer use in the country while urea is considered to be most favourable fertilizer because of its climate conditions, the other nitrogenous fertilizer such as ammonium sulphate and sodium nitrate have virtually become non existent as compared to urea. Such is not the case with DAP; because of its high analysis (contributing both nitrogen and phosphorus) it has not been able to displace other phosphoric fertilizers. Various NPK complexes and SSP will continue to have prominent shares in the market. While NPK complexes have found their niche with respect to the need-based crops and soils, SSP is widely acceptable fertilizer for sulphur deficient soils and crops like pulses and oilseeds, which require more of it. SSP in addition to phosphorus contributes 12 percent sulphur and 20 percent calcium to the soil.

Desirable Product Mix

While urea and DAP will continue to dominate India's product pattern in the years ahead, there is need for further modification or search for new materials to improve their keeping quality and efficiency to use. In this context, newer material like urea super granules (USG) and urea ammonium nitrate (UAN) liquid fertilizer could find future applications. In the case of liquid fertilizers, the problems related to their distribution, storage, and application need to be tackled. Urea coated with any of the available cheap indigenous material to improve its keeping quality and/or to import slow-release characteristics would also find application. All new material can be tailored to the needs of certain areas they would be most effective and serve the purpose for which they have been developed in the first place.

The use of SSP needs continued encouragement because it also provides sulphur and calcium which is equally important for plant growth. Nitrophosphate (NP) needs accelerated thrust as it avoids use of imported sulphur. Water solubility of available P_2O_5 in NP has always been a contentious issue, but there are several situations wherein nitrate nitrogen and 50% water soluble P_2O_5 are entirely suitable for application. Production and application of Nitrophosphate has so far not attained its due share and an expanded role of this product is foreseen in the coming years. In this context, it may be emphasised that the role of direct application of phosphate rock and partially acidulated phosphate rock (PIPR) is being encouraged in the overall supply of P_2O_5.

The use of zincated urea and boronated SSP is being encouraged in the country. Recently, these fertilizers have been brought under the purview of the Fertilizer Control Order.

Fertilizer Production

India has now emerged as the third largest producer and consumer of mineral fertilizers in the world (Table 8).

The total production of nitrogen (N) and phosphate (P_2O_5) fertilizers during 2000-2001 was 11.757 million tonnes and 5.056 millions tonnes respectively. This represents a capacity utilization of about 89 percentage and 92 percent, respectively. Thus, total production of N and P_2O_5 reached the high point of 10.54 millions tonnes (Tomer, 1996). During 2000-2001 domestic production of nitrogenous and phosphatic fertilizer was 10.942 and 3.734 millions tonnes respectively.

Table 8: Fertilizer Production, Imports, and Consumption (India)

Year	Production	Imports	Consumption (,000 mt)				
			Nitrogenous (N)	Phosphatic (P)	Potassic (K)	Total Consumption	Consumption of Cropped Area
1951/52	27	52	59	7	-	66	0.6
1955/56	89	63	108	108	13	10,131	0.9
1960/61	166	419	212	53	29	294	1.9
1965/66	357	413	575	133	77	785	5.1
1970/71	1,061	630	1,479	541	236	2,256	13.6
1975/76	1,828	1,635	2,149	46	278	2,894	16.9
1980/81	3,005	2,759	3,678	1,214	624	5,516	31.5
1985/86	5.756	3,399	5,661	2,005	808	8,474	47.4
1990/91	9,044	2,758	7,997	4.221	1,328	12,546	70.0
1991/92	9,863	2,769	8,046	3,321	1,361	12,728	69.7
1992/93	9,751	2,976	8,427	2,844	884	12,155	66.6
1993/94	9,045	3,167	8,789	2,669	908	12,366	67.8
1994/95	10,438	3,117	9,642	3,006	1,182	13,830	75.8
2000/01	14,679	2,617	10,861	4,212	1,557	16,630	87.2

However, since there are no known potash reserves in India, the country is totally depended on imports. During 1996-1997 fertilizer production increased to 14.31 millions and food production touched 198.17 millions tonnes. It was reported that for nitrogen, the shares of urea is 81.9 percent, while for phosphatic production the share of DAP is 50.7 percent, of SSP, 18.9 percent and of others, 30.4 percent. (Tomer, 1996).

Fertilizer Import

The demand supply gap of fertilizer in the country is met through imports. The entire domestic requirement of potassium fertilizer is imported as there are no known reserves of potash in the country. The availability of phosphatic fertilizer is also largely dependent on imports, either in the form of finished fertilizer or raw materials and intermediates. For nitrogenous fertilizer, the present indigenous production meets approximately 80 percent of the country's requirement.

Urea, DAP, and MOP (muriate of potash) are the only fertilizer currently being imported in the country while imports of urea are exclusively on government account, import of DAP and MOP are largely on private trade account.

Fertilizer Consumption

Consequent upon the expansion of irrigation and area under HYV's, fertilizer consumption has increased subsequently from 1.5 million tonnes of plant nutrients in 1967-68 to abount 11 millions tonnes during 1988-89. The consumption of nitrogen at 9.51 million tonnes during 1994-95 was an all-time high. However, the consumption of P_2O_5 and K_2O at 2.95 and 1.06 million during 1994-95 is yet to reach the record level of 3.32 and 1.36 million tonnes respectively achieved during 1991-92. After decontrol of phosphatic and potassic fertilizer during August 1992 sale, their price rose steeply. Thus consumption of phosphatic fertilizer went down resulting in deterioration of NPK ratio to 9.6:28:1. The total consumption of N + P_2O_5+ K_2O was 16.631 million tonnes during 2000-2001.

Balanced fertilizer use or rather balanced crop nutrition, ensures an optimum supply of all essential nutrients. Balanced fertilizer application is also a must for sustaining optimum yields and profits. Many Indian soils are deficient in all major plant nutrients. Higher consumption of P_2O_5, and K_2O has therefore to be encouraged to achieve the desirable consumption ratio of 4:2:1 to maintain the crop productivity and soil health.

In India, about 80 percent of the fertilizer consumption is confined to the irrigated areas which account for about 35 percent of the total cropped area and which cover four major crops primarily rice, wheat, sugarcane and cotton.

With expansion of rainfed areas, the production of oilseeds, pulses and course cereals is also gaining importance. The rainfed areas which constitute about 70 percent of the cultivated areas utilized only about 20 percent of the total fertilizer consumed. Efforts need to be made to increase fertilizer consumption in these areas as well.

Although there is positive growth in the consumption of fertilizer, the per hectare consumption of fertilizer nutrient (N + P_2O_5 + K_2O) was only 74 kg during 1994-95 against world average of 87 kg, the Asian average of 129 kg and China, Pakistan and Bangladesh average of 308 kg, 102 and 111 kg respectively. Similarly, productivity of rice in China is about 3.8 tonnes against 1.7 tonnes per hectare in India. The picture in respect of wheat is no different. Therefore, there is still a long way to go before India can claim to have a level of performance comparable with that of other nations.

Fertilser Self-sufficiency

Of the three main nutrients required in large amounts for various crops nitrogen, phosphate and potash-indigenous raw material are available mainly for nitrogen. Thus, the government policy has been to achieve maximum possible degree of self-sufficiency in the production of nitrogenous fertilizer based on utilization of indigenous feedstock. As of now, the country is self sufficient to the extent of about 80 percent for nitrogen. Prior to 1980, nitrogenous fertilizer plants were based mainly on naphtha feedstock. A number of ammonia-urea plants were also set up between 1978 and 1992 based on fuel oil as feedstock. In 1980 two coal based plants were set up for the first time in the country. With associated and free gas becoming available from offshore Bombay high and South Basin, a number of gas-based ammonia-urea plants have been set up since 1985.

In the case of phosphatic fertilser, however, domestic raw material constrains do not permit self-sufficiency in production. Recoginising this, the deliberate policy-mix involving the following strategies has been adopted.

- Domestic production based on indigenous rock phosphate and imported rock phosphate and sulphur.
- Production based on imported intermediates i.e. ammonia and phosphoric acid.
- Import of DAP, has been adopted.

Roughly 70 percent of the requirement of phosphatic fertilizer is met through the first two options. However, since indigenous rock phosphate fertilizers supplies are only 5 percent of the total requirement of P_2O_5 phosphatic fertilizers produced in the country are substantially based on imported raw materials and intermediates.

As regard potassic fertilizers, there are no known and commercially exploitable reserves of potash In the country. Thus, per force, the entire requirement of these fertilizers is met through imports. Even in the case of complex fertilizers having potash, the potassic content is entirely import- based.

VI. Fertilizers Use and Environment Problems

To meet the increasing demand for food and other basic agricultural commodities, the increased use of fertilizers and other chemicals is imperative in a developing country like India. This may raise question regarding the contribution of fertilizers to pollution. The deterioration of the environment due to fertilizer application may primarily be caused from poor fertilizer use efficiency of around 40 percent (45 percent in N, 20 percent in P and 55 percent in K), and from losses occurring through runoff, leaching, volatilization denitrification etc. Detrimental effects are evidenced by (1) Eutrophication of lakes and water resources due to high concentration of P and NO_3 (2) NO_3 pollution of ground water through leaching (3) gaseous losses of N particularly NO_2 and (4) metal toxicities from phosphorus sources.

Nitrate (NO_3) content may increase due to high soil erosion in areas of high rainfall and sandy soils with low exchange and buffering capacities of soil. In general, with the current level of fertilizer nutrients use of about 60 kg/ha per annum (40 Kg N/ha per annum) in India, there is no cause of anxiety regarding fertilizer pollution since no evidence has been documented thus far, except for a high concentration of NO_3 in well water in some parts of northern and central India. Nevertheless, considering the volume of fertilizers to be consumed in the future and the low level of recovery by crops, close monitoring is required to take appropriate precautionary measures.

Constraints Impeding Agricultural production and/or Fertilizers use: Policy Measures need to be considered

Despite significant achievements, Indian agriculture faces serious challenges from an ever-increasing population, limited available land, depleting soil fertility, and shrinking of non-renewable resources. Crop production in India is subjected to the vagaries of weather and the resulting drought and floods. The challenge comes particularly from the fragile ecosystem of arid regions (which represent 12 percent of the geographical areas) which are characterized by low and erratic rainfall, poor availability of other inputs (Irrigation water, fertilizer, seeds and agricultural credit).

In the areas where large scale application of chemical fertlilser are made, has started showing decreasing trends of fertilizer response to food production with an average decline in fertilizer use efficiency (FUE) from 17.1 Kg/Kg

NPK during 1970-71 to 8.1 Kg/Kg NPK in 1988-89 with possibility of its going down to 6.5 Kg NPK by 2000 AD (Sankaram, 1990). Although a part of this behaviour may be attributed towards diminishing return of increased fertilzer dosage, a major cause of such reducing response is due to lesser application of organic manure/biomass in comparison to chemical fertilizers resulting in deterioration of soil properties/ health. The situation may be more serious for vegetable cultivation where higher application of mineral fertilizers are recommended (Singh, 1996).

Agriculture and its development are far from sustainable in many parts of the world, both in developed and in developing countries. Pollution, very high rates of energy use and consumption of non-renewable resources need to be addressed in farming systems of the world. Traditional agriculture in many developing countries is threatened by the increasing population at rates higher than 2% per year, especially if plant nutrients removed from the fields or from community territories exceed the natural replenishment of soils by these nutrients (dry deposits and rain deposits, mineral weathering, irrigation water, nitrogen fixation) and external applications. In most developing countries, for the next 50-100 years agricultural production will need to be increased by at least 3% per year to maintain food security levels and to enhance the self-reliance of resources-poor marginal farmers. This situation will drastically increase the removal of plant nutrients from the soils. The most important issue is how to increase world agricultural production, while preserving the quality of the environment and the productive capacity of land and water resources.

An agricultural ecosystem differs from a natural one in which plant nutrients are constantly being removed. A natural ecosystem will generally be closed cycle and at equilibrium in terms of plant nutrient and soil organic matter content; nutrients taken up by plants will eventually return to the soil as plant and animal residues and be recycled through the activities of soil fauna and micro-organisms. In an agricultural system this will very rarely be so because the produce is removed from the land for food or other uses. There is, almost inevitably, a net removal by crop, plus those lost in other ways (eg. Leaching, runoff, gaseous losses) must be replaced if the agricultural system is to be sustainable. Therefore the agricultural system is like an open cycle. If replenishment is not done the system is, in effect, 'mining' the soil reserves of plant nutrients and this will inevitably lead to a decline in soil fertility and productivity.

Low-input sustainable agriculture (LISA) is workable if the balance between removal and supplies of plant nutrient is equilibrated. Organic agriculture can not create plant nutrients except by fixation of atmospheric nitrogen through micro-organisms. Organic agriculture recycles plant nutrients, although with low profitability, as with every biological process, and the area providing these nutrients has to be much larger than the areas receiving organic material in order to balance the system. However in many cases, creating favourable conditions for insects, worms and micro-organisms in soils, intensive organic agriculture can develop conditions for high soil productivity in well-supplied field.

To ensure higher yield, crops require an instant flow of nutrients at different growing stages which can not be supplied by natural weathering of minerals and organic materials. Organic sources release these plant nutrients slowly. This can be limiting factor, and the chemical fertilizers can provide sufficient plant nutrient flow to the crops. On the contrary, at the early stage of the crops much faster or with low-demand condition, organic sources can limit plant nutrient losses by leaching or gas, or avoid chemical fixation of nutrients by soil components, especially in very acidic situation. Thus, plant nutrition management should not be conceived only in terms of balance sheets.

The countries with an urgent need to bring about the quantum jump in agricultural production also suffer by the vicious circle of poverty, low- input agriculture and soil degradation. This cycle can only be broken by transforming resource-poor, low-input farming into science-based The latter must be based on judicious use of off-farm inputs because increased and sustained crop production will not be possible without improving and sustaining soil fertility and productivity.

With low input, the minimum dietary requirement can only be met from about 0.5 ha land *per output.* The greatest challenge facing mankind in the 21st century is to produce the basic necessities of food, feed, fiber, fuel and raw materials from 0.14 ha land *per output* or less. Technological options for sustainable management of soil, water and plant nutrient resources in the 21st century must address this issue. The *per caput* land requirement to meet the basic needs depends on the inputs.

While the use of mineral fertilizers is the quickest and surest way of boosting crop production, their cost and other constraints frequency deter farmers from using them in recommended quantities and in balanced proportions. The cost escalation resulting from a shift in the policies of many governments, through reduction or withdrawal of subsidies from fertilizers without correspondingly enhancing produce prices has further aggravated the situation.

These limitations - associated with either source of plant nutrients - are often overcome when they are used in judicious combination providing a mixture which in the long terms, is not only complementary but synergistic. Moreover, environmental hazards attributed to prolonged and sometimes heavy rates on mineral fertilizer application can be reduced by optimizing fertilizer use efficiency through a judicious mixture of organic sources, and mineral fertilizers at levels sufficient to replenish the soil nutrients removed by the crop. Specific eco-friendly bacterial fertilizers depending on the nature of crop can also be used to supplement plant nutrients such as nitrogen and phosphorus, and improve soil fertility.

VII. Concept of Integrated Plant Nutrient Supply (IPNS) System Biological Resources: Crop Science

Swaminathan (1974) stated what we hence need is an IPNS system involving fertilizer, organic manures of different kinds and biological nitrogen fixation through introduction of legumes in crop rotations. Commenting further he stated that legumes will help in maintaining soil fertility. It is based on this approach that Biovillage programme of M.S. Swaminathan Research Foundation is being operated.

It is quite clear that no single source of plant nutrient be fertilizer, organic manure/crop residues or biofertilizer can meet total nutrient requirement for sustainable agriculture taking care of increasing human needs. India is third largest mineral fertilizer country in the world. However, the harvested crop remove about 12-13 million tonnes more NPK every year from renewable sources than what is supplied through additional chemical fertilizer showing this contribution from organic and biological resources. Crop Science Society of America (1993) unanimously endorsed that decreasing FUE and declining soil organic matter (SOM) levels are serious threats to sustainability.

The main objective of IPNS system is to maintain, improve or adjust soil fertility and supply optimum levels of plant nutrients for sustainable agriculture by integrating possible sources of plant nutrients. It is also being increasingly recognized that IPNS system does not have scope to promote extreme approaches such as chemical farming, organic farming, natural farming etc. It aims to make the use of organic manure/compost of all kinds, recycling of crop residues, green manures and biofertilizers in an integrated or blended way to get higher use efficiency of inputs for sustainable crop yields. This requires a synthesis of traditional wisdom and ecological prudence of tropical agriculture with modern advances in technology.

Components of IPNS

Soil: Agricultural productivity depends largely on the top soil which is about 15-20 cm deep for the most soils. This upper most soil layer provides many functions such as rooting zone for the plant, supplies a balanced mix of plant nutrients and absorbs, retains and supplies moisture for plant growth. Soil is a biodynamic system because it supports a large population of diverse beneficial micro-organisms such as bacteria, actinomycetes, fungi, algae, protozoa etc.

that decompose biomass, mediate in cycling of plant nutrients and protect plant from pest and diseases. All these functions of soil microorganism are essential for maintaining soil fertility and productivity *of* agricultural soils. Soils by these capabilities are considered by many as "healthy soils". The various physical, chemical and biological properties interact in complex ways that determine the potential of soil' fitness for sustained production of healthy and nutritious crops.

The integration of growth-enhancing factors that make a soil productive has often been referred to as "Soil quality". The soil quality has traditionally focused on and has been equated with, soil productivity. Recently the concept of soil quality has been broadened to include three other attributes such as food safety and quality, human/ animal health and environmental quality.

The capacity of soils to produce safe and nutritious crops in a sustained manner over a long period of time to improve the human and animal health without impairing the natural resources base or polluting the environment are important criteria. Some believe that good quality soils are essential to produce healthy and nutritious crops which can improve human and animal health but the research data in this relationship is lacking.

There are several causes for a decline in the soil quality such as nutrient losses due to runoff and leaching, depletion and soil reduction in microbial biomass and deterioration of physical condition of soils. It can also happen through the accumulation of toxic substances from excessive use of chemical fertilizer and pesticide and atmospheric pollution (Sampson, 1981; Hornic and Parr, 1987; Gaur *et al.*, 1990).

Soil Organic Matter: The maintenance or restoration of soil quality depends on organic matter addition and an array of beneficial micro-organisms, it supports. A regular addition of organic matter can effectively offset many of the degradation processes. It is also the best way to develop a biologically active soil (Table 9 to 11), that require less energy for production of crops (Bhardwaj and Gaur, 1985), increases resistance of pest and diseases and enhances decomposition of toxic residual pesticides (Allison, 1973). The soil biological properties as an indicator of soil quality has been neglected largely because of availability of single reliable index for predicting biological properties of soil. The chemical and physical properties can be measured by using relative simple and standardized methods of measurements. Therefore, more attention should be given to soil biological properties to establish their interaction/relationship with chemical and physical properties of soil, the plant health and food quality are important but poorly understood. Plant health and nutritional quality may prove to be useful and reliable indicator of soil quality. (Hornick, 1992).

Table 9: Average bacterial population in differently treated soil during organic matter decomposition (Population expressed in 106/g soil).

Soil	Control (Soil)	Farmyard manure at therate		Wheat/paddy straw at the rate		Bajra/Maize stalk at the rate	
		0.5% C	2.0% C	0.5% C	2.0% C	0.5% C	2.0% C
Hill Soil	28.63	26.66	52.93	57.46	88.82	-	-
Alkali Soil	3.39	8.79	29.60	17.37	28.71	-	-
Heavy clay soil	11.66	16.54	17.56	24.69	28.71	20.09	28.39
Acid red loam soil	5.06	20.51	23.51	12.77	17.41	24.97	28.11
Sieronzem Soil	31.10	51.80	73.33	-	-	54.48	78.10
Alkali Soil	0.35	0.51	1.13	-	-	0.62	1.03
Alluvial Soil	32.08	74.83	141.18	92.12	197.76		-
Red Soil	16.38	20.52	39.29	-	-	228.89	367.41
Medium Black Soil	3.85	2.45	2.51	1.78	3.35	-	-
Acid lateritic Soil	5.12	4.56	3.07	1.63	3.61	-	-

Table 10: Average actinomycetes population in differently treated soil during organic matter decomposition (Population expressed in 106/g soil).

Soil	Treatment						
	Control (Soil)	Farmyard manure at the rate		Wheat/paddy straw at the rate		Bajra/Maize stalk at the rate	
		0.5% C	2.0% C	0.5% C	2.0% C	0.5% C	2.0% C
Hill Soil (HP)	8.93	11.23	20.79	20.51	30.23	-	-
Alkali soil (Delhi)	0.67	2.60	16.27	3.74	6.56	-	-
Heavy clay soil (Patna)	2.26	3.10	3.46	4.41	5.06	4.21	4.79
Acid red loam soil (kanke)	1 .52	11.84	2.21	1.78	2.29	2.38	2.74
Sieronzem Soil (Hissar)	13.47	37.44	55.83	-	-	52.26	102.75
Alkali Soil (Rajthal)*	15.98	15.51	30.41	-	-	15.29	14.19
Alluvial Soil (Cuttack)	3.22	9.74	16.08	8.77	13.24	-	-
Medium Black Soil (Pune)	1. 62	1.10	1.56	1.09	1.94	-	-
Acid lateritic sell	3.29	4.26	3.60	2.98	1.72	-	-
Red Soil (Hebbal)**	8.44	5.94	9.39	-	-	12.47	11.69

* Number X 10^3, ** Number X 10^5

Table 11: Average fungal population in differently treated soil during organic matter decomposition (Population expressed in 106 /g soil).

Soil	Treatment						
	Control (Soil)	Farmyard manure at the rate		Wheat/paddy straw at the rate		Bajra/Maize stalk at the rate	
		0.5% C	2.0% C	0.5% C	2.0% C	0.5% C	2.0% C
Hill Soil (HP)	11.57	14.67	17.57	17.11	28.96	-	-
Alkali soil (Delhi)	4.64	6.26	5.23	12.25	27.14	-	-
Heavy clay soil (Patna)	12.39	13.54	13.66	18.26	28.34	55.91	73.32
Add red loam soil (kanke)	10.82	12.35	12.84	29.57	57.02	72.69	99.39
Sierozem Soil (Hissar)	28.69	25.46	24.98	-	-	43.36	40.30
Alkali Soil (Cuttack)	27.15	39.00	48.23	75.54	134.15	-	-
Alluvial Soil (Hissar)	1.63	1.84	1.82	-	-	1.93	2.56
Red Soil (Hebbal)	6.74	4.06	4.89	-	-	42.43	95.55
Medium Black Soil (Pune)	13.29	11.71	13.36	13.71	12.68	-	-
Acidic Lateric Soil	18.07	14.41	14.54	17.54	26.29	-	-

Biodiversity: Soil micro-organism and invertebrates play a vital role in decomposition of organic matter and nutrient cycling and could be important indicators of soil quality. The various species, numbers and functions of these organisms often are sensitive to environmental stresses and changes in soil properties associated with cultivation and crop production. The changes in biodiversity of soil organisms (microbes, earthworms, insects) may provide indications of soil degradation or rehabilitation (Stork and Eggleton, 1992; Juwarkar *et al.* 1992; Visser and Parkinson, 1992)

Recycling of organic matter supplements plant nutrients. Plant depends on organic matter as well as minerals to support growth and high yields. However, most cultivated soil in tropics and subtropics due *to* high temperature and intensive cultivation are subjected to rapid decomposition of organic matter. Most soils of Indian plains have less than 1 percent organic matter. They are also often poor in more than one or two mineral nutrients.

Cultivated soils are very good in recycling of wide range of organic residues if the farm wastes/residues are regularly returned back to soil. Thus it can become an important component with chemical fertilizers in IPNS system. The added organic matter is decomposed in the soil with activities of soil flora and fauna resulting in production of CO_2, H_2O, minerals and humus substances which become part of the soil. The added fertilizers and minerals react with both soil minerals and organic compounds. Recycled wastes thus in no way are alien to the soils natural make up or process. However, these must not bring and pollute the soil with excess quantity of heavy metals, pathogens and un-decomposed highly resistant organic materials which may adversely effect or spoil the soil quality.

Recycling of farm and other organic residues bring in much needed organic matter and plant nutrients and thereby improve the soil fertility a key component of soil productivity. Organic material can be recycled by the following different methods.

Cropping/Farming System- Biomass Production

The cropping system rather than an individual crop, and the farming system rather than an individual field, are the focus of attention in this approach for developing IPNS practices for the main agro-ecological zones of a country and for the various categories of farms. IPNS identify the best associations of various types of plant nutrients in the different fields for balanced plant nutrition and higher yields, at the same time sustaining soil fertility and controlling nutrients losses. The farming system approach helps the farmer in the choice of the most adapted IPNS techniques to any situation and in increasing farm production and labour productivity. It is envisaged that locally available material of plant or animal origin as by products of agricultural activities be used or where such materials are not abundantly available, *in situ* production of organics be attempted. In area and farms where non-cultivated areas are available (forest, pastures, fallows), development of biomass production is attempted for a better contribution to plant nutrient transfer to cultivated areas in a sustainable system. In areas and farms where land is scares and limited to the fields, increasing biomass production in fallows and on the borders/bunds of the plots can be attempted. Biological sources which have considerable potential, are quick-growing leguminous crops grown as a part of the cropping system and incorporated into the soil at an appropriate stage as green manure, leguminous trees grown in hedgerows and their loppings used as mulch materials or incorporated into the soil of the cropped alleys between them, forage or grain legumes properly inoculated with *Rhizobia* grown in the cropping sequences; and the use of *Azolla* or blue-green algae with wetland crops and bacterisation of cereals with non-symbiotic nitrogen fixing bacteria such as *Azotobacter or Azospirillum*. Phosphate nutrition of plant can be improved by including phosphate solubilizing bacteria or fungi in the cropping system.

The following process must be understood in the context of organic recycling and IPNS (i) in the recycling of crop residue will bring back certain amount of nutrients to same fieid. However, they are not new sources and are deficient in the same nutrient as those of the soil (ii) *in situ* growing of leguminous crops/shrubs and their biomass incorporation will bring atmospheric N into the system and occasionally, through their deep roots, nutrients leached beyond root zone and subsoil, mineralized nutrients are brought to the surface layer for the use of annual crops; (iii) when biomass is brought from outside the plot or farm and/or pasture land on uncultivated soil it is a transfer of nutrients from one place to another. So, except for nitrogen where biological fixation can introduce nutrient into these soil- crop system, IPNS is a process which limits the losses in the system and increases the efficiency of nutrients, but not a process which supplies plant nutrients for the whole system. However, such a system can transform biomass with low economic values such as paddy straw or stubble into biomass with a higher value of food grains.

Contrary to the 'Low External Input' (LEI) and 'Organic Farming' approaches, the IPNS involves a low to medium external input approach, taking into account a holistic view of soil fertility and plant nutrition management for a targeted yield based not only on cropping and farming systems but also on distinct geographical areas or villages as a dynamic system.

Plant nutrient cycles, plant nutrient transfers and plant nutrient productivity are usefully analysed with clear reference to the original natural ecosystems, when it is possible and within the long-term evolution of the agricultural eco-systems. The analysis of plant nutrient management is related to an area, and the concept and the methods depend on the size of the area. This analysis must be arrived out at field, farm, village or human group territory level, at the level of the medium-size watershed. The challenges of crop intensification, as far as plant nutrient management is concerned, consists of the increase in the plant nutrient reserves and their productivity and the reduction of losses from the cropping system. These challenges must be dealt at the level of the whole farm, at least and in many cases at the village level. The IPNS programme should develop innovations and promote their extension of the best use of all sources of plant nutrients in order to upgrade their productivity and to reduce plant nutrient losses. If the fertility of the soil and plant nutrient content of the soil/crop system have already been lowered to a higher degree by inappropriate practices, one major task of IPNS will be at least to stop the unfavourable ongoing evolution and if possible to upgrade the plant nutrient content of the soil/crop system. This needs better understanding of the constraints and of the efficiency of practices required to rebuild more productive system than through the simple recycling of organic materials. The role of mineral and bio-fertilizers must be clearly defined within the range of practices needed to increase their efficiency and within the complete cycle of plant nutrients involved within the soil/crop system.

Promotion of Integrated Plant Nutrition System (IPNS)

IPNS aims at maximizing the efficiency of plant nutrient supply to crop through better combination of available farms inputs and external sources of plant nutrients ensuring sustainable agricultural production through improved productive capacity of soils. Such system may significantly reduce needs for mineral fertilizer, because they provide timely and sufficient supplies of plant nutrients, according to targeted crops yields and reduce as far as possible plant nutrients losses in cropping system.

Progress towards achieving these broad objectives must be viewed in the various ecological and economic contexts of agriculture in developing countries. Firstly, there are situation where plants nutrients are mined from the soil because of uptake by crops and losses of plant nutrient through erosion, leaching and volatilization exceed the low level of plant nutrient supply. Thus IPNS will assist in attaining a better balance of nutrients, initiate the intensification of cropping systems with limited use of chemical fertilizers coupled with better recycling of local organic and biological sources of nutrients and, above all with a drastic reduction of nutrient losses.

On the other hand there are situations where plant nutrient efficiency is low, even with significant supplies of nutrient from various sources. In these cases, IPNS will improve the combination of plant nutrient sources and cropping techniques and improve efficiency. In most cases, this low efficiency is due to an unbalanced supply of plant nutrient (too much nitrogen relative to other nutrients) or to another limiting factor such as secondary or micronutrient deficiency, physical or physico-chemical constraints of the soil and climatic risks. IPNS can be used to relieve such constraints, profitability depending on the availability of resources such as plant nutrients, equipment, energy, adapted varieties and irrigation.

There are serious problem where losses of plant nutrient are polluting the environment because of excessive or improper management of plant nutrient supplies, such as nitrates in surface and ground water, phosphates in surface water and nitrous oxides and ammonia in the atmosphere. Here IPNS will better balance or even reduce the supply of nutrients, maintain yield but focus efforts on both increasing plant nutrient efficiency and reducing the plant nutrient losses.

There are arid and semi-arid areas where the maintenance of soil organic matter is crucial for efficient plant nutrient management, maintaining soil permeability and water holding capacity, and promoting development of deep rooting system capable of exploiting water stored in the soil. To promote IPNS it is necessary to produce sufficient biomass in order to restore in the soil at least the quantity of organic material equivalent to that mineralized during the crop rotation. In mining the reserves of soil plant nutrient, farmers reduce the capacity of their soils and thus produce less of bio-mass. However, the mineralisation rate of soil organic matter is rapid when the soil temperature is high, and the biomass production is directly related to the availability of water and availability of plant nutrients. Thus, in the semi-arid tropics it may be difficult to restore the organic matter content of a degraded soil unless a sumptuous organic manuring is done.

In the humid tropics, leaching of plant nutrients, erosion, acidification and immobilization of plant nutrients in the soil may hamper the efficient supply of plant, nutrients. Additional competition from weeds and pressure of pests decrease this efficiency. However, crop as well as biomass production and crop diversity are higher but the mineralisation of soil organic matter is generally lower, than in the semi-arid tropics. The climatic risk is also low as compared to semi-arid areas and the natural conditions are generally more favourable for agricultural intensification. IPNS will then address rather diversified level of intensification depending upon the condition of the market for agricultural products and farmers' production interests. The upgrading of the soil fertility parameters is easier in the semi-arid or arid tropics because of comparatively higher bio-mass production. Here checking of plant nutrient losses is more complex than in semi-arid tropics because the turnover of plant nutrients is higher and the pressure of the factors involved in the losses is also greater.

In the irrigated areas, the removal of drought risk greatly increase the potential for plant nutrient efficiency. However, the economy of plant nutrients is often quite low because of lesser control on nitrogen losses in the cropping system or unbalanced fertilization. IPNS in irrigated areas is rather special as the decomposition of crop residues has to carefully regulated for timely release of nutrients and to check the leaching losses of plant nutrients. However, biological fixation of nitrogen (blue green algae, *azolla*) is possible through traditional flooding irrigation. IPNS will then mostly involve in improving plant nutrient efficacy, as maintaining soil organic matter content is easier than in rainfed areas because of moisture and higher biomass production.

The economics of nitrogen is a major problem in IPNS. The biological fixation of atmospheric nitrogen may provide signification quantities to the cropping system if water as well as phosphorous and sulphur are available. However, nitrogen fixation can not me*et al.*l nitrogen requirement above moderate level of intensification. There is a wide range of free living soil bacteria that fix nitrogen from the atmosphere and make it available for plant growth. There are other bacteria, notably *Rhizobia* and actinomycete viz. *Frankia,* which live symbiotically in plant root nodules obtaining energy and nutrient from root exudates and providing in return the fixed atmospheric nitrogen. These have been exploited by farmers for many years and sustained crop production in Asia, Europe and other parts of the world before the discovery of mineral fertilizers in the latter part of the last century.

In China, India, Thailand, Vietnam, and other Asia Countries, the alga *Anabaena, azolla*e, which lives symbiotically with the water fern *Azolla*, has sustained rice cropping for centuries by providing much of the nitrogen required. In India blue-green algae and *Azolla* are being used as bio-fertilisers in rice cultivation.

The challenges now are to use conventional techniques combined with methods of genetic engineering to improve the extent of nitrogen fixation and widen the range of plant and environments that can benefit. At present naturally or managed plant/microbe systems in good rainfed areas can provide 20 to 60 kg of nitrogen per hectare, sufficient to sustain cereal yields of around one tonne. Technology could conceivably raise this by 25 percent by the year 2010. Conventional plant breeding techniques may assist to obtain better efficiency through varieties using plant nutrients better to raise yields, or with stronger rooting systems to avoid losses by leaching. Varieties tolerant to soil constraints such as salinity, lack of oxygen, free aluminum etc. will also have more benefits from plant nutrient supply than traditional varieties.

IPNS is likely to make a significant contribution to the growth of foodgrain production during the period ending 2010 and in the achievement of sustainable agricultural systems. It would be unwise, however, to underestimate the challenges facing IPNS in short to medium term, and to overestimate the gains in the long term. Lack of livestock and labour will be major constrain in some areas: eg. many small holders can not keep sufficient livestock to generate the required amount of manure (up to 10 tonnes or more per hectare) or can not provide the large labour inputs for collecting, carting and spreading it. Where cultivated area can not be extended and the national target of food self-sufficiency is on higher side, it may be difficult with current or foreseeable technologies to achieve or sustain the required high yields with recycling of farm residues and use of microbial inoculants (bio fertilizers) alone for plant nutrient supply that some ecologists argue for (Norse, 1988).

China has been in this situation for several decades. It has efficient systems for organic residue recycling and biological nitrogen fixation. Yet, since 1950 food grain production has become increasingly dependent on mineral fertilizer, and China is now the largest user of mineral fertilizers in the world. In spite of the present efforts in China to develop further the science of ecological agriculture, this trend seems likely to continue and even intensified due to increasing shortage of labour for the collection and application of organic manures. Several other countries e.g. Senegal, Burundi, Rwanda, Malawi, India and Thailand are facing similar problems.

References

Allison, F. E., 1973. Soil organic Matter and its Role in Crop Production, Elsevier Sci. Publ. Co. New York.

Anonymous 1927. Forest destruction and its effects. Nature, 119 :37-39.

Bhardwaj, K. K. R. and Gaur, A. C. 1985. Recycling of Organic Wastes, ICAR, New Delhi, pp. 104.

Blum, W.E.H. 1994. Sustainable Land use and environment. Proc. Ind. Soc. Soil Sci., New Delhi, pp 21-30.

Consultative Group on International Agricultural Research (CGIAR), 1989. Sustainable Agricultural Production: Implications for International Agricultural Research, FAO, Research & Technology Paper No. 4. pp 131.

Gaur, A. C. Neelkantan, S. and Dargan, K. S. *et al.* 1990. Organic Manures, ICAR, New Delhi, pp 159

Gaur, A.C. 1998a. Integrated Plant Nutrient Supply System, Blending Organic, Bio and Chemical and Resources In. Soil-Plant-microbe Interactions in Relation to Integrated Nutrient Management, IARI, New Delhi pp 14-27.

Gaur, A.C. 1998b. Organic and Biofertilizers in Sustainable Agriculture. Natl. Seminar on Biosystems for sustainable Agriculture U.P. Council of Agri. Res. Lucknow pp 8-9.

Gigou, T. 1986. Proc Second IMPHOS Regional Seminar, New Delhi, pp 379-404

Hayani, Y. and Otsuka, K. 1991. Beyond the Green Revolution: Agricultural Development in the New Century. Paper presented to the World Bank on Agricultural Technology: Issues for the International Community and the World Bank, Oct. 1991.

Hazra, C.R. 1991. Nutrient Balance and Sustainable Agriculture in Central Plateau and Hill Region. Fert. News 36(6): 35-42.

Hornick, S. B. 1992. Factors affecting the nutritional quality of crops. Amer J. Alternative Agric. 7 : 63-68.

Hornick, S.B. and Parr, J.F. 1987 Restoring the productivity of marginal soil with organic amendments. Amer J. Alternative Agric. 2:64-68

Juwarkar, A.S. *et al.* 1992. Restoring of manganese spoil dump using press mud. In: Environmental Issues and Management of Waste Energy and Mineral Production (eds: R.K. Singhal *et al.*) pp 827-830, Balkema, Rottordam, Brookfield.

Nambiar, K.K. and Ghosh, A.B. 1984. Highlights of Research of a long Term Fertilizer Experiment in India (1971-1982), IARI, New Delhi pp 100.

Norse, D. 1988 Policies for Sustainable Agriculture: Getting the Balance Right Paper prepared for the IFAD. International Consultation on Environment, Sustainable Development and the Role of Small Farmers Rome, Oct. 11-13, 1988.

Pingali, P., Moga, P.F. and Velasco, L. E. 1990. The Post-Green Revolution Blues in Asian Rice Production: The Diminishing Gap between Experiment Station and Farmer yields. IRRI Social Sciences Division, Los Banos.

Randhawa, N. S. and Tandon, H. L.S. 1982 Fertilizer News 27(2) :11-26.

Sampson, R.N. 1981. Loosing Soil quality In: Farmland or Wasteland, Rodale Press, Emmaus, Pennsylvania pp 132-152.

Sankaram, A. 1990. The Hindu survey of Agriculture pp 163-169 Saxena, S. K. 1993 Indian Country Report. Fertilizer Policy in Asia and the Pacific (Eds: Saleem Ahmed and Allen Clark) APO-East West Centre. Pp. 198.

Shantaram, M. V. 1998. Environment Assessment for Agricultural Development in Asia and Pacific. Report APO study Meeting June 25 to July 5, 1996, Tokyo) Asian Productivity Organization, Tokyo. Pp 188-208.

Singh, D.K. 1996 Important Tip for cultivation of Leafy vegetables in Winter. Indian Farmers Digest. 30: 17-22.

Stork, N.E. and Eggleton, P. 1992. Invertebrates as determinants and indicators of soil quality. Amer. Alternative Agric. 7:38-47. Swaminathan, M.S. 1974. Proc. FAI-FAO Seminar on Fertilizers. pp 61-63.

Tandon, HLS 1987 Phosphorous Research and Agricultural Production in India: FDCO, New Delhi pp 160.

Tandon, H.L.S. 1991a. Sulphur Research and Agricultural Production in India, TSI, Washington, D. C. USA. Pp. 140

Tandon, H.L.S. 1991b. Secondary and Micronutrients in Agriculture .FDCO, New Delhi. pp 122.

Tandon, H.L.S. and Sekhon, G.S. 1988. Potassium Research and Agricultural Production in India FDCO, New Delhi. pp 122.

Tomer, K.P.S. 1996. Country Report, India: In: Appropriate use of Fertilizers in Asia and Paciic. Proc. Seminar, Taipei, Taiwan, ROC (Ed. Saleem Ahmed) Asian Productivity Organization Tokoyo, pp. 190-205.

Visser, S. and Parkinson, D. 1992. Soil biological criteria as indicators of Soil quality: Microorganisms. Amer J. Alternative Agric. 7: 33- 37.

Von Uexkull, H.R. 1988. Proc Seminar on small Holder Development in Pacific, IBSRAM, Bangkok, pp 121-132.

World Commission on Environment and Development 1992. Our common future Oxford University Press, Oxford.

2

Animal Residues

Type and Potential of Organic Residues

A number of diverse sources of organic materials are available for recycling in agriculture. These are mostly of plant origin and recycled either directly or after having being used as feed or food by animals and humans. In view of high organic content and mineral nutrients in the biomass, these are considered as valuable inputs to improve the soil fertility and crops production. Organic resources are low cost inputs but are valuable based on its potential for nutrients and organic matter content.

Attractive Features of Organic Resources

- Organic matter is the basic resource of several essential plant nutrients.
- It buffers the effects of pH dependent changes on nutrient availability.
- They are generated close to their use.
- They are renewable as they are generated regularly.
- With increasing demand for increasing food production, the potential of organic resources is expected to rise.
- Farmers are familiar with organic resources from time immemorial.

The differences in the nature of organic resources such as between a succulent green manure and animal dung are quite clear. Leguminous green manures fix atmospheric nitrogen and mineralize faster are thus superior for plant nutrition as compared to stable organic manure such as FYM/compost and crop residues which are important from organic recycling, disposal and plant nutrients points of view.

Crops remove substantial quantities of plant nutrients. Harvested grains remove 75 percent of nitrogen and phosphorus and 25 percent of potassium of the total removed by the crops and rest is left in the crop residue. In case of animal dung, a crop plant has already served as a source of nutrient twice, a major input in FYM or rural compost production. If entire urine and dung are conserved the nitrogen in FYM is about 17 percent of what was originally present in the growing plant.

Normally, 95 percent of nitrogen and sulphur are present in soil organic matter. Nearly 70 percent of zinc and copper are in organic form (Hodgson, 1969). Accordingly to Sanchez (1976), 60-80 percent of soil phosphorus is of organic origin while Dusburg *et al.* (1989) reported it to be 20 to 75 percent. Thus a reduction in soil organic matter level multiplies the nutrient deficiencies.

I. Animal Wastes, Cattle Dung, Urine and FYM

The handling of animal wastes includes, collection, storage and spreading or disposal on land. The term livestock excretions mean fresh excrements including solid or liquid portions. Animal wastes can be divided in three main categories.

i) Farmyard manure where straw or litter is decomposed with animal wastes in windrows, pits or heaps.

ii) Slurry, which is a mixture of excreta/urine often containing varying quantities of rain water or wash water.

iii) Animal wastes are separated into solid and liquid portion. This third category of manure is obtained with the use of mechanical slurry separators where liquid part is used for irrigation of land and the solid part is handled separately as a solid. In dry climate conditions, manure dries out quickly due to evaporation of heaps and used as solid material.

Quantities of Animal Wastes: The characteristics and quantities of animal wastes depends on several parameters such as digestibility of animal feed, its composition such as protein and fibre content of feed, environmental conditions like temperature and humidity and species and age of the animals and their physiology. The composition of wastes from ruminants such as cattle, buffalo, goat and sheep are different than wastes obtained from pigs and poultry, which are highly digested. The livestock dung consists of mainly undigested food which is left over after microbial action. It is also contains residue of digestive fluids, waste mineral matter, withered cells from intestine linings, mucus bacteria and other undigested matter like dirt residues consumed along with food.

Animal wastes differ from city and industrial waste because they are usually highly concentrated and relatively low in volume. Biological oxygen demand (BOD) may vary from 1000 ppm for dilute slurry to over 50,000 for slurry for silage effluent and concentrated pig slurry. BOD refers to the quantity of oxygen required to oxidize organic matter within 5 days of incubation.

The excreta can be grouped based on the solid content. The excreta having 4% solids are in liquid form which can be easily be collected and directly used for irrigation purpose by flushing or pumping device. The excreta having

4-15% solids are in semisolid state and require special device for pumping in biogas plants or irrigation purposes. The excreta with 20% solids can be easily collected with the help of spades/rakes and transported in a cart or wheel barrow to the site of composting or for use in biogas plants. Livestock wastes are semisolids and have to be handled and utilized in this form.

Characteristics of Bovine Excreta: The result of nutritional trial conducted at NDRI, Karnal and IVRI, Izatnagar on Sahiwal and Tharparkar cattle and Murrah buffaloes provide accurate quantitative evaluation of dung and urine excretion and their properties (Daniel *et al.*, 1967, 1968; Gopal Krishna, 1971, Krishna Singh and Mudgal, 1967, Mudgal and Ray 1967, Sebestian, 1967). The dry matter intake, digestibility, excreted quantity of dung and urine, and nitrogen, phosphorus and potassium balance were recorded to determine the qualitative and quantitative characteristics of cattle excreta.

The lactating cattle and buffalo showed maximum intake as well as greater excretion of dung and urine. This was followed by dry animal, heifers and male calf (Gaur *et al.*, 1995). The quantity of excretions were dependent on the body weight of animal plus feed intake.

Dung consisted about 75 to 85 percent moisture, 15 to 25% organic matter and 2 to 5 percent mineral matter. Organic matter of dung mainly comprised of 78 to 90% of total carbohydrate (Crude fibre + nitrogen free extract), 9 to 18% crude protein and 2 to 5% ether extract (Neelkantan and Singh, 1975).

The average excretion of wet dung from different types of cattle and buffaloes in different states, of India was documented by Gaur *et al.* (1995). The highest excretion of about 15 Kg per animal was reported in Delhi, Haryana, Himachal Pradesh, Punjab, Maharastra and the lowest rate of bout 3.6 to 5 Kg/animal reported from Kerala. The all India daily turnover of dung trend in Kg for cattle and buffalo groups were registered by Neelkantan (1975) which showed varying rates. In case of cattle, young stock excreted 4.90 Kg, where as adult female and male excreted 11.3 Kg and 15.6 Kg respectively. Buffaloes produce 5.7 and 14.8 Kg in case of young stock and adult respectively. However, low excretion rates of 2.9 to 7.0 Kg in different states of India was reported by Goel *et al.* (1973) based on IARS survey which did not take into account, dung evacuated in pastures, fields and road sides.

The average excretion of urine/day/cattle or buffalo was 7.4 and 8.0 litres respectively. The variation depending on the size and type of animal in urine production has been documented (Gaur *et al.* 1995).

Constrains in their Optimum Use: Though its fertilizing value in agriculture has been well recognized more than 50% of the cattle dung produced in the country at present is burnt as fuel and thus lost to farming. The traditional method of preparing and storing farmyard manure is generally faulty and sub-standard. The cattle dung waste is first collected in open backyard and when a cartload is full, it is removed to another heap or to an uncovered pit in a site outside the village. The heaps lie exposed to the sun, without proper maintenance of moisture etc. hence the material does not decompose properly and losses of nutrients and the material takes its toll due to rains or strong winds. Cattle urine, a valuable nutrients resource either is not conserved or is stored in a defective manner, resulting in losses of plant nutrients and the poor quality of the final product.

Undigested protein is excreted in faeces and the excess nitrogen is excreted in the urine as uric acid or urea. Potassium is absorbed during digestion but eventually most of it is excreted through urine. Calcium, magnesium, iron and phosphorus are excreted mainly in the faeces. US studies on distribution of plant nutrients in faeces and urine of dry cows have shown that 95 percent of potassium, 63 percent nitrogen and 50 percent of sulphur are found in urine. The wastage of nitrogen and potassium rich urine and losses of nitrogen through ammonia volatilization and washing away of the waste reduces the fertilizing value of FYM in India to a great extent. The percentage of N, P_2O_5 and K_2O in Indian dung is 0.30, 0.15 and 3.09 percent against the values of 1.0, 0.3, and 1.09 percent in European conditions. About one-half of this nitrogen, one-sixth of P_2O_5 and more than half of K_2O are readily soluble and subjected to losses.

Fertilizing Values: Animal residues and the byproducts form important feature of recyclable organic resources in agriculture. Their potential and fertilizer value (Table 12) show that their role in improving plant nutrition and to reduce the requirement of mineral fertilizers would be significant.

Table 12: Potential and Utilizable Quantities of Animal Residue

	Plant Nutrients (Million Tonnes)		
	1991*	2011**	2025*
Total	12.40	18.02	21.64
Utilizable	7.40	10.81	12.98
Fertilizer Equivalent Value	3.70	5.40	6.49

Source: *Bhardwaj and Gaur (1985) **Sekhon (1997)

The chemical composition of animal wastes excreted is influenced by the feed, body weight and age of the animal. The nitrogen and C/N ratio of dung excreted in different age group of Sahiwal cattle varied from 2.0 to 2.6 and 21.7 to 28.8 respectively. Similarly in case of Murrah buffaloes, the nitrogen and C/N ratio varied from 1.42 to 2.96 and from 49 to 19.5 respectively (Gaur and Singh,

1982). These result show that the quantity of excretion is influenced by nature of feed stuffs and the age of the animals.

The nutrients contents of animals wastes/by products and human wastes have been presented in table 13. This table also includes the meal group of organic manure such as blood meal, meat meal fish meal etc. Their nutrient content varies depending on the nature or group of manure. Meal group of manure are richer in plant nutrient, but the availability (quantity) is much less as compared to animal dung and human wastes.

Table 13: Nutrient content of animal and human wastes.

Organic matter	**Percent**			**Content (Kg/tonne)**		
	N	P_2O_5	K_2O	N	P_2O_5	K_2O
Cattle Dung	0.35	0.12	0.17	3.5	1.2	1.7
Sheep + Goat	0.65	0.50	0.65	6.5	5.0	6.5
Blood Meal	11.0	1.20	1.00	110.0	12.0	10.0
Meat Meal	10.0	2.50	0.50	100.0	25.0	5.0
Tankage	7.5	7.00	-	75.00	70.00	-
Hair/Wool Waste	12.30	0.10	0.30	123.0	1.0	3.0
Fish Meal	7.00	6.00	2.00	70.00	60.00	20.00
Raw Bone Meal	3.5	22.50	-	35.0	225.0	-
Steamed bone meal	1.50	27.50	-	15.0	275.0	-
Humans	1.35	0.75	0.50	13.5	7.5	5.0

The theoretical potential and availability of cattle dung is 2000 million tonnes per year. It contains 3.44 million tonnes N, 1.31 millions tonnes P_2O_5 and 2.21 millions tonnes of K_2O (Gaur and Singh 1995).

Cattle accounts for about 90 percent of the total animal dung and nutrient. The proportion of cattle dung available for manure purpose has decreased to 30 percent. Its major competing use is in the form of dung cakes. Wherever possible, it is better to route it through biogas plants in order to get both fuel and manure. Organic manure/compost is also being commercially produced from rural and city wastes and sold to farmers for Rs 2000 per tonne.

Farm-yard Manure: Good quality farmyard manure is considered as the most valuable organic manure applied to soil. It is most commonly used organic manure in India. It consists of mixture of cattle dung, the beddings used in the stable and household wastes. Though, the value of FYM has been well recognized from time immemorial, more than 50 percent of the cattle dung produced in the country is burnt as fuel and thus is lost to agriculture. Not only this is a substantial loss, but also the traditional method of preparing and storing FYM is generally faulty. The loose heaps lie exposed to sun without maintenance of proper moisture, aeration etc. resulting in preparation of poor quality manure. Cattle urine is generally not conserved and finds its ways into the drains.

About half of the nitrogen, one-sixth of phosphorus and more than half of the potash are readily soluble and subject to losses. However, the loss of nitrogen and mineral elements caused by careless handling can be reduced by using an absorbent bedding for cattle dung and storing dung in stone or brick lined pits, mixing with straw and other vegetable materials in proper proportion and use of some amendments such as rock phosphate. Thus, if urine is properly conserved, the losses of soluble minerals nutrients through seepage and ammonia volatilization are prevented, microbial decomposition of raw organic is encouraged and nitrogen losses are minimized. The relative absorbent capacity of different farm wastes such as wheat straw, pea straw, dry leaves, saw dust, soil etc. varies. Sawdust has a good water absorbent capacity followed by farm residues and soil.

Conservation of Urine: If urine is not conserved in the bedding used for cattle it must be collected in a pucca pit and then added to manure mixture in the pit or heap, Nitrogen in the urine is mainly in the form of urea, which readily changes to volatile ammonium carbamate through bacterial action and may be lost as ammonia if the process is not controlled. This loss can be reduced to a great extent if the manure and urine-absorbed litter are put together or in layers. Application of alout 2-3% rock phosphate and 0.5% soil may check the losses of nitrogen and improve the quality of manure. The losses through escape of ammonia can also be prevented by applying calcium sulphate, gypsum or superphosphate. Gypsum has been found effective as an ammonia absorbing agent. Super-phosphate besides absorbing ammonia, supplies additional phosphorus and thus improves the bioprocessing of compost and its quality.

Maturity of FYM: The maturity of the compost is important from the quality point of view. Mature compost should be applied just before sowing the crop. However, partially decomposed FYM should generally be applied to soil about three to four weeks before the sowing of a crop. In case there is sufficient moisture in the soil, the FYM will be further decomposed and improve the physical, microbiological and chemical properties. In any, case after FYM is carted to the field, it should be evenly spread and worked into the soil soon to avoid the loss of nitrogen. The practice of leaving the manure in small heaps scattered in the field for several days before incorporating into the soil, results in serious deterioration of its quantity and quality particularly if strong winds blow. In vegetable and fruit cultivation, the application of mature compost in conjunction with fertilizers to young plant can give best results.

II. Biogas and Digested slurry

The disposal of wastes from animal and human population under anaerobic digestion results in better and cleaner environment. Biogas is a terminal product which consists mainly of 50 to 80% methane (CH_4) and 15 to 45 % carbon dioxide (CO_2) with traces of hydrogen sulphide (H_2S) and ammonia. The methane gas generated in biogas digestors is used as smokeless fuel for cooking, for lighting and can also be used as engine fuel in combination with liquid fuels. The residual slurry is valuable source of organic manure and water (Idnani *et al.*, 1969, Nagar, 1978; Gaur *et al.*, 1995, Dierk, 1990)

Microbial Process: Biogas is produced by microbial breakdown of organic wastes such as animal, human, farm and agro-industrial wastes under anaerobic conditions, i.e. in the absence of oxygen at a highly negative redox potential. The various groups of bacteria perform a series of complex biochemical processes. Generally, it involves four groups (1) Hydrolytic bacteria: These bacteria convert insoluble or partially soluble complex polymers into waste soluble compounds (ii) Acetogenic bacteria: These bacteria form acetic acid and hydrogen from soluble complex organics (iii) Hydrogen Oxidising homoacetogenic bacteria: These bacteria form acetic acid (iv) Methanogenic Bacteria: They catabolise alcohol and organic acids to methane and carbon dioxide. However, methanogenic bacteria are not able to catabolise alcohol other than methanol or organic acid other than acetate and formate (Bryant *et al.*, 1967, Bryant *et al.*, 1977).

The methanogenic bacteria represent a rather diverse group of microorganisms belonging to Archaebacteria (Balch *et al.*, 1979, Jones *et al.*, 1987) and *Methanosarcina barkeri*, *Methanococcus mazei* and *Methanothrix soehngenii* have been isolated in pure cultures. Various investigations have indicatated that the population of methanogens in anaerobic digesters range between 10^6 and 10^8 per ml (Hobson *et al.*, 1981, Rajasekaran *et al.*, 1986)

Methanogenic bacteria are most strict anaerobic bacteria known. Studies on methanogenic bacteria revealed the presence of F420. This compound is fluorescent under long ultraviolet light. This property was suggested by several workers for tentative identification of methanogenic bacteria (Doddema and Vogels 1978; Kataoka *et al.*, 1991).

In the domestic sewage digesters, hydrogen utilizing, *Methanobacterium formicicum* is very common (Smith 1966). Hobson and Shaw (1974) reported that the population of hydrogenotrophic bacteria were in the range *of* 10^5-10^8 per ml sewage sludge and 2 X 10^5 - 10^6 in piggery waste digesters.

Most of methanogenic bacteria can utilize hydrogen and carbon dioxide as their sole carbon and energy source. The need for hydrogen in acetate utilisation by some strains was also reported. Some genera utilized acetate or formate as their sole carbon or energy source (Ferry, 1992). *Methosarcina* sp. utilized methanol and *Methanospirillum* sp. utilized 2- propanol (Widdel, 1986).

Evaluation of Cultures in Biomethanation

The anaerobic roll tube technique was followed for isolation of Clostridium sp. (cellulolytic culture) and methanogens identified as *Methanobacterium* sp. and *Methanosarcina* sp from biogas digester (Hari and Gaur-unpublished). These isolates were evaluated as inoculants in biomethanation of rice straw and grass (Table 14) (Plates 1-5). The results showed that the maximum biogas was produced from the grass based digester as compared to rice straw based digester. The use of mixed inocula marked CM^3 gave the maximum biogas followed by CM^2, CM^1 and MBS.

Table 14: Biogas production from rice straw and grass inoculated with pure culture.

Inocula	**CM_1**		**CM_2**		**CM_3**		**MBS**	
Detention period (Week)	**Rice Straw**	**Grass**	**Rice Straw**	**Grass**	**Rice Straw**	**Grass**	**Rice Straw**	**Grass**
1	-	-	-	18.22	-	22.22	-	-
2	22.22	13.00	23.89	23.67	36.67	38.89	12.22	8.56
3	16.89	26.44	14.89	8.33	23.78	10.00	10.89	10.22
4	13.56	10.89	18.33	22.22	11.78	18.33	14.44	22.22
5	7.22	53.33	35.89	41.89	20.00	52.89	27.33	41.89
6	1.11	33.33	11.11	11.11	16.67	18.89	14.44	13.33
7	-	22.33	5.89	3.56	8.78	11.11	10.000	6.67
8	-	12.23	1.33	1.10	3.56	7.44	5.22	4.67
Total (Liters)	61.00	171.56	113.13	130.11	121.22	179.78	94.56	107.56

CM^1: *Clostridium* sp. + *Methanobacterium* sp.
CM^2: *Clostridium* sp + *Methanosarcina* sp.
CM^3: *Clostridium* sp. + *Methanobacterium* sp. + *Methanosarcina* sp.
MBS: Manicipal Biogas slurry.

Plate 1: Gram stained cells of *Clostridium* sp. (12.5x 100.0 x)

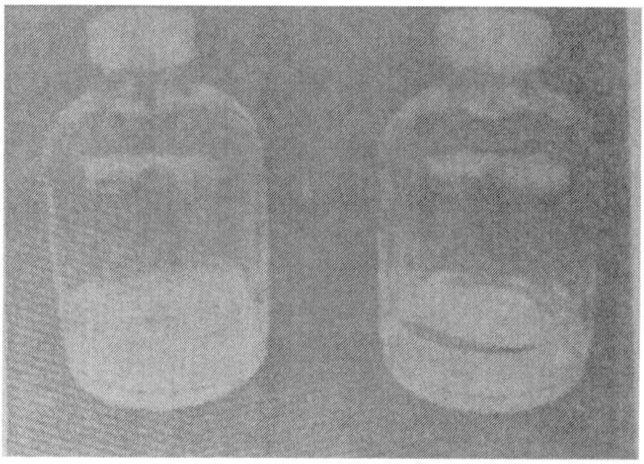

Plate 2: Cellulose paper degradation by *Clostridium* **sp.**

Plate 3: Autoflorescence of *Methanobacterium* sp. (12.5x 45.0x)

Plate 4: Autoflurescence of *Methanosarcina* sp. (12.5 x 45.0 x)

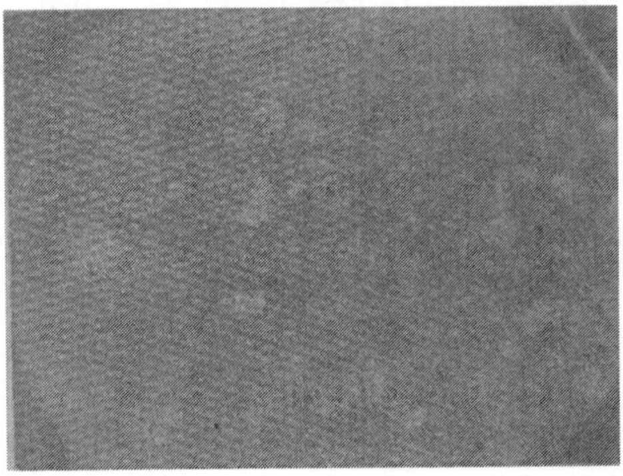

Plate 5: Long rod shaped bacteria in MBS inoculated rice straw digestor

Table 15: Chemical composition of spent rice straw and grass after bio methanation by mixed inoculation

Constituents % dry wt	CM[1]		CM[2]		CM[3]		MBS	
	Rice Straw	Grass	Rice Straw	Grass	Rice Straw	Grass	Rice Straw	Grass
Volatile Solid %	76.5	74.50	76.0	74.50	75.00	76.00	76.0	76.0
Organic Carbon %	45.9	44.7	45.6	44.7	45.3	45.6	45.3	45.6
Ash Content %	23.5	25.5	24.0	25.5	24.5	26.0	24.0	26.0
Cellulose %	25.3	19.3	23.3	16.7	20.0	16.0	22.0	16.3
Hemicelluloses %	26.9	28.8	27.2	27.2	25.2	25.7	23.1	24.4
Lignin %	15.2	15.9	17.9	14.5	20.5	18.6	9.0	16.0
Total phenols %	3.3	2.8	3.0	3.3	3.3	3.3	4.0	2.8
Nitrogen %	1.5	1.6	1.6	2.1	1.5	1.9	1.8	2.0
Phosphorus %	0.41	0.5	0.4	0.4	0.4	0.6	0.5	0.6
C/N ratio	29.8	26.8	28.3	21.3	31.2	24.1	25.1	22.6
C/P ratio	111.9	99.3	111.2	106.4	107.9	77.3	97.0	80.0

CM[1]: *Clostridium* sp + *Methanobacterium* sp

CM[2]: *Clostridium* sp + *Methanosarcina* sp

CM[3]: *Clostridium* sp+ *Methanobacterium* sp+ *Methanosarcina*

MBS: Municipal biogas slurry

The analysis of raw materials (rice straw and grass) is as follows:

Rice Straw: Volatile solid 86.0%, organic carbon 51.6%, Ash content 14%, Cellulose 34.5%, Hemicellulose 40.3%, Lignin 10.4%, total phenols 0.3875%, total N 0.64 %, total P, 0.17%, C/N ratio 80.6, C/P ratio 303.5.

Grass: Volatile solids 83.1%, organic carbon 49.8%, Ash content 14.9%, Cellulose 31.2%, Hemicellulose 39.0%, Lignin 12.5%, total phenols 0.395%, total N 0.78%, total P_2O_5 0.14%, C/N ratio 63.9, C/P ratio 356.

The result on the spent rice straw and grass (table 15) clearly showed that there were considerable reduction in the content of organic carbon, cellulose and hemicellulose in all the digesters. Apart from the reduction of organic component, the nitrogen and phosphorus contents were augmented. The C/N and C/P ratios were favourably influenced.

Biogas technology: The technology of biogas production in quite old, at present, more than 3.2 million small and big size plants are in operation in rural India. Feed stock used in these plants is mainly cattle dung. However, an enormous quantity of other potential organic wastes like crop residues, fruit peelings, plant biomass, agro-industrial wastes etc remain grossly under utilized. Moreover, some of these wastes create environmental pollution which can be hygienically recycled through this route.

The physico-chemical properties of different substrates for biogas system (Table 16) were reported by Oblisami (1992).

Table 16: Physico-chemical Properties of different feedstocks.

Feed Stock	Moisture	T.S%	V.S.%	pH	COD mg/L	Protein mg/ml	Cellulose
Cow dung	89.5	11.5	8.6	7.2	13,120	0.52	16.4
Buffalo dung	81.4	18.6	13.8	7.9	11,760	0.60	13.2
Dry animal dung	83.3	16.7	11.8	7.3	14,950	0.42	10.0
Stray Cattle dung	79.3	20.7	18.2	7.1	15,520	0.47	7.2
Goat dung	80.4	19.6	15.7	8.0	13,120	0.57	10.4
Poultry dropping	65.8	34.2	20.1	7.8	9,160	0.65	6.8
* Human waste	89.4	10.6	9.6		-	-	-

**Human Waste, Carbon 45%,* N6.5%, *P4.8%, K 1.4%*

The total solid content of the wastes varied from 10.6 to 34.2%. The total solid content was highest in case of poultry manure. The volatile solids was highest in poultry droppings (20.1%) and the least in case of cow dung (8.6%). The moisture content of wastes ranged from 65.8 to 89.5%. It was lowest in the case of poultry dropping. The pH of the wastes ranged from 7.1 to 8.0.

The gas production was highest in case of poultry droppings (308 ml/day) where as with stray cattle dung, it was lowest (78.0 ml/day). Goat dung, buffalo dung, cowdung and dry animal dung produced 130, 128,103, and 84 ml of biogas/day respectively. The population of methanogens was also found to be highest in poultry dropping digester followed by buffalo, goat, cowdung, dry animal dung and stray cattle dung digesters.

Factors Affecting Biogas Fermentation

The rate and extent of biogas production is affected mainly by environmental and operational factors:

- Biodegradability and quantity of the substrate.
- The seeding material
- The environment
- The design of digester

Substrate

Substrate that cannot be degraded or degraded slowly even after pre-treatment are not suitable for biogas production. The composition of the organic matter determine the yield of methane i.e quantity of methane in ml produced per Kg of biomass or total solids or volatile solids. Reduced compound, such as fats and protein produce more methane than substance relatively oxidized such as carbohydrate. Animal dung, municipal sludge, food and vegetable processing wastes, distillery effluent etc. are easily amenable to biogas fermentation.

The carbon to nitrogen ratio (C:N) of substrate is an important factor to determine digestibility. The C:N ratio of 30:1 in a substance is desired to achieve optimum gas production especially in system using cattle dung. It is however, important that both the carbon and nitrogen contained in a substance are in available form for the growth of the bacteria.

The concentration of available N and P can determine the nutrient limiting condition of a digester. Wastes which are low in nitrogen content may therefore require supplementation with N source. similarly, supplementation with phosphorus may be required in case of wide C: P ratio materials like rice straw. It was reported that when rice straw (C: P ratio 333:1) was amended with P source like rock phosphate or super phosphate to C: P ratio 200:1, maximum amount of biogas was produced. However, rock phosphate was found better than superphosphate (in this case) Bardiya and Gaur 1997). Further, Fe, Mg, K, Na, Ca, Co, Ni, Se, are reported to stimulate methane production. Some of these micro-nutrients are however, detrimental to gas production when present in higher concentration.

Seeding Material/Starter Culture

Biogas fermentation can be initiated quickly if the substrate is inoculated with adequate good quality inoculum such as old digested slurry or sludge. The key to starting and maintaining a successful digester lies in providing a balanced distribution of both acid former and methane formers. So far, as animal dung is concerned, the requirement of seeding material is not felt as these substrate are already full of requisite bacteria. However, in case of crop residues, sewage sludge, distillery effluent etc. the seed material at the time of startup should be at least twice the volume of the substrate.

Environmental Factor

It is essential to provide a favorable environment for all group of bacteria responsible for biogas production. Methanogens have unique nutritional and environmental requirement. Besides maintaining strict anaerobic conditions, the major environmental factor which influence gas production are pH, temperature and toxic compounds.

pH

Acid forming bacteria can tolerate a pH as low as 5.5, but the methanogens are sensitive to low pH values. Biogas production proceeds well within a pH range of 6.6 to 7.6 with the optimum range being 7.0 to 7.2 (Dague, 1968). The pH pf the digester may drop to 6 and below if there is an excessive accumulation of volatile acids. This may occur when the organic loading rates are excessively high and/or when toxicity is caused in the digestor. When the digestor becomes sour, the problem can be overcome either by stopping feeding of the digester for a few days or by adding lime to adjust the pH. Alkalinity is measure of the buffering capacity of the contents and acts as safeguard against pH fluctuations due to volatile acids. A good buffering is generally ensured in the digester if the alkalinity were maintained within 2500 to 5000 mg/litre of the digesting medium.

Temperature

Temperature of the digester is an important factor since the activity of bacteria is dependent on it. There are two optimum temperature ranges for biogas production. These are the mesophilic, i.e 30-45°C and the thermophilic, i.e 50-60°C (Acharya, 1958, Chawla, 1986). Gas production rates in unheated digester are lower during winter months as compared to the rates observed in summer months. As a rule of thumb doubling of gas production rate takes place for every 10°C increase in the digestion temperature in the mesophilic range. Thus, the temperature determines the digester capacity and period of detention/retention. Longer periods will be required at lower temperature and shorter for higher temperature, that is stable. Thermophilic digester can be run for shorter detention times as compared to mesophilic digesters.

Toxic Compounds

Toxicity to biogas production may result due to various factors. Accumulation of volatile fatty acids and undissociated ammonia are commonly associated with failure of digestor. Molecular oxygen also inhibits gas production. Inhibition is also caused by excess concentration of certain ions such as sulphide, calcium, magnesium, potassium, sodium, copper, cadmium, iron, chromium, nickel, etc. (Hayes and Theis, 1978). This problem can be counteracted by using antagonistic ions (sodium for overcoming ammonium-toxicity) and synergistic ions (calcium, magnesium or potassium) for overcoming ammonium toxicity. Heavy metals as an ion can adversely effect the bacteria.

The various substrates available in rural areas are, however, generally free of these toxicants. The inhibition of gas production caused by ammonia and volatile acids in digester based on cattle dung and crops residue can be corrected by proper management, dilution and operation, biogas production from industrial effluents requires consideration to be given to the presence of toxic compounds.

Indian Design of Biogas Plants

The designs of biogas plants developed in the country are mainly based on cattle dung as feedstock. These models can also be appropriately used for production of biogas from night-soil and other organic material which can be mixed with water to form a slurry.

There are two well tested and field worthy design of biogas plants. These are: i) Floating gas holder type (KVIC type gobar gas plant), and ii) Fixed dome type (Janata biogas plant). Few new models have been field- tested and approved for extension. These new models are as follows:

i) Floating gasholder type plant having digester made of angle iron and polythene sheet, called the Ganesh model.

ii) Floating gasholder type plant having digester made of pre-fabricated ferro-cement segments.

iii) Floating gasholder type plant having gas holder made of fibre-glass reinforced plastic.

iv) A plant having conical shaped base and shell shaped digester made with a provision of a gas holder, called, 'pragati model'

v) Fixed dome model dispension with the shuttering work for construction of dome portion, called the 'Deenbandhu model'

Hydraulic retention periods have been standardized for different areas according to temperature zones as indicated below:

Zones	Mean Temperature during winter month (°C)	Recommended retention period for a biogas plant (days)
I	More than 20	30
II	15-20	40
III	10-15	55
IV	Less than 10	Not Suitable

Common sizes of family type biogas plants are 1 to 10 m3 gas production per day. Floating gas holding type plant of large sizes are used for community and institutional purposes, their ranges being from 60 to 145 m^3 gas production per day.

Miscellaneous Organic Resources

For introduction of successful biogas programme, regular and abundant availability of organic feedstock is required besides animal and human wastes. Other feedstock such as rice straw, pressmud, spent wash, algae and water hyacinth can be utilized for methanogenesis. The type and availability of wastes varies from place to place due to variation in agro-climate condition. In future, one has to go in for cultivation of energy crop which will supply that required feedstock for biogas digester. Such energy plantation can be terrestrial, fresh water or marine plants. Common seaweeds can be also be used as feedstock for anaerobic digestion.

Algae was incorporated at 3% dry basis with cattle dung for anaerobic digestion (Venkatraman and Kaushik, 1975). They reported that at the end of 10 days, total gas produced from algae amended dung digester was 3314 cm3/kg feedstock as compared to dung alone digèster (1900 cm3/kg feedstock).

Water hyacinth is another aquatic plant weed which is available in large quantities in certain area (360-480 tonnes/ha). Water hyacinth that is large scale nutrient remover and plays an important role in water purification . and pollution control.

Wolverton and McDonald (1979) carried out studies with water hycinth *(Eichhornia crassipes)* and duck weeds. They reported that maximum volumes of methane was produced from water hyacinth. In India, Deshpande *et al.*, (1979) reported that incorporation of water hyacinth with cattle dung (1:1) enhanced the volatile fatty acids in the digester. The studies further showed the water hyacinth and cattle dung when used as a feedstock (1:1) were 16% more efficient over cattle dung alone (Sushil Kumar, 1980-Per comm).

Manurial Aspects

The recovery of plant nutrients from spent slurry is an economic return to farmers using the biogas technology. The slurry is rich, a source of water, humus substances and plant nutrients. The nutrient content in spent slurry is reported to be higher on dry weight basis than in the feedstock due to conservation of nutrient during digestion and gasification (Badger *et al.*, 1979, Gaur *et al.*, 1995, Jeet Singh *et al.*, 1987).

Tripathi (1993) reported that Ministry of Non-conventional Energy Sources (MNES) had launched a countrywide demonstration programmes in farmers' field through difference agencies such as state nodal departments, FORRAD, New Delhi and Gujarat State Fertilizer Corporation (GSFC) etc. A number of field demonstrations have shown an over all increase of 24.2 percent in crop yields of cereals, pulses, oilseed, potato, vegetable crops etc if biogas slurry is utilized. Besides in agriculture, its utilization in aquaculture, vermicomposting, mushroom cultivation and *Azolla* production was reported (Tripathi, 1993, Tripathi *et al.* 1993, 1993a, Senapati, 1993, Garcha and Sharma, 1993, Singh et *al.*, 1993).

The benefit of the biogas plant has been recognized at the national and individual levels. It has more relevance to the nation in terms of environmental and sanitation improvement, reduction of deforestation, etc. but these consideration do not influence the individual to adopt the technology. So far, over 3.3 million domestic biogas plants have been set up in the country. Assuming that 85 percent of these plant of an average capacity of 4 cubic meters gas production per day are working at a given time for 300 days in a year, these plants are estimated to produce gas equivalent to 11.32 million tonnes of firewood per year valued at Rs. 452.8 crores. Besides these plants are estimated to produce 56 million tonnes, of enriched biogas manure valued over at Rs. 500 crores per year. Thus, biogas plants generate sustained savings in terms of firewood and fertilizers to the nation.

Prospects

Biogas technology has started revolutionizing the life of rural masses in many areas. The demand for biogas plant has now started coming from the people living in remote areas. A large number of voluntary organizations and private entrepreneurs are involved in the promotions, extension, installation and maintenance of biogas plants. It has created employment for both skilled and semiskilled workers. It is estimated that the country has a potential of setting up about 13 to 14 million biogas plants based on cattle dung as feedstock. With improvement in the technology to process other kinds of feedstock, this potential is likely to increase further. With a view to capture the potential already available in the country, new strategies have been developed for implementation of a programme aimed at installation of about 1,50,000 to 1,75,000 family size and 500 large size biogas plants per annum. Emphasis needs to be laid on quality construction and post-installation servicing. Research and development efforts should also be intensified to develop design based on diversified feedstocks and reduced retention periods. A new market strategy involving pre-fabricated structure including institutional finance have also ensured. With these measures, it is hoped that the programme will make still faster progress in the coming years.

III. Poultry Manure

There are about 60,000 poultry farms in India and ranks 5th in the world in egg production and 22nd in poultry meat production. Besides, these they generate poultry manure and hatchery wastes of high economic value. It is rich in plant nutrient since liquid and excreta are ejected together without any loss of urine.

It is estimated that nutrient value of manure from 3 birds on deep litters is superior than that of one cow. Average daily of fresh manure from broilers is around 70-80 kg/100kg live weight. The excrete from hen, 1 to 6 months old ranged from 229 to 888 g/head/day depending on age etc. (Morino, 1975). The poultry excreted about 5% of its body weight/day. Chicken upto 13-weeks age on an average excreted about 274 g/head/ day.

The nutrient content of poultry manure fresh, dried floor litter and deep litter with saw dust and straw has shown variable composition in respect to major plant nutrients (Table 17).

Table 17: Composition of Plant nutrients in different type of poultry manure

Type	Percent			
	Moisture	N	P_2O_5	K_2O
Fresh	57	1.07	0.40	0.50
Dried	7	2.33	0,86	1.08
Fresh	75	1.47	1.15	0.48
Floor Litter	24	3.03	2.63	1.40
Fresh	80	0.76	0.63	0.22
Dried *	30	3.04	2.52	0.88 *
Kiln Dried *	10	3.80	3.15	1.10 *
Deep Litter with saw Dust + straw*	-	1.80	2.30	1.40

*Garner (1967)

The result showed that content of N, P_2O_5 and K_2O was related to moisture content, method of drying and the method of collection as floor litter or deep litter (Garner, 1967, Jai Kishan and Hussain, 1977, Gaur *et al.*, 1995). It has been reported to contain 1400 ppm of iron, 90 ppm of zinc, 210 ppm of Mn, 7.1 ppm of Cu, and 5 ppm of boron.

Nitrogen excretion by adult layers ranged between 0.65 and 0.75 g against nitrogen uptake of 2.65 and 2.90 g/head/day where as phosphorus excretion showed narrow range (0.36-0.39 g) against intake of 0.62-0.70 g per head per day. In this manure, 60% of nitrogen is present as uric acid, 30% as more stable organic nitrogen forms and the balance as mineral nitrogen (Srivastava, 1988). Most of nitrogen in poultry excreta is in the form of uric acid which is converted to urea by uricase enzyme and NH_4- N in a short time. This leads to substantial losses (10-80%) of the excreted nitrogen (Sims and Wolf, 1994).

Poultry manure ferments very rapidly and thus it may lose up to 50% of its nitrogen within 30 days. Therefore, poultry manure can be applied to soil directly as soon as possible. After application it should be properly incorporated in the soil. The fertilizing value of poultry marure can be improved by co-composting or anaerobic digestion with wice C/N ratio residues.

Poultry excreta contains a large load of bacteria, fungi, virus and protoza. Total viable counts of 10^8 to 10^9 (CFU/g dry wt) in poultry waste have been reported (Nodar *et al.,* 1990). Some of the microorganism may be pathogenic and their direct application to soil may cause health problems. The most frequently reported pathogens are *Clostridium* spp. and *Salmonella* spp. (Alexander *et al.*, 1968). Therefore co-composting of poultry litter/ excreta is an effective system to eliminate pathogens from these wastes. Poultry wastes mixed with other wastes can also be subjected to anaerobic digestion to produce biogas for cooking and spent material can be used as organic manure. Poultry droppings have been useful in producing enriched compost for producing of mushroom.

Singh and Srivastava (1971) and Singh *et al.* (1973) attributed the higher efficiency of poultry manure to narrow C/N ratio and higher content of readily mineralisable nitrogen.

Uses of Poultry Waste

Organic manure : Poultry manure can be dried and pulverized. It can be used on lawn, garden, greenhouse soil and farms. Fresh poultry manure can be also applied and incorporated into soil as soon as possible since it ferments quickly.

Compost: Gaseous ammonia liberated from fresh poultry wastes and its higher salt content show phytotoxic effect.

The losses of nitrogen and phytotoxic effect can be minimized by composting of poultry wastes with wide C/N ratio material such as maize stalks, ragi husk, deep litter poultry waste and cowdung slurry in layers of 30, 30, 15, and 5 cm respectively. This order may be repeated. Optimum moisture and aeration as required in composting should be taken care of. In this way the losses of nitrogen and phytotoxic effects can be reduced resulting in a good quality organic manure.

The field experiment conducted in black soil of Tamil Nadu Agricultural University farm, Coimbatore (Table 18) has shown the efficiency of poultry manure in comparison to other types in increasing the grain and straw yield of rice (Budhar *et al.,* 1991).

Table 18: Effect of Organic Manure including poultry manure on yield of rice

Type of Manure	Grain Yield (Kg/ha)	Straw Yield (Kg/ha)
FYM	5338	7570
Biogas Slurry	5493	7110
Poultry manure	6634	8760
Crotalaria Juncea	5922	7870
Sesbania rostrata	6436	7480
Control (NPK)	5166	7810
CD at 5%	7.79	1.60

IV. Meal Group of Manure

The dead or slaughtered animal supply two type of wastes such as dead carcasses and abattoir waste. In India nearly 8-10% of livestock die due to various causes. From dead animal, many commercial products such as hides, meat, fats, bones, horns, hooves, hair etc. can be obtained. The wastes from the abattoir where livestock and poultry are slaughted for human consumption are available in large quantities. The plant nutrient content in various animal wastes, have been presented in table 19.

Table 19: Nutrient Content of Meal Group Manures

Nature	N%	P_2O_5%	K_2O%
Blood-meal	10-12	1-2	1.0
Meat-meal	10-11	2.5	0.5
Horn & hoof meal	11-13	1.0	—
Raw Bone-meal	3-4	20-25	-
Steamed Bone meal	1-2	26-30	—
Fish meal	4-10	3-9	1.0

These are quick mineralizing organic manure suitable for all type of soils and crops. This group includes blood meal, meat meal, horn and hoof meal, bone meal, fish meal etc. some of them have been discussed below:

Blood meal: This is a major slaughterhouse waste product which contains 10 to 12% N, 1 to 2% P_2O_5 and 1% K_2O. Its C/N ratio ranges between 3 to 4:1. The availabilities of blood meal is estimated at about 55000 tonnes per year and only a part of it is utilized. It decomposes quickly in the soil and can be used as fertilizer any time during the growth of the crop.

Meat meal: Solid Slaughterhouse wastes consist of wastes, intestines etc and have good manurial value. Meat meal is prepared by sterlising and rapidly reducing the moisture content of the meat of the fallen animals in wet rendering plants. It contains about 10% N, 2.5% P_2O_5 and 0.5% K_2O. It has a C/N ratio ranging between 2 and 3:1.

Tankage: It consists of refuse meat that accumulates in packing plants and meat market together with meat of dead animals. The material is cooked in steam pressure, which liberates fats, most of which is skimmed off. The meat is next subjected to pressure to remove as much as possible of the remaining fat to make it suitable for fertilizer use. A better quality tankage is also used for feeding purpose.

Horn and hoof meal: The horns of cattle, buffalo, and sheep and hoofs of cattle, buffalo, sheep, goat, horse and pig are of considerable value. The horns and hoff are pressed for recovery of glue. The ground horn and hoof meal is a high grade nitrogen carrier and is in great demand for use as manure in tea gardens and coffee plantations. It is obtained by grinding of hoof and horns after drying. It contains 11-13% N, 1% P_2O_5 and 2.5% lime.

Fish meal: Sea food canning is an important industry in almost all the maritime states of the country. The wastes arise mainly during processing due to low market value of non-edible varieties etc. Prawn shell and head, trash fish and frog legs are the main by products available in the country. Prawn waste is used either as manure or for supplementing cattle and poultry feed. Trash fish which constitutes 15 to 30% of the total marine harvests are generally not preferred for table purpose because of their small size, bony structure and poor taste. These are converted into fish- meal or fish-manure.

Fish-meal processed by drying non-edible fish or wastes from fish industry is well balanced organic manure containing appreciable quantities of nitrogen and phosphorus. Fish-meal is also produced by processing in steam digesters and then drying. The total N varies from 4 to 10%, P_2O_5 from 3 to 9% and K_2O about 1.0%. It has C/N ratio of about 4 to 5:1. The use of fish manure is common in the coastal areas of India.

Fish tankage: Fish tankage, fish scraps, dried ground fish meal (fertilizer grade) is dried. The products consists of 6.5% to 10% N and 4-8% P_2O_5.

Acid fish: It is similar to fish tankage except that it has been treated with sulphuric acid to prevent its decomposition. Acid fish contains 4 to 6.5% N and 3-6% P_2O_5.

Fish meal/tankage: It consists of offal from fish canneries and entire body of non-edible fish such as menhaden and dog fish. This waste is low in N but high in P_2O_5 as compared to whale fish. Its composition is variable. In Manhaden oil industry, raw fish is streamed under pressure, oil is removed, the pulp is dried and ground. Whale meat may be treated and used in the same way.

Tannery waste: Hair is a tanney waste which contains 10.5% N. Leather meal is obtained from scraps and other remains from the manufacture of shoes and other leather goods. The leather meal is prepared by processing and grinding of scrap and has 6 to 12% N. It also contains small percentage of P_2O_5 and K_2O.

Wool waste: It consists of short wool and 'dirt' from raw wool during cleaning and contains 3 to 8 percent N. The processing or treatment of wool wastes and similar nitrogenous material such as leather and hairs waste with strong acid make the nitrogen in these material available to crops. This may be accomplished by mixing of rock phosphates and sulphuric acid for producing of wet base manure.

Bone meal: It consists of bones with small quantities of adhering material. Bones accumulate as a by product in packing and rendering plant. Bone meal is obtained as powdering by crushing of bones and is used as a fertilizer.

Uses: Sterilised bone-meal is an important mineral supplement in livestock feed, yet it is used chiefly as phosphatic fertilizer. Small quantities of nitrogen are also added when applied to soil. The availability of phosphorus from bone meal depends on the particle size, the finer the particles, the greater the phosphorus availability. It is available in two forms (1) raw bone-meal (2) steamed bone-meal.

Raw bone-meal: It consists of finely ground and dried material. It contains from 3 to 4% nitrogen and from 20 to 25% P_2O_5. The percentage of nitrogen and phosphorus varies with the quality of bones and age of animals. According to the standards laid down by Indian standard institute, the raw bone meal must fully pass through 2.3 mm I.S. Sieve of which not more than 30% shall be retained on 850 micron I.S. Sieve.

Steamed bone-meal: Steamed bone meal is obtained by boiling raw bones under steam pressure and then it is finely ground. During boiling, much of the fat and some of the nitrogenous material are removed.

Steaming increases the percentage of phosphorus and reduces nitrogen content in the steamed bone-meal. Due to removal of fat from the bones, steamed bone-meal become porous, brittle and easy to grind.

Steamed bone-meal also decomposes more rapidly in the soil than raw bone-meal. Accordingly to I.S.I. Standard, not less than 90% of the material should pass through 1.18 mm I.S. sieve. The other specifications regarding the chemical composition of raw and steamed bone-meal have been laid by I.S. I. as follows:-

Raw bone-meal: Maximum moisture 8%, minimum total P_2O_5 by weight, 20% minimum available P_2O_5 (2% Citric acid) 8% and minimum N by weight 3%.

Steamed bone-meal: Maximum moisture 7%, minimum total P_2O_5 by weight 22% available P_2O_5 16% nitrogen nil.

Bone tankage: This term is applied to tankage containing a high proportion of bone which is too much to class it as bone meal. Nitrogen content varies widely.

Uses: Sterilized bone-meal is an important mineral supplement in livestock feed. It is also used mainly as phosphatic fertilizer. It is considered useful in all soil having good drainage. It is less effective on heavy soils and calcareous. In such soil, use of phosphate solubilizing bacteria such as *Pseudomoas striata* or *Bacillus polymyxa,* as seed or soil inoculants can result in solubilization of bone-meal phosphorus and improve its efficiency (Gaur, 1990). It has particularly notable effect on soil which are well supplied with organic matter. The effect of organic matter also stimulates the soil microbial population including that of nitrogen fixing bacteria and phosphate solubilizing microorganisms (Bhardwaj and Gaur, 1985, Gaur, 1990, 1999). Paddy, wheat and other cereals crops respond very well to bone-meal particularly in acidic soils and other soils inoculated with phosphate solubilizing microorganisms. Sugarcane, vegetables, fruits, leguminous plants, pastures, and grasses are all benefited by bone-meal.

Bone meal is applied to the soil at sowing time or just before it. It is preferably drilled in the soil. A dose of 112-224 Kg bone-meal/ha is sufficient for most cereal crops. For vegetable and fruit crops about 500-600 Kg bone-meal/ha is applied.

Nitrogenous tankage: It is made from wastes of wool, silk, hair, feather, fur scrap, leather, felt which are treated with dilute sulphuric acid under pressure. The product is dried and ground. It is good nitrogen fertilizing material containing 6-10 percent nitrogen.

Garbage tankage: City waste containing both plant and animal refuse waste are cooked under pressure. The water is pressed out grease is removed and finally ground for use as organic manure. The water extract portion can be used for spraying the plants.

V. Seafood Wastes as Manure

The use of organic materials such as scrap fish and fish wastes as fertilizer was popular practice until the early twentieth century when inorganic fertilizer technology developed. By 1930 low cost inorganic fertilizers had replaced most organic materials. From 1950 to 1970, the dominance of inorganic fertilizers on the commercial market was nearly complete. Fish waste fertilizers were produced only by isolated small plants for speciality application. In recent years, energy and environmental concerns and rapid price increases for inorganic fertilizers have renewed interest in the use of waste organic materials of all kinds for fertilizers.

Processing and Application

Seafood wastes may be applied as fertilizer either unprocessed (fresh) or processed (dried, composted, etc.). Unprocessed seafood wastes have undesirable characteristics, such as:

1. High water content (60 to 90%)
2. Extreme liability to becoming putrid, due to the high lipid content.
3. Difficult application, requiring specialized equipment.
4. Seasonal production.

The high water content makes economical transportation possibly only for very short distances (less than 25 km) because of their putrescibility, unprocessed seafood wastes have to be removed from the processing plant and applied immediately. Surface applied wastes can create odors and attract flies and rodents; therefore, these must be immediately tilled into the soil. Other possible application techniques are the plough furrow cover method, in which wastes are deposited in a ploughed furrow and immediately covered beneath the soil surface.

Seafood wastes can generally be applied uniformly with a manure spreader. However, grinding of the waste at the processing plant will result in a more homogeneous material that is easier to apply.

Table 20: Typical composition of Oregon Crab and Shrimp Processing Wastes.

Constituent	Fresh (%)		Dry basis (%)	
	Shrimp waste	Crab waste	Shrimp waste	Crab waste
Nitrogen	1.3	1.6	5.6	4.4
Phosphate (P_2O_5)	1.0	1.2	4.3	3.3
Potash (K_2O)	0.05	0.30	0.22	0.83
Sulphur	0.06	0.18	0.26	0.50
Lime ($CaCO_3$ equivalent)	6.0	15	26	42
Magnesium	0.18	0.33	0.78	0.92
Water	77	64	—	-

Source: Costa and Gardner, 1978

The typical crab and shrimp processing waste characteristics are shown in table 20. The $CaCO_3$ content of shellfish wastes acts similarly to lime and is very beneficial for acid soils, shellfish wastes, particularly shrimp wastes, are relatively deficient in potassium.

The seasonal nature of the sea foods processing industry may conflict with the farmer's requirement for fertilizer. The practicality of using seafood wastes in a particular locality will depend upon the demand for fertilizer during the seasonal seafood processing periods. However, the wastes are organic and will be mineralized slowly.

Where sufficient local agricultural land is available, fertilizer demand is sufficient during processing seasons, and cooperative arrangements between seafood processors and farmers can be developed, the use of unprocessed seafood wastes as fertilizers may be a very economically attractive method for processors to dispose of wastes and for farmers to supply nutrients to their crops. This method is especially applicable to small, isolated seafood processing plants, where shipment to a centralized waste processing plant is not practical.

Processing of seafood wastes may be practiced to increase the stability, reduce the water content, increase nutrient concentrations, and produce a fertilizer which is convenient to apply. In a dry, stable state, storage and transportation over relatively long distances to available market is economically practical. The primary methods by which seafood wastes may be processed included drying, composting, and extraction of valuable components.

The most direct approach for producing a dry, stable fertilizer is drying. Drying and grinding of seafood wastes into a fertilizer-grade meal would be very similar to a fish meal process. Fryer and Simmons (1977) list this product as a 10:4:1 fertlizer. Composting is a method of converting organic matter into a drier, non-odiferous form through bacterial action.

References

Acharya, C.N. 1958. Preparation of fuel gas and manure by anaerobic fermentation of organic materials. Res. Ser., ICAR, New Delhi pp 58.

Alexander, D.C., Carrier, J.A.J. and Mckay, K.A. 1968. Can. Vet. J. 9: 127-131

Balch, W.E., Fox, G. E., Magrum, L. J., Woeso, C. R. and Wolfe, R. S. 1979. Methanogens: Reevaluation of a unique biological group. Microbiol Rev. 43: 260-96

Badger, D. M., Bogue, M. J. and Stewart, D. J. 1979. Biogas production form crops and organic wastes N. Z. J. Science. 22: 11-20.

Bardiya, N. and Gaur, A.C. 1997: Carbon and Phosphorus ratio for methane production from rice straw in batch fermentation. Indian Jour Microbiol. 37: 81-84

Bhardwaj, K. K. R. and Gaur, A. C. 1985 Recycling of organic wastes. ICAR, New Delhi pp 104.

Bryant, M. P. Wolin, E. A., Wolin, M. J. and Wolfe, R. S. 1967. Methanobacillus omelianski a symbiotic association of two species of bacteria. Arch. Microbiol 59:20-31.

Bryant, M. P. Campbell, L. L., Reddy, C. A. and Crabill, M. R. 1977. Growth of Desulfovibrio in lactate or ethanol media low in sulphate in association with H2 utilizing methanogenic bacteria. Appl. Environ. Microbiol 33:1162-69.

Budhar, M. N. Palanippan, S. P. and Rangasamy, A. 1991, Indian J. Agron. 36(2): 251.

Chawla, O. P. 1986. Advances in Biogas Technology, ICAR, New Delhi.pp 144

Dague, R. R. 1968. Water Pollution control Federation 40: 2021- 32

Daniel, S. J., Bhorsekar, M. R. and Mullick, D. N. 1967 studies on utilization of home grown roughage by dairy cows. I. Use of leguminous fodder during winter season. Indian J. Dairy Science. 20:181

Daniel, S. J., Bhorsekar, M. R. and Mullick, D. N. 1968 studies on utilization of home grown roughage by dairy cows II. Use of different roughage during summer season. Indian J. Dairy Science. 21:43.

Deshpande, P., Saranaik, S. Godbole, S. H. and Wangle, P. M. 1979. Use of water hyacinth as an additive in biogas production. Curr.Sci. 58: 490-492.

Dierk, D. 1990. Biomanure from small biogas plants GATE questions and answers informations 2/90:15-21.

Doddema, H. J. and Vogels, G. D. 1978. Improved identification of methanogenic bacteria by fluorescence microcopy. Appl Microbiol 36:752-54.

Dusburg, J. M, Smith, M. S. and Doran, J.W. 1989. In: Dynamics of SOM in Tropical Ecosystems (Eds. Coleman *et al.*) University of Hawaii, USA p 245.

Ferry, J. G. 1992. J. Gen. Microbiol 139: 371-378.

Garner, H. V. 1967. Manures and fertilizers. Bull 36 H.M.S.O. London pp 104.

Garcha, H. S. and Sharma, P. K. 1993. Effect of biogas effluent slurry for mushroom cultivation. In: Biogas Slurry Utilisation. Consortium of Rural Technology New Delhi 131-134.

Gaur, A. C. and Singh, R. 1982. Integrated nutrient supply system. Fertilizer News 27:87-98.

Gaur, A.C. 1990. Phosphate Solubilizing Microorganisms as Biofertilizer. Omega Scientific Publishers, New Delhi pp.176.

Gaur, A. C. 1999. Microbial Technology for composting of Agricultural Residues by Improved Methods, ICAR, New Delhi, pp

Gaur, A. C. Neelakantan, S and Dargan, S. 1995. organic Manures, ICAR, New Delhi pp 159.

Gaur, A. C. and Geeta Singh 1995. Recycling of rural and Urban wastes through conventional and Vermicomposting. In: Recycling of Crops, animal, human and industrial wastes in agriculture (Ed. HLS Tandon) FDCO, New Delhi.

Goel, B. B. Singh, K. B. and Singh, K. P. 1973. Availability and disposal of dung in India: Indian J. Anim. Sci. 43: 671.

Gopal Krishna 1971. studies on energy and protein requirement for milk production in India dairy animals Ph.D Thesis, Agra University, Agra.

Hayes, T. D. and Theis, T. L. 1978. J. Water Pollution Control Federation 50:61-72.

Hobson, P. N. and Shaw, B. G. 1974. The bacterial population of piggery wastes anaerobic digesters. Water Res. 8:507.

Hobson, P. N. Bousfield, S. and Summers, R. 1981. Methane production from agricultural and domestic wastes (Ed. Andrew Porteous) Applied Sciences Publishers Barking, England pp 269.

Hodgson, J. F. 1969. Soil Sc. Soc. Am. Proc. 33: 68-75.

Idnani, M. A. Laura, R. D. and Chawla, O. P. 1969. Indian Farming 19 (1):35.

Jai Kishan and Hussain, K. Q. 1977. Utilization of poultry excreta as sources of nitrogen for the growing calves. Indian J. Anim sci. 47: 790.

Jeet Singh, Jain M. C. Sushil Kumar and Singh, K. D. 1987. Effect of biogas slurry applied in conjunction with chemical fertilizers on wheat yield and fertility status of soil. Proc. Bio-Energy Soc. Fourth Convention Symp. (BESI) Udaipur, pp 246-250.

Khan, G., Gupta, S. K. and Banerjee, S. K. 1981. J. Indian soc. Soil Sci. 29: 19

Krishna Singh and Mudgal, V. D. 1967. The comparative utilization of feed nutrients from Lucerne hay in buffalo and cros-bred Zebu heifers. Indian J. Dairy Sci. 20:142.

Kataoka N, Tokiwa, Y and Tekeda, K. 1991. Apppl. Environ. Microbiol. 57:3671-73.

Morino, K. 1975. Engineering Characteristics on animal wastes as manures, fuel and feed. All India Faculty Programme on waste and feed handling for animal production. PAU, Ludhiana Jan, 1975.

Mudgal, V. D. and Ray, S. N. 1967. Studies on the nutrient requirement for growth III calcium and phosphorus metabolism in claves with three levels of protein intake Indian J. Dairy Sci. 20: 5.

Nagar, B. R. 1978. J. Scientific and industrial Research 37:445- 48.

Narahari, D. 1999. Fertilizer value of poultry excreta. Agro India 4 : 4-5.

Neelakantan, S. 1975: Assessment of the availability of cowdung for potential gobar gas production. Agricultural situations in India 30:633.

Neelakanta N, S. and Singh, K. 1975. Assessment of dairy cattle wastes and efficient utilization Indian Dairy m 27: 233.

Nodar, R., Acea, M. J. and Caraballas, T. 1990 Biol. Waste 33: 95-105.

Oblisami, G. 1992. Enhanced biomethavation employing appropriate inocula for alternative feed stocks In: Alternative feed stocks for biogas. Consortium on Rural Technology. New Delhi pp 1-9.

Rajasekaran, P. Murugesan, R and Palnisamy, A. 1986. Influence of temperature on microbial members and biogas production of some anaerobically digested wastes. Agric. Wastes 17: 83-89.

Rajsekaran, P. 1981. Effect of fish and chicken wastes on microbial population and biogas generation. Madras Agric. J., 68: 444-449.

Sanchez, P. A. 1976. Properties of Management of Soil in Tropics. John Wiley and Sons, New York, USA pp 618.

Sebastian, L. 1967. The comparative utilization of feed nutrients for milk production by Zebu and buffalo. MSc. Thesis, Punjab University, Chandigarh.

Sekhon, G. H. 1997. Nutrients needs of irrigated food grain crops in India and related issues. Proc. Symp on plant Nutrients Needs, Supply, Efficiency and Policy Issues. 2000-2025. Nat Acad. Agric. Sci. New Delhi pp78-90.

Senapati, B.K. 1993. Vermitechnology in integrated waste biomass management: Concept, prospects and problems. In Biogas slurry utilization. Consortium of Rural Technology, New Delhi. pp 57-74.

Sims, J. T. and Wolf, D. C. 1994 Adv. In Agron. 52:2-72.

Singh, K. and Srivastava, O. P. 1971. International Symp. On Soil Fertility Evaluation Proc. 1: 815.

Singh, K. Minhas, M. S. and Srivastava, O. P. 1973. Indian J. Hort. 30: 537.

Singh, P. K., Singh, D. P. and Satpathy, K. B. 1993. Use of cattle slurry and animal dung for *Azolla* production and its utilization as a biofertilizer for rice. In: Biogas slurry utilization Consortium of Rural Technology, New Delhi pp 135-143.

Srivastava, O. P. 1988. Indian J. Agric. Chem. 21: 1.

Smith, P. J. 1966. The microbiology of sludge methanogenesis. Development in Industrial microbiology 7: 156-161.

Tripathi, A.K. 1993. Biogas slurry a boon for agricultural crops. In : Biogas slurry Utilization. Consortium on Rural Technology, New Delhi. 11-14.

Tripathi, S.D. Ayyappan, S, Madhumita Das and Sandhya Bhandari. 1993 use of Biogas slurry in Aquacuture, In: Biogas slurry utilization. Consortium on Rual Technology, New Delhi pp 97-98.

Venkatraman, G. S. and Kausik, B. D. 1975. Stimulation of gas production from cowdung by algal supplementation. Curr. Sci. 44 :807-808.

Wolverton, B. C. and Mc Donald, R. C. 1979. The water hyacinth from prolific pest to potential provider AMBIO, 8(1): 2-9. Royal Swedish academy of sciences.

Widdel, F. 1986. Growth of methanogenic bacteria in pure culture with 2-prapanol and other alcohols as hydrogen donors. Appl. Environ. Microbiol. 51: 1056-62.

3

Agricultural Residues, Tree Wastes Green Manure and Urban Wastes

I. Crop Wastes

Crop residues are the main sources for recycling of the organic and plant nutrients removed from the soil during the growth of crops. These residues have competitive uses such as fuel and cattle feed. There was an estimated demand of 583.6 million tonnes of dry fodder against the supply of aljout 398.8 million tonnes (Singh, 1997). Besides the use of crop residues as animal feed, industrial uses of crop residues also pose competitive demands. The potential of crop residues of some principal crops was worked out by Bhardwaj (1995). It was assumed that not more than one third of the total crop residues (273.63 million tonnes) was available for recycling as plant nutrients (Table 21). Out of a total 5.61 million tonnes ($N+P_2O_5+K_2O$) available in the form crop residues, only 2.47 million tonnes can be mobilized either through *composting* or *in situ* incorporation or mulching. Based on these estimations, Sekhon (1997) projected that approximately 4.82 and 5.79 million tonnes of $N+P_2O_5+K_2O$ will be available from recycling of crop residues by the year 2011 and 2025 with utilizable value of 2.21 and 2.26 million tonnes respectively (Table 22).

Table 21: Availability of some crop residues in India and their plant nutrient potential.

Crop	Residue to economic yield ratio	Residue yield (Thousand tonnes)	Nutrient potential as NPK (Thousand tonnes)	
			Total	Recyclable
Rice	1.5	1,10,495	2,398	799
Wheat	1.5	82,63-1	1,504	501
Sorghum	1.5	12,535	262	67
Maize	1.5	11,974	252	84
Pearlmillet	1.5.	6,967	121	40
Barley	1.5	2,475	51	17
Fingermillet	2.0	5,351	118	39
Sugarcane (Striped Cane)	0.10	22,736	423	423

Potato tuber	0.5	7,867	141	141
Groundnut pods	1.5	10,598	339	339
Total	——	2,73629 or	5609 or	2470 or
		273.63 m.t	5.61 m. t.	2.47 m. t.

Damodar Reddy and Subba Rao (1997) also estimated that about 376 million tonnes- of crop residues are available annually in the country. They assumed that about 50 percent of available plant residues are utilized as cattle feed and fuel. The remaining residues, should be recycled in organic farming which can meet about 25% of total annual NPK requirement in India.

Table 22: Total and utilisable amounts of plant nutrients from crop residues.

Plant Nutrients (million tonnes)			
Crop Residues	1991*	2011**	2025**
Total	3.32	4.82	5.79
Utilisable	1.52	2.21	2.66
Fertilizer equivalent value	0.76	1.10	1.33

Sources: *Bhardwaj and Gaur (1985) *Sekhon(1997)

The resources surveys regarding the availability of farm wastes in different agro-climatic zones by research / extension workers in general revealed as followed.

1. Major amounts of crop residues are used as animal feed and fuel in rural India. Only a small quantity of the crop residues remain unutilized or wastes. Even the left over organic wastes are not being used scientifically for application to cultivated land or for converting it into compost. The general trend is to bum off the left over crop residues. Farmers generally rely mostly on animal dung for making farmyard manure.
2. Considerable amount of cattle dung in used as fuel in the form of dried dung cakes. This practice is highly intensified in the Indo-gangetic plain comprising the states of Bihar, Haryana, Punjab and Uttar Pradesh where about 40% of the bovine dung is used for making fuel dung cakes. The reason is dearth of alternative sources of fuel. However, in hill area such as Himachal Pradesh, about 95% of the total bovine dung is used for making farmyard manure. The reason is the availability of wood for fuel purpose. Nevertheless, the surveys revealed that the farmers are fully aware about the value of animal wastes, but do not follow proper or improved methods for making compost from animal dung or other available organic resource.

3. The regional conditions of farming such as mechanisation of agriculture has an influence on the availability of crop residues as . wastes. Rice straw is a complete waste in Punjab, Haryana and western Uttar Pradesh where as it is largely used as animal feed in other parts of India. On the contrary, in states like Maharashtra and Kamatka, wheat straw is not preferred as animal feed.
4. Some of the crop residues such as sugarcane trash, sunflower and safflower straw and residues, potato haulms, groundnut shells and groundnut haulms have been found to be exclusive wastes for recycling purposes.

II. Tree Wastes

Plantation Crops: Plantation crops such as coconut, arecanut, cashew, oil plam, tea, coffee, rubber and cocoa are important high value commercial crops grown in the states of Kerala, Tamil Nadu, Karnataka, Andhra Pradesh Assam and West Bengal. The total area under these crops excluding cocoa cover an area of about of about 3.43 million hectares in India. The area in hectares under different crops is as follows: coconut 1 6,68,000, arecanut 2,38,000, cashew 3,81,000 rubber 4,66,000, tea 4,21,271, coffee 2,42,201. Plantation crops occupy the land for decades and yield different types of tree residues throughout the year in contrast to field crops which generate biomass on season-basis. Hence large amount of plantation organic wastes in the form of leaf fall, stems, twigs and other plant parts becomeavailable in the plantation area. Because of their perennial nature, plantation crops remove very high quantities of plant nutrients from the soil. Leaves, sheath, husk and bunch wastes of arecanut nemove 95,10 and 110 kg of N, P and K per hectare (Bhat and Mahapatra 1 989). Therefore, utilization of all kind bf wastes and byproducts of plantation crop is essential to return the plant nutrients to soil removed by the crops.

Nair *et al.* (1997) reported that 1,65000 tonnes of $N+P_2O_5+K_2O$ can be supplied through wastes of plantation crops (Table 23)

Table 23: Plant nutrients supplied by some plantation crops (OOOt.)

Crop	2000 AD			2025 AD		
	N	P_2O_5	K_2O	N	P_2O_5	K_2O
Coconut	79.2	7.6	49.5	460.7	44.2	287.9
Arecanut	6.8	3.7	9.2	15.8	8.6	21.0
Cocoa	1.0	0.32	2.87	16.2	4.9	44.8
Coffee	4.7	0.23	1.22	7.9	0.39	2.1
Total	91.71	11.85	62.79	500.6	58.09	355.8

Coconut Plantation: Coconut is cultivated in a very large area. The possible uses of coconut organic wastes have been discussed in the following pages.

Botanically, the coconut shell is the endocarp of the coconut fruit. Nearly 67 per cent of coconut production is concentrated in Kerala only, followed by Tamil Nadu (16.0%), Mysore (10.5%) and Andhra Pradesh (2.6%). The annual availability of shells from the coconut produced in the country is estimated to be about 0.88 million tonnes. The world production of shells amounts to about 3 million tonnes, concentrated mostly in Netherlands, Philippines, India, Ceylon, Malaya and New Guinea.

Chemical Composition: Coconut shell contains about 50 per cent cellulose and 35 per cent lignin. The ash from the burnt shells contains a fair amount of potash (about 45%), sodium (15%), calcium (6%), phosphorus (4%), and other elements in small quantity. The moisture content of shells varies widely according to maturity and storage. The percentage composition of ash varies with the habitat, variety of coconut and maturity of the drupe. It does not contain any oily or resinous substance.

Domestic and Ornamental Uses: Shells are generally used for domestic and ornamental purposes-such as drinking cups, bowls, smoking pipes, lamps, spoons, sugar and tea pots, flower vases, paper weights, bangles, buttons, combs, ladles, cuff-links, musical instruments and a variety of other fancy articles. The chief advantage of coconut shells for-which these are favoured for the manufacture of artistic articles is that these can be carved beautifully, decorated with lacquer and inlaid with ivory, silver or other metals so as to exhibit an ornamental effect. Moreover, these are available in different sizes and shapes so as to serve various purpose.

Fuel: The shells are used as fuel in south India and Ceylon mostly in kitchens and to a lesser extent in laundries, smitheries, bakeries, brickyards and limekilns. It is not favoured as a boiler fuel because of the corrosive nature of its vapours on metal parts mostly due to the acidic nature of the products of combustion when air intake is insufficient.

Industrial Uses

Shell Charcoal: Coconut shell contains about 52 per cent carbon, 42 per cent oxygen and (*per cent hydrogen. On burning these shells in suitable kilns in a limited supply of air, charcoal is obtained. A good quality charcoal does not contain more than 10 per cent moisture, 2.0 per cent ash, 0.1 per cent chloride, 15 per cent volatile matter and the particle size not more than 5 per cent shall pass through 0.63 cm mesh sieve. As it contains a smaller percentage of ash, so it can be used as a fuel for automobile gas generators if provided in better shape and form. Shell charcoal is used mostly in laundry, smithy, bakery etc. Well powered shell charcoal is used as a dentrifice. A substance based on coconut shell carbon for the defluoridation of drinking water has also been reported.

Activated Charcoal is extensively used for making active carbons which are of gas absorbent grade. These carbons are used in gas masks, fertiliser, rayon industries and for the recovery of solvent. Regional Research Laboratory, Hyderabad has developed a process to manufacture active carbon from coconut shell (NRDC process No. 232 102). The process consists of crushing the coconut shell to the required size, treating with

zinc chloride and activating in a rotary kiln. The activated material is washed, pelletised, dried and packed. The product obtained conforms to the following specifications: benezne adsorpotion at 30°C, 45 - 55 per cent; ash content, not more than 2 per cent; surface area, 1100-1200 m^2/g and pore volume, 0.7-0.08 m^2/g.

Destructive Distillation Products: By destructive distillation of coconut shell (say 1 ton), the primary products obtained are uncondensable gases (300 kg) pyroligneous acid (350 kg), settled tar (80 kg) and charcoal (300 kg) the portion which remains in the carbonising apparatus. By redistillation of pyroligneous acid, wood naphtha- a mixture of acetone, methanol and acetic acid are obtained. Settled tar, on fractional distillation, yields phenols and cresote. The remaining portion called pitch can be used for road making.

The carbonised shells from the first still when treated suitably with steam and acid will give active carbon which has got great commercial importance. It is used in large quantities in the refining of vegetable oil, sugar, molasses, fruit juices, glycerine syrups and also for controlling flavour and odour of water especially in the beverage industry. Further, it is used in pharmaceutical industry for its absorbent action on alkaloids, enzymes, toxins and poisons of various kinds. Active carbon is also used in the natural gas industry for the recovery of gasoline and solvent vapours such as petroleum ether, sulphuric ether, benzene, alcohol, ethylacetate and similar materials from air. It is also used for the recovery of fermentation products such as alcohol, butanol etc. and for the purification of gases.

Furfural Yield: Coconut shell contains about 30 per cent pentosans which can yield about 15 per cent furfural. Furfural is a chemical of great importance for a variety of rapidly developing industries, like petroleum, nylon, vegetable oils, plastics, synthetic rubber, resins and for the preparation of many other economic products (Anonymous, 1975).

Coconut Shell Flour: It is used as a compound filler for synthetic resin glues and as a filler and extender for phenolic moulding powders. It gives lustrous finish to moulded articles and improves their resistance to moisture and heat. It is obtained by breaking shells into pieces and then grinding into fine particles of 100-500 mesh. It has a great potential particularly in .the plastic industry but some problems are acting as bottlenecks in its maximum exploitation. Lack of suitable machinery is one of the main impediments. In the existing process, the resultant powder of 100 mesh is hardly to the extent of 60-70 per cent and as such becomes uneconomical. Secondly, availability of other flours such as wood flour, walnut shell flour, coirpith ends etc. are acting *as* competitors in its export potential.

It is evident from the above account that coconut shell is the most versatile part of the world's most versatile tree - the coconut palm.

Coconut Pith

Coir fibre is extracted from coconut fruit husk and in the process unusable coir dust is generated particularly near fibre extraction plants. Coir dust is thus the non-fibrous light fluffy tissue of coconut husk which is a waste product of the coir industry. This product constitutes nearly 60 per cent of weight of the husk. It accumulates in abundance in the coir extraction centres in the coastal areas of Kerala. Considering the vast quantities of coconut pith available in factories manufacturing rubberised coir mattresses and also the disadvantages it offers to the fibre industry, it is highly worthwhile to explore the avenue for its utilization.

Chemical Composition of Coir Pith

On an average, coconut pith contains about 35 per cent cellulose and 25 per cent lignin. The composition of samples obtained from various centres would vary depending upon the proportion of fibre shorts present as well as the source from which it is obtained i.e. retted or dry decorticated husks. (Table 24)

Ash content amounts to about 8.7 per cent with very low nutrient contents, N 0.30 per cent; P_2O_5 0.05 per cent; K_2O, 0.90 per cent and CaO, 0.40 per cent and sulphur 0.20 per cent.

Coconut pith can be utilized as building material *as a* sources of furfural, in agriculture and other purpose.

Table 24: Composition of Raw Coir Pith (%)

Moisture (dried at 105°C)	11.90
Ash	8.70
Fats and Resins	1.80
Cellulose	35.00
Lignin	25.00

Pentosans	7.45
Loss of material on boiling with caustic soda	12.00
Fixed carbon	37.10
Sulphur	0.20
Posphorus	0.11

Source: Anitha D.R (1991)

In Building Industry: Particle boards prepared from wood, begasse etc. require the normal proportion of adhesive (6 to 9 per cent by weight) for bonding the particles; but coconut husk / pith is unique in that it has considerable self-bonding property. This can result in greatly reduced cost of ptoduction of particle boards. Other inherent advantages of boards from this waste are resistance to fire and decay.

Tropical Products Institute, London developed a process of making building slabs or hard boards from coconut pith but it was found that the presence of small proportion of soluble substances in this waste inhibits the setting of cement, thus making it necessary to use a rich mix. Therefore, it seems unlikely that coir dust can be used satisfactorily in the manufacture of hard board, for any application involving the use of water, unless the manufacturer is prepared to use extreme compressive pressures and still greater proportion of resin.

The possibilities of partial substitution of phenol in phenol formaldehyde resins by various lingo-cellulosic materials was investigated. Usually the plywood boards are made by using the product obtained through the condensation of phenol (100 parts), formalin (185 parts) and sodium hydroxide (5 parts) but replacement of a part of phenol in this formula by coconut pith even gave satisfactory results (Annonymous, 1968).

Hume (1949) described a process for preparing hardboards from a mixture of coconut husk-shorts and coconut pith. It was claimed that durable unlaminated boards of any thickness capable of being sawed, nailed and glued could be produced. In addition, this board could be subjected to a process of petrification which renders it completely insect and weather proof as well as imparts stony appearance capable of taking colour finishes.

Coir dust in combination with rubber can form composition flooring, ceiling boards and similar material. Joint experiments carried out by Coconut and Rubber Institute in Ceylon showed that these products can keep well without deterioration for a period exceeding five years.

Source of Furfural: The available analytical data shows that the organic matter in coconut pith contains about 8.1-11.4 per cent of pentosans. On this basis a recovery of 4-5 per cent of furfural from coconut pith could be expected.

Furfural is an important industrial chemical used in most of the expanding industries like petroleum, nylon, vegetable oil, plastic, synthetic rubber, resins and for the preparation of many other economic products (Anonymous, 1975).

Concrete: Coconut pith is a durable light weight material having a very high thermal insulating value. The Central Building Research Institute, Roorke has developed a process to manufacture concrete from retted coconut pith and portland cement. The novelty of the product lies in the fact that compared to the conventional insulating material-like concrete, the cement-coconut pith concerete is much lighter, easier to apply and gives better thermal insulation for equal cost.

Expansion joint filler finds wide applications in all such construction work where joints in slabs, beams, shells, prefabricated components etc. are to be filled up.

For the Manufacture of Gaskets: At present the rubberised cork sheets are prepared from imported cork granules. National Chemical Laboratory, Pune has developed a process for the manufacture of gaskets from coconut pith (NRDC Process No. 83091). This process consists of mixing the pith with neoprene and other ingredients and vulcanising the mix in a press under pressure and temperature in suitable moulds. These gaskets could be used in automobile engines and similar equipment as a substitute for rubberised cork sheets.

From this brief account, it is evident that coconut pith holds a great potential particularly as a partial substitute for phenol in phenol formaldehyde resin adhesives, as a filler for resin adhesives, as "a raw material for moulded products, and in the rubber and electro-chemical industries as well; but the main problems in the use of this bulky material is the difficulty of its drying to the desired moisture content. The material as available is very moist, containing over 500 per cent of moisture. So drying is an important and necessary operation before it can be economically exploited.

Use of Coir Pith in Agriculture

These wastes are available in abundance in South India. According to Arumugham and Damodharan (1993), there are approximately 84,000 retting and coir extracting units in Kerala producing white fibre and about 650 brown coir units located in Tamil Nadu, Kamataka and Andhra Pradesh. Direct use of unretted coir dust should be discouraged as these contain 8-12 per cent soluble tannin-related phenolics which may act as plant and microbial inhibitors including immobilization of soil nitrogen.

The composition of unretted coir dust contains moisture 11.9%, lignin 30%, cellulose 22% and organic carbon 26.0% among its major constituents. Among the plant nutrients, it contains 0.2% N, 0.08% P, 0.8% K, 0.2% S, 0.9% Ca and 0.7% Mg (Nagarajan *et al.*, 1990)

Composted or biodegraded coir dust have been recycled by application to soil with beneficial effects on the crop yield and physical properties of soil (Nagarajan *et al.*, 1990)

Coir pith compost was prepared using *pleurotus caju* or *pleurotus* platypus as lignolytic fungal inoculant by the special technique (Singaram and Pothiraj, 1991) *Pleurotus* (*Basidiomycetes* spp.) have been found to possess lignolytic enzymes such as laccase and peroxidase that bring about faster decomposition of the coir pith (Mani and Marimuthu, 1992). The chemical characteristics of composted coir pith and the raw coirpith have been compared in table 25.

Table 25: Chemical Composition of Raw and Composted Coir Pith

Parameters	Raw coir pith	Composted coir pith
Lignin (%)	30.00	4.80
Cellulose (%)	26.00	10.10
Organic carbon (%)	29.00	24.50
Nitrogen (%)	0.26	1.06
Phosphorus (%)	0.01	0.06
Potassium (%)	0.78	1.20
C/N ratio	111.50	23.10
Volume (M3)	1.00	0.52

Composted coirpith was enriched with N,P,K and has a low C/N 23.1:1. It also contained higher amount of calcium (0.5 ppm) magnesium (0.48 ppm) manganese (0.09 ppm) zinc (25.00 ppm) and copper (6.20 ppm) as compared to raw material due to 48% reduction in the volume of the raw material during the biocomposting. This material can be used as a substitute to farmyard manure where the latter is not available in adequate amounts.

Several studies conducted in Tarnil Nadu, using 10 tonnes composted coirpith / ha increased the crop yields. (Veerabadran, 1991) Ammal and Durairajamuthiah (1996) found coirpith as a good source of potassium for rice crop.

The yield of groundnut crop was increased when a red sandy loam soil was amended with 5t / ha sawdust plus coir dust for two successive seasons (Loganathan era/. 1979). There was also a significant improvement in soil structure due to use of these amendments. Coir dust incorporated in a clay loam soil at the rate of 10t / ha resulted in appreciable improvement in a number of physical properties in comparison to FYM and press mud (Mayalagu

1983). There was a strong correlation between the pod yield of groundnut and improvement in infiltration rate, hydraulic conductivity and water holding capacity of soil.

Singaram and Pothiraj (1991) obtained significant response in maize and finger millet with the application of coir pith at the rate of 10 tonnes / ha in a tannery effluent degraded soil. The good response obtained for these two crops may be due to absorption of heavy metals by coir pith and improvement in the infiltration property of soil. In a sodic soil (pH 10.1) application of coir pith compost with gypsum and fertilizer together resulted in 33 per cent increase in grain yield of rice. Since the coir pith is slightly acidic, it has an additive effect with gypsum in lowering the pH of the soil along with an increase in exchangeable Ca and K content.

Addition of rock phosphate to acidic coir pith material (pH 5.8) favoured availability of phosphorus, while Zn SO_4 + $FeSO_4$ addition helped in formation of natural organic complexes of Zn and Fe (Savithri and Rani Perumal, 1993, Ramasamy and Son, 1993)

It is also excellent soil mulching material and has been used particularly in coconut nurseries and citrus culture.

It can be concluded that plant residues which are very slowly biodegraded act mainly as soil ameliorant and thus improve physical condition of soils for needed crop productivity. Hence, depending on the availability and economics of use, such comparatively highly lignified and wide C/N ratio materials also have their recycling potential.

Coir Pith for Biogas Production

Accumulation of this organic waste in industrial yards causes environmental pollution and fire hazards. At present biogas is produced mainly from cow dung. A trial was undertaken in the Central Plantation Crops Research Institute (CPCRI) Kasaragod to examine the feasibility of generating methane from coir pith by using either alone or in combination with cowdung. In this trial 5 KVIC model biogas plants were used (Singh *et al.*, 1998)

Initially, cowdung and coir pith were mixed in a mechanical mixer using a 3 hp, 3 phase electrical motor, a reduction gear box with output rpm of 10. The shaft was then coupled with a pipe of 75 cm having horizontal angle iron pieces of 20 cm length welded for churning and proper mixing of cowdung and coir pith. This unit was placed over a round cement tank of 1 m diameter and depth about 1.2 m. Initially, cowdung and coir pith was put in the required ratio on volume basis and water was added in the ratio of 1:1 (i.e. cowdung + coirpith: water). Four combinations were used in the trial besides the control of

100% cowdung. The combinations were 80% cowdung + 20% coirpith, 60% cowdung + 40% coirpith, 40% cowdung + 60% coirpith and 20% cowdung + 80% coirpith.

The gas output from individual biogas plants was connected to individual gas flow meters for monitoring the quantity of gas produced on per day basis. The output of all the five biogas plants was then connected to a 5 hp dual fuel engine. The gas flow was utilised to test the efficiency of biogas run engines. Initial results indicated that the combination of 80% cowdung and 20% coirpith released 10% more methane. A mixture of this gas and diesel could replace diesel fuel to run 5hp water pumps.

Tea Plantations: The major tea growing states are Assam, West Bengal, Kerala and Tamil Nadu. Tea plantations require good knowledge as tea is grown exclusively for its tender shoots. Besides, the production of commercial tea, plant provids considerable amount of prunning waste (Ranganathan 1975).

Table 26: Nutrient Content in Tea Prunnings.

Plant Part	Dry matter (t/ha)	N	P	K	C	C/N ratio
		(kg / ha)				
Foliage	7.2	252	30	72	2880	11
Twigs	3.6	85	10	21	1440	17
Secondary Wood	4.2	44	13	13	1764	40
Primary Wood	9.6	101	28	28	4220	42

The C/N relationship of ihe different parts (Table 26) and nutrient content indicate that it would decompose at a faster rate and also show high manurial value particularly in the case of the foliage residues. Therefore, the incorporation of tea prunnings in soil is commonly practised.

Tea Wastes: The dust which arises from manufacture of tea can be used as organic manure. It has a composition similar to tea leaves (Eden 1958) and would be able to supply useful amounts of nitrogen and potassium and small amount of other essential plant nutrients. In India during the course of tea production, processing and storage about 10 million kg of tea wastes become available in the form of fluffs, stalks and sweepings. It is an important raw material for extraction of caffeine. The decaffeinated tea waste can be used as organic manure or animal feed.

The consumption of instant tea has increased and spent tea leaves which contain 50 percent moisture has been utilized as soil mulch. Croyle *et al.* (1974) have examined the potential of this product as an animal feed and for composting. The chemical analysis showed that the dry matter in the spent leaves has a relatively high protein content (26.4%) but the material is fibrous (64%) and much of the fibre is lignified (lignin 36%). Furthermore, much of the nitrogen appears to be bound to the cell wall constituents which would make the crude protein fraction less available for enzymatic digestion. In view of the above reasons and since the waste may contain appreciable residues of tannins and caffeine it may not be suitable as animal feed but can be composted to a make it good quality product for nursery or garden crops. Spent tea waste has 0.28 to 3.5% N, 0.4%, P_20_5 and 1.5% K_20 with a C/N ratio of 9 to 11.

Leaf Litter of Shade Trees Associated with Tea Plantations: Succulent loppings of *Albizzia* and *Chinensis* are another source of organic matter associated with tea plantations. The data from Tocklai experimental station indicate that 2.50 to 5.0 tonnes of leaf litter from shade trees are added to a hectare of land. *Albizzia* is a nitrogen fixing leguminous tree, nutrients get mineralised faster and provides a good amount of nitrogen and other nutrients of P, K, Ca and Mg.

Venkata Ram (1978) reported a loss of 89 tonnes of organic matter from a hectare of tea garden soils during a four-year cycle in south Indian conditions. The nutrient balance (4 year cycle) was maintained by addition of 94 tonnes organic matter/ha containing 20 tonnes of leaf litter, 22 tonnes of tea prunnings, 5 tonnes of weeds and 47 tonnes of shade tree litter and loppings.

In recent years, the demand for tea grown through organic manures is also growing. Therefore efficient management of on-farm and off-farm resources of organic matter is desirable in tea gardens.

Coffee Plantations is confined to states of Karnataka, Kerala, Tamil Nadu and Andhra Pradesh in India. Plant litter, the husk obtained in processing beans for commercial coffee and litter from shade trees are the main organic residues of coffee plantations. Nitrogen fixing leguminous shade trees are invariably grown in coffee gardens. *Erithrina* spp. are of common occurrence. Aranguren *et al.* (1982) reported that coffee plantations in Venezuela received little or no fertilizers for 10 years when were grown in association with shade trees as *Erithrinaj* sp. and *Inga* sp. There was substantial reserves of nitrogen in 0-60 cm depth which was associated with litter fall dominated by shade tree litter accounting for a contribution of 86 Kg N/ha/year out of 189 Kg N/ha year. The decomposition of leaf litter was rapid. Therefore, it seems that the shade tree litter plays a very significant role in maintaining soil fertility in coffee plantations.

Coffee Waste

Coffee and pulp coffee hulls are two major main by products which are generated from the production of coffee beans from coffee cherries. Their uses in agriculture have been widely investigated over several years. Another coffee by product, coffee grounds has become available in increasing quantities due to extensive use of instant coffee which has been examined as an animal feed.

Coffee Production and Processing

Coffee cherries are preferably picked when they are fully ripe and red in colour for production of best quality coffee. Normally, each cheery contains two seed or beans covered by four layers. These layers are pericarp (fruit wall). Pulp, a slimy matter, a stiffer outer covering of the beans themselves known as the parchment, and a thin almost transparent inner lining covering the seed which is called the seed testa. All these coats have to be removed to prepare the coffee beans. The method to remove the coats such as older or dry method or wet method which is often referred as west Indian process.

In dry method, the cherries are dried either on the trees or on the stone drying grounds (barbercues). The process of drying takes almost 15 to 25 days.

In the wet method, the cherries are pulped as soon as possible after picking, within 24 hours to avoid fermentation. They are placed in large tanks full of water immediately after harvest to separate immature from ripe cherries. The formers are light and float on the surface and latter sink to the bottom from where they are sucked through pipes leading to pulpers. There are various types of pulpers, all designed on much the same principle .The cherries can not pass though them without being reduced to a pulp by rasping action of revolving cylinder or other moving surface. The beans and pulp mixture is passed into a vat full of water, where mechanical stirring causes the heavy beans to settle to the bottom and lighter pulp to be carried the away by an overflow of water. The beans are drawn off in a stream of water and then separated by being passed through a form of sieve. The slimy layer which still adheres to bean is removed by fermentation with yeast and bacteria in special vat. This takes about 18 to 24 hours, but the process can be hastened by use of enzyme preparations or 2% NaOH solution. Washing of beans with water together with stirring and agitation several times makes it completely free of mucilage. They are dried until yellow-green in colour and then either sold as such which are still covered by the last two coats or dehulled or polished before being sold. For dehulling or threshing, a variety of machines are available which are designed to crack the parchment without injuring the beans. This is quickly achieved if the material was thoroughly dry since it is easily broken. Finally simple rubbing and winnowing removes the light pieces of parchment from the heavy beans.

Instant coffee is made by treating beans usually a mixture of different varieties with boiling water. The solution obtained is freeze dried to give instant coffee.

However, a considerable amount of residue is left containing about 80 percent water. The residue is dried by heating to give dried coffee residue.

About 430 kg of coffee pulp are obtained from 1000 kg of cherries which is about 30% of original cheery. Dehulling at a later stage yields about 60 kg of coffee hulls equivalent to approximately 40 kg of dry wastes which is about 12 per cent of the cherry. The yield is about 390 kg of beans containing about 190 kg of dry matter. In the preparation of instant coffee about 67 per cent of dry matter of the beans is left in the residue.

Coffee Pulp

The high water content in coffee pulp causes problems in its transportation, handling, preservation and direct use as animal feed. Bressani *et al.* (1975) studied ensiling for storing of pulp as animal feed. The pulp is spread on a clean surface in sun to reduce its water content to about 65 per cent. The process of dehydration is achieved with 8 to 10 hours. It is then ensiled in a silo trench along with 3 to 5 per cent sugarcane molasses. The finding although did not show any significant effect on the nutritive value of residue but there was significant decrease in cafffein and total tannin content. The pulp silage is accepted by animal if fed immediately after removing from the silo. However, its exposure to air causes it darken in colour and becomes less palatable to animal. The dark colour is due to exidation of polyphenols present in coffee pulp which can possibly be controlled by using polyphenol oxidase inhibitors. Murillo *et al.* (1976) ensiled coffee pulp with equal proportions of either Napier grass (*Pennisetum purpureum*) or maize fodder. Molasses at the rate of 16 per cent was added to each mixture and the products were considered of better quality than coffee pulp silage alone.

The handling and preservation of coffee pulp is facilitated if the pulp is dehydrated in the sun or by heat. The pulp is spread in the sun on a clean surface to thickness of 5 to 10 centimeters and it is turned 3 to 4 times daily to 12 per cent moisture content in about 24 to 32 hours. During drying the colour changes form red to black but the product stores well and loses some of its caffeine. Sun dehyradation is economical but large area of surface is needed and the dark colour decrease the intake by animals.

Composition of Coffee Pulp

The mean analysis of dry coffee pulp is presented in table 27. Fresh coffee pulp generally contains 60 to 45 per cent water and dehydrated pulp has moisture of about 12 per cent.

Table 27: Chemical composition of coffee pulp

Component	Range (g/kg dry matter)
Crude Protein	90.1-128.1
Ether extract	21-29
Crude Fibre	146-240
Ash	64-96
N-free extract	534-678
Calcium	6.34
Phosphorus	1.33
Potassium	20.20
Magnesium	trace
Sodium	1.14
Zinc	4.60 (mg/kg)
Copper	5.70 (mg/kg)
Manganese	7.20 (mg/kg)

Source : J.H. Topps : 1993

There appears to be some loss of N free extract, which may be due to removal of a portion of soluble carbohydrates either by fermentation or seepage.

In general, the data show that the coffee pulp has a nutritive value equivalent to that of an average quality hay or grass silage.

The total fibrous constituents in pulp approximately half is in the form of lignin, therefore the digestibility of pulp in ruminants in not high and very low in pigs. It appears that 25 per cent of total protein is associated with lignin and may not be completely available. The availability of protein may also be affected by tannins in coffee pulp. It is known that tannins form complex with protein and render them less available for microbial digestion in the animal gut. The work in needed to examine the adverse effect, if any of tannin. Caffeine, about 1.5 per cent of the dry matter is also present in the coffee pulp and is known to have a diuretic effect that is likely to adversely affect the nitrogen retentions of animals given appreciable amount of this by product. The storage of pulp either ensiled or in dehydrate form for several months reduces the caffeine content. A reduction of 70 per cent after 15 months of storage has been registered by Bressani *et al.* (1972). The almost complete removal of caffeine is possible either by successive washing with water or treating the residue with *Penicillium crustosum* (Schwimmer and Kurzaman, 1972)

The amino acid composition of protein in coffee pulp shows that it has relatively tiigh quality for nonruminants. It appears to compare favourably with that of soybean meal, but is more deficient in sulphur-containing amino acid. This may further limit if the tannins in the pulp make some of the amino acids partly unavailable to the animal.

Coffee Pulp as Animal Feed

Studies have shown that weight gain and food intake of cattle decreased as the level of coffee pulp in the feed increased. These effects were marked when the level of coffee pulp was greater than 20 percent (Bressani *et al.*, 1975). Similar results were reported earlier with lactating cows. (Choussay 1944; Work *et al.*, 1946)

The inclusion of coffee pulp at level below 20 per cent did not affect "milk production. The depressing effect of coffee pulp on the weight gain of cattle is less marked when the level of dietary protein is high. Therefore, it appears that the value of pulp is limited by poor utilization of protein contained in it due to presence of tannins in the pulp. The coffee pulp may be considered as average quality of roughage and its inclusion in production feeds should be less than 20 percent. Ground coffee pulp as feed was not found favorable in growing pigs.

Coffee Pulp as Organic Manure

The use of coffee pulp as organic manure in coffee plantation for meeting the plant nutrients, need of the coffee cherries is well known and is obviously important in sustainable agriculture. The pulp has been used to make coffee molasses and has been examined as a part of supplementary feed for fish, *Tilapia aurea.* Coffee molasses is made by drying the sugar from the mucilage and pulp.

Composting Coffee Pulp

Where coffee is processed on the farm, the pulp may be composted with other farm waste and applied to the land. Centralization of the coffee processing for economic and marketing reason has largely disrupted this pattern of utilization of the coffee pulp as a soils amendment. Other use for the pulp have not been developed, so it has become a disposal problem and a source of pollution. The small scale methods of composting employed on the farms were not suited to composting the volumes of coffee pulp generated at central processing plants and large scale composting method had not been adapted to the characteristics of the coffee pulp. (Wilson *et al.*, 1980). It appeared that aerated pile method for coffee pulp was well suited to composting without any conditioning treatment or addition of other material since it has an initial texture that presents little resistance to air movement and has a favourable carbon : nitrogen ratio.

Preliminary tests, however, indicated that a pile of coffee pulp will settle and lose its porosity and that any air movement will be hampered.

Thus, some preliminary conditioning of the pulp appears to be essential for rapid aerobic, thermophilic composting. Conditioning methods that were considered included solar drying, dewatering with a screw press, and blending with another material.

Most of the coffee in central America is produced in mountainous area of volcanic origin; thus pumice gravel which seemed to have the essential characteristic for conditioning the coffee pulp would be readily available to most coffee processing operations. In trials of the pumice gravel as a conditioning agent for the coffee pulp, the compost piles with pumice settled less than those without. Observations of the colour and odour of the pulp also indicated that the treated piles had maintained aerobic condition through out while untreated piles were mostly anaerobic. Ratios of 3 and 4 parts coffee pulp to 1 part pumice gravel were tried with apparent success.

Four piles surface treatments were tested to evaluate the attractiveness of the coffee pulp to insects, when no cover was used, the coffee pulp dried quickly resulting in surface that did not seem to attract insects. However the dried coffee pulp would not be composted. Application of 1 cm of pumice sand had much the same effect. A burlap cover kept the coffee pulp moist, but would be difficult to keep in place if exposed to wind. One cm of dry coffee pulp compost remained dry and unattractive to insects on its outer surface and was moist with fungal activity as evident at its inner surface. Thus, it appeared that any of the surface treatments could be used for insect control and that the compost treatment would produce the least amount of dried uncomposted material.

Coffee Hulls

Coffee hulls are a dry, stable material that is easy to store and handle. Its rate of decomposition is very slow because it is covered by waxy film and has a high lignin content. The chemical composition range (g/kg dry matter) of coffee hulls (dry basis) is as follows:

Crude protein 26.3-101.7, Ether extract 6.5-20.2, Crude fiber 297-757, Ash 5.4-75.3, N free extract 204-515, Calcium 1.61, Phosphorus 0.30

Coffee hulls contain between 5 and 10 per cent water. The protein content is very low comparable to wheat straw and fibrousness of the product is pronounced, of this half is cellulose and approximately on-fourth and one-fifth are hemicelluloses and lignin respectively. The high lignin content renders the by product indigestible and it may be considered as a filler in animal feed. The value of hull may be improved by delignification using sodium or calcium hydroxide solutions (Bressani *et al.,* 1975).

Use of Coffee Hulls

The very high fiber and very low protein content make coffee hulls as unsuitable feed for non-ruminants. In case of ruminant, hulls may be included in diet for older cattle upto the level of 20 percent without appreciably affecting their performance. For young cattle, the inclusion of hulls may be more detrimental and may give a perceptible decrease in performance.

Coffee hulls have been used as fuel. In coffee plantations it is used as mulch where it slowly decomposes and acts as a soil conditioner.

Dried Coffee Residue

This is a by product from the manufacture of soluble or instant coffee. The chemical composition of the residue varies depending on the differences in the original beans and drying method. The dry residue is

stable, brown powdery material which has good storage property and is relatively easy to handle.

The range of chemical composition of dried coffee residue (g/kg dry matter) based on several chemical analyses are as follows:

Crude protein 102-130, Ether extract 225-283, Crude fiber 466-510, Ash 7-8, N free extract 143-168, Calcium 0.55-0.65, Phosphorus 0.50-0.70, Magnesium 0.15-0.25, Sodium 0.40.060, Potassium 0.35-0.45.

It also contains zinc (9-12ppm) copper (25-29ppm) and manganese (26-36 ppm). The residue contains about 5 to 10 per cent moisture. It has an average protein content of about 10 percent of dry matter but a high fiber and a relatively high fat content. Its total ash content is very low and the major elements are present in small amount.

Uses of Dried Coffee Residue

Most of the work carried out to examine the use of dried coffee residue as a feed relates to ruminant. Conflicting results have been obtained when the digestibility of by product was examined using sheep and to lesser extent cattle. In majority of these studies with sheep and cattle, inclusion of dried coffee residue in more than very small amount in the feed caused a marked diuretic effect. Campbell *et al.* (1976) reported that inclusion of 10 or 20 per cent residue more than trebelled or quadrupled the urinary output in Holstein steers respectively. This resulted in increased losses of urinary nitrogen and sodium. Another problem with this residue is that it is not readily acceptable or palatable to cattle or sheep.

Dried coffee residue has not been reported to be suitable food for non ruminant farm livestock.

Coffee residue can be used as soil conditioner. If economically / feasible, it could be small industrial source of oil and fat. Due to high calorific value, it can be used a fuel.

Arecanut is cultivated mainly in states of Karnataka, Assam and Kerala. Arecanut garden wastes include leaf litter, leaf sheath, arecanut bunch waste and arecanut husk. The NPK content of different parts (table 28) has been documented by Bhatand Mahapatra (1989).

Table 28: Nutrient content and total addition from Arecanut

Type	Av. dry biomass/ Palm (tonnes)	Nutrient content (%)			Nutrient added/ Palm/Year (kg)		
		N	P	K	N	P	K
Leaves	4.7	1.59	0,16	1.01	63.6	6.4	40.4
Leaf Sheath	1.7	0.66	0.08	1.51	11.2	1.4	25.6
Husk	1.7	1.00	0.12	2.40	17.0	2.0	40.0
Bunch	0.3	1.04	0.17	1.10	3.1	0.2	3.6

These residues have nutrient potential and are good raw materials for preparation of manure. The manuring of arecanut gardens is normally practised by using *Gliricidia* and *Mimosa* loppings, forest litter, compost, cattle manure and fish manure at the rate of 15 Kg/ palm/ year. Considering the nutrient potential of arecanut wastes, it may be advised that all the manure/ compost requirement of arecanut gardens can be met by recycling of these organic sources either by composting or *in siiu* decomposition in the soil.

Oil Palm

Major organic wastes of oil palm plantations are palm trunk, palm fronds, empty bunches, palm press fibre and mill effluent. The latter two products are wastes when palm shells are processed into palm oil in the mills. Oil palm cultivation in India is of recent origin, hence the technology for the use of oil palm wastes is not much developed. Potentially, it appears that sufficient quantity of organic wastes are being generated by oil palm cultivation. The chemical composition of palm oil mill effluent (Pome) is presented in table 29. It is rich in nitrogen 1194 mg N/L with solid matter 67000 mg/l.

Table 29: Characteristics of Pome before and after treatment

Parameters	Untreated	Treated
pH	3.86	7.10
BOD*	35,836	1,912
COD*	83,356	27,934
Total solids*	67,932	31,791

Suspended solids*	39,233	22,906
NH_4-N*	50	138
Total N*	1194	1004

*values in mg/l Sources: Urn, 1998

Lim (1998) reported that in Malaysia and Thailand application of Pome and rubber effluents to soil by the furrow system improved yields in palm oil (Table 30). The improved yields were caused due to the nutrients in the effluents and improving of the soil properties.

Table 30: Optimum rates of land application of Pome on Oil Palms

Effluent Type	Application system	Optimum rate (rain equivalent) per year (cm)	Application frequency (times/year)	Yield increase (%)
Untreated	Flatbed	3.3	4	8
	Tractor/tanker	4.6	12	12
Treated	Flatbed	6.7	4	23
	Sprinker	2.5	6	18
	Flatbed	6.6	4	14

Table 31: Impact on water quality in tube-wells following oil palm effluent applications

Parameter	Concentration (mg/l)	
	Control plot	Applied plot
pH	5.2	4.9
BOD	5	5
NH_4-N	—	—
NO_3-N	8	5
K	45	17

The soil system is effective in removing nitrogen from the effluent (Table 31). The field monitoring results also show little or nil ground water contamination from the application of moderate amounts of Pome. Uncontrolled excessive application, however can result in higher amourjt of leachates in the ground water and in surface runoff.

Mulching with prunning and wastes of plantation crops at the rate of upto 40 tonnes/ha improved oil palm growth and their yields. The yield increases due to mulching were higher than fertilizer alone.

Thamburaj and Kuthubutheen (1989) produced good compost by decomposing palm press fibre by inoculating with efficient cellulolytic fungus. Palm oil plantations are generally established by providing a legume cover.

Once established, these legume covers persist for several years. *Pureria phasiofoides Calapogonium muconoides, Centrosena pubescens* and *Mimosa invisa* are useful cover crops for palm oil plantations. These nitrogen fixing tree legumes annually contribute significant amounts of carbon and nitrogen to soil by leaf litter generated annually. Proper management of organic waste materials available from oil palm plantation and cover legumes can contribute significantly to soil fertility.

Cashew Plantation

The organic wastes from cashew plantations are dry leaves, vegetable parts, apple wastes and shells. The vegetable part consists of twigs and dead branches, which are generally used as fuel. The shells are used for extraction of cashew nut shell liquid for industrial use. Cashew apple is produced in significant quantities but mostly it is wasted, though it is a source of fermented beverage, other by products and plant nutrients. The published literature on recycling of cashew wastes is quite rare. Dry leaves of cashew plants are the most convenient organic wastes. It is estimated that 20-25 Kg dry leaves are produced by 12-15 yrs old tree. The leaf litter is a good material for composting studies, however, are needed to evaluate the manurial value of the leaf litter.

Cocoa

It is a shade-loving crop. It is therefore, grown as mixed crop in the plantations of arecanut and coconut. Pure p'antations of cocoa are rarely found in India. The wastes generated are leaf litter, twigs and dried branches which directly fall on the ground. The pod wastes (husk) is also recyclable after extraction of cocoa beans.

Cocoa is grown in a single hedge and double hedge in the interspaces of coconuts. Cocoa leaf litter in double hedge system could contribute 55 Kg N, 11 Kg P and 35 Kg Kof plant nutrients per hectare (CPCRI, Kasorgod). Cocoa is also grown successfully as a mixed crop in the interspaces of arecanut. Manikandan *et al.* (1987) examined an arecanut-cocoa system on a laterite soil after 13 years of planting. It was reported by them that leaf fall was well distributed in the system and provided more organic matter in arecanut and cocoa basins than in the interspaces. The pH of soils of cocoa basins was lower than that of the arecanut basins, showing therefore, that cocoa leaf fall is more acidic than arecanut leaf litter.

Rubber Plantations

An immature rubber plantation removes about 728 Kg N/ ha during its first six years of growth where as fertilizer N applied to rubber is about 200 Kg /ha during this period. The balance is possibly made up by the cover crop (Potty and Mannotra 1993). Generally rubber plants do not respond to fertilizer application after 6 years of growth.

Cover crops and the leaf litter are local sources of organic matter in rubber plantations. Suitable legume cover crops recommended in case of oil palm plantation can be grown in rubber plantation as well. Besides, *Macuna bracteata* is another introduction to south Indian rubber plantations. Usually a cover crop is established ahead of the rubber planting, but it can also be planted along the planting of rubber plants. Potty and Mannotra (1993) observed considerable improvement in organic carbon, total N and available K in the soil of a rubber garden after 3 years of establishment of the cover crops of Purer/a *phasiotoides* and *Macuna bracteata.*

The review in respect of the utlisation and recycling of the residues and by products of post-harvest processing, leaf litter of shade trees and cover crops grown in the gardens of most of the plantations make an important source of biomass and plant nutrients. Most of the information is projection based. It is therefore, needed to make quantitative estimates of residues entering the system, rates of nutrients mineralisation, soil organic matter maintenance and nutrient balance. An integrated approach is required to evaluate the contribution of all sources of organic matter an over all effect on soil fertility including the contributions of nutrients fo replacement of fertilizer use.

III. Wild and Aquatic Weeds

Several wildly occurring waste plants constitute a potential source of nutrients and organic matter for cultivated lands. Ipomoea *(Ipomoea carnea),* Water hyacinth *(Eichhorniacrassipes),* wild sage *(Lantanacamara)* and Eupatorium *(Eupatorium adenophorum)* are a few of such plants. These plants are usually unwanted in nature as they adversely affect environmental quality in various ways. At the same time, at succulent stage of growth, these plants serve as effective manurial resources because of their ready decomposability in soil. In view of the findings that organic residues having equal to or greater than 1.74% nitrogen content cause net mineralization of nitrogen (Frankenberger and AbdeLmagagid, 1985). The green biomass of these plants can be incorporated in soil shortly before crop sowing or even at the time of sowing without any chances of nitrogen immobilization.

The limited research information that is available regarding its manurial utility of the wild plant sources, it may be assumed of their utility for nutrient cycling and land improvement.

Ipomoea biomass consists of leaves and tender stem parts which contain a good concentration of plant nutrients such as N, P and K and were found comparable to nitrogen fixing legume, *Sesbania bispinosa* (Chatterjee *et al.* 1979). The nitrogen mineralization pattern in both the plants was also similar. A field trial using 10 tonnes *Ipomoea* showed that the green biomass could supply 50Kg of nitrogen to rice crop (Jha *et al.,* 1980) Mandal *et al.* (1993, a, b) compared different manurial sources for summer as well as winter rice in sandy loam soil at Kalyani (West Bengal). Organic residues were incorporated in soil one week before transplanting rice. The data (Table 32) reveals that only Ipomoea applied in combination with the half recommended dose of fertilizer N could compare with the full recommended dose of fertilizer nitrogen on the yield of winter rice. The effect was even greater than *Azolla* while for summer rice fertilized with lower level of nitrogen, Ipomea, *Azolla* and farmyard manure performed equally well. Nitrogen uptake by the rice crop from the applied organic materials was found effective in proportion to the effects on crop yields in both the growing seasons. The fertilizer nitrogen was applied in split dosages as indicated in the table 32. The other organic materials though relatively less effective also proved to be good manures.

The shrubs like *Lantana* and *Eupatorium* are wild plants with extensive growth in favourable environments, particularly in humid hill tracts, causing harm to sflvi-pastora! vegetation and plantation crops. These produce luxuriant nutrient rich biomass which can be profitably utilized for direct land application or composting. Bhardwaj and Kanwar (1991) observed that application of 10 tonnes/ ha of *Lantana camara* in silty clay loam soil in combination with fertilizer N could substitute 30 Kg N/ ha for wheat crop.

Table 32: Comparative effectiveness of different organic materials on the yield of rice (2 year-mean)

Treatment	**Winter rice (Kg /ha)**	**Summer rice (Kg/ ha)**
Azolla	6474	5346
Azolla + 60 Kg N (30+30)/ 30 Kg N (15+15)	7346	5605
Ipomoea + 60 Kg N (30+30)/ 30 Kg N (1 5+1 5)	8141	5629
Water hyacinth + 60Kg N (30+30) 30 Kg N (15+15)	6057	4552
Water lettuce + 60 Kg (30+30) 30 Kg I\I (15+15)	6755	4851
FYM + 60 Kg N (30+30) 30 Kg N (15+15)	6749	5547
120 Kg N (60+30+30) 60 Kg N (30+15+15)	8020	6149
Control (-N)	4320	3360

Organic material was applied @ 10 tonnes/ha

in summer rice the recommended dosage was applied at 60 Kg N/ ha in 3 split doses (30+15+15) and reduced dosage with organic matter was (30 Kg / ha in 2 split doses (15+15).

The shrubs like Lantana and Eupatorium are wild plants with extensive growth in favourable environments, particlularly in humid hill tracts, causing harm to silvi-pastoral vegetation crops. They produce luxuriant nutrient rich biomass which can be profitably utilized for direct land application or composting. Bhardwaj and Kanwar (1991) observed that application of 10 tonnes/ ha of *Lantana camara* in silty clay loam soil in combination with fertilizer N could substitute 30 Kg N/ha for wheat crop. The total and mineral nitrogen content of soil at harvest of crop was improved due to treatment with *Lantana camara.* The decomposition rate of *lantana* was lower than that of *Sesbania aculeata* but it was higher than wheat straw. However, it was observed in a separate field study that for better effect on plant *Lantana camara* needed about 15 days of precropping decomposition in the soil.

Bhardwaj *et al.,* 1988). *Lantana camara* contains some alkaloids. It is probable that the certain intermediate products may show toxicity to wheat seedlings when the crop is sown immediately after its incorporation. In another two-year field trial, at Palampur, (Velayutham and Bhardwaj, 1994) a combined application of 20 tonnes/ha of Eupatorium *adenophorum* plus 60 Kg fertilizer N / ha produced as much yield of the crop as was recorded by the application of 100 Kg. fertilizer N (Table 33).

Table 33: Effect of the green biomass of *Eupatorum adenophorum* on wheat yield and properties of an acid silty clayey loam soil

N source (P&K)	Wheat grain yield (t/ha)	Soil properties at crop harvest				
		Mineral N (kg/ha)	Avail P (kg/ha)	Avail K (kg/ha)	Water holding capacity (%)	Bulk density (g/cm3)
Control (-N)	2.62	25	11	185	45	1.29
Eupatorium 1	2.90	42	10	268	48	1.24
Eupatorium 2	3.19	41	13	287	50	1.18
Eupatorium HN60	3.70	56	17	270	47	1.24
Eupatorium 2+N60	4.12	62	20	281	50	1.18
N100	3.98	34	13	179	46	1.30
CD (P=0.05)	0.26	NS	4.8	26.2	3.60	0.08

Eupatorium l: 0 t/ha, Eupatorium 2:20 t/ha (green foliage), Basal application of P & K.

Incorporation of *Eupatorium residues* improved the soil nutrient status and its physical properties.

It is concluded that wild plants, when applied at the rate of 10-20 tonnes /ha can meet from one-third to one half of the nitrogen required for the crops such as wheat and rice with appropriate management.

Aquatic Weed- Water Hyacinth

Water hyacinth *(Eichhornia crassipes)* is a free floating plant, originally a native of Brazil. It was introduced over a century ago as an ornamental plant in the South-East Asia and has become established throughout the tropics and propagates explosively. Temperate zones are free from this plant as it is killed by severe frost. In India its prevalence assumed the proportion of serious menace in Bengal in the year 1914 and spread to Assam, Bihar, Orissa, Eastern Uttar Pradesh and other states also. It was recognised that this plant affects seriously the health and prosperity of the people unless prompt action is taken to eradicate and check it from spreading. It was found that it has begun to grow in cultivated areas as a result of which paddy and jute could not grow. Further, fishes can not live or thrive in the water in which water hyacinth exists. The problem of eradication of water hyacinth was so severe that legislation was enacted in Bengal in 1919 and later in parts of Orissa and United Provinces. However, inspite of lot of research and eradication work a suitable formula for getting rid of this weed could not be evolved.

Area Infested with Water Hyacinth

Though exact figures as to the area covered by this weed in India are not available yet, the approximate figures from Bengal as 12,000 ha., and Bihar as 80,000 ha indicate the extent of its distribution. It is estimated that the total acreage under its domain to be about 200,000 ha, including the affected areas in Assam, Eastern U.P., Andhra Pradesh, Tamil Nadu, Orissa and Kerala. The adverse effects of such uncontrolled growth on agriculture, fisheries, transport and human health are obvious and the necessity for its collection, consequent disposal and prevention of further growth is required.

Problem of Eradication

Water hyacinth being aquatic is easy to collect by manual labour or by mechanical appliances but its disposal is more difficult due to its high water content. It takes some time to dry before it can be burnt. Most difficult of all is prevention of further growth since the seeds remain dormant for at least one year and maintain their viability for several years. The plants are themselves able to resist considerable drought upto 57% of the saturated value of the water content of soil due to corresponding change in osmotic strength, decrease in the rates of total leaf area to total root length and the presence of mucilage at the leaf base. The plant produces long roots with progressive drying of mud and thrives at pH range between 6.0 to 8.0. However, seeds lying at the bottom of the tank can not germinate in the absence of oxygen hence keeping the tanks flooded possibly may prevent their germination. Spraying with chemicals, such as copper sulphate solution (0.018%) was reported to kill the plant, but the chemicals are not consistently effective since calcium oxalate crystals escaping from disintegrating leaves, antagonise the effect of copper sulphate. .Other chemicals had also *a* rather limited effect. In fact, the inevitable conclusion arrived at by certain earlier scientists including Sir J.C. Bose (1928) and P. Parija (1934) was that the only effective means of controlling this weed was in its mechanical collection and destruction.

Mechanical Devices for Harvesting and Dewatering of Water Hyacinth

Manual method, though quite suitable for removal of water hyacinth from small and shallow water areas, is not suited for medium and larger water spreads. The central Institute of Fisheries Technology at Cochin has developed a machine for harvesting of both submerged and floating weeds. It is recorded that 1 to 1.5 ha. can be cleared in 8 hours. The machine can also be used for fishing and fish transport in inland waters.

As water hyacinth contains about 90-95 per cent water, it can seldom be used for feeding animals or other purposes. The common practice is drying by exposing a thin layer of the plant to the sun. The other method could be sophisticated solar collector- heat transfer system.

Other devices developed are different types of roller-belts, cone-and screw-presses. Therefore, attempts may be made to develop and identify simple and economic devices for dewatering of water hyacinth for adoption in the country.

Water Hyacinth as a Source of Manure and Biogas Production

Water hyacinth can be used as surface mulch, green manure or compost. It can also be used after drying as fuel and the ashes subsequently added in the soil, although both nitrogen and organic matter are lost during burning. Recently, attempts have been made to use water hyacinth for production of biogas without loosing its manurial value.

As a mulch, water hyacinth is used in tea gardens during dry period for the conservation of soil moisture. On rotting, they act as manure. The plants can also be ploughed or buried into the soil and allowed to decompose for a short duration before sowing of crop. However, the problem with these methods is that as the plants are bulky, it is difficult to handle and transport to long distances.

Compost

Water hyacinth is a good material for compost making. The fresh plant contains approximately 95.5% moisture, 3.5% organic matter and 1% ash. The percentage of N, P_2O_5 and K_2O is 0.04, 0.06 and 0.20, respectively. As yield of about 250 tonnes of water hyacinth per hectare per year can be obtained, there is a potential of producing about 2 million tonnes of compost annually in the country. In terms of manorial value, one tonne of water hyacinth compost contains on dry basis about 20.5 Kg. N, 11.0 Kg P_2O_5 and 25 K_2O. This compost contains higher concentration of potash than farm-yard manure or town compost.

As fresh plants contain about 95% water, the chief obstruction in its use as manure is the difficulty in its transport to any distance. The rotting of fresh water hyacinth, on the other hand, involves a considerable loss, to the extent of 70% K_2O and 60% of N. in the form of liquid which squeezes out in large quantities. In order to prevent this the wilted or dried plants are more suitable for composting.

Practices Developed in China

Traditionally Chinese have been culturing aquatic plants in a systematic way in areas with networks of waterways to provide material both for compost making and fodder for pigs. Special techniques have been developed for localized cultivation to avoid, obstruction in navigation.

Thirty to forty tonnes/ ha. of plants are used to seed the water surface. The seeding is done in April or early May. The plants reach maturity in 2-3 months. A long-handled rake is used to pull out the plant which are loaded into boats and ultimately taken to wherever needed. The second harvest is taken after 3 months. A harvest of 87.5 tons/ ha. of open water per year and 250 tons/ha, of open sewage water/ year have been reported.

Chinese Method Composting

A typical technique for making compost from water hyacinth, was developed in China which is as follows. The approximate quantities of different materials put in a circular or rectangular pit are aquatic plants or green manure crops-0.75 tonnes, rice straw-0.15 tonnes, river silt-7.50 tonnes and pig excreta or animal dung-1.0 tonne. The materials are put in alternate layers, first layer of water plants, second layer of silt and straw mixture and third layer of stable manure ; each layer being 15 cm. thick. The top layer is made of mud and a water column 3-4 cm. deep is kept at the hollowed surface to create anaerobic conditions which help in reducing the nitrogen losses to a minimum. The contents of the-pit are turned over three times, the first turning is done after a month of filling of the pit. During this turning 20 Kg. superphosphate is added and thoroughly mixed with the composting material and water is added to the mass to provide optimum moisture. The second turning is done after another month and the third after about 2 weeks. The manure is ready for application in about 3 months time. Each pit gives about 8 tonnes of manure which is sufficient for 0.1 ha. of land. The composition of such a manure is organic matter -7.8 to 10.3%, N -0.3%, P_2O_5-0.2% and K_2O-0.3%.

Work Done in India

The process of composting of plant material was first developed in India and is commonly known as "Indore Method". This method involves putting of water weeds and cow-dung layer by layer to form a heap or pile. The material is allowed to decompose for a period of about Smonths. A simple and quick method of composting of water hyacinth in pits has been developed by Assam Agricultural University. The steps involved in this method are mentioned below:

1. Choose elevated and levelled ground. Dug enough number of pits of desired sizes tappering towards bottom. The dimension of a pit may be 3.0 m. in length at the bottom and 3.5 m in length at the top, 2.0 m breadth at the bottom and 2.5 m breadth at the top and 1.5 m height. Each such pit can accommodate about 8 tonnes of compostable material.
2. Chop wilted water hyacinth plants into pieces of about 15-20 cms. Length.
3. Place bamboo poles vertically in the pit at certain distances.
4. Put the plant material in layers of 1 m. thickness.
5. Spread a thin layer of old decomposed water hyacinth plants collected from the bottom of dried up ponds and water lagoon, as a starter to speed up the composting process.
6. Then pits are covered with bamboo mates or palm leaves.
7. Sprinkle water on the plant mass periodically
8. After a week or ten days when plant mass sank down, add successive layers of wilted plants until the whole pit is packed.
9. Cover the material with soil to check evaporation of moisture.
10. Remove bamboo poles, leaving empty cylindrical space as such to act as ventilation chimneys for keeping aerobic condition.
11. Compost gets ready in about 3-4 months.

It is found that water hyacinth compost is good for crops like rice, potato, maize, jute and vegetables. The recommended doses of compost for rice, maize and jute is 20, 5, 7.5 tonnes per hectare, respectively.

Biogas Production

The recovery of fuel from water hyacinth and other aquatic weeds, even if on a small scale, has interesting implications, especially for rural areas. While dried water hyacinth could be burnt to produce heat-energy, attempts were made to capitalize upon one of nature's processes of anaerobic decomposition to produce biogas. The high moisture content of aquatic weeds is an advantage in this process as it is needed for proper decomposition. In a pioneering work conducted on bio-coqversion of water hyacinth into methane gas in U.S.A., it was estimated that each kilogramme of water hyacinth on dry weight basis yields about 370 litres of biogas with an average methane content of 69 per cent. The biogas thus produced can be used as fuel for cooking, lighting, running of diesel engines and as a source of power. The residue left after the decomposition is used as manure.

Although large number of biogas plants are in operation in the country, these, however, process mainly cattle dung. The present conventional design of 'gobar gas plant' cannot as such be used for processing of water hyainth as choking of inlet and outlet pipes and scum formation due to floating nature of the material have been observed.

Laboratory experiments conduced at Central Mechanical Engineering Research Institute, Durgapur (West Bengal) have indicated that chopped and dried water hyacinth mixed with water in 1:2 proportion or mixed with cattle dung in 1:1 proportion and then slurry prepared, yield biogas. The gas generation from water hyacinth and dung mixture is less dependent on change in temperature than gas production from cattle dung alone. Experiments conducted at the University of Agricultural Sciences, Bangalore, showed that water hyacinth alone as well as in combination with cattle dung is a good material for biogas production. In similar experiments carried out at Sri Parasakthi College for Women at Courtallarh, it was observed that water hyacinth with night-soil yielded more biogas than water hyacinth and dung in the same proportion. A combination of [dung, urine and water hyacinth gave more biogas and nitrogen-rich manure! than individual material. At Gobar Gas Research Station, Ajitmal (U.P.), it was found that water hyacinth to dung in 3:2 ratio was a good combination for obtaining high quantity of gas containing low concentration of carbon dioxide (24%). However, trials conducted by using these material in dome-[shaped biogas plant were unsuccessful as removal of slurry under force of gravity was not found and the material formed scum layer.

A "Scale-up" model plant has been constructed at Central Mechnical Engineering Research Institute, Durgapurto determine design parameters [for conduction of actual working plant of 2-3 cu.m. gas prodution per day 'capacity. One such plant has been constructed by using bricks, eement-imortar and cement-concrete. The plant needs periodical cleaning, say twice in one year. Further work to develop a suitable design of biogas plant based on water hyacinth is in progress at afore-mentioned institutions.

China has made remarkable progress in this field. The unique feature of biogas plant is that the gas holder and the digester are combined in one unit and the size of inlet and ouj let openings is large. In the construction of these plants local materials are used.

Common combination of raw materials used in China are: a mixture of 10% human excreta, 30% animal dung and 10% vegetable material and 50% water. Vegetable materials include aquatic plants, crop residues, green grass etc. which are decomposed for more than 10 days prior to their being put into the digester.

It is estimated that the amount of water hyacinth that can be grown on 1 ha. of water surface can be converted to 70,000 cu.m. biogas; corresponding to 150 MWH/year, which is more than what rural electrification schemes offer. In addition, the sludge obtained from the fermented water hyacinth is rich in potassium and also contains other plant nutrients. The annual sludge yields from a 2 cu.m. capacity plant could be about 5 tonnes/year. As there is a qreat scope of culturing water hyacinth in India, attempts should be made to standardize design of biogas plants suited to the need for the material and' such plants may be set up on pilot basis to / asses their economic feasibility.

Water Hyacinth as Animal Feed

Acquatic and semi-aquatic plants may be utilised as feed for cattle, sheep, goat and pigs. The water hyacinth which is available abundantly in ponds, lakes, rivers and other water resources in the different parts of the country is some times consumed by animals in the time of scarcity. Although, some times animals eat water hyacinth, it is not very palatable and it is also not possible to maintain health of cattle, buffaloes and pigs when the later is fed as such as sole diet.

Leafy parts of water hyacinth contain as much as 25-35 per cent protein. The individual amino acid constituents are present in about the same proportion as in land forages of similar crude protein content. However, the methionine and lysine level, generally considered to be the limiting amino acids in plant protein are often lower than in high quality terrestrial crops. Lysine content of water hyacinth is equal to that found in milk. Besides fiber content of such plant can make potential source of animal feeds for the ruminants. Although nutrient composition is a guide to feeds' value, but high digestibility is crucial if the nutrients are to be available to the animal. When water hyacinth is used as a roughage experiments indicate that although some time cattle eat, it is not palatable and it is not possible to maintain normal health on sole feeding of this plant, however, when water hyacinth is used as a roughage, experiments indicate that the nutrients are satisfactorily available to ruminants.

Practices in Other Countries

In Indonesia, pigs are fed rations containing water hyacinth after certain treatments like boiling and mixing with other feeds. In China pig farmers boil chopped water hyacinth with vegetable wastes, rice bran, copra cake and salt to make a suitable feed. In Malaysia, fresh water hyacinth is cooked with rice bran and fish meal and mixed with copra meal as feed for pigs, ducks and fish in ponds. Similar practices are common in Philippines and Thailand also.

Studies Carried Out in India on Water Hyacinth

Many attempts have been made to make satisfactory uses of water hyacinth as feed for animals. In 1921 a water hyacinth Committee was set-up in Bengal at the scarcity of grass. Studies on the use of water hyacinth as animal feeds have been carried out at the Indian Veterinary Research Institute, Izatnagar and the Eastern Regional Station of the National Dairy Research Institute, Kalyani for improving the nutritive value and palatability of water hyacinth and have developed methods by which this weed can be fed to animals.

The work carried out at Ranchi under the All India Coordinated Research Project on the utilisation of Agricultural by-products and Industrial Waste Materials it was observed that when water hyacinth was fed as a sole feed or supplemented upto 60% in a grain mixture the growth of the pigs was retarded. It indicated that there is some incriminating factor present in the water hyacinth which needs to be removed. The work carried out at Bhubneshwar indicated that when the pigs are fed with water hyacinth as sole diet, this showed symptoms of diarrhea. The problem could be, however, minimised by removal of water from the water hyacinth by pressing and by feeding at about 20% level in the pig ration. The addition of 1% of molassess and urea helped in maintaining the growth and nutrition of the cattle and pigs.

The studies carried out at the Indian Veterinary Research Institute has indicated that the intake of the buffaloes and cattle was only 1 Kg dry matter per 100 Kg body weight when water hyacinth is used as a sole diet. This caused depression in the growth of the livestock. Chaffing and mixing with wheat or paddy straw in a ratio of 3:1 was found to increase the intake of the cattle and buffaloes to 1.5 kg per 100 kg body weight. It was also found possible to dry the water hyacinth without any loss in its feeding value and it can be utilised in the cattle feed by mixing with chaffed wheat or paddy straw and molasses in the ratio

7:2:1 in a silo for 8 to 10 weeks and then could be fed as a source of energy to cattle and buffaloes. It may be observed that it is possible to replace the part of the livestock ration to the extent of 25% by using water hyacinth on a dry matter basis provided the following methods are adopted.

1. The water hyacinth collected may be pressed to remove the water and then fed to the livestock upto 25% level in their feed on dry matter basis.
2. Water hyacinth can be converted into hay by drying it in the sun or oven dried which will help its preservation for the longer period and could be mixed with livestock feed as and when required.
3. Green water hyacinth after chopping can be ensiled by mixing with chaffed wheat and paddy straw and molasses in the ratio of 7:2:1 in a silo for 8 to 10 weeks and then fed as a source of energy to the cattle and buffaloes.

Treatment of Waste Water and Industrial Effluent by Using Water Hyacinth

It has been observed that water hyacinth can scavenge inorganic and some organic compounds from water and industrial effluents. Effluent renovated by the plants is stripped of its pollutants and, whon released into waterways, causes less environmental damage. The clean water produced is, in most situations, suitable for reuse in irrigation and industry. Furthermore, the plants themselves can be harvested and used, thus providing additional benefit.

Upgrading of Sewage Lagoons

In an experiment conducted by National Aeronautics and Space Administration at Florida in U.S.A., it was observed that when sewage passed through a pond at a rate of 2.2 million litres per ha per day water hyacinth growing in the pond removed 80 percent of the nitrogenous compounds and 40 per cent of the phosphorous compounds in 2 days. Under ideal conditions water hyacinth can recover the following elements (Table 34) from waste-water.

Table 34: Nutrient uptake from waste water by water hyacinth

Elements	Amount recovered (Kg/ ha/day)
Nitrogen	22-44
Phosphorus	8-17
Potassium	22-44
Calcium	11-22
Magnesium	2-4
Sodium	18-34

Water hyacinth culture also removes algae and fungal bacteria, greatly reduces suspended matter, and removes odor-causing compounds. Reportedly, the resultant effluent is clear, odourless, and contains little nitrogen.

IV. Green Manuring

Green manuring refers to incorporation of fresh green matter in soil to supply the plant nutrients contained in the biomass. About 67 million hectares in area is under green manuring (Saxena, 1995). It can be defined as a practice of ploughing or turning into soil fresh green plant residues for improving the soil fertility and scale of plant nutrition. The importance of green manuring in agriculture has been recognised from early times in many countries. Although green manuring has been a traditional practice in India, this practice declined after the introduction of high yielding crop varieties and use of chemical fertilizers and pesticides. In the past few years, there is revival of interest in green manuring due to concern for soil health, demand for organic food and increasing cost of chemical fertilizers. Green manure is a cheaper alternative to chemical fertilizer nitrogen. The process also contributes to maintenance of soil organic matter and enhances the availability of other plant nutrients, besides nitrogen.

A large number of plants both legumes and non-legumes are used as green manure. The choice of a particular green manure crop has to be made based on the soil type, the time available to raise the crop and the facility for irrigation. The legumes are preferred for green manuring because they are capable of fixing atmospheric nitrogen in symbiotic association with micro-symbiont, *Rhizobiumor Azorhizobium* which becomes available to beneficiary crop on its decomposition in the soil. Root exudates of legume crop also benefit the crop, it was reported that decomposition of legume residues, was much faster as compared to cereal residues and farmyard manure (Gaur *et al.*, 1971,1973).

Panse *et al.* (1965) reported the results of 65 trials comparing different green manures used in rice culture. Most trials included sunnhemp, *Sesbania* acu/ eatefcand *Vigna* species. Sunnhemp showed better response in low rainfall areas than *Sesbania* but both performed better than other green manures, sucrVas mungbean (*Vigna radiata*) and urid (*Vigna mungo*). Under good moisture conditions, *Aeschynomene americana* showed a higher response than *Sesbania.* Indigeneous tall-growing rice cultivars were used and the average response of 0.24 t/ha was reported. The response in south India was of relatively higher magnitude. Uppal (1955) reported that among *sesbania, sunnhemp* and clusterbean, *sesbania* was the best green manure for rice in alkali soils. Singh (1963) found that even in normal rice soils, *Sesbania* has been preferred over sunnhemp because of its ability to withstand prolonged moist conditions.

Agarwal *et al.* (1950) recommended based on several field trials, *dhaincha* (*Sesbania aculeata*) as an appropriate green manure for rice crop in U. P. particularly the areas which receive more precipitation and possess hard clayey soils in bunded paddy fields. *Sesbania speciosa* is a perennial plant, has slow initial growth but is more tolerant to drought than S. *aculeata* (Patnaik *et al.*, 1957). Leaves and prunnings of perennial plants are used at the time of planting or somewhat later.

Besides legumes several other plants have been used as manure, water hyacinth grows extensively in standing water and contains about 2-3% nitrogen on a dry weight basis. Sea weeds, with 1-2% nitrogen are used for green manuring in coastal areas. Mirchandani and khan (1952) also reported the use of *Ipomea cornea* (grown as hedge), mango leaves and *cannabis sativa (bhang,* widely growing) as green manures.

Khare and Rai (1968) compared the effect of P on nitrogen fixation by leguminous crops and found that maximum nitrogen was excreted by sesbania, followed by soybean, cowpea and green gram. A net increase of soil N occurred with sunnhemp. It shows that the contribution of *Sesbania* to soil N is actually more than what is added as roots and tops. *Sesbania* is also more efficient in mining P from lower soil depth. (Subbiah and Mannikar1964).

An Ideal Green Manure Crop

An ideal green manure crop should possess the following important agronomic and physiological characters (FA01977, IRR11988).

1. Show an early establishment and fast growth with potential for production of large biomass.
2. Be photoperiod insensitive.
3. Possess efficient nitrogen fixation capacity to accumulate large quantity of nitrogen within 4-6 weeks.
4. Is tolerant to adverse soil and weather conditions.
5. Is tolerant to pests and diseases.
6. Is easy to incorporate.
7. Is easily decomposable.

Benefits of Green Manure

1. It stimulates the microbial biomass and the biocycle processes in soil.
2. It improves physical, chemical and biological properties of soil.
3. It helps to maintain organic matter status of cultivated soils and adds nitrogen in particular.

4. It increases the availability of other plant nutrients like phosphorus, calcium, potassium, magnesium and iron besides addition of nitrogen in significant amounts as a result of fixation of atmospheric nitrogen.
5. Green manuring improves aeration in rice soils by stimulating the activity of organised surface film of algae and bacteria.
6. Many green manure crops have additional uses as sources of food, feed and fuel.
7. Green manure increases the yield of crops.
8. Green manure brings about favorable changes in the physico-chemical properties of soil and in reclamation of saline-sodic soils. *Sesbania aculeata* is particularly useful because it can withstand drought ard waterlogging and well adopted to sodic soil. At pH higher than 8.5. amendment of soil with gypsum may be necessary to ensure germination and high biomass production. Both treatments are complimentary in reclamation of sodic soils.
9. Green manure also acts as a mulch and helps in conservation of soil moisture.

Methods of Green Manuring

Sunnhemp, *Sesbania,* clusterbeans, sen, cowpea. *Vigna radiala, Vigna mungo* and fodder legumes are used as green manuring crops. Sesbania and sunnhemp are most popular, in India, two methods are being practised for green manuring.

1. *In situ green manuring:* Green manuring crops such assunnhemp, sesbania, cluster beans etc are grown in the field and after 4-8 weeks of their growth, are incorporated in the same field before the sowing or transplantation of the main crop. This method is generally followed in northern India.
2. *Green leaf manuring :* This refers to incorporation of green leaves and tender twigs collected from perennial shrubs and trees grown in waste lands, on field borders and adjacent forest areas. The commonly used shrubs and trees such as *Glyricidia, Sesbania specioss,* subul *(Leucaena leucosephala) karanj(Pongamiapinnata)* etc. This practice is common in southern and eastern part of India.

Types of Green Manure Crops

The green manure crops can be grouped as cultivated annual legumes, perennial legumes and wild annual legumes. The characteristics of some green manures is given in table 35.

1. *Leguminous green manures:* The plants which are generally used in rice cultivation as green manures are quick growing, accumulate greater biomass of green matter quickly, easily decompose and contain considerable quantity of nitrogen. Some of such common leguminous green manure crops, their biomass and nitrogen potential are also presented in table 36 (Abrol and Palaniappan 1988).

Table 35: The characteristics of some annual and perennial green manures

Cultivated	**Green matter G (t/ha)**	**Moisture (%)**	**N content (% dry basis)**
Annual legumes *Cassia mimosoides*	47	74	2.99
Crotalariajuncea	16.5	73	3.01
Cyamopsis tetragonoloba	7.0	60	3.53
Indigofera anil	6.8	74	3.66
Sesbania aculeata	14.8	78	2.67
Sesbania speciosa	7.8	78	2.43
Vigna unguiculata	10.0	85	2.63
Vigna radiata	7.7	82	2.96
V. tribolus	5.3	79	2.47
Perennial legumes *Cassia hirsuta*	5.0	81	2.52
Desmodium gyroides	1.4	72	3.35
Gliricidia maculata	3.0	75	3.46
Sesbania punctata	3.7	73	2.42
Tephrosia Candida	2.3	67	3.20
Wild annual legumes *Aeschynomene americana*	8.9	78	3.14
Calapogonium mucunoides	4.5	74	3.02
Cassia tora	5.2	71	2.13
Lathynjs sativus	—	82	4.67
Tephrosia purpurea	3.5	70	3.46

Table 36: Some common leguminous green manure crops for rice

Name	Botanical name	Growing season	Out put in 45-60 days		
			Green Matter (t/ha)	Nitrogen %	Nitrogen added (Kg/ha)
Sunnhemp	*Crotalaria juncea*	Wet	21.1	0.43	91
Dhaincha	*Sesbania aculeata*	Wet	20.2	0.43	86
Pillipesara	*Phaseolus trilobus*	Wet	18.3	1.10	201
Green gram	*Vigna radiata*	Wet	8.0	0.53	42
Cowpea	*Vigna sinensis*	Wet	15.0	0.49	74
Cluster bean	*Cyamopsis tetragonobba*	Wet	20	0.34	68
Senji	*Melilotus alba*	Dry	28.6	0.57	163
Khesari	*Lathytvs sativus*	Dry	12.3	0.54	66
Berseem	*Trifoliuma lexandrinum*	Dry	15.5	0.43	67

Adapted from Abrol and Palniappan (1988)

They differ widely in biomass production and nitrogen content ranging from 2.0 to 4.9% (Vachhani and Murthy, 1964, Ghai *et al.,* 1985). In rice *Sesbania,* sunnhemp and wild indigo *(Indigofera tinctoria)* depending on the soil type and agroclimatic conditions are recommended. Sunnhemp is recommended for sugarcane, wheat and finger millet. Lupin, sunnhemp, cowpea, Guar etc. are recommended for potato crop (Krishnamurthy, 1978, Sharma *et al.,* 1991)

2. Non-*grain legumes:* The exotic stem nodulating *Sesbania rostrata* has much promise for lowland rice especially with adequate irrigation. Another promising stem nodulatirjg introduction from Madagascar is *Aeschnomene afraspera* which is capable of withstanding water stress to some extent. Its potential under Indian conditions has not been fully explored.

Though *Tephrosia purpurea* and *Phaseolus tribolus* grow relatively slowly and accumulate much less nitrogen than *Sesbania,* they are more adapted to drought conditions.

Tephorosia purpurea has the additional advantage of self-sowing and is not browsed by cattle. Hence in the deltaic areas where a single rice crop is grown due to limited water availability, *T. purpurea* is grown in the long rice follow season without irrigation. After 3-4 months, it sheds seeds in the field which germinates after rice harvest in the next year. The green manure is incorporated 7-10 days before rice transplanting. Wild indigo is another drought tolerant green manure crop for rainfed rice which can contribute 62-182 Kg N/' ha to the succeeding rice crop, depending on soil and agroclimatic conditions (Garrity and Flinn 1988).

3. *Grain legumes:* Some annual grain legumes crops are also used as green manure after harvest of grain. Mungbean stover incorporated in the soil after the pod harvest contributed about 60 Kg. N/ ha to the succeeding rice crop (Rekhi and Meelu 1983). Evaluation of several grain legumes showed that mungbean was the best yielding 1 tonne grain and 2.5 tonnes dry matter/ ha equivalent to 50 Kg N/ ha (Meelu *et al.,* 1986). Mungbean, black gram and cowpea could contribute about 50-60 Kg N/ ha for the succeeding rice crop (Kulkarni and Pandey 1988). Cowpea was best among the grain legume tested contributing about 65 kg N/ ha in addition to 500 Kg grain/ ha (John *et al.,* 1989).

   Phaseolus tribolus can be raised with limited irrigation in summer and a part of biomass can be used as fodder and another part as green manure.
4. *Perennial shrubs and trees:* The prunning from the perennials are collected and used as green manure. Some of the perennials grown on bunds and waste lands are used as green leaf manure which were listed by Gaur *et al.* (1995). *Thespesia populinea, Cassia auriculuta, Pongamia glabra, Meliaazadirachta, Calotropis gigantea* etc. are used for green manuring in south India. Some wild legumes viz. *Glin'cidia meculata, Tephrosia purpurea, T. Candida, Ipomoea carnea, Cassia tora, Sesbania* spp. etc can be grown on bunds and waste lands for utilizing their vegetative part for green leaf manuring.

Characteristics of Some Important Green Manuring Crops

1. *Sesbania aculeate (Dhaincha)*

It is quick growing succulant green manure crop which adapts itself to different agro-climatic conditions. It can be grown under adverse conditions of waterlogging, drought and salinity etc. The roots get profusely nodulated by *Fthizobium* and thus are capable of fixing atmospheric N. The plant has a soft stem and makes good growth in about 6-8 weeks. It is the most commonly grown green manuring crop in India. It performs better in low rainfall and limited soil moisture regimes. Recommended seed rate is 20 to 25 kg/ha. It produces abundant green biomass ranging from 10 to 20 tonnes per hectare depending upon the age at harvest.

2. *Sesbania rostrate*

It is an exotic leguminous crop which forms nodules both on stem and root. It was introduced in India about a decade ago from I.R.R.I., Philippines. It is a tropical legume and thrives well under flooded and water logged conditions producing aerial nodules on the stem. This can be grown either prior to rice crop or in between two rice crops. Though naturally propogated by seeds, seedlings and rooted stem can also be used as planting material. The normal seed rate is 30-40 Kg/ha. To get early, uniform and vigorous seedlings, seeds have to be scarified with concentrated sulphuric acid for 15 minutes. Summer season, the period of April to July is best season for producing a large biomass and good seed production. The photosensitive nature of this crop testricts its usage during winter and short day periods.

3. *Sesbania speciosa*

It is adaptive to different soil types such as sandy, loamy, clayey and alkaline soils. Initially the growth during 30 to 40 days is very slow, it. picks subsequently making rapid growth. It withstands to salinity to some extent. It has no serious pests or diseases. The plant has greyish appearance with soft hairs on the stem and leaves. The stem is pithy but if allowed to grow for more than four or five months it makes it woody and becomes difficult to be cut by sickle or pulled out.

There are different methods of growing *Sesbania speciosa* in rice field. Three or five days prior to harvesting of rice crop, seeds at the rate of 50 Kg/ ha are broadcasted in the field. The seeds get into the soil while harvesting of the rice is carried out by farmers. With the available soil moisture, the sesbania seeds germinate. This method is very easy to follow, as it involves no preparatory cultivation for raising green manure crop.

The other method involves broadcasting of sesbania seeds at 35-50 Kg/ ha after ploughing the field. A good stand of the crop can be obtained by irrigation. Where two crops of rice are taken, three weeks old seedings of Sesbania can be grown along the borders of the field and utilised as green manure for the second crop. Such border planting of this crop at a spacing of 5-10 cm in one hectare of land will give about 5000 to 8000 kg of green biomass for the second crop.

Each plant of Sesbania gives about 400-600 g of seeds. For growing green manure in one hectare, 50-60 Kg seeds will be necessary. Therefore, about 125 to 150 vigorous plants are left among the border plants to get sufficient seeds from these plants. A 60 days crop growth will yield about 10,000 kg/ha of green matter. For one hectare of rice crop 6,500 Kg of green matter will be sufficient.

4. *Crotalaria Juncea* (Sunnhemp)

It is a fast growing green manure-fibre crop. It grows up well in loamy and heavy soils. It does not withstand heavy irrigation or continues waterlogging.

There are a number of varieties varying in duration ranging from 75 to 150 days. The general appearance of the crop is greyish to greenish. The tall, robust and late duration varieties are used for retting and fibre extraction also. The seed rate is 25 to 40 Kg/ ha and the yield of green matter may range from 12000 to 25000 Kg /ha depending on the agro-climatic conditions and duration of the crop. The crop even can be cut when it is 45 days old.

5. *Tephros/apurpurea* (vn\d Indigo)

It is highly drought tolerant. It is a slow growing green manure crop. It is not browsed by cattle. It becomes self seeding and it is suitable for single season rice lands, where water is scarce during the off-season.

The seeds are sown as broadcast in the standing crop of rice just a week before harvest. The seeds have a waxy, impermeable hard seed coat and do not quickly germinate. The seeds are to be pounded with sand or steeped in hot water at 55°C for two to three minutes. The seed rate is 25 to 40 Kg/ha while the green manure yield varies from 3500-6000 Kg/ha.

6. Indigofera tinctorfa

This is a perennial shrub. It is found wild as well as cultivated. There are two species which closely resemble each other and generally grown as indigo (Madras Indigo and Bengal Indigo). The seed rate is 25 to 30 kg/ ha and the yield of green matter varies from 10,000 to 12,000 kg/ha

Green Leaf Manure Shrubs and Trees

Green leafy material is gathered from all sources for manuring purposes by farmers. In addition, loppings from miscellaneous trees are also gathered for use as leaf manure crop. Green leaves have the same effect as *in situ* green manure on the soil fertility and crop yield.

The common shrubs growing in waste lands are *Cassia auriculate Dodonia viscosa, Calotropis gigantea etc.*

Leguminous trees like *Pongamia glabra* and *Cassia siamea* can be planted in waste lands for augmenting the supply of green leaves. The trees do not require any attention after they get established and start growing. A brief description of some common shrubs and trees utilized for the collection of green leafy material is given below:

1. *Glyricidia (Glyricidia maculate)*

A shrub that comes up well in moist situations under favorable agro-climatic conditions, it quickly grows and develops as a tree and used for shade and green leaf manure in tea, coffee and cocoa plantations. It can be planted on alternate field bunds of wetland, 1 to 2 m apart pr as a thick hedge by close planting in three or four rows at 0.5 m spacing or along field border or along farm roads for green biomass production. The shrub is prunned two to three times a year and it withstands repeated loppings.

Glyricidia can be propagated by planting stem cuttings or seedlings 30 to 60 days old raised in nurseries. The establishment of seedling is better compared to stem cutting. Within two years after planting the plants are ready for lopping. Each plant gives 5 to 10 Kg of green leaves annually. About 375 to 400 plants on bunds in one hectare will yield 2500 to 3500 Kg of green leaves annually.

2. *Ipomoea cornea*

It is a fast growing with extensive branching drought resistant plant. It produces abundant green matter in short time. It is multipled by use of mature stem cuttings. Stems cuttings of about 0.3 m long with three to four nodes and axillary buds are planted at a distance of 1 to 2 m all along the wide field bunds, irrigation channels and fencs. About 1800-2000 cuttings can be accomodated in one hectare as border planting and two to three cuttings can be taken in a year. Each plant will give about 5 to 7 Kg of green matter/lopping.

3. *Cassia auriculata*

It is a very common growing plant on waste lands, hill tops, plain sea shores, almost in wild condition. It is a hardy plant. The plant is propogated through seeds as the seeds get dispersed and plant grows without any efforts. When the plants are in flower especially in off-season, they are cut and applied to field.

4. *Azadirachta indica*

It is profusely branching large ever-green tree with profuse foliage. It grows in different types of soil and can be grown on field borders, roads, waste lands also in parks and gardens. Trees are planted at a spacing of 5 to 6 metres. One or two loppings per year are taken in favorable seasons, each lopping giving 150-200 Kg of green biomass.

5. *Deiris indica (Syn. Pongamia glabra)*

It is a leguminous moderate size evergreen tree. It grows is coastal forest, on river banks, tank bunds mostly along the streams, waste lands and road sides. Trees are established by planting two or three months old seedlings at a distance of 4 to 5 metres. Two loppings can be taken in a year, yielding about 150-200 Kg/ lopping.

6. *Thespesia populnea*

It is also an evergreen tree which grows in all soil types. The trees are grown in gardens, parks and also in waste lands at a distance of 4 to 5 metres. It is propagated by stem cuttings. It establishes fast and produces numerous branches. Two or three loppings of green leaves can be taken in a year during favorable environments. Each lopping may provide 100 to 150 Kg. of green matter.

Green Manure in Cropping System

Rice - wheat is the most important cropping system in northern India. There is a gap of about 40 to 70 days between the transplanting of rice crop and harvesting of wheat crop. Short duration legume crops such as mungbean and cowpea can be grown in the zaid season and biomass can be incorporated in the soil after picking the pods. Forage legumes such as cowpea can also be grown during this period using a portion as fodder and the rest as green manure. The fallow period may extend upto 90 days when 'basmati' rice is grown after wheat. Similar strategy of green manuring can be adopted in maize -wheat crop rotation.

Green manuring can also practised in Northern India in potato-sunflower and potato-wheat rotation. There is a fallow period of about 120-130 days after wheat crop and before planting of potato crop in late September. Potato is known to respond markedly to green manuring. Green manuring can also be included in oilseed based cropping system, particularly *toria* - sunflower cropping system. After harvest of sunflower in late May, there is a fallow period of around 100 days before sowing of toria in mid - September. The fallow period can be used to grow grain legumes for dual purpose with minimum input of water, the period being a part of the rainy season. Similarly green manuring can be done in toria-wheat rotation by growing a grain legume after harvesting of wheat crop.

Green manure can be interplanted with sugarcane and cotton crops and incorporated into soil after 50 to 60 days of growth. In rainfed areas where monocropping with wheat is practised, green manure plants, particularly sunnhemp can be raised in kharif and incorporated into the soil before the recession of monsoon.

Rice - based cropping systems occupy a major area in irrigated land in south India. Single or two rice crops are grown followed by pulses, cotton and vegetables. With such intensive nutrient mining, multiple nutrient deficiencies are occurring leading to decline in soil productivity. A number .of green manure crops have been evaluated for their biomass production and nitrogen contribution so that they can be grown for the system in the off-season (Table 37).

Table 37: Evaluation of green manuring in rice-based cropping system.

Green manure	Biomas (t ha-[1])	N accumulation (Kg ha[1])	Salient features
Crvtalaria juncea (sunnhemp)	16.8	159	Quick growing, easy seed production
Sesbania aculeata (daincha)	26.3	185	High biomass production wide adaptability
Sesbania rostrata	24.9	219	Stem nodulafing, tolerant to waterlogging
Tephrosia purpurea	16.8	115	Drought tolerant, self seeding
Phaseolus trilobus	17.6	126	Green manure-cum-fodder cover crop

1.Rice varieties matured in 60 days, *Source*: Palniappanera/. (1990).

Intercropping of *Sesbania rostrata viMh* Rice

Stem nodulating *Sesbania rostrata* can be intercropped in lowland kharif at 1.5m intervals taking on row for every 10 rows of rice. Thirty-day old S. *rostrata* can be cut at about 30cm height and incorporated in the field. The cut portion can be ratooned and final growth ploughed in soil before the transplanting of rabi rice. Studies conducted at Coimbatore showed the benefit of intercropping S. *rostrata* and rice, incorporating of cut portion at 45 days after sowing of rice and the ratoon growth was incorporated before the transplanting of the second crop (Table 38)

Table 38: Effect of intercropping the green manure tree *Sesbania rostrata on* rice yield.

Treatment	Biomas added (t ha-[1])	N contribution (Kg ha-[1])	Grain yield (t ha-[1])	
			Kharif	Rabi
Green manure plant (whole) incorporated at 45 days of growth	6.2	53.8	4.50	495
Green manure plant cut and incorporated at 45 day of growth	5.2+	82.6	4.14	5.64
Ratoon incorporated after rice harvest	5.4			
No green manure	-	-	4.60	3.88
CD (5%)	-	-	0.18	0.13

Decomposition of green manure

Green manures are rich in water soluble constituents such as nitrogen and other minerals and comparatively low in cellulose and lignin. The nutrient content and C/N ratio of some green leaf manures is presented in table 39

Table 39: Nutrient content of some green leaf manures

Crop	N (%)	P (%)	K (%)	C: N ratio
Conventional green manures Annuals *Sesbaniaaculeata*	3.97	0.37	4.80	21
Crotalariajuncea	1.90	0.34	3.60	44
Perennials (leaves and twigs) *Gliricidia septum*	2.76	0.28	4.60	31
Pongamiaglabra	2.78	0.27	0.19	33
Non-conventional green manures trees (leaves and twigs) *Azadirachtaindica*	2.83	0.28	0.35	70
Delonix elate	3.51	0.31	0.13	27
Detonixregia	2.76	0.46	0.50	32
Phettophorwn ferugenum	2.63	0.37	0.50	32
Weeds *Adathoda vesica*	1.32	0.38	0.15	60
Parthenium hysterophous	2.68	0.68	1.45	30
Echomiacrassipes	3.01	0.90	0.15	29
Trianthema portulacastnim	2.64	0.43	1.30	32
Ipomoea cornea	2.01	0.33	0.40	43
Calotropisgigantia	2.06	0.54	0.31	64
Cassia pistula	1.60	0.24	1.20	120

Source: S.P. Palaniappan

Green manuring crop residues are low in lignin and C/N ratio as compared to straw or mature dry plant residues. The rate of decomposition of green manure is faster and plants nutrients are converted from organic to inorganic forms due to microbial activity in the soil. The main objective of green manuring is to supply plant nutrients especially nitrogen.

According to Daji (1934) when young plant matter is incorporated in soil, there is a danger of nitrogen loss, since the formed nitrate is likely to leach out unless the succeeding crop is sown timely, to utilise the nitrate formed. The process of green manuring may also have a priming effect on the soil organic matter. When mature green crop is ploughed into soil, due to comparatively slow process it would also contribute to the formation of humus substances. The age of the plant will play an important role in the kind.of benefits which may accrue to the crop due to green manuring.

Traditionally green manure crops are grown in rainfed fallow fields and incorporated 2-4 weeks before sowing of the main crop. This practice is not feasible in intensive agriculture system because the fallow period before sowing of kharif season crop is 40 to 60 days only. The results (Table 40) showed that delay in transplanting of rice after green manuring was not only disadvantageous but also decreased the yield responses.

Table 40: Effect of interval between incorporation of green manure and rice transplanting on rice yield as reported from three sources.

Decomposition period (days)	Rice grain yield (t ha^{1})		
	1	2	3
0-1	5.9	5.9	6.2
5-7	—	6.1	5.3
10	5.7	6.1	—
14-15	—	5.4	4.8
20	4.4	—	—

Source: 1= Bhardwaj (1982), 2= Ghai et al. (1988), 3= Beri et al. (1989)

The decomposition products have a marked effect on biological, physico-chemical and physical properties of soil. Effect on pH is complex but tends towards neutrality. Redox potential is reduced under water logged conditions but electrical conductivity tends to increase. Physical conditions are improved primarily due to decomposition products of organic matter, synthesis of microbial polysaccharides and development of fungal network in soil. This improves the proportion of water stable aggregates. Bulk density is lowered and porosity, water holding capacity and infiltration rate of soil are enhanced. It checks soil erosion and nutrient losses. Green manuring brings about favourable changes in the physico-chemical properties of saline-sodic soils and helps in their reclamation.

Crop Responses to Green Manuring

In rice, *Sesbania aculeata* and *sunnhemp* have been reported to increase the yield by over one hundred per cent. Increase in yield of 1-2 tonnes/ ha are commonly documented (Table 41).

Table 41: Rice grain yield response to green manuring (GM) in North India.

Green manure crop	Age (days)	Rice grain yield (t ha^{-1})		Response to GM (%)	Reference
		No GM	With GM		
S. aculeata	66	2.64	5.64	114	Dargan *et.al* (1975)
S. aculeata/ Sunnhemp	60	3.22	3.95	23	—
S. aculeata	50	1.55	2.69	74	Tiwari *et al.* (1980)
S. aculeata	50	3.30	6.10	85	Ghai *et al.* (1985)
S. aculeata/ Sunnhemp	56	2.62	3.68	41	Sharma and Mittra (1988)
S. aculeata/ Sunnhemp Cowpea	60	2.70	5.50	105	Beri *et al.* (1989a)
S. *aculeata*	60	2.70	5.80	115	Beri *et al.* 1989b)

Dargan *et al.* (1975) reported higher rice yields due to green manuring with *sesbania aculeata* both in control without any applied fertilizer and treatments containing fertilizer nitrogen. The other studies with green manuring also showed high responses in the productivity of rice (table 41).

Singh and Brar (1985) reported that cowpea green manure at 10 and 20 tonnes ha^{-1} supplying 60 and 120 Kg N ha^{-1} increased the grain yield of maize by 64 and 155 per cent over the control respectively. Green manure provides benefits beyond the nitrogen content because the increase in the yields of rice reported can not be achieved with any amount of fertilizer nitrogen (Dargan *et al.*, 1975, Tiwari *et al.*, 1980, Beri *et al.*, 1989a).

Such responses are due to cumulative and legume effect of the green manure crop on soil properties and plant growth.

Substantial response of potato crop to green manuring *(Sesbania, Cluster bean,* Mungbean) has been reported by Singh *et al.* (1975), Sharma *et al.* (1988) and Sharma and Sharma, (1988) (Table 42).

Table 42: Response of potato to green manuring (GM) in India.

Green manure crop	Age (days)	N addition	Potato yield			Reference
			-GM (t ha^{-1})	+GM (t ha^{-1})	Increase (%)	
Cluster bean	56	—	13.0	18.6	43	Singh *et al.* (1975)
Sesbania	—	90kg N ha^{-1}	13.7	18.5	36	Sharma *et al.* (1988)
Sesbania/ Mung bean	—	70-90 kg N ha^{-1}	9.6	14.7	53	Sharma and Sharma (1988)

Several field experiments showed a grain yield increase of rice equivalent to 26-50 Kg N/ ha (Rajendra Prasad and Goswami, 1991). The organic carbon and total nitrogen of green manured plots were 0.67% and 0.054% respectively against 0.58% and 0.021% in control plots.

The superiority of inclusion of green manure crop *(Sesbania aculeate)* in rice-wheat cropping system to sustain crop yield and a saving of chemical fertilizers was shown at Ludhiana (Brar *et al.* 2005) The results clearly showed that green manuring before rice plantation not only saved 50% of NPK fertilizers but it also improved rice crop yields. However, there was no residual effect of green manuring on suceeding Rabi wheat crop Yields.

Under dry land conditions it was observed that 59 to 120 Kg N / ha were fixed by green manure crops (McDonagh *et al.* 1995). The short duration groundnut and a mixed *Sesbania rostrata* and multipurpose cowpea (MFC) produced 4 tonnes / ha oi dry residues. Groundnut yielded 1910 Kg / ha whilst intercrop yielded 1000 Kg of fresh pods of MFC.

Synergic Effect of Green Manure

Experiments conducted at Ludhiana have shown that application of green manure and crop residues together helped in alleviating the negative effect of crop residues on the yield of rice (Table 43). The grain yield of rice in soil amended with wheat or rice straw combined with green manure crop was higher or similar than the yield obtained with application of 150 Kg N/ ha in control plots. The synergy interaction of green manure and crop residues increased the soil organic carbon and soil fertility besides the favorable influence on other soil properties.

Table 43: Effect of fertiliser N, green manuring, and crop residues on grain yield of rice (t ha-1)

Treatment 1	1988	1989	1990	1991	1992	1993	1994
Control (no N)	4.0	4.6	3.7	4.3	4.1	3.4	2.8
150 Kg N ha^{-1}	6.3	6.6	6.2	6.5	5.7	5.6	4.0
180 Kg N ha^{-1}	6.6	6.9*	5.8	6.7	NT2	5.3	3.8
GM	6.6	6.5	6.2	6.5	5.8	5.5	3.9
GM + wheat straw	6.9	6.9	6.4	6.8	5.6	5.5	3.9
GM + rice straw	6.9	6.9	6.7	7.0	5.9	5.3	3.9
Wheat straw + 1 50 Kg N ha^{-1}	NT	6.8	5.8	6.1	5.2	4.8	
FYM	NT	6.3	4.6	5.7	6.0	4.8	
*FYM + GM	NT	7.7	6.1	6.9	6.5	6.1	
LSD (P=0.05)	0.53	0.59	0.46	0.45	0.32	0.37	

1. Gm= green manure, FYM= farm yard manure
2. NT= not tested

In maize-wheat-mung rotation, incorporation of mungbean straw (after picking of pods) along with wheat straw markedly increased maize yield over no residue or the treatment with wheat straw alone. Maize yield with 60 Kg N/ ha in mungbean residue + wheat straw plots was similar to that obtained with 180 Kg N ha'^{1} on unamended plot, thus saving 120 Kg N/ ha. The grain yield of the following wheat crop was also increased significantly on the amended plots.

Biological nitrogen fixation through green manuring legume crops converts atmospheric dinitrogen into plant-usuable form and helps to maintain soil nitrogen status as a substitute for fertilizer nitrogen to obtain large crop yields (Peoples and Craswell, 1992)

Economics of Green Manuring

The contribution of leguminous green manure crops to N nutrition of the following rice or maize crop has been studied throughout north India. Apparent recovery of green manure N in different studies in north India varied from 32 per cent to 50 per cent (Bijay Singh and Yadvinder Singh, 1997). In general, recoveries of nitrogen from green manure was similar to those from fertilizer. Many studies provide an indirect method of N contribution by green manures by comparing rice yield responses to N fertilizer with and without green manure. The parameter nitrogen fertilizer equivalence (NFE) of a green manure is often calculated as the quality of fertilizer N that must be applied to obtain grains yields equal to those obtained with green manures alone. This approach could be best regarded an approximation because the green manure effect is not due to N contributidn alone but due to other benefits as well. The NEF values of different green manures in wetland range from 45 to 148 N/ ha from 45 to 60 day- old green manure crop.

The leaching losses of green manure N are normally expected to be smaller than chemical fertilizer N because mineralization of green manure during two weeks after incorporation ranged from 12 per cent to 38 per cent under upland condition (Yadvinder Singh *et al.* 1992).

Denitrification losses in green manure amended soils are affected by the supply of nitrate N rather than the availability of organic carbon (Yadvinder Singh *et al.* 1992). The interval between the incorporation of green manure and application of fertilizer together with irrigation method during crop growth will markedly influence N losses through denitrification. Green manuring with *Sesbania* reduced losses of fertilizer N by 30% mainly through denitrification although plant uptake of fertilizer N was also reduced (Azam *et al.*, 1985).

Green manuring reduces losses of nitrogen by ammonia volatilization through temporary microbial immobilization of N and production of organic acids. Khind *et al.* (1989) reported ammonia volatilisation loss of about 1 per cent of applied total nitrogen in a sodic soil (pH 10) green manured with 6 tonnes *Sesbania I* ha containing 2.5% N and 100 Kg urea N/ ha. The loss of N from urea treated soil was 4.5 per cent.

Substantial information is available on the contribution of green manure towards increasing crop yields and improving soil properties resulting in conditions favorable for sustainable agriculture. A critical issue for wider adoption appears to be the practical and economical feasibility of the technology at the farm level. Tillage and seeding operations are major component of labour and cash required by farmers to grow green manure crops. Irrigation cost is an important aspect in several part of India.

Constrains in Green Manuring

In spite of several advantages with BNF and green manuring, there are some limitations which are hampering the adoption of this technology on a wider scales particularly in a country such as India. The constrains are:

1. In intensive farming, farmers do not wish to set apart about 8 weeks exclusively for growing green manure crop. Moreover, in north India, planting green manure crop may delay rice plantation beyond the normal transplanting period.
2. Farmers opt for fertilizer N because typical agronomic constrains such as intense heat during May and June common to all green manures remain as real barriers for adoption.
3. The costly item in green manuring is the seed. Inadequate availability of quality seeds and desired species at lesser cost is one of the major problems in the adoption of green manuring by the farmers.
4. Production of leguminous green manuring crops is adversely affected by insects and pests when grown before or during rainy season in north India.
5. Many farmers do not know that legumes fix nitrogen in their root nodules and thus benefit the crop. However, proper ,extension and education of farmers will help in adoption of the practice. *Efficient Rhizobium* cultures for seed treatment should be made available to farmers at convenient places.
6. Non-availability of water resources may restrict the raising of green manure crops.

V. Urban Solid Wastes

The growing population, increasing urbanisation and expanding industrialisation have created problem both for agriculture and environmental sanitation. The generation of urban wastes has definitely created problems of pollution of land, water and air. The only solution under the present situation lies in considering these foul smelling obnoxious, highly polluting material not as wastes but as potential new resources of immense value for meeting the plant nutrient needs.

In general "solid waste" is the term used internationally to describe non-liquid waste materials arising from domestic, trade, commercial, industrial, agricultural, mining activities and from the public services. The disposal creates lots of health and environmental problems.

Approximately 500-1000g of solid wastes per person is generated in cosmopolitan cities of India (Table 44) (Plate 6).

Table 44: Solid wastes potential in some Indian cities

City	Tonnes /day
Mumbai	6050
Delhi	4000
Calcutta	4000
Bangalore	2000
Chennai	4000
Lucknow	1500
Hyderabad & secunderabad	1300
Ahmedabad	1280
Surat	1000

Plate 6: Composting of city wastes

It is being realised at various levels that hygienic disposal of city waste is of great importance for abetment of environmental pollution problems and conservation of resources. About 60 MT of municipal solid waste from 300 class I cities calls for our immediate attention as it is daily collected and lies in accumulated form at dumpsites leading to problems of environmental contamination and an unhygienic disease oriented living conditions. It is surprising to note that India is the second largest country of the world having agriculture as the most dominating way of livelihood. Only a few small scale (partially-run) composting plants were operating during the last 1-2 decades (Plates 7-8). In the process, over 500 MT of precious organic matter got wasted which otherwise would have replenished at least 50 million hectares of cultivated land. In the same period, we have concentrated on increasing the food production only through chemical means thereby facing the problem of chemical residues in food, ground water contamination, loss of organic carbon of soils, loss of biological equilibrium in soils, deterioration of physical condition of soil, increased soil erosion and deterioration of food quality in terms of taste, flavour, shelf life etc.

Nature of City Solid Wastes

Indian Waste has 35 to 40% organic matter alongwith moisture content of 40 to 50% makes it 75 to 90% component of waste stream. Balance fraction of 10 to 25% comprise of plastic, syn. textile, sand, stone, ceramic, hard wood etc. This 75 to 90% component is ideal for recovery as organic manure (Table 45). If, it is not done and instead waste is just dumped or land filled it leads to putrefaction, production of toxic leachates, foul smell, pathogens and creates many problems.

Plate 7: Semi-mechanized compost plant system

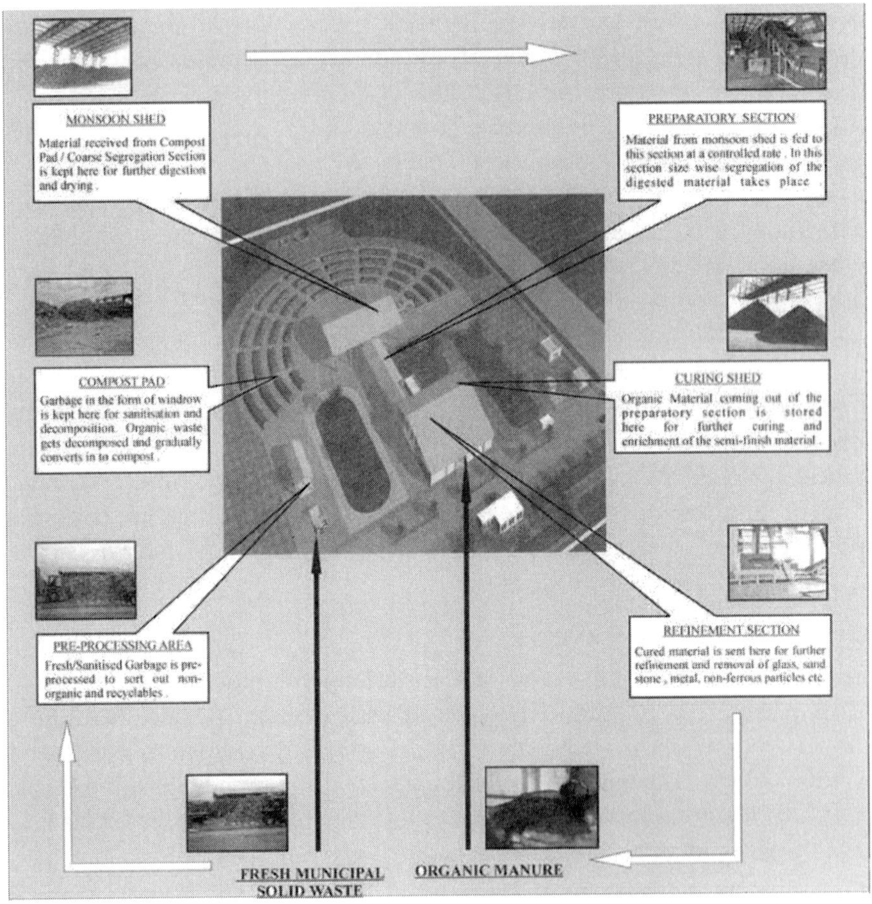

Plate 8: Compost prepared at semi-mechanized system

Table 45: Components urban solid wastes

Biodegradble material	Content (By weight) (%)	Non-Biodegradble material (%)	Content (By weight) (%)
Organic matter*	78.20	Plastic	3.71
Coconut shells	2.70	Leather	3.21
Jute	0.95	Rubber	0.75
Cotton textile	2.2	Syn. textile	0.69
Paper products	3.12	Metals	0.37
		Glass/Cereamics	1.45
		Sand etc.	2.65
Total	87.17	Total	12.83

*Composting of leafy biomass, fruit, vegetable peels, prunes from lawns, straw, fodder etc.

Source: Trivedi and Gurdeep (1992 a and b). Khan et al. (1994)

The calorific value of Indian waste is only 800 to 1400 KJ/ Kg. Therefore energy recovery from the same is costly and is a non-viable project. Indian waste is all mixed and segregation at source is still not done in any city in the country. This not only makes handling and processing difficult but also rules outapplication of foreign technologies and machines as the same is based on segregated wastes. Due to hot humid condition, putrefaction and rotting cycle of organic wastes starts very quickly leading to disease and foul smell problems.

The solution of the above problems lies in quick bioconversion of organic wastes.

Technology Options

In view of land degradation and demand for organic food, it is essential that the biodegrable organic wastes have to be utilized for meeting the soil health and plant nutrients need. The following technology options are available for organic waste recycling and management:

1. Aerobic Composting
2. Anaerobic composting and / or methane gas production
3. Vermiculture
4. Fuel brickets/ pellets
5. Power generation

Aerobic Composting

This is a key technology and is the most practical and commercially viable option in many parts of the world. With scientific advancement several innovative parameters have been incorporated of better efficiency, speed of fermentation and quality of the final recovered productions. Salient features of different products have been described by Gaur (1999). Organophos (phospho compost) enriched compost, microbially hastened compost etc can be prepared using appropriate microbial technology.

Anaerobic Digestion

This technology can be practised with source segregated wastes combined with cattle dung and selectively picked up wastes from agromarketing units and market yards. The project cost for 150 TPD model is between 600-900 lacs thereby making the gas as well as manure costly. Requires recurring subsidisation.

Vermiculture

It is most eco-friendly technology and ideal process for small scale decentralised composting plant. It can handle upto 20 MT wastes/ day. Selected market yard and food wastes are required. The activity is best suited to NGO and large orchardists due to paucity of space and large number of labour requirement.- Investment level is low

Fuel Brickets/ Pallatisation

This option has not been successful in India due to soft succulent waste, higher drying cost and efforts and low calorific value output at a higher production cost. This activity may turnout to be useful if combined with composting operation where some woodly and lignin rich materials get rejected. If such materials are used for fuel bricketing, there can be operational and commercial synergy. However fuel bricketing plant put by Excel Industries Ltd. with assistance of G.O.I, is existing in Mumbai with recurring losses for the last 8 years.

Power generation

Power generation from mixed municipal solid waste is a difficult task. One plant with huge investment by G.O.I. (Delhi Municipal Corporation) was put which did not work. The plant was based on foreign technology and the government has subsequently lost the case in the International court.

Microbial process for conversion of city wastes into compost

Urban solid wastes containing energy and nutrients for crops need to processed by biological and mechanical methods into good quality enriched organic manure. The process H is carried forward with the growth and activity of bacteria, fungi, actinomycetes and micro and macrof-auna. The process can be hastened by use of efficient cellulolytic, lignolytic, proteolytic and other type of organisms producing different enzymes for breakdown of long complexes at mesophilic (25M5°C) and thermophilic (50°-75°C) condition. It is most effective and economical method (Plate 9). It was reported that dehydrogenase and alkaline phosphates were much higher in crop residues compost whereas cellulase and protease were notably higher in city compost (Gaind and Lata, 2004)

Plate 9: Ligno-cellulolytic fungi *Phanerochate crysosporium*

Steps in Biocoftversion

Selection of efficient microbial strain → culture multiplication, inoculation of waste with efficient culture → Enzymes (production) multiplication and growth in organic biomass → Energy + CO_2+Nutrients, At high temperature... Pathogens are destroyed or Inactivated Maturation and Stabilization phase (Temperature decreases to mesophilic and ambient temperature).

Secondary maturation= Synthesis of hurmic substances takes place

Non-mechanical methods were used earlier when volume of city wastes to be handled was small. Mechanised composting is the only alternative for handling enormous quantities of wastes generated in bigger and metropolitan cities. The setting up of mechanical compost plants involves large capital investment. The financial sources of various agencies like municipal corporation committees, agro-industrial corporations which undertake setting up of such plants are not large enough to meet the capital cost of these projects. No doubts, the ministry of Agriculture provides subsidy to the extent of 50 lakhs to meet the capital cost of the plant, machinery and civil construction works. The rate of interest charged by the banks pushes the cost of construction and the production costs to such an extent that the marketing of compost is not so easy and requires an appropriate action plan for the marketing of the product. The main problem in making such plants economically viable is the sale of product. The only solution to make compost acceptable to farmers at reasonable cost is through an improvement in its quality. This can be possible by making organo-biofertilizers of varying grades such as phoscompost (organophos) rock phosphated or bone meal compost or microbially enriched organic manure. The use of efficient microorganism and amendment with relevant by products for enriched product is essential.

Solid waste management in Indian metropolitan cities

Mumbai generates about 6000 tonnes of garbage per day during peak season from April to October. The physical components of the solid wastes of Mumbai are 38% ash and soil, 20% wood and grass, 10% Paper, 2% plastic and 30% others. It has three measure chemical components of 40% moisture, 22% combustible and 38% ash content. The wastes are dumped at four dumping sites at Deonar, Mulund, Malad and Gorai areas. The garbage is disposed off untreated after only preliminary levelling and debris layering. This creates unhealthy environment.

The amount of solid wastes generated in Delhi is around 4000-5000 tonnes/day. The solid waste is dumped in landfil-sites at Ghazipur, Bhalswa, Tugtakabad, Mastthal and Mandawli. Some of wastes around 200 tonnes are composted at NDMC compost plant Okhla. Recently during April, 2001 a compost plant has been promoted by APMC to convert vegetable and fruit wastes generated at Azadpur Subzi mandi. The APMC compost plant is located at Tikri Khurd, G. T. Karnal Road, Delhi and making "Organophos" manure. The project seems to be a failure because proper techniques/ technology are not being followed.

About 4000, tonnes of solid wastes accumulate daily in Kolkata which are collected and transferred to dumpyards. Many scientists have evaluated the fertilizing value of Kolksta City wastes from different landfill sites of various durations and also city wastes prepared at compost plants at Bantalla, Kolkata. Several workers (Khan *et al.*, 1981, Datta Gupta *et al.*, 1985, 1986a,b,c, Chattopadhyaya *et al.*, 1992) reported favourable effect of application of decomposed city wastes on soil fertility and physical condition of soil. No toxicity of heavy metals has been reported. Favourable effect of city wastes on growth, yield and nutrient uptake by vegetables and rice has been reported. In garbage farms of East Kolkata, 25 varieties of vegetable are grown in a year without addition of chemical fertilizers (Christine 1989).

Bangalore generates over 2500 tonnes of wastes daily some of the solid wastes is converted at partially mechanised compost plant at Bangalore. The collection and transporation of city garbage to dumping yards and compost plants need to be improved a great deal. Chennai generates about 2000 tonnes of garbage daily besides debris around 500 tonnes daily. The garbage is dumped at two low-level areas viz. Kodungaiyur and Perungudi which are almost full.

Compost Making - A Key Technology

Composting is a microbiological and ecofriendly technology for disposal and conversion of solid organic wastes into manures. Composting is being practised in bigger cities such as Delhi, Mumbai, Kolkata and many others. In many of these cities, manure produced from city garbage and wastes are sold to farmers and individuals for use on their farms and kitchen gardens.

Mechanical Composting Plants

The wastes are generated from various sources such as vegetable and fruit wastes from central markets, hotel, restaurants, canteens, hostels, kitchen wastes from residential areas and apartments besides the wastes from leaf litter and tree prunnings, street sweeping etc. The Nitrogen content of such wastes is found to be around 2 per cent which is favourable for rapid composting. The different types and categories of waste suitable for bioprocessing into enriched organic manure / compost are follows:

1. **Crop residues** (straw, fruit and vegetable, leaves, stover etc.) Banana, Papaya, date palm stubs, wild and aquatic weeds.
2. **Animal and slaughter house wastes** Animal dung, poultry droppings and piggery wastes, Slaughter house wastes such as meat meal, fish meal, bone meal, hair and wool waste.
3. **Agro-industrial products** Rice husk, rice bran, bagasse, molasses and pressmud, cotton litter, cotton seed hull, ginning wastes, Jute mill waste, Tabacco wastes, Fruits/ vegetable processing wastes (mango kemal, mango peels, banana peels, citrus, apple, pineapple, custard apple wastes). Soya sauce and Tomato ketchup wastes Potato, Tapioca, Corn industry's wastes, Non-edible oilcakes, coirpith, Leather wastes, fish, prawn and lobster shells, sericulture wastes, cocoons, coffee and tea, spent mushroom meal. Cold storage and market yard wastes.

The ideal size of compost plants may be those with a daily intake capacity of about 120-150 tonnes of good quality-wastes. The extent of mechanisation should be tailored to local conditions due to variations in the characteristics of urban wastes from city to city and location. Vegetable and fruit market wastes are of good quality as compared to conventional garbage. The non-compostable wastes are discarded before or after composting according to the available facilities.

Sanitisation of Compost Material

Solid wastes from urban and rural areas may be contaminated with a variety of pathogenic microorganism such as bacteria, viruses, protozoa etc causing outbreak of several type of diseases. Therefore the san'rtasation requirement for composting should be strictly monitored. The maintenance of high temperature in the range of 55-65°C in compost heaps for 3-4 days is required to kill the pathogenic organism if any present in the raw material in the compost (Table 46).

Table 46: Sanitisation requirement for composting in some European Countries

Country	Temperature (°C)	Exposure (Days)
Austria	65	6
Belgium	60	4
Denmark	55	14
France	60	4
Italy	55	3
Netherland	55	2

One of the merits of composting is that the pathogens are killed by the high temperature which prevails in the heap during active thermophilic decomposition phases.

Quality of Compost

Generally the compost produced at mechanical compost plants from conventional garbage are poor in plant nutrients and generally contain 0.5 to 0.8% N, 0.5 to 0.7 phosphate and 0.5 to 0.8% potash. To improve the quality of compost in city plants better quality -garbage containing high percentage of organic matter should be used. This will reduce the transporation of inert materials, helps in effective capacity utilization and produce good quality compost. Blending of sewage waste waters or sewage sludge during composting as source of plant nutrients and water will be beneficial (Gaur 1983). Frequent analysis of the compost samples to monitor the quality should be done and enrichment with other by products such as rock phosphate, flyash, basic slag, pressmud, fish meal, non-edible cakes etc in judicious amounts should be done to scale up the quality and reduce the cost of production per unit of plant nutrient.

VI. Urban Liquid Wastes - Sewage and Sludge

In many developing countries, recycling of human wastes is dependent on the installation of pipelines and water treatment plants unless its direct handling is socially accepted as in china. In a large country like India the amount of sewage sludge is barley 0.5 million tonnes (Juwarker *et al.*, 1992). In India sewerage system exists in big cites where sufficient households are connected to sewerage system or make agricultural use of sewage, a convenient and economic proposition. The system provides both water and plant nutrients to crops. The chemical composition of sewages sludge as reported by Juwarker *et al.*, (1991) is as follows:

Organic carbon 19.5-31.80%, Total N 1.08-2.34%, total P_2O_5 0.84-2.14% K_2O 0.53-1.73%, Zinc 700-2146 ppm, copper 194-535 ppm, manganese 176-465 ppm.

Status of Pathogen

The presence of pathogenic bacteria, viruses, *ascaris* ova and protozoa cysts in the sewage and sludge and the likely problems in agriculture associated with their need to be examined. A large number and variety of pathogenic bacteria, viruses and protozoa are found in the domestic and industrial wastes which can cause innumerable health problems. The work in India on this aspect has been a low-key affair but NEERI, Nagpur is likely to pay greater attention to these aspects.

The relative number of enterobacteria in sewage and sludge is regulated by the type of sewage treatment. Primary treatment mostly consisting of settling process remove 35-45 per cent pathogens while more than 95 per cent pathogens are removed by secondary treatment. Thu use of treated sewage thus minimize the health risk. The potential risk due to pathogens to the farm workers and consumers depends on the survival of pathogen in soil and on crops. Their survival is affected by several environmental factors such as soil temperature, pH, moisture and antagonistic interaction between organisms of soil and sewage waste waters.

The survival of enteric pathogens coli forms, *E-coH, E. streptococci* was higher is soil treated, with 1:1 diluted and primary treated sewage as compared to secondary treated sewage and fish pond waste water upto 120-150 days. A high degree of health risk persists in the use of untreated and 1:1 dilute sewage and to some extent in the case of primary treated sewage (Juwarkar, 1987). Effect on the vegetables like spinach, cabbage and cauliflower grown in soil treated with untreated sewages were reported to be contaminated with *Salmonella.* However *E-coli* forms and *E. streptocci* were observed even with treated sewage use (Jurwakar *et al.* 1990). The survival of virus on crops surface is shorter than in soil because they are exposed to deleterious environment. Survival of polio virus type 1 on vegetables such as lettuce and radish was observed for 14 to 36 days after irrigation.

To minimize health hazard to farm workers and consumers due to sewage handling and use, it is desirable to treat the sewage at least up to secondary level prior to irrigation particularly for the crops which are consumed raw.

The primary treated sewage can be used to grow crop such as cereals, pulses, oilseed and fiber crops.

Major Plant Nutrients

Plant nutrients content of sewage varies from city to city and on seasonal basis. The average concentration of NPK in sewages of a few Indian cities in presented table 47 (Kale *et al.,* 1973).

Table 47: NPK concentration (ppm) in sewage

City	N	P	K
Mumbai	47.7	12.2	n. a
Kolkatta	40.0	5.5	15.9
Delhi	28.5	13.7	41.0
Hyderabad	37.0	14.7	26.0
Kanpur	73.0	2.5	6.8
Chennai	59.2	25.3	19.2
Nagpur	58.3	7.0	41.6

The average concentration of N, P and K in Indian sewages is estimated 60,20 and 40 ppm respectively (Shende 1988). Considering an estimate of 40,10, and 30 K g/m3 NPK respectively effluent applied @ 10,000m^{3}ha/year will add NPK @ 400,100 and 30 kg/ha/year. On such fields, additional fertilization is not needed (Shuval 1990, WH01989)

Normally domestic sewage contains 99.9 percent water and 0.10 percent suspended colloidal and dissolved solids containing plant nutrients and organic matter.

Sewage sludge from highly industrialized wastes can pose heavy metal contamination of foods: While copper, zinc, iron and manganese are essential micronutrient for growth of plants, cadmium, chromium, nickel and lead have no such role to play and are pollutants. There is considerable variation in heavy meta'fcontent of sewage sludge (Table 48) produced in different Indian eilies (Juwaekar *et al.,* 1992).

Table 48: Total and available heavy metals in sewage sludge

City		Heavy Metal (ppm)						
		Cu	Zn	Mn	Cd	Cr	Ni	Pb
Ahemdabad (Pirana)	Total	535.2	2145.7	215.0	3.5	60.4	32.3	76.8
	Available	50.3	201.2	11.0	0.8	-	2.3	8.3
Delhi (Okhla)	Total	307.2	1260.6	190.7	1.5	82.0	191.5	41.7
	Available	34.6	146.3	9.2	0.5	-	7.6	6.7
Chennai	Total	210.3	935.0	465.2	8.3	38.5	60.5	16.5
	Available	17.2	77.2	24.5	1.3	-	2.2	1.9
Nagpur	Total	272.5	832.5	242.5	1.5	49.2	14.8	24.3
	Available	45.7	70.3	9.2	0.2	-	1.6	3.5
Jaipur	Total	265.3	1720.2	255.6	7.3	176.2	37.5	66.9
	Available	29.1	79.0	26.0	1.4	-	0.3	6.4

These elements are converted into various torms after application of sewage sludge in soil. The available form is usually a small part of the total content but can accumulate to toxic levels.

Processing of Raw Sewage

Application of raw sewage to soil for growing crops in not recommend due to pathological and biochemical reasons. Appropriately treated sewage sludge and use of sewage water for irrigation are commonly used for recycling of human wastes in agricultural land. Composting of human wastes can be practised and used as organic manure for land application.

In absence of sewerage systems, implementations of on -site sanitation method should be carried out where excreta is stored over a long period(up to 2 years) as in the case of twin pit pour flush pit latrines which are being used in India. When the pit get filled up with night soil, it is disconnected from the latrine and elements allowed to get decomposed. During the process, all pathogens including the helminthic eggs are killed. The organic manure thus produced can be safely used for land application. Thermophilic composting or composting for long period are appropriate technologies for making the waste, pathogen free before use.

Apart from conventional primary and secondary treatments of sewage, composting offers another technology for producing safe end product for recycling on agricultural land.

Composting Technology

In many cities, drainage system has been provided and the same is being extended regularly to provide the hygienic conditions to the inhabitants. The sewage treatment plants are used to purify the water for its reuse and sludge is dried in the tanks and used as organic manure.

Where the sewage treatment plants have not been installed, it is utilized as source of nutrients and irrigation water on organic sewage farms.

Besides the necessity of sewage farming from point of view of utilization of resources, it is also an effective method to avoid pollution. Unrestricted discharge of city liquid wastes in rivers and streams and on land results in pollution of environment and is a public health hazard.

Sludge from sewage treatment plant has become another resource to be managed by composting. Composted sludge has been found to be a good material for agricultural use. On an average sludge has been found to contain 1.6% N, 1.0% P and 0.2% K on dry weigh basis (Parr *et al.*, 1978). Sludge is not being composted in India although it is not used in developed countries without composting. Sludge in India is dried in open tanks near sewage treatment plants and sold to Govt. organizations and farmers for use in gardens lawns and parks in the city regardless of possible heath hazard.

The composting of sewage sludge and city wastes can improve nutrient value of the compost and sanitization process will kill the pathogens in sludge. The value of compost thus produced will be more than raw sludge.

Upto 1965, town refuse composting was an organized process in India, the practice was discontinued due to handling problems of increasing volumes of city refuse and greater interest in chemical fertilizers. Mechanical compost plants were installed in few cities in India during 1970's to convert the organic wastes into organic manure for agricultural and horticultural use. Few of the plants continued functioning while others were closed down due to operational, maintenance and marketing problems. Generally the compost produced at these compost plants was found useful for land application (Gupta *et al.*, 1986. Velayutham and Bhardwaj 1994). Compost produced from city garbage and refuse are mostly, low in nitrogen and phosphorus content. However, the quality of the city compost can be improved by composting sewage waste water or sludge with the city organic wastes (Gaur 1983).

Effect on Crops

The use of night soil, waste water or sewage sludge saves water and provides plant nutrients. Continued application of sewage or sludge will increase the soil fertility resulting in luxuriant growth, delayed maturity and lodging thus adversely effecting the grain filling and its harvest index. Judicious use of sewage farming system in many countries such as Australia, China, Germany, Guatemala, India, Mexico, Tunisia and USA has been successful in cultivation of food grain, vegetable, fruit and fooder crops (Shende, 1988, WHO, 1988). The use of waste waters in India increased the yields of vegetables, fruits, groundnut and tobacco crops (Table 49).

Tree plantations irrigated with sewage have produced lush green trees. Forest trees irrigated with sewage resulted in fast growth without any harmful effect on the environment. Water stagnation, foul smell and mosquito breeding were never noticed in these places. Nutrient deficiency, toxicity or salinity problems were also not reported. Eucalyptus (hybrid), *Leucaena leucocephala* and poplar have been successfully grown with sewage irrigation at Karnal (India) and any type of trees can be grown utilizing human wastes (Chabbra, 1969).

Table 49: Yield of Crops Irrigated with sewage waste water, at Poona Annual Crop Yield (t/ha)

Crop	Canal Water(X)	Diluted (1:1) Waste water	Undiluted Waste water.
Beetroot	8.75	15.60	16.27
Carrot	9.71	8.72	11.75
Raddish	7.26	6.14	8.33
Potato	6.12	7.00	9,33
Ginger	6.04	9.18	9.80
Papaya	26.72	27.91	37.10
Knolkhol	9.70	11.76	16.57
Cabbage	9.27	11.32	12.13
Cauliflower	6.96	7.08	9.09
French Beans	6.63	8.20	8.06
Tomato	10.01	—	13.38
Tobacco	1.12	1.25	1.25
Groundnut	2.88	2.90	3.17

X: Land manured before planting. Source: Mahida (1981)

The chemical composition of crops fed on sewage treated soil in Calcutta area showed high levels of nitrogen, moderate levels of P, S, and Mg but exceptionally high level of calcium (Adhikari *et at,* 1992). The concentration of K in mustard and gourd stems and N in mustard roots as well as in spinach leaves was above the toxic limit as set by WHO (1989).

The analysis of different types of crop such as cauliflower, mustard, spinach and radish showed exceptionally high concentration of Fe and Zn. The concentration of Cu was above the toxic limit in the leaves of most of the crops. Manganese concentration was below the optimum level. In most cases, Cd, Pb and Cr were present in high concentrations in the plants. Lead concentration was above the critical toxic level in cauliflower, mustard, spinach and radish, Cadmium and chromium concentration was below the critical toxic level.

Hazards from heavy metal accumulation are relatively less in grain crops than in the leafy produce due to lower concentration of these metals in the grains than in the vegetative parts (Adhikari *et at.,* 1992)

The effect of waste water irrigation on soil properties depend on soil type, diluted or raw irrigation etc. The soil productivity can be sustained for several years if agricultural operations are optimum and sewage water is used judiciously considering the soil texture. Fine textured soils have more surface area and greater porosity and can retain waste water for longer periods for plant use and loss of water through seepage is also less as compared to coarse textured soils, therefore finer textured soil need to be irrigated less frequently than the coarse textured soils.

The sewage treated soil had higher contents of available N, P and K as compared to untreated soil (Adhikari *et at.,* 1992). The concentration of micronutrients viz, Zn, Cu and Fe was above toxic limit but that of Mn was within the limits. The concentration of non - essential heavy metals such as Pb, Cd, Ni, Cr and Co was higher than normal. Such accumulations were due to repeated application of sewage water as the water themselves don not carry high concentration of these elements. Except for Pb and Cd, all others accumulate mostly in the surface of soil.

Method of Application

There are about 212 sewage farms in India covering more than 50,000 hectares. Hydraulic loading on these farms ranges farm 22.4 m^3 to 336.8 m^3/ ha/day which is several times higher than the normal loading (Jurwarkar *et al.*, 1992). The present system is very much inclined towards disposal of sewage rather its proper utilization for plant nutrients.

Application of sewage on land is quite easy and usually practiced in the outskirts of cities. The raw or treated sewages is pumped and carried through pipes in different areas of farms and made available to plants through ridges or furrows or by flooding.

Experience with sewage farming in India shows that excessive application of sewage is common. Farmers generally do this as they are charged the price on area basis and not based on the amount of sewage used. This may result in very high rates of nitrogen application which are not only above the crop requirement but also beyond the retention capacity of soils. Such practice can lead to ground water pollution and in certain cases result in soil sickness as well. Thus there is a need to optimize management practice for recycling of sewage waste water in farming. Optimum rates and interval of sewage application to agricultural soil should be based on the crop needs and soil health considerations and not merely as an outlet for waste disposal.

The minimum degree of treatment needed to eliminate health risk and raise suitable crops with treated and untreated wastewater is presented in the Table 50.

These should be used according to the degree of treatment for different crops. These essentially indicate that the plant products, which are directly consumed by humans should be grown with sewage which has at least received secondary treatment.

Table 50: Suggested crop for Irrigation with treated sewage

Treatment level	Suggested crops in order of preference
Primary treated	Cash crops (Cotton, Jute, Sugar-cane, and Tobacco)
	Essential Oil Crops (Citronella, Mentha and Lemon grass)
	Cereals and pulses (Wheat, rice, green gram, black gram, sorghum pearl millet)
	Oil seeds (Linseed, sesame, castor, sunflower, soybean and groundnut)
	Fruit Crops (Coconut, banana, citrus, saputa, guava, grapes, papaya and mango)
	Vegetables (Brinjal, beans, ladyfinger etc.) These should be properly cooked before eating
Secondary treated	All crops listed above
Secondary treated and disinfected	All crops without restriction

Source: Juwarkar et al. (1992)

Utilization of Sewage in Aquaculture

The nutrients in the sewage waste waters are of immense value in aquaculture for growth and multiplication of bacteria which promotes the growth of phytoplankton and zooplankton which are essential for growth of fish (Mandal and Chattopadhyay, 1992)

The sewage wastes is applied in ponds and other water bodies to culture fishes such as carps *(Cyprinids carpis)*. Indian major carps (Calta, marigol and rohu) and Tilapia *(Oreochromis mossambicus)*. According to WHO (1989) reports about two third of world yield of farmed fish is obtained from ponds fertilized with excreta and animal manure.

In India, there are about 132 waste water fertilized fish ponds covering 120 sq Km. Most of these are located in West Bengal. The largest amongst covering 4400 ha exists in Kolkatta and is stocked with Indian major carp and tilapia, The ponds are fed with raw waste water from Kolkatta and this is probably the largest example of waste water based aquaculture (Bose 1944, Edwards 1985, Strauss 1986). The mean annual yield are approximately 1000-1400 kg/ha. The fish ponds supply 10-20 percent of the fish consumed in greater Kolkatta. Fishes have been raised very successfully in waste stabilization ponds with an annual yield upto 3000 kg/ha/year (WH01989)

The use of excreta in agriculture has the hazard of passive carriage of pathogens. For safe use of fish, it is essential that ponds are fed only with treated sewage or stored night soil or sludge.

A well designed compost system can produce almost pathogen free compost and can be safely used in agriculture. Evacuation of the digestive track content usually occurs rapidly due to shock or capture. It is suggested that fish be held in fresh water immediately after capture for at least few hours for them to evacuate their intestines and thus reduce contamination.

Pathogens, Toxicities and Precaution

The presence of pathogens and heavy metals are two major problems in the recycling of sewage waste. The application of proper technology can overcome some of these problems to a large extent.

Several types of pathogens are present in excreta and persist if sewage is not properly treated (Gaur *et al.,* 1995). Communicable disease caused by pathogenic bacteria, Viruses, protozoan cyst and helminthes eggs occur when these infectious agents remain viable and are not eliminated or killed. Another potential hazard of applying sewage in agriculture is the pollution of soil. Usually soil functions as filter and the microorganisms in the applied sewage are retained in its upper layer. They don't move to low horizons of soils except in certain conditions such as limestone formation where microorganism can be transported to lower horizons. Under most conditions, hook worm juveniles usually remain viable in the upper layer of the soil and infect the host's skin.

The protozoan cyst and helminthes remain viable in the top soil and cause infections. The proper sewage treatment and composting of sewage and sludge before use would destroy most of these pathogens.

Heavy Metal

Sewage contains a variety of heavy metals which may be useful or harmful to plant and animals depending upon the nature and concentration of the metal.

The essential elements such as copper, iron, manganese and zinc can be harmful and toxic if present in higher concentration either in the soil or the waste due to repeated application. The studies conducted at Kolkatta area showed that repeated application of sewage to soil can result in build up of very high concentration of certain heavy metals in soil and plants. (Adhikari *et al.,* 1992). This may possibly be caused due to the discharge of industrial wastes into the sewage system.

The heavy metals are present in higher concentrations in the solid when the sludge are separated than in the liquid phase. This is due to metal' sewage attraction for negative charged colloidal particle. Therefore, the plants growing in sludge amended soils absorb very little heavy metals, even if their contents in the sludge may be on higher side. This trend has been reported in case of chromium, copper mercury, nickel and lead. However in case of cadmium and

Zinc, uptake of higher concentration of these heavy metal by com leaf and grain was reported from sludge amended soil (Leeper, 1978)

In studies conducted in sierozem soil (Hisar) showed that non essential heavy metal such as lead and nickle applied at the rate of 4 prri had the least adverse effect on the yield of green gram and pigeon pea while cadmium and mercury were most toxic. (Gupta *et al.* 1994) Experiments indicate that the plant can be harmed when their levels in the soil are greater than 3 ppm water extractable boron, 30 ppm EDTA -extractable Zn and 20 ppm acetic acid extractable Zn and 20 ppm acetic acid extractable cadmium (Purves, 1977).

From human health point of view, cadmium is considered a potential hazard and excessive cadmium intake through rice has been associated with condition, known as Itai Itai in Japanese. Maximum permissible weekly limits of cadmium intake by humans has been placed at 400-500 mg cadmium (Nordberg 1974)

Precaution

Sewage waste should be treated to such an extent that all infectious agents are killed or eliminated before use. This can be accomplished by combination of time and temperature although other environmental factors such as sunlight and oxygen tension can also affect their viability. At temperature higher than 65°C no excreted pathogen can survive for more than a few minutes, except spore forming bacteria e.g. *Clostridium perfringens* and Hepatitis A virus. The only process which will produce pathogen - free material are batch thermophilic digestion, thermophilic composting or drying with detention period of 2 years. At 50°C for 13 days. all pathogens will be killed (Strauss, 1985, Pereira 1986). Therefore it is desired that sewage or sludge applied to land should be subjected to any of these processes.

A chain of stabilization ponds well maintained can give effluent free from pathogens. This effluent may be pumped or allowed to flow by gravity and can be applied on the land by flooding through furrows, by subsurface irrigation or localized irrigation in which individual plant is irrigated.

Farmer should take precautionary measure to save themselves from possible pathogens in water or soil. The regular use of gumboots and rubber gloves while at work is advisable. The vaccination against the likely disease such as typhoid, gastroenteritis, cholera, poliomyelitis etc. may be the part of precautionary measures. The maintenance of good personal hygiene and cleanliness, use of soap and antiseptics immediately after work for washing and covering the mouth and nose loosely with a wide cloth while at work to prevent inhalation of aerosols and dust is essential.

References

Abrol. I.P. and Palaniappan, S.P. 1988. Prod. Symp. On Sustainable Agriculture: The role of green manure crops in rice farming systems. IRRI, Philippines. pp71-82.

Adhikari, S.,Mitra, A. and Gupta, S.K. 1992. Post-irrigation effect of sewage application on soil micronutrients and some heavy metals in north eastern fringe of Calcutta. Proc. Workshop on micronutrients, IBFEB-HFC, Calcutta pp. 280-293.

Agarwal, R.R., Man Singh and Chandra Pal 1950. Dhaincha as a green manure in U.P. Agri. Dept. Bull 104. Supdt. Printing and Stationery, Allahabad.

Ammal, U.B. and Durairajmuthiah, N. 1996. J. Indian Soc. Soil Sci. 44:445.

Anitha, D.R. 1991 Coir Pith - Potential wealth in India. In. Proc. Seminar on utilization of coir pith in Agriculture . Tamil Nadu Agriculture University, Coimbatore p1.

Anomymous 1968. Paintindia. October issue.

Anomymous 1975. Yojna 19(12): 221-222.

Aranguren, J, Escalante, G. and Herrera , R. .1992 PI. Soil 67:47.

Arumugham, C and Damodharan, A,D. 1993 Coconut Based Coir Industry in India . Oxford and IBM Publishing Co. Pvt. New Delhi, pp 663-640.

Azam, F, Malik, K.A. and Sajjad, M.I. 1985.PI.Soil 86 : 3-13.

Beri,V.Meelu, O.P. and Khind 1989a Trop. Agri(Trinidad) 66:11-16.

Beri.V.Meelu, O.P. and Khind 1989b Trop. Agri(Trinidad) 66 :209-212.

Bhardwaj, K.K.R. 1982. Indian J. Agron. 27:284-85.

Bhardwaj, K.K.R. 1995. Recycling of crop residues, oil cakes and other and plants products in Agriculture. In Recycling of Crop Animal, Human and Industrial wastes in Agriculture (Ed: HLS Tandon) FDCO, New Delhi pp 9-30.

Bhardwaj, K.K. R. and Gaur, A.C. 1985. Recycling of organic wastes (Fifteen years of Research 1968-1982) ICAR, New Delhi, pp 104.

Bhardwaj, K.K.R. Dev, S.P. and Kanwar, K. 1988. Proc. Seminar on control of Lantana camara and Ageratum. Himachal Pardesh Krishi Vishwavidyalaya, Palampur. pp 36-41.

Bhardwaj, K.K.R. and Kunwar, K. 1991. Indian J. Agri. Sci. 61:898-903.

Bhat, N.T. and Mahapatra, A.R. 1989 J. Plantation Crops 16:443.

Bijay Singh and Yadvinder Singh 1997 Green Manuring and Biological Nitrogen Fixation : North Indian Perspective In: Proc. Symposium Plant Nutrient Needs, Supply Efficiency and Policy Issues 2000-2025. Nat. Acad. Agri. Sci., New Delhi.

Bose, P.C. 1944 Calcutta Sewage and Fish Culture Proc. Natl Inst. Sci. India.10:443-454.

Brar, B.S., Benipal, D.S. and Dhillon, N.S. 2005 Green Manuring for better crop yields. Indian Farming (In Press)

Bressani, R. Cabezas M.T., Jarquin R and Murillio, B. 1975. The use of Coffee Processing waste as animal feed. Proc. Conf. on animal feeds of tropical and subtropical origin. Tropical Products Institute, London, pp. 107-117.

Bressani, R; Estrda, E. and Jarquin, R. 1972. Coffee pulp and husk. 1 Chemical composition and amino acid content of protein of the pulp. Turrialba 22:299-304 147). Agricultural Residues, Tree Wastes, Green Manure and Urban Wastes

Cambell.T.W. Bartely, E.E. Bechtle. R.M. and Dayton. A.D. 1976 Effects of coffee grounds on ration of digestibility and diuresis in cattle, on in vitro rumen fermentation and on rat growth. J. Diary Sci. 59:1452-1460

Chhabra, R. 1989. Sewage Water. CSSRI., Karnal

Chatterjee, B. N. Singh, K. I. Pal. A and Marti, S. 1979. Indian J. Agri. Sci. 49:188-192

Chattapadhyaya, N. Datta Gupta, M and Gupta, S.K. 1992 J. Indian Soc. Sci. 40:464

Choussy, F. 1944. Coffee pulp in the diet of dairy cow. Anales del instituo technologico de EL Salvador 1:265-271

Croyle, R.D. Wilson. LL and long T.A. 1974 Potential of spent tea leaves for animal feed and composting. Compost Sci. 15 :28-30.

Christine Furedy 1989. Appropriate Technology for Urban wastes in Asia. Biocycle. July 1989, pp. 56-57.

Daji, D.A. 1934. The Decomposition of green manure in soil J. Agri. Sci. 24:15-27.

Datta Gupta, M Chatopadhyay, N and Gupta, S.K. 1985. Indian Agri 29:294.

Datta Gupta, M. Chattopadhyay, N and Gupta, S.K. 1986a. Intl. Rice Research Bull. 11 :30.

Datta Gupta, M Chattopadhyay, N and Gupta, S.K. 1986b. J. Indian Soc. Agri. Sci. 1 : 30.

Datta Gupta, M Chattopadhyy, N and Gupta, S.K. 1986c. J. Indian Soc. Agri. Sci. 7 : 346.

Damodar Reddy and Rao, S. 1998. Farmer and Paliament 35:38

Dargan, K.S. Chillar, R.K. and Bhardwaj. K.K.R. 1975. In alkali soil green manuring for more paddy. Indian Farming 25(3) :13-15.

Eden, T. 1958 Tea. London. Longmans.

Edwards, P. 1985. Aquaculture: a component of low cost sanitation technology. World Bank Technical Paper no 26, Washington, D.C.

FAO 1977. Recycling of organic wastes in Agriculture. Report on FAO/UNpP Tout t6 Republic of China. Soils Bull. FAO. Rome

Frankenberger, W.I. Gr. and Abdelmagagid, H.M. 1985. PI. Soil 87 551-271.

Gaind, S. and Lata 2004. Quantitative evaluation of exo-enzyme during composting. Indian Jour of Microbiol. 44(3): 5-9.

Garrity , D.P. and Flinn, J.C. 1988. Proc. Symp. Sustainable Agriculture: The role of green manure crops in rice farming System. IRRI, Philippines 111-129.

Gaul, B.L. Abrol, I.P. and Dargan K. S. 1976. Note on irrigation needs of Sesbania aculeata for green manuring during summer. Indian J. Agri Sci. 46:436-38.

Gaur, 1999. Microbial Technology for Composting of Agriculture Residues by Improved Methods. ICAR, New Delhi pp 78.

Gaur, A.C. 1983. Enrichment of Urban Compost. Proc. Int. Symp.on Biological Reclamations and Land utilization of Urban wastes (Eds. F. Zucconi et al) Universita degli studi di Napoli, Naples pp 531-536.

Gaur, A.C., Neelkanthan, S and Dagran, K.S. 1995. Organic Manures ICAR, New Delhi pp 159.

Gaur, A.C., Sadasivam, K.V. Vimal. O.P. and Mathur, R.S. 1971. A study on decomposition of organic matter in an alluvial soil: CO2 evolution, Microbiological and Chemical transformation PI. Soil 35: 17

Gaur, A.C. , Sadasivam , K.V. Vimal. O.P. and Mathur, R.S. and Kavimandan SK. 1973. Studies on humification of organic matter in red ra/carsoil Zbl. Bakt. II Bd. 128 :149

Ghai, S.K. *et al.* 1985. Comparative study of potential of Sesbanias for green manuring. Trop Agric (Trindad) 62:52-56

Ghai, S.K. Rao, D.L.N. and Batra, L. 1988. Biol. Fertil. Soils 6:22-25

Gupta, M. D. Chattopadhyaya, N and Gupta, S.K.1986. Ann. Agri. Res. 7:346

Gupta, V.K. *et al.*, 1994. 25 years of Micronutrient Research in Soils and Crops c/f Haryana, Dept, of Soils, HAU, Hisar. Pp 99

Hume 1949. Economic Bot. 3 : 42-45.

IRR11998. Symp. Recommendations Proc. Symp. On. Sustatnable Agriculture. The role of green manure crops on rice farming Systems IRRI Philippines pp 1-4

Jha, K.P., Chandra P and Nanda, B.B. 1980. Oryza 17.-18-23

John, P.S., Pandey, R.K., Buresh, J. and Prasad, R. 1989. Fertile. News34(9):16-26.

Juwarkar. A.S. 1987. A case study on the use of sewage for crop irrigation - Indian experience, FAO, Rome.

Juwarkar, A.S. *et al.* 1987. Yield and cadmium accumulation by gram and barley grown in calcareous clayey soil. IAWPC Tech. Annual 14-79-85

Juwarkar, A.S. *et al.* 1990.Exploitation of nutrient potential of sewage and sludge through land application. Report of Expert consultation of the Asian Network on Bio and Organic Fertilizers. FAO -RAPA 1991/7, Bangkok, Thailand

Juwarkar, A.S. *et al.* 1991. Proc. Int. Conference Mineral Development on Environment.

Juwarkar, A.S. *et al.* 1992. Biological and industrial wastes as sources of plant nutrients. In: Fertilizers, Organic manures, Recyclable wastes and Biofertilizers (HLS Tandon, Ed) FDCO, New Delhi pp 72-90

Kale, C.K., Mishra, R.P., Rao, N.U. Shende, G.B. and Siddiqi R.H. 1973. Sewage Farming Nagpur, a Course manual, NEERI., Nagpur

Khare, N.K. and Rai, M.M.1968. J. Indian Soc. Soil.Sci. 16 :11-114

Khan, G. Gupta, SK and Bannerjee , S.K. 1981. J. Indian Soc. Soil Sci. 29:19

Khan, S.A.. Rao, C.V. and Bandyopadhyay, M. 1994. Indian J. Environ. Hlth. 34(4): 248

Khind, C.S. Garg, A and Bajwa, M.S. 1989. Int. Rice Res. Newsl 14(2): 31-32

Krishnamurthy, R. 1978. A. Manual on Compost and other Organic Manures Today and Tomorrow Publishers, New Delhi: 85-119

Kulkami, K.R. and Pandey, R.K. 1988. Annual legume for food in a rice based cropping system. Proc. Symp. on sustainable Agriculture.

The role of green manure crops in rice farming systems. IRRI, Philippines, 289-299

Leeper. G.W. 1978. Managing the heavy metals on land. Marcel Dekkar, Inc. New York

Lim, K. H.. 1998. Proc. International oil palm/Palm oil Conf. Agriculture pp. 575-595

Loganathan, N.S. Narsimhan, V. Lakshminarasimhan C.R. and Surendran, R. 1979. Madras Agri. J. 66:278-280

Mahida. U.N. 1981 Water Pollution and Disposal of Waste Water on Land Tata McGraw Hill, New Delhi.

Mandal, LN. and Chattopadhyay, G.N. 1992. Fertilizers Management in Fish Ponds, In. Non traditional Sectors for Fertilizers Use (Ed: HLS Tandon) FDCO, New Delhi 1-17

Mandal, B.K. Das, N.C. and Ghosh, R.K. 1993a Indian J. Agri. Sci. 63:195-199

Mandal, B.K. Das, N.C. and Ghosh, R.K. 1993b Indian J. Agri. Sci, 63:340-344

Manjkandan, P. Joshi, O.P. Hameed Khan, H. Mahapatra, A.R. and Bidappa, C.C. 1987 Trap. Agri. (Trinidad) 64:13

Mani, M.T. and Marimuthu, T. 1992 Mushroom Res. 1: 49

Mayalagu, K. 1983. Influence of different soil amendments on the physical properties of a heavy black soil and yield of groundnut TMV 7 in the Periyarvaigai command area. Madras Agri. J. 70 /304- 309

McDonagh, J.F. Tobmsan, B., Limpinuntana, V and Giller, K.E. 1995. Grain legumes and green maures as pre rice crops in North East, Thailand. PI. Soil 177:111-126

Meelu O.P. 1986. Green manuring in low land rice. Paper at planning meeting - Workshop of International network on soil fertility and evaluation Research Hangzhon, China

Mirchandani. J.J. and Khan, A.R. 1952. Green Manuring, ICAR Bull. No 2. ICAR, New Delhi. P. 84.

Murillo, B.L Daqui, L Cabezas, M.T. and Bressani, R. 1976 coffee pulp and hulls II. Chemical composition of coffee pulp ensiled with Napier grass and maize. Archives Latinomericanos de Nutricion 26 .-33-45.

Nagarajan,- R. Ramasamy, K.Savithri, P and Manickam, T.S. 1990 Coir Waste in Crop Production. Bull. TNAU, Coimbatore.

Nagarajan, R. Manickam, T.S. Kothaandaraman, G.V. Ramsany, K and Palanisamy, G.V. 1985. Madras Agri J. 72 :553.

Nair, M.K. Bridappa, C.C. Palanisamy S.P. and A.K. Upadhyay 1997. Proc. Symp. Nutrients Need, Supply, Efficiency and Policy Issues 2000-2025. Nat. Acad. Agri. Sci. New Delhi pp 157-168.

Nordberg, G.F. 1974. Health Hazards of Environmental Cadmium Pollution Ambio 5:55-66.

Palniappan S.P. etal 1990. Proc. Int. Symp. on Natural Resources Management for Sustainable Agriculture.

Panse, V.G. Abraham., T.P. and Leelavathi. C.R. 1965 Green Manuring of crops. ICAR Bull (Agri) No 2, New Delhi pp 83.

Parr, J.F. Espstein , E and Wilson, G.B. 1978 FAO Soils Bull. 36 FAO, Rome pp 301.

Patnaik, S., Ramarao, A.S. and Nanda, J.S. 1957. Sesbania speciosa comes to Orissa. Indian Farming 7:20.

Peoples, M. and Craswell, E. T. 1992. Pl. Soil 141 :13-29.

Pereira, N. 1986 Pathogen survival in a refuse /sludge forced aeration compost system. In. Effluent Treatment and Disposal. Pergamon Press Oxford, pp 373-391.

Potty, S.N. and Mannotra, J.M. 1993 In. Fertilizer Management in Commercial Crops. (Ed. HLS Tandon) FDCO, New Delhi. P. 133.

Purves, D. 1977. The contamination of soils and crops by toxic elements normally found in municipal wastes and their consequences for human health. In Waste Water Renovation and Reuse (Ed. D'ltri) Marcel Dekkar Inc. N.Y.

Ramaswamy, P.P. and Son, T.T.N. 1993 National Seminar on Development of Soil Sci. Indian Soc. Soil Sci. New Delhi. p66.

Rajendra Prasad and Goswami, N.N. 1991 Soil Fertility Restoration and Management for Sustainable Agriculture in South Asia. Adv. in Soil Sci. 17: 37-77.

Ranganathan.V. 1975 Potash Rev. 27 (8): 1.

Rekhi, R.S. and Mellu, O.P. 1983 Oryza 20:125-129.

Resales, F. 1973. The use of dehydrated coffee pulp in feeding of ^ pigs. MSC thesis Universidad de san Carlos. Guatemala.

Schwimmer, S. and Kutrzamn, R.H. 1972. Fungal decaffeination of roast coffee infusions. J. Food Sci. 37: 921-924.

Sharma, A.R. and Mittra , B.N. 1988 J. Agri Sci. Cambridge 110: 605-608.

Sharma R.C. and Sharma H.C. 1988, J. Agri. Sci. Cambridge 111: 193-195.

Sharma, R.C. GovindaKrishnan P.M. Singh, R.P. and Sharma, H.S. 1988. J.Agri. Sci. Camb. 110 :499-504.

Sharma R.C. era/1991. Nitrogen Management in Potato. Tech. Bull 32, CPRI, Simla pp 74.

Savithri, P. Subhiah, S., Nagarajan, R., Manim, S. and Gopalswamy, S. 1991. Seminar on utilization of coipith in Agriculture, Tamil Nadu Agri. Leniv., Coimbatore pp 129.

Savitri, P and Rani Perumal 1993. National Seminal on Development in Soil Sci. New Delhi, p 64.

Saxena, S..K. 1995 In. 24th IPI colloquiim, Chiang, Thailand. Feb 21-24:1995.

Sekhon, G.H. 1997. In Proc. Symp. on Plant Nutrients Need, Supply, Efficiency and Policy issues 2000-2025 Nat. Acad. Agri Sci New Delhi pp 78-90.

Shende, G.B. 1988. Status of wastes water treatment and agriculture reuse with special referende to Indian experience and research and development needs. Proc. FAO Regional Seminar on the treatment and use of sewage effluent for Irrigation . Nicosia 7-9th October, Butterworth, London.

Shuval, H.1.1990. Waste Water irrigation in developing countries. UNDP-World Bank Water and Sanitation Programme.

Singh Bhajan and Brar, S.P.S. 1985 J. Res. Punjab Agri. Univ. 22: 243-252.

Singh N.T. 1963 Effect of sugarbeet cultivation in the reclamation of saline - alkaline soils in Punjab M. Sc. Thesis PAD Ludhiana.

Singaram, P and Pothiraj, P. 1991-Proc. Seminar on utilization of coripith in Agriculture. TNAU, Coimbatore pp 203-207.

Singh N.T., Verma, S.M. and Josan , A.S. 1975 J. Indian Soc. Soil sci. 23:259-262.

Singh, T.V. Mathew. A.C. and Bosco, S.T.J. 1998. Tree world 8(3):3

Singh Punjab 1997. In Proc, Symp on Plant Nutrients Needs, Supply .Efficiency and Policy Issues 2000-2025.Nat. Acad. Agric. Sci., New Delhi pp 217-230.

Somasundram, E. 1991. Studies on the direct and residual effect of applied and intercropped Sesbania in Rice. M.Sc. (Ag) Thesis, TNAU. Coimbatore.

Strauss, M. 1986. Waste Water and excreta use in India. Dubendorf International References Centre for Waste Disposal.

Subbiah, B,V, and Mannikar, N.D. 1964. Selection of green manure crops for uptake of subsoils phosphorus - studies with p-32, Indian J. Agric sci. 34:21-27.

Thamburaj, J.J. And Kulhutheen, N.J. 1989. Biol Wastes 27:257.

Tiwari, K.N. Tiwari, S.P. and Pathak, A.N. 1980. Studies on green manuring of rice in double cropping system in partially reclaimed saline sodic Soil Indian J. Agron 25 :137-146.

Trivedi, P.R. and Gurdeep, R. 1992a Disposal of Solid Wastes. In Solid Waste pollution. Akashdeep publishing House, New Delhi, pp 116-186.

Trivedi, P.R. and Gurdeep, R. 1992b Management of Solid Waste. In Management of pollution Control. Akashdeep Publishing House, New Delhi.

Topps, J.H. 1993. Coffee and Tea Wastes. In: Handbook of organic waste conversion, Van Nostrand Rhunhold Co.

Uppal, H.L. 1955. Green manuring with special reference to Sesbania aculeata for treatment of alkaline soils. Indian J.Agric. Sci. 25:211-217.

Vacchahi, M.B. and Murthy, K.S. 1964. Green manuring for rice Research. Report No 17ICAR, New Delhi pp 50

Veerabadran, V. 1991. Proc. Seminar on utilization of coir pith in Agriculture TNAU, Comibatore p 91

Venkata Ram, C.S. 1978 Proc. 1st Ann. Symp. Plantation Crops. Central Plantation Research Institute, Kasargod (Kerala) p 438

Velayutham, M and Bhardwaj, K.K.R. 1994 Indian Farming 44 (9): 39-41.

W.H.0.1989. Health Guidelines for use of Wastes water in Agriculture and Aquaculture. World Health Organization. Report Series 778, Geneva.

Wilson, G.B., Parr, J.F. and Sikora LJ. 1980. Experience with organic composting in developing countries. Biol Waste Management and organic Resources Laboratory, U.S. Dept. of Agriculture, Beltsvile, M.D. 20705

Work, S.H., Van Severen, M.L. and Escalon L. 1946. Preliminary information on the value of coffee pulp substituted for maize in the rations of milking cow. Cafe De El Salvador 16:773-780

Yadvinder Singh, Bijay Singh and Khind, C.S. 1992. Ady in Soil Sci. 20:237-309.

4

Agro-Industrial Byproducts

1. Oil Cakes

Several varieties of oil seeds are produced in India. Oil seeds contain 40 to 50 percent oil and residues left after extraction are known as oil cakes. Oil seed cakes are generally referred as concentrated organic manure because of their high content of plant nutrient, nitrogen in particular as compared to conventional farmyard manure or compost. Oil cakes also invariably contain 2 to 15 percent oil depending on the method of extraction. Oil seeds are further demarcated into edible and non-edible oil seed cakes depending on their palatability to animals. Both type of oil seed cakes can be used as organic manures, but edible oil seed cakes are generally used as animal feed as warranted by socio-economic reasons (Bhardwaj, 1995). Non-edible cakes are generally used as organic manure to supply major plant nutrients like nitrogen, phosphorus and potassium.

Potential of Non-edible Cakes

The authentic estimates of quantity of non-edible cakes produced in the country are not available because the oil yielding trees are grown and scattered our vast tract of the country. Only a part of oil bearing seeds of these trees are being collected and processed. According to an estimate, annual production of non-edible oil seed cakes was of the order of 3.8 million tonnes annually which on average can provide 87, 25 and 57 thousand tonnes of N, P_2O_5 and K_2O respectively (Singh and Gurumurti 1984).

The manurial value of oil seed cakes is mainly due to their nitrogen and organic matter content although they contain fair amounts of phosphorus and potassium and micronutrients. The chemical composition of major plant nutrients of some oil seed cakes produced in the country and presented in Table 51.

The rate of decomposition of oil seed cakes is much faster due to their narrow C/N ratios compared to plant residues and farm wastes. The C/N ratios of different oil seed cakes such as groundnut, neem, mustard and mahua cakes were reported to be 6.8, 10.6, 11.8 and 21.3 respectively (Mukherjee *et al.*,

1991). It was reported that edible cakes of groundnut and mustard mineralized faster than non-edible cakes of neem and mahua indicating better value of edible cakes as compared to non-edible cakes. However, edible cakes are preferred as animal feed for milch cattles. Both type of cakes have been extensively reported to be used in sugarcane crop and betel leaves. (Gaur *et al.*, 1995)

Oilseed cakes are applied in powdered form Just at the time of sowing or as top dressing (Saxena and Singh 1980). However, oil cakes should not be applied in a manner so as to directly come in contact with the germinating seeds or young plants, as they may prove harmful.

Table 51: Nutrient Contents of some oil cakes

Name of the oil cake	**Percent**		
	N	P_2O_5	K_2O
Non - edible cakes Neem *(Azadirachta indica)*	5.20	1.05	1.48
Karanj *(Pongamia glabra)*	3.98	0.93	1.32
Mahua *(Madhuca indica)*	2.90	0.81	1.85
Castor	4.20	1.85	1.42
Edible cakes Cotton Seed (decorticated)	6.45	3.15	1.68
Cotton Seed (undecorticated)	3.90	2.45	1.60
Coconut	3.45	1.50	2.00
Groundnut	7.25	1.70	1.45
Mustard	4.52	1.65	1.40
Rape Seed	4.95	1.84	1.30
Sesame	6.00	2.10	1.26
Safflower (decorticated)	7.90	2.23	1.89
Safflower (undecorticated)	4.89	1.44	1.20
Linseed	5.45	1.50	1.30

Non-edible Cakes on Nitrification Process

Non-edible cakes are not used as animal feed and thus are mostly used as concentrated organic manure. Besides their utility to supply plant nutrient, these have been found to regulate nitrification process.

Powdered Neem-cake mixed with medium black soil at Pune alongwith 100 Kg N/ha, exerted appreciable inhibitory effect on the number of nitrifying bacteria (Table 52). The decline in the population of *Nitrosomonas* and *Nitrobacter* was evident due to application of different rates of neemseed cake. The most significant inhibition took place in the initial stage of growth and it was more to

Nitrosomonas than *Nitrobacter.* The population decreased progressively over 4 weeks time. The trend of recovery of *Nitrobacter* population was noted at 4th week of incubation but the rise was also higher in untreated soil and treated with lower rate of *neem* cake. The trend of these results were supported by the data on ammoniacal and nitrate nitrogen in soil. (Table 53). While ammoniacal nitrogen was increased nitrate nitrogen approximately proportionately decreased under the influence of *neem*-cake. This relationship between these two forms of nitrogen was intensified with increasing quantities of neem-cakes in soil. From regulating point of view, it is advantageous to retain ammoniacal nitrogen in soil because this form is less prone to leaching losses.

Similar trends were recorded in sierozem sandy loam soil of Hisar where 100 ppm fertilizer nitrogen alongwith different quantities of *neem* cake were applied (table 54 and 55).

Table 52: Effect of Neem seed-cake on nitrifying bacteria in medium black soil

Neem Cake (% of urea N)	Incubation period (weeks)							
	Mtmsomonas (MPH X 10^4)				*Nitrobacter* (MPN X 10*)			
	1	2	3	4	1	2	3	4
0	2400	130	92	17	50	20	3	92
20	800	110	54	24	30	20	3	92
50	200	240	17	5	50	10	2	11
100	100	50	16	5	20	10	2	30
200	50	20	11	35	20	10	1	13

Table 53: Effect of *Neem* seed-cake on nitrification in medium black soil

Neem Cake (%of UreaN)	Incubation period (Weeks)													
	NH_4-N(ppm)							NO_3-N(ppm)						
	1	2	3	4	6	8	10	1	2	3	4	6	8	10
0	71	79	82	85	81	70	60	102	100	131	74	64	70	59
10	78	81	84	86	84	73	64	85	89	104	70	57	66	54
20	80	83	85	86	82	75	63	84	87	99	74	68	70	65
30	83	85	87	88	84	75	64	60	86	92	74	71	70	55

Table 54: Effect of *Neem* seed-cake on nitrate formation in sierozem sandy loam soil

Neem cake applied (%C)	Incubation period (days)								
	0	3	7	14	21	28	42	56	70
	NO_3-N (ppm)								
0.00	14	26	66	82	96	91	99	102	104
0.02	11	9	48	58	86	91	95	98	102

0.05	9	7	41	44	65	88	93	95	98
0.10	9	9	29	44	66	86	91	102	104
0.20	9	9	8	9	12	42	86	95	102
0.50	1	1	4	2	2	16	71	112	115
2.00	5	4	7	6	6	5	6	19	61

Table 55: Effect of *Neem* seed-cake on the population of nitrifying bacteria in sierozem sandy loam soil

Neem Cake (%of ureaN)	**Incubation period (weeks)**							
	***Ntrosomonas (MPN* X 10⁴)**				***Nitrobacter (MPH* X 10⁴)**			
	1	**2**	**3**	**4**	**1**	**2**	**3**	**4**
0	2400	130	92	17	50	20	3	92
20	800	110	54	24	30	20	3	92
50	200	240	7	5	50	10	2	11
100	100	50	16	5	200	10	2	30
200	50	20	11	35	200	10	1	13

The results on the effectiveness of neem-seed cake as nitrification inhibitor was also confirmed by the studies conducted at Ranchi, Jharkhand in acid red loam soil (pH 5.5) which was treated with *Pongamia* (*Karanj)* cake (Yadav and Srivastava, 1987). From the incubation studies conducted in different soils, with neem-seed cake and *Karanjseed* cake, it was clearly shown that the use of nonedible cakes checked the losses of nitrate from soil and improved the fertilizer use efficiency of urea (Bhardwaj and Gaur, 1985)

Several laboratory and field studies have confirmed the role of nonedible seed cakes in improving the efficiency of added fertilizer nitrogen (Bains *et al.*, 1971) and regulating the growth and activity of nitrifying bacteria when urea was coated or blended with *neem* seed or *Karanj* seed cakes than when urea was applied alone. Commercial production of oilseed cake coated nitrogen fertilizers will be of practical importance to rice farmers.

Ahmad *et al.* (1978) reported that alcohol extracts of karanj seed cake lipids applied at the rate of 20ppm reduced the population of *Nitrosomonas* and *Nitrobacter* than simple application of karanj cakes. It was observed that alcohol extract of non-edible cakes contain compounds inhibiting nitrification process (Sahrawat and Parmar, 1975). The active principle subsequently characterised as karinjin which was comparable to N-serve, a nitrification inhibitor (Sahrawat and Mukerjee, 1977). In further studies alcohol extract of *karanj* and *Neem* cakes were effective as nitrification inhibitor at par with Karanjin (Sahrawat, 1982).

Control of plant parasitic nematodes

Castor cake has been reported to be a good vermicide against termites. Non-edible cakes are known to act as biopesticides. High dosages of fertilizer nitrogen increased the population of soil nematodes. Since the soil fumigation to control plant parasite nematodes is very expensive and not ecofriendly, the effect of different types of organic matter including neemseed cake with and without chemical fertilizer was studied (Gaur and Prasad, 1970)

The results (Table 56) clearly showed that application of NPK in absence of organic matter amendment markedly increased the population of parasitic nematodes associated with wheat crop. Addition of organic matter even in presence of applied NPK suppressed the plant parasitic nematode population and *neem* cake was found to be most effective treatment in lowering their numbers.

Table 56: Effect of organic matter and inorganic fertiliser on nematode population in soil

Treatment	Nematode population/250mg soil	
	Plant parasitic	Saprozoic
Control	56	628
Wheat straw	63	2183
Wheat straw + neem cake	25	1090
F.Y.M	98	1203
Dhaincha	83	963
N.P.K	170	400
N.P. K+ wheat straw	85	725
N.P. K+ wheat	55	985
Straw +neem cake N.P. K+F.Y.M	125	529

The findings were further confirmed from the subsequent studies conducted under field conditions (Prasad *et al.*, 1972).

Neem (Azadirachta indica) in the biocontrol of crop pests

Neem tree has emerged as single most important source of botanical insecticides. All parts of the tree are biologically active though the maximum insecticidal activity is observed in seed Kernel. Neem tree is a native of Indian subcontinent. Neem derivatives affect many crop pests by disturbing their growth and inhibiting feeding, reproduction, oviposition (Shmutterer, 1988) and hormonal system of pests (Rembold *et al.*, 1984). In addition neem is also reported to cause toxicity to eggs and impair egg hatchability.

Pradhan *et al.* (1962) reported about the feeding deterrent property of neem seed kernel suspension against desert locust. Subsequently several biologically active components were isolated from various parts of the tree, notable being the isolation of meliantrol and azadirachtin. Both these ingredients at very low concentrations inhibited the feeding of locust, Azadirachtin is effective at a very low concentration compared to meliantrol.

Crop pests constitute a major problem in sustainable food production. The use of pesticides is not only causing soil and water pollution but their \residues are finding their way in the food chain. The use of biopestecides will be safe and can play an important role in sustainable agriculture. There is need to strengthen R & D in this area.

Crop yields: It was found that mixing of *neem* cake and urea in 3:10 ratio was effective in increasing rice yields in a number of experiments conducted in Maharashtra (Ketkar, 1974). In rice field amended with urea and neem cake treated urea, significantly higher levels of NH-N were recorded upto 30 days of crop growth under low land rice conditions (Subbiah and Kothandraman, 1980). In neem cake blended urea, applied to rice, higher levels of NH,-N in soil even at 35 and 65 days of crop growth was reported (Tiwari, 1989). In Andhra, in farmers fields, it was found that oil cake blended urea was more effective than urea alone even in under poor water management conditions (Reddy and Shinde, 1981). Rice yields, nitrogen uptake and nitrogen use efficiency were significantly enhanced with oil cake blended urea, neem oil and neem oil extracts (Surve and Daftardar, 1985) neem oil and its extracts were as effective as oil cakes as such.

Application of oil cake blended urea increased the yield and nitrogen uptake by maize crop besides it reduced the leaching of NO2-N by 9-3 per cent (Singh and Singh, 1989).

The beneficial effect of *neem* seed cake plus wheat straw in reducing the population of plant parasitic nematodes during the growth of maize and mungbean was reported earlier (Prasad *et al.*, 1972). The effect of organic materials including neem seed cake in particular on the grair yield and nitrogen uptake by maize crop (Ganga-5) under field conditions with normal package of practices was investigated (Gaur and Mathur, 1979). The treatments tested and data obtained in the field trial are presented in table 57. A basal dressing of single superphosphate (80 kg P_2O_5/ha) and muriate of potash (50 kg K_2O/ha) was applied. Different types and combination of organic matter was incorporated in soil, 15 days prior to sowing except FYM with optimum moisture for allowing decomposition of wide C/N ratio material. A combination of straw (13.5 tonnes/ ha) and *neem* seed cake (4.5 tonnes/ha) in the proportion of 3:1 was included as one of treatment to examine the effect of *neem* seed cake on the rate of decomposition of straw and its effect on the yield and nitrogen uptake on the crop.

Table 57: Effect of organic materials on yield and nitrogen uptake by maize crop.

Treatment	Grain Yield (Q/ha)	Straw Yield (Q/ha)	Nin grain (%)	N uptake in grain (Kg/ha)
Control	48.0	160.0	1.58	73.4
N (200 kg N/ ha)	62.2	200.0	1.73	107.7
Straw (0-15 cm depth) + N (200 kg/ha)	62.6	220.0	1.68	105.1
Straw (0-7.5 cm depth) + N (200 kg/ha)	55.6	230.0	1.73	96.2
Straw +neem cake + (015cmdepth) +N(200kg/ha)	76.0	270.0	1.75	133.5
FYM (0-15 cm depth) + N (200 kg/ha)	70.4	240.0	1.63	114.8
CD 5%	11.0	52.0	-	-

The results (Table 57) showed that the application of 200 kg N/ha increased the yield and nitrogen uptake of maize crop by 29.2 and 46.7 percent over the control treatment with basal application of P and K only. There was no additional gain in yield of the crop when straw was incorporated in 0-15 cm soil depth showing a balancing effect in the rate of mineralization and immobilization processes in the soil. But the incorporation of straw in shallow layer (0-7.50 cm) decreased the grain yield and nitrogen uptake by the crop by 10.5 and 10.8 percent respectively showing nitrogen immobilization due to increasing the organic biomass load in the shallow layer and a short period of decomposition of 15 days only. It is interesting to highlight that the use of non-edible neem seed cake can play on important role in surmounting the unfavourable effect of wide C/N ratio material even during short period and nitrogen uptake was increased by 22.2 and 23.9 percent due to application of straw and *neem* seed cake in the proportion of 3:1 alongwith the fertilizers (NPK) over NPK fertilizer treatment. The maximum yield was obtained with *neem* seed cake treatment which was even more than with the standard FYM treatment. It is concluded that combined use of straw and non-edible oil seed cakes in optimum proportions would not only augment the soil heath but also increase the economic benefits in higher food production.

Non-edible oil seeds- oils and cakes

Non-edible oil seeds like *neem,* karanj, *mahua,* castor, etc. which are found in forest and non-forest areas, are one of the most potential sources for augmenting and collection of the oil seed supplies. Non-edible seed oils can be utilized in various industries particularly soap industry. Edible and non-edible oils are consumed mainly by the soap, paints, varnish, toilets, hair oils and lubricants producing units. These items consume substantial quantities of oil.

Apart from soap industry, non-edible oils are also used in leather tanning, for medicinal purpose and household uses like illumination. Some of them such as *neem, karanj,* castor and mahua etc. are already being exploited while others like sal etc. are of potential importance. General description about some oil seeds trees found in the country is given below.

Neem Tree (*Azadirachta indica*)

It is also called Margosa tree or Indian lilac. During the vedic period great importance was given to plants as sources of many essentials in life such as food, clothing, medecines and shelter. Rigveda and Atharvaveda described medicinal plants. Ayurveda has a tradition of using herbals and considered all plants as potentially medicinal. The traditional classes of medicine, Ayurveda, Siddha and Unani system incorporate several well organised system of diagnosis and cure.

Neem is one of the such versatile trees which is natural pesticide and is used for multiple purpose.

Neem is medium sized to large evergreen tree often reaching a height of 25 to 30 metres. The bark is somewhat rough and grey and the leaves are bright green consisting of 9-15 leaflets. Each leaflet is oblique or slightly curved, coarsely toothed, shining on the upper surface and reaching a length of 30 cm. Flowers are rather small, whitish, scented and emerge from march to may. The fruits are oval shaped, about 1.5 cm long and 0.75 cm in thickness, greenish yellow when ripe and one-seeded soft and pulpy. The seed has a brown kernel, embedded in a hard white shell. The maturity of fruit and their fall varies from state to state.

Habitat: The neem tree in found all over India and Myanamar either wild or planted as shady trees in villages and towns, on roadside, in agroforests and waste lands but thrives best in the drier climate regions. In the areas, where the rainfall varies 400-800 mm, having an extended dry, season and maximum shade temperature as high as 120°F, this tree performs very well. In India, this tree occurs largely in states of Uttar Pradesh, Tamil Nadu, Maharashtra, Karnataka, Rajasthan, Madhya Pradesh and Delhi. It is very sensitive to frost and damaged young plants do not recover easily. For this reason it is not a common tree in Punjab and Himalayan Tarai belt. According to an estimate by Khadi and village industries commission (K.V.I.C), there are about 180 lakhs neem trees all over India. Presently it also grows in tropical and subtropical regions of Africa, America and Australia (Rawat, 1993-94).

Modern concept of Neem tree

Neem bark: The bark is a good bitter tonic, astringent and antiperiodic. It contains a resinous bitter principle and usually prescribed in the form of a tincture or an infusion. It is also considered as beneficial in-malarial fever. The bark is also considered useful in skin diseases.

Neem leaves: The best known parts are the leaves which are used to keep insects out of wollen clothes, books, food grains etc. The leaves are bitter and have a faint but characteristic unpleasant smell. Leaves are dried in shade before their use as moth repellant. The leaves are applied to boils in the form of poultice and a decoction is recommended on ulcers and eczema.

Neem oil: The oil content of kernel is about 50 percent and can be extracted by improved, bullock driven or power driven *ghanis* (wooden oil mills) hand screw presses or hydraulic cage presses. There are two methods of processing neem fruit. One method consists of depulping the fruits and subsequent decortication of dried depulped fruit and other consists of drying of whole fruit and then decortication. Pressing of kernel yields comparatively better quality oil than pressing of whole fruit or undecorticated neem seed.

Neem oil is a brownish yellow, non-drying with an acrid and bitter taste and unpleasant odour. The quality of oil differs depending on the quality of seed and the method of processing it. The fatty acids and composition range consists of palmitic acid 13.6-14.9%, stearic 14.4-19.1%, arachidic 1.3-2.4%, oleic 49.1-61.9% and linoleic 7.5-15.8%.

Uses: Neem oil is used in the manufacture of brownish coloured soaps, in medecine and as insecticides. Medicated soaps (Margo) with neem odour are available in the market. It can also be used for production of oleic and stearic acid.

The bitter principles of neem oil have been investigated by several workers. Nimbidin can be extracted with alcohol, which is highly bitter and contains sulphur. Nimbidin preparations of the oil are reported to be efficacious in a variety of skin diseases, septic sores and ulcers due burns and useful in bleeding gums and pyorrhoea.

The kernel is the most effective part of the neem so far as the biological effect is concerned. As many as seven distinct chemicals having insecticidal properties have been isolated from it. The utility of *neem* lies in some biologically active chemicals called limonoids; Scientists have isolated several limonoids from seeds, bark and leaves of neem. Each of these limonoids has a different mode of action. It has been possible for neem to resist more than 200 species of insects, mites and nematodes. Neem based products are medium to broad spectrum insecticides for a variety of plant eating or phytophagous insects. The larvae of moth pests are particularly sensitive to neem derivatives and therefore, the main targets. Also, insects find difficult to develop resistance to neem products.

The most extensive studies on properties of neem products are antifeedant activity, growth disruption and toxic effects. The antifeedant property of neem kernal against desert locust was observed in 1962 (Proadhan *et al.*, 1962). Since then number of studies on the insecticidal properties and biological effect of oil, kernel, cake and their extracts have been discovered. As many as about 70 insects of different orders have been attacked by the antifeedant property of neem. The most active part of neem, the kernel can be used for this property.

Neem Cake: This is used as organic manure in agriculture. It has multiple uses in agriculture, poultry and animal husbandry. When neem cake is added to soil, it reduces to population of insects. It supplies organic matter to soil and plant nutrients to crop plants and protects crops from termites and plant parasitic nematodes, Mixing neem cake (3%) with nitrogenous fertilizers suppresses the population of *Nitrosomonas* bacteria involved in conversion of ammonia to nitrate and thus enhances nitrogen- use efficiency of fertilizers.

Mixing of neem cake with soil has been effective against nematodes infesting tomato, okra, brinjal, cole and rice crops, guava, citrus, banana and cardamom.

Karanj or Pongam *(Pongamia glabra)*

This is an evergreen tree found in villages, roadsides, banks of rivers and in forests. In localities where it is known to have been cultivated successfully, the absolute shade temperature varies from 100 to 120° F, the absolute minimum 30-60° F and the rainfall from 50 to over 250 cm. The fruit or pod is 7.5 cm long and encloses reddish seeds each about 2.5 cm long and weighing about 1 gm. The seed contains about 27 to 39 percent oil and resinous substance. Villagers collect the seeds and decorticate them with the help of wooden hammers or by thick sticks. Separation of kernels is done by winnowing processes, the seed is chiefly pressed in expellers through *ghanis* are also being used. In expellers, 24 to 26 percent of oil and 68 to 72 percent of the cake are reported to be obtained.

Karanj oil is an acrid, reddish brown, non-drying oil having about 61.30% oleic, 9.72% linoleic, 6.06% palmitic, 4.30% arachidic, 3.22% lignoceric, 4.36% dihydroxystearic, 2.19% stearic and less than 1.0% of myristic, and linolenic acid together.

Uses: Oil is chiefly used for leather tanning, lightening and to small extent in soap making, medecine and lubrication.

The cake is used as organic manure for supply of plants nutrients and organic carbon to soil to improve its soil fertility. This also acts as a nitrification inhibitor and improves the efficiency urea in rice cultivation.

Mahua *(Madhuca latifolia)*

Mahua is largely found in sub-Himalayan tracts particularly concentrated in the states of Uttar Pradesh, Tamil Nadu, Maharashtra, Gujrat, Bihar, Orissa, Madhya Pradesh, Karnataka and West Bengal. *Madhuca latifolia* is replaced in Tamil Nadu and Karnataka by *M. longifolia*.

Both the species are frequently cultivated in and around village, as avenue trees, etc. The seeds (1 to 4) are contained inside fleshy fruits. The kernels weigh about 75 percent of the whole seed and contain about 50 percent of the fat. Mahua oil is pale yellow with semisolid fat having oleic (43.3%), palmitic (23.7%), stearic (19.3%), linoleic (13.7%) and lignoceric (1%). The fat is extracted by hot expression in screw presses and hydraulic presses, *ghanis* and sometimes in oil expellers.

Mahua fat is somewhat near to tallow in its properties and fatty acid composition than many fats. Almost the entire production of mahua seed oil is used in soap making. However, soap made from this oil are prone to get rancid and hence its use in superior quality washing or toilet soap is limited. It can be used in manufacture of candles and for preparation of fatty acids. It contains soponin, a poisonous ingredient and thus it is not used as cattle feed.

Mahua is used as organic manure by farmers. The cake has about 2.7% nitrogen which is less compared to nitrogen content of *neem* or *karanj* cake. The rate of mineralization of *mahua* seed cake is slower as compared to other non-edible cakes used as concentrated organic manure due to presence of saponin in it. Therefore it is incorporated in soil much before sowing or transplantation of crops. It is frequently used as organic manure in sugarcane plantations, in grass lawns and parks on account of vermicidal and insecticidal properties.

II. Rice Milling Industry

Rice Husk

Rice husk is the largest product of rice milling industry comprising 20 to 25 percent of paddy. The availability of rice husk is about 15 mt. annually (Gaur, 1978). It is often burnt which is not acceptable from environment point of view. It is a poor source of plant nutrients. It generally contains 42.6 percent cellulose, 20 percent lignin, 18.6 percent pentosans and 18.7 percent ash, N 0.3 to 0.4 percent, P_2O_5 0.2 to 0.3 percent and KO 0.3 to 0.5 percent. The ash contains 90 percent silica. Occasionally it is used for soil application for improving the physical condition of saline and alkaline soils. Incorporation of rice husk @6 tons/ha in a sandy loam soil reduced the yield of first wheat crop even in combination with chemical fertilizers, but the yield of second crop in the next year increased significantly in rice husk treated plots (Prihar *et al.*, 1972). Application of husk @10-15 tons/ha to an alkaline soil having low water permeability brought about significant increase in its infiltration rate (Khosla, 1976)

Crop Responses

The effect of various organic materials including paddy husk was studied on grains yield, seedling emergence, soil hardness, bulk density and root length of black gram and groundnut (Shanmugam *et al.,* 1995). They reported that rice husk decreased the hardness of soil and reduced the bulk density.

The results from coordinated research project showed that application of paddy husk @ 5 tons/ha increased the yield of maize and wheat crops grown in alluvial soil from 18.6 to 22.1 and 30.6 to 35.0 q ha1 respectively. The results of experiment conducted in red sandy, loam soil on groundnut, bajra and wheat where paddy husk was applied @ 5 tonnes/ ha showed favourable effect on groundnut or bajra but no favourable or adverse effect was recorded on wheat crop yields. Shanmugam *et al.* (1995) compared various organic and inorganic material with regard to their effects on yield of black gram and groundnut grown in lateritic soil. They reported that application of paddy husk @10 tonnes ha1 increased the root length and yield of both the crops than the control. Rice husk was comparable to pressmud cake but was better than FYM.

Rice Bran

Rice bran is pericarp or outer covering of rice seed, has a relatively high fat content. Rice bran is a valuable by-product of rice milling industry. Paddy yield about 5 to 7% bran. The production of rice bran is about 4 million tonnes, annually although, it is produced in substantial quantities, it remains unutilized commercially in developing countries.

It has limited use as organic manure since this is utilised for production of rice bran oil. The rice bran oil has many useful characteristics, both as edible and industrial oil. It also provides useful raw materials for various industries. Mostly industrial grade oil is being produced from the rice bran. To produce edible grade oil, it is necessary to have bran of low free fatty acid content and this is a serious constraint under Indian conditions.

The economic extraction of oil from bran depends mainly on its quality. The oil content of bran varies with the type of rice mill. The oil content of parboiled rice is higher than from raw rice. Parboiling produces some movement of oil from lower layers to the surface so that the top layers get enriched in oil.

The main problem in the availability of rice bran are large number of huller mills and lack of centralised collection. The bran available from hullers is mixed with finally ground rice particles, which can find use only as an animal feed or fuel.

The chemical analysis of rice bran shows wide variations:- moisture 8-12%, protein 10-14%, carbohydrates 38-44%, oil 10-24%, fibre 9-12%, ash 9-12% and pentosans 8-11%. Unlike other oils, the crude rice bran oil contains free fatty acids, gummy colouring and odorous materials as well as organic impurities. Rice bran oil has unsaturated fatty acids like oleic (41-45%) and linoleic (28-37%) as having characteristics of a high grade edible oil.

However crude rice bran oil is mostly used for non-edible purposes particularly used in the manufacture of soaps, detergent, washing powder and paints. Edible grade rice bran rice oil can be used like other vegetable cooking oils. However, there are several constrains in refining rice bran oil such as high free fatty acid content, large refining losses, high silica, sand and hull content, presence of wax, unsaponifiable matter, colour and odour etc.

III. Byproducts of Sugar Industry

Sugar industry in India is the second largest to textile only consuming about 50 percent of the total sugar cane production. There are à large number of sugar factories around 416 functioning and also are being established. The by products of sugar industry are molasses, bagasse, pressmud and spent wash. Molasses serves as raw material for production of alcohol and rectified spirit with production of one litre alcohol, nearly fifteen litres of spent wash (effluent) are being discharged in the environment.

Byproducts

Bagasse: It is an important by product of sugar factories and about 5 million tonnes of dry bagasse is annually produced in the country (Gaur *et al.*, 1995). Bagasse is a fibrous residue left over after extraction of sugarcane juice. The quantity of bagasse depends on the fibre content which varies from 33 to 36 percent in Northern India and 26 to 30 percent south India. The production of bagasse has increased by about 4 times over a period of 50 years. The composition of bagasse depends on the variety of sugarcane, degree of premilling preparation and the efficiency of crushing plant. The average composition of moist bagasse (%) is as follows: fibre 48.5, moisture 48.0, sugar 3, and minor constituents 2.

The average composition (%) of bone dry bagasse is as follows: Cellulose 45, pentosans 28, lignin 20, ash 2, and sugar 2. Bagasse ash is rich in silica, potassium and iron.

Conventional Uses of Bagasse

1. **As Fuel:** The bagasse produced in India is almost entirely used as a fuel in sugar factories for generating steam in boilers to drive the prime movers as well as for boiling and concentrating sugarcane juice. The wet mill bagasse contains about 48 percent moisture but it is used after drying in special type of furnaces. The reason is that it is readily available and cost effective. The average gross calorific value (GCV) of bagasse with about 48 percent moisture is about 4,000 Btu/lb and that of furnace oil and coal are 18,000 and 11,000 Btu/lb respectively.

One tonne of bagasse generate about 2.5 tonnes of steam. Attempts have been made to compress bagasse into briquettes directly as fuel substitute for charcoal. However, due to high potassium salts, these could not replace coke in the firing of lime- kiln. At national sugarcane institute (NSI); Kanpur, bagasse has been used for production of methane (200 cu. metres of combustible gas/ tonne of bagasse)

2. **As Pulp and Paper:** This kind of its use is well recognised. The advantages of using bagasse as compared to bamboo in paper manufacture are as follows:-

   i) Bagasse can be readily collected in large quantities from sugarcane factories at less cost.

   ii) The paper made out of high percentage of bagasse has good opacity and printing charactristics.

   iii) Pulp produced from bagasse fibre has good bursting strength and also requires less refining power.

The disadvantages of bagasse to bamboo are:

i) The cellulose content of wet bagasse is about 20 percent only as compared to 90-95 percent in soft woods and 60-70 percent in hard woods.

ii) Bagasse pulp is short-fibred and long fibred pulp like bamboo pulp, waste paper, cotton linters or rags to the extent of about 25 percent has to be admixed for making good quality paper with good strength. Pith; is highly absorbent and improper removal of pith would mean high consumption of chemicals, drastic cooking and bleaching process.

The problems of sugar factories while sparing bagasse are

i) The regular supply of alternate fuel at economic price is not assured. One thousand tonne capacity factory would require about 100 tonnes coal or 50 times furnace oil per day. The requirement would be 150 tonnes per season.

ii) The use of coal instead of bagasse is costly.

Small paper plants situated near sugarcane factories have some advantages

i) The cost of transportation of bagasse is nominal

ii) Pith separated from bagasse can be burnt in the boilers. Thus the working of integrated sugar and paper plants with an installed capacity of 5000 tonnes per annum would be successful.

3. **Furfural and activated carbon:** Bagasse contains about 21 percent pentosans which can yield 17 percent furfural on dry weight basis. The factors affecting choice of bagasse as raw material are its availablity and cost, cost of conversion and the value of the products obtained.

   The production of activated carbon from bagasse involves low temperature, carbonisation and simultaneous chemicals activation. It has higher capacity for colour removal compared to coconut shell or filter mud (press mud).

4. **As cattle feed and fertilizer:** Bagasse can serve as cattle feed. It has good quantity of lignin, cellulose, and pentosans but with low nutritive value. Ammoniation can increase the digestibility of the fibre. In this process, bagasse absorbs about 2 percent nitrogen. For adult cattle bagasse can be impregnated with urea-molasses mixture after fine chaffing.

5. **Use in composting :** Bagasse can be composted admixed with farm residues and used as organic manure. Decomposition sets in soon and the pith is degraded quickly than the fibre. The average N and P_2O_5 content of bagase is about 0.25 and 0.12 percent, respectively. Compost prepared from bagase has N and P2O, content of 1.4 and 0.4 percent, respectively. It is estimated that about 14 Mt. of compost can be produced annually from bagasse.

Pressmud

The raw sugarcane juice obtained during milling from sugarcane contains 77 percent water, 8 to 20 percent sucrose, 0.3 to 3.0 percent reducing sugars, 0.5 to 1.0 percent organic compounds and 0.2 to 0.6 percent inorganic compounds (Manohar Rao, 1979). The raw juice contains suspended and dissolved impurities. The suspended impurities include the dispersed soils, bagasse particles, wax, fats, proteins, gums, pectins, tannins and colouring matter extracted from the cane during milling. These remain in colloidal form while dissolved impurities in the juice are glucose, fructose and inorganic salts such as N, K, P and others which are absorbed by sugarcane plant from the soil during its growth.

It is necessary to remove suspended and dissolved impurities from the juice in order to obtain sugar crystals. The processes such as sulphitation and carbonation are generally used to purify the cane juice. The juice clarification by sulphitation process is carried out with the help of lime and sulphur dioxide. $CaSO_3$ is obtained as precipitate alongwith impurities is termed sulphitation press mud (SPM). In carbonation process lime and calcium carbonate alongwith impurities are precipitated in the form of carbonation press mud (CPM). Press mud and filter cakes are synonyms and suffix cake after press mud carries no extra meaning.

Pressmud is a waste material obtained as by product of canesugar mills which is of the tune of 3 percent of total cane crushed. It was estimated that during the year 1995-96 about 8.0 million tonnes of press mud were generated which would rise to 8.6 mt. during 2001-02. The statistics showed that Maharashtra produced 1380 thousand tonnes of press mud followed by U.P. 1150 thousand tonnes followed by Tamil Nadu 643 thousand tonnes followed by other states. It is estimated that about 5 million tonnes per year of press mud cake is produced in India (Jambhekar, 1997).

The production of SPM and CPM is about 3 percent and 7 Percent of the quantity of cane crushed in the sugar factory respectively. There are two major potential press mud producers such as *khandsari* units and sugar industry. The bulk of press mud is obtained from sugar factories.

Properties and chemical composition of press mud

Press mud is soft, spongy, light weight, dark brown to brownish white in colour containing sugar, fibre, coagulated colloids, cane wax, inorganic salts etc. The method of juice clarification and the quality of cane crushed affects the composition and properties of press mud such as pH and content of plant nutrients. It is fairly a rich source of N, P, K and some micronutrients (Kumar and Mishra, 1991 and Yadav, 1995). The press mud derived from sugar factories using sulphitation process (SPM) contains 9.3% $CaSO_3$ and 35.8 organic matter (Chand *et al.*, 1977). SPM contains 1.1 to 3.1 percent N, 0.6 to 3.6 percent P, 0.3 to 1.8 percent K in addition to secondary and micronutrients, Fe (880-2500 ppm) Zn (153-272 ppm) Mn (111-1500 ppm) Cu (45-126 ppm) and acidic in pH (Table 58). On an average 1000 kg SPM contains 17KgN+36kg P+14 kg K+23kg S. On the other hand CPM has been found richer in $CaCO_3$ (60%) and contains 11% organic matter. It contains 0.6-0.9 percent N, 0.4 to 2.0% P, 0.36 to 1.6% K, 2100 ppm Fe, 275 ppm Zn, 2150 Mn and 250 ppm Cu (Table 58). It is alkaline in nature.

Tabel 58: Comparative chemical composition of SPM and CPM

Parameters	Oven dry-basis	
	SPM	**CPM**
Organic carbon (%)	26.0-43.2	13.1-15.1
Nitrogen (%)	1.1-3.1	0.6-0.9
Phosphorus (P%)	0.6-3.6	0.4-2.0
Potassium (K%)	0.3-1.8	0.36-1.6
PH	acidic	alkaline
$CaCO_3$ (%)	0.2	60
$CaS0_4$ (%)	0.4	traces

Effect on crop yields

Press mud has been mainly used as a source of plant nutrients, an ameliorant for acidic and saline-sodic soils, as a medium for raising sugarcane seedling and as a carrier for preparation of bacterial inoculants such as *Rhizobium*. Acid soils around sugar mills could benefit from carbonation press mud (CPM) since most acidic soils are less productive due to low water retention, low fertility and poor physico-chemical conditions as farmers can not afford to use commercial limestone as amendment.

Carbonation press mud (CPM) from sugar factories may be economical alternative to lime. Datta and Gupta (1983 a, b) reported that application of 2 tonnes/ha CPM to soil increased the yield of maize crop by 129.4 percent and of wheat crop by 100.8 percent. The residual effect of press mud on pea crop was double that of limestone.

It is often believed that a year old press mud is better as organic manure for crop plants. On the other hand, the studies have shown that fresh press mud can be used without any adverse effect on soils and crops (Yadav, 1992). However, some workers do not advocate direct application of press mud to soils because it contains 8-10 percent wax and prefer it to be composted before use. (Patil and Kale, 1983). It can be composted with distillery spent wash and utilised as an enriched organic manure than press mud alone (Thopate *et al.,* 1991).

Mohite and Balsaraf (1992) reported the beneficial effect of press mud @ 5 tonnes/ha with iron pyrite (1.0 to 1.5t/ha) on the grain yield and phosphorus uptake by maize crop and followed by wheat crop in a black calcareous soil. The residual effect of press mud was found equal to the effect obtained with 20 tonnes FYM/ha.

Yeast sludge: Yeast is added to fermentor for production of alcohol from molasses. After fementation is over, yeast sludge is separated as a waste product. The dead hydrolysed yeast cells are present in the raw spent wash in the storage tanks and lagoons. Yeast sludge can be composted with press mud and spent wash. It contains more proteins, vitamins and minerals. Yeast sludge is also used as feed supplement for cattle, poultry and fishes. However, many of the distilleries use this yeast sludge alongwith E.T.P. sludge as an organic manure to augment the soil fertility level. The usage of yeast and ETP sludge are also used for composting to enhance the nutrient value of manure and also can be used by farmers in pit or heap composting, process. Yeast sludge after sun drying can be supplied to the farmers since there is a good demand to utilize the same as organic manure.

Spent wash

Distillery spent wash is a byproduct of sugar cane. It contains all the minerals whatever the sugarcane crop has been taken up during its growth from the soil. During the processing for sugar, these minerals are left out in the molasses that forms raw material for the alcohol industry. The minerals in the molasses find their way in the spent wash left out after alcohol distillation.

The effluent termed as spent wash is dark in colour and is essentially an extract of sugar cane and microbial residues. Being a plant product, spent wash is non-toxic and does not contain any heavy metals or other toxic residues but it is highly environment polluting because it has a very high BOD (45,000 to 50,000 ppm) and COD (100,000 to 200,000 ppm) is highly acidic (pH around 4) and has high dissolved solids and organic matter concentration. Organic matter present in spent wash is the cause of high BOD. Most of monosaccharides and disaccharides during alcoholic fermentation are degraded. The lefts over residues are some pentoses and derivatives, peptides, proteins, phenolic compounds etc and by products of fermentation. The other organic sources are yeast and autolysed yeast cells.

The organic mixture has already passed through a series of degradative steps during fermentation and is now enriched with compounds which are difficult to breakdown. One of the methods is to pass it through methanogenic conversion via biogas plants with long residence time and special microbes are inoculated for increasing the efficiency of anaerobic system. The other method is cocomposting of spent wash with press mud or other suitable organic wastes. It can handle recalcitrant substrates. It permits surface growth of filamentous fungi and actinomycetes to facilitate decomposition. The handling of material becomes easy and investment is low. The final product is solid compost and its suspension in water hardly shows any spent wash colour.

Components of Composting

Press mud: The typical composition of press mud is as follows: org carbon 20-24%. N 1.26%, P 3.38%, K 1.46%, Ca 3.2%, Mg 1.65%, S 0.23%, Cu 52ppm, Zn 69 ppm, Mn 898 ppm, Fe 24 ppm.

Spent wash: It is very valuable resource and source of multiple plant nutrients. It is free from toxic and harmful components. The BOD is very high due to presence of organic and inorganic nutrients. Hence the direct application of spent wash to soil is harmful because it damages porosity and oxygen transfer capacity of soil. It also leads to germination disorder in crops and contamination of sub-soil water. So the conventional method is biomethanation to reduce the BOD value by secondary and tertiary aerobic treatments. But even after primary digestion the spent wash has high BOD (5,000 mg/l) very dark colour and repulsive odour. The characteristics of spent wash before treatment is given below:

pH 4.0 to 4.5, Total solids 2,20,000 to 2,30,000 mg/l, BOD 80,000-10,0,000 mg/l, COD 1,80,000 to 2,00,000 mg/l, Dissolved solid 1,80,000-2,00,000mg/ 1, Suspended solids 15,000 to 25,000 mg/l, Acidity (Ca CO,) 16,000 to 18,000 mg/l, Sulphate (as SO) 11,000 to 15,000 mg/l, Chlorides (as Na*) 3,000 to 5,000 mg/l, Calcium (as CaO) 4,000 to 6,000 mg/l, NH-N (as N) 500 to 700 mg/l, Spent wash quantity (lit/ lit alcohol) 5-7 litres.

Composting of Press Mud and Spent Wash

The method of converting spentwash into organic manure is based on the principle of organic recycling where solid wastes such as press mud is utilised. This would also check the pollution of surface and ground water with spent wash.

The work has been reported (Table 59) on the quality of compost prepared cut of spent wash and press mud (Khambe *et al.*, 1992). The findings show that manurial value of spent wash compost was higher than farmyard manure. Moreover, spentwash compost contained considerable amount of micro-nutrients which are deficient or limiting in many soils. The essential elements such as Zn, Fe and Mn were present in spentwash press mud compost in desired concentrations.

Table 59: Analysis of spent wash-press mud compost

Parameters	Compost 1	Compost 2
PH	7.1-8.0	7.1-7.6
Organic carbon %	30-35	30-40
Nitrogen %	2.0-2.5	2.7-3.5
pa*	2.0-2.5	3.0-3.5
K_20%	2.3-2.7	3.0-3.5
C/N ratio	15-14	10-11.4
Fe%	0.99-1.05	1.6-1.8
Cu%	0.72-0.85	0.72-1.0
Zn%	0.07-0.26	0.19-0.42
Cd%	0.012-0.017	0.04-0.14
Pb%	0.010-0.029	0.010-0.013
Na%	0.52-0.54	0.52-0.54
Mn%	0.11-0.12	0.21-0.22
Cl%	2.4-2.0	3.9-6.2

Subba Rao (1990) reported that spent wash can be composted either with press mud, bagasse or sugar cane trash taken in proper proportion and inoculated with organic matter decomposing cultures. Murkute *et al.* (1992) further undertook the biocomposting of spent wash either by mixing with press mud, bagasse or sugar cane trash. The initial analysis of the raw materials used in composting is presented in table 60.

Table 60: Chemical analysis of organic raw materials

Organic material	Org. C%	Total N %	C/N ratio	PH	BOD
Spent wash	25.0	1.87	15.5	4.3	45000
Press mud	33.0	1.50	22.0	6.5	—
Sugarcane trash	45.0	0.4	112.5	—	—

Microbial Inoculant

Compost was prepared by heap method (10*3*1 meter). Microbial culture was added @ 1kg/ tonne of wastes at the time of heaping. For composting press mud spent wash was added periodically to press mud by maintaining 60% moisture after every turning.

Enrichment of Compost

Azotobacter (nitrogen fixing bacteria) and phosphorus solubilizing bacterial culture each were added to compost biomass @ 500 g/tonnie of compost after the thermophilic stage after 4-5 weeks of initial composting was over at the time of turning. The results (Table 61) showed that nitrogen content with the use of culture was increased to 1.92% as compared to control, 1.21%, the maximum nitrogen was recorded in the treatment having press mud plus spent wash in 1:2 ratio. The C/N ratio was also favourably influenced by the use of cultures against no culture (control). The BOD of compost biomass was decreased to 62 ppm due to use of the cultures as compared to control (181 ppm).

Table 61: Chemical analysis of press mud-spent wash compost

Treatment	org. C%	Total N%	C/N ratio	PH	BODppm
	Press mud -	spent wash (1:2)			
Culture	33.6	1.92	17.5	7.3	62
No culture	35:9	1.30	27.7	7.4	181
	Press mud-	Sugpr cane trash-	Spentwash		
Culture	16	1.42	11.2	7.8	55
No culture	19	1.08	17.5	7.7	162

The composting of spent wash with press mud has been taken up successfully in some distilleries in Tamil Nadu on a large scale. However there is a scope to further improve the quality and maturity of the finished product by making amendments and use of the microbial cultures.

Biocomposting at Rajshree Sugar and Chemical Ltd (RSCL) Distillery

Process Description

1. The composting process is carried out on concrete platform lined with LDPE sheets to prevent any percolation. Windrows: The press mud is used to form the windrows of 3 to 4.5 metres wide, 1.5 metres height and length minimum of 120 m.
2. With the help of aerotiller, the material will be moderately mixed with spent wash and inoculated with special culture to enhance the decomposition rate, raising the temperature of the windrows to 65° C. The culture solution is sprayed uniformly on windrows and mixed immediately with the help of aerotiller.
3. Application of spent wash : In view of thermophilic decomposition process, moisture drops to 35-40% and decomposition is slowed down and temperature is reduced to less than 60°C. Spent wash is sprayed to adjust and maintain 50-60% moisture. Spent wash contains 20% solids and 80% moisture. It is essential to maintain the temperature at around 65°C for faster rate of decomposition.
4. The maturity of compost is attained within 10 weeks. It is free from weeds, nematode cysts, plant palthogens etc. The composting technology

is relevant for disposal and use of spent wash by all distilleries and help in converting the wastes into ecofriendly enriched organic manure.

5. The well-lined storm water collection are provided to collect the rain water run off and leachates if any during the rainy season. The leachates collected from the composting yard is reused in biocomposting process. Hence there is no possibility of ground water pollution due to leachates from the composting yard.

Distillery Effluent Used as Ferti Irrigation

Many distillery units have set up biomethanation plants to generate methane and by this primary treatment, the COD is reduced by 65% and BOD by 90%. The distillery effluent can be returned to the soil in an effective manner without causing any pollution problem. The treated effluent is neutral to slightly alkaline pH (7.6) and dark brown in colour with a BOD of around 5000 ppm. It is fairly high in nitrogen (1800 ppm) phosphorus (420 ppm) and high in potassium (11,500 ppm). It contains fairly high amounts of calcium (1050 ppm) magnesium (2,200 ppm) and sulphate (2,440 ppm). It also contains appreciable amounts of micro-nutrients viz. Zinc (8 ppm) manganese (5 ppm) iron (85 ppm) and copper (6 ppm). Each tonne of treated effluent, on an average contains 1.8 kg N, 0.4 kg P, 11.5 kg K, 1 kg Ca, 2.2 kg Mg, 2.5 kg SO_4, beside 30 kg of organic matter and some micronutrients. The analysis of the effluent showed that it does not contain any toxic heavy metals (Devrajan *et al.*, 1994). The results of various trials showed that treated distillery effluent can be considered as a liquid fertilizer and can be applied to soil in an effective manner with proper soil and crop management practices for increasing crop yields.

IV Sawdust and Cotton Wastes / Byproducts

Saw Dust

Sawdust, a by product of saw mill and plywood industry which is available in huge quantities is mainly used as fuel, packing material and in ice factories. It has good water absorption capacity (table 62). This characteristic can make sandy soils retain more irrigation or rain water for the crops under water stress conditions. Its application to heavy soils shall make them easier for cultivation by improving the physical condition of the soil. Sawdust is rich in lignin (13-14%) and a high C/N ratio material (500:1) low in nitrogen (0.12%) and also low in phosphate (0.20%) and is biodegraded at a very slow rate. This will be useful in tropical soils in retaining organic matter for longer as compared to FYM or compost which is mineralized at a faster rate and build up of organic carbon is much less.

The application of sawdust in the soil may temporarily immobilize primarily the nitrogen and phasphorus, resulting in stunted and poor growth of the plants. This can be surmounted by application of nitrogen fertilizer to meet the nitrogen requirement of the microbial biomass participating in the decomposition of sawdust (Naphade 1981). Sawdust can be used as bedding in the cattle shed to absorb urine. The enriched sawdust can be dried and applied to soil as N source (Idnani, 1966) or composted alongwith farm wastes.

White *et al.* (1959) reported that sawdust application immobilized nitrates and caused temporary deficiency of nitrogen which can be overcome by application of adequate amount of nitrogen. Bollen and Lu (1979) reported that 6 to 12Kg N/ tonne of sawdust was sufficient to counteract N deficiency during the first year. Its value as mulching material and for control of plant parasitic nematodes has also been documented.

Table 62: The relative water absorbent capacity of different materials

Material	**Quantity of water (in kg) retained 1 kg of material after 24 hrs. of soaking**
Sand	0.25
Soil	0.50
Dry leaves	2.00
Wheat straw	2.20
Sawdust	4.35
Peat	6.00

Source: Handbook of *Agriculture (1997)* I.C.A.R. *New Delhi, Page-217*

Cotton Wastes / Byproducts

Cotton is an important fibre crop of India. The products of cotton seed are oil seed cake, linters and hulls. One tonne of cotton yields 48% cottonseed cake, 26% hulls and 15% oil and 11% is composed of fibre and shrinks during ginning. Cotton seed oil cake can be both used as feed and manure and the main waste by products arising out of cotton cultivation are as follows:

1. Cotton stalks
2. Cotton linters and hulls
3. Cotton leaves and other plant parts
4. Cotton dust.

Rainfed cotton crop yields two and a half tonnes stalks per hectare and under irrigated condition, the yield of cotton stalks is about 5 tonnes ha1. Total production of cotton stalks on all all-India basis is about 14.0 million tonnes which are generally used for fuel purpose.

Cotton Dust

Cotton textile mills mainly in their blowing rooms produce a large quantity of this waste. Textile mills in India produce about 35,000 tonnes of this waste per year. Cotton dust contains 8.0% moisture, 70% organic matter, 41.0 organic carbon, 1.4% total N, 0.6% P2O, 1.2% K2O. This waste has greater amount of plant nutrients than routine city refuse. Its C/ N ratio of 29.4:1 is ideal for composting and compost can be prepared within 3 weeks (Gaur, 1978).

Cotton Seed Meal

Although it is usually high in phosphorus, cotton seed products are low in lysine and vitamins A and D. Dairy cattle fed on cotton seed products produce milk with a higher fat content since cotton seed oil consists of saturated fat. Raw cotton seed contains gossypol, a compound which is toxic to swine, poultry and young cattle (Morrison 1959). However, adult cattle were fed as much as 4.5 kg cotton seed meal daily for three years but did not show any adverse effect.

V. Paper Mill Sludge

Paper mill sludge is an industrial by product with high content of organic matter which is slowly biodegradable. Since there are limited alternatives for its practical use, it can be utilized as an organic manure, either as raw or after composting. The effect of combined pulp and paper mill effluent used for irrigation of sugar cane crop increased the soil pH, organic carbon, total nitrogen, phosphorus and potassium over the control (Table 63) (Kannan and Oblisami, 1990).

Table 63: Effect of irrigation with pulp and paper mill effluent on soil chemical properties

Irrigation period (Year)	pH	Organic C%	Total N (mgkg[1])	Available P (mgkg-1)	Available K (mgkg-1)
Control	7.5	0.33	110	21	190
1	7.9	0.55	160	30	190
2	8.0	0.83	170	31	190
3	8.3	0.91	180	35	240
15	8.4	1.20	220	38	240

*Mean of intervals (March, April, June and August)

The cumulative effluent irrigation also increased the population of bacteria, actinomycetes and rhizobia in soil. The bacterial population was highest after 15 year of irrigation followed by 3, 2 and 1 year of its application (Table 64). The numbers were directly proportional to the improvement in soil properties.

Table 64: Total microbial* population in soil irrigated with paper and pulp mill effluent

Irrigation period (year)	Number/gram dry soil		
	Bacteria (10^5)	Actinomycetes (10^3)	Rhizobia (10^3)
Control	117	33	10
1	135	37	12
2	145	37	13
3	168	39	17
15	196	59	18
CD (5%)	2.05	1.27	0.50

*Mean of 5 intervals in an year

The effect of paper-mill sludge in another experiment was carried out for two years on the yellow lupin plants (Vasconselos and Carbal, 1993). The composition of pulp-mill sludge was as follows:

Moisture 56.9%, pH 7.3, organic matter 81.31%, N 0.65%, P 0.05%, ash 18.7%, C/N ratio 75.5, Ca 5.30%, K 0.90%, Mg 0.05%, Na 0.25%, Cu 25 ppm, Mn 262 ppm, Zn 33 ppm, Ni 75 ppm, Cd. 3 ppm, Pb 77.3 ppm and Hg 0.86 ppm.

The results of the first year showed that application rate greater than 50 tonnes ha-1 of pulp mill sludge has depressing effect on the crop yield but it was not so in the second year (Table 65) The main cause of depressive effect on the yield may be attributed to high C/N and/or C/P ratio of the applied sludge. The sludge significantly increased the soil pH and improved the organic matter status. There appears to be significant liming potential in the sludge since a change in the pH of 1 to 1.5 units was recorded due to application of 10 to 50 tonnes of sludge ha $^{-1}$.

Table 65: Effect of pulp-mill sludge on yellow Lupin

Qty.of mill pulp Sludge (tha-^{1})	Soil analysis at the end of 1 year		Yield		
			I Year		II year
	PH	Organic natter (%)	Leaf area (cm^2)	Dry matter (9 pot^1)	Dry matter (gpot^1)
0	5.10	0.59	4.65	16.00	19.60
10	5.50	0.61	4.27	13.53	21.40
30	6.13	0.76	5.18	15.20	22.00
50	6.70	0.78	3.37	11.07	21.30
70	7.00	0.79	2.60	9.00	21.90
90	7.33	0.79	1.55	5.30	23.20
110	7.60	0.87	0.92	3.07	20.80
130	7.70	0.88	0.72	2.90	19.90

This sludge has a higher amount of calcium and low amount of sodium. This is of interest for acidic soils poor in organic matter content. Nitrogen uptake by the lupin crop which is a leguminous crop should not have been affected but it was strongly decreased in presence of sludge applied at higher rates.

References

Ahmad, N, Rai, M, Yadav, N.L. and Sahi, B.P. 1978. Studies on efficiency of Karanj cake and its lipid associates on nitrification regulation under submerged rice cultivation in a calcareous soil in Bihar. J.Indian Soc. Sci. 26: 304-05.

Bains, S.S, Prasad, R and Bhatia, 1971 Fert News 16 (3): 30-32

Bhardwaj, K.K.R 1995. Recycling of crop residues and other plant products in agriculture. Recycling of crops, animal, human and industrial wastes in agriculture (Editor HLS Tandon) FDCO, New Delhi, pp 9-30

Bhardwaj, K.K.R and Gaur, A.C. 1985. Recycling of organic wastes, ICAR, New Delhi, pp 104

Bollen, W.B. and Lu, K.C 1957, Proc. Soil Sci. Soc Am. 21:35

Chand, M., Abdul, I. P. and Bhumbla, D. R. 1977. Indian J. Agric. Sci. 47: 348-354.

Datta, M and Gupta, R.K. 1983 a. J. Indian Soc. Soil Sci. 31:511

Datta, M and Gupta, R.K. 1983b. J. Indian Soc. Soil Sci 31:635

Devrajan, L, Rajannan, G, Ramanathan, G asad Oblisami, G. 1994 Kissan world

Gaur, A.C. 1978 Recycling and utilisation of organic wastes as fertilizers. Proc. India-FAO-Norway seminar on Development of Complementary use of mineral fertilizers and organic materials in India Ministry of Agriculture, Dept. of Agriculture (Fertilizer division) New Delhi. Pp 109-126.

Gaur, A.C. and Prasad, S. K. 1970, Effect of organic matter and inorganic Fertilizer on soil and plant nematodes Indian jour. Ent. 32 : 186-188

Gaur, A.C. and Mathur, R.S. 1979. Effect of straw, neem cake, and farmyard manure on yield and nitrogen uptake by maize crop. Indian J. Agron 24: 449-50

Gaur, A. C, Neelkantan, S and Dargan, K. S. 1995, Organic Manure, ICAR, New Delhi pp 159

Idnani, M. A.1966. Indian Farming Jan. Issue, p.44

Jambhekar, H. A. 1992, In Proc National Seminar on Organic Farming. M P K V, Pune, pp 52-53

Kannan, K and Oblisami, G. 1990. Influence of Irrigation with pulp and paper mill effluent on chemical and microbiological properties. Biology and Fertility of Soils 10: 197-201

Ketkar, C. M. 1974. Neemcake blended urea on nitrogen economy Fert. News. 19(2):25-26,

Khosla, B. K. 1976. Infiltration and water distribution in sodic soil as affected by rice husk incorporation. J. Indian Soc. Soil, Sci. 24 :436-38

Khambe, S.D., Salunkhe S.S. and Subba Rao B., 1992 Proc. Natl. Seminar on Organic Farming. MPKV, Agricutlure college, Pune. PP 29-31.

Kumar, V and Mishra, B.1991, J. Indian Soc. Soil, Sci 39:109

Manohar Rao, P. J. 1979. Sugar Technology for Administrators in Indian Sugar Industry Maharashtra Sugar Foundation, Pune.

Mohite, A.V. and Balasaraf, M.D. 1992, Proc. National Seminar on Organic Farming, Agri. College Pune PP 26-28

Morrison, F. B. 1959. Feeds and Feeding, The Morrison Publishing Co. Clinton, Iowa, U.S.A.

Mukherjee, D. Mitra, S. and Das, A.C. 1991, J, Indian Soc. Soil, Sci 39:457-62.

Murkute, S.B. Thopate, A.M., More, B.B. and Hapase, D.G. 1992. Proc. Natl. Seminar on organic Farming. MPKV, Pune, 67-68.

Naphade, K.T. 1981. Ph.D. Thesis, Agchem Soil Sci, PG School, I.AR.I., New Delhi

Parihar, S.S., Verma, K.S, and Bhajan Singh 1972. Direct and residual effect of rice husk on crop growth on a sandy loam soil. Indian J. Agron 17:344-47

Patil, S. P. and Kale, S.P. 1983. Indian sugar 32: 905-08

Pradhan, S., Jotwani, M.G. and Rai B.K.1962. The Neem Seed deterrent to locust. Indian Farming 12: 7-11

Prasad, S.K., Mishra, S.D. and Gaur, A.C. 1972. Ind. J. Ent. 34:307- 311

Rawat N.S. 1993-94. Every man's science pp 194-199.

Reddy, M.N. and Shinde, J.E. 1981. Fertilizer News 26(2) 21-22.

Rembold, H.,Forster, S. Egoppett, C., Rao, P.J. and Siber, K.P.1984, The azadirachtin, a group of insect growth regular of the neem tree, PP 153-162 In: Natural pesticides from neem tree (Azadirachta indica A. juss) and other tropical plants (Eds. Schumutterer H. and Ascher, K.R.S) Proc. Second Int. Neem conf. May, 1983, Ravischholzhausen.

Sahrawat, K. L. 1982, J. Indian Soc Soil Sci. 30: 156-59

Sahrawat, K. L. and Mukherjee, S. K. 1977, Plant Soil 47: 27-36

Sahrawat K. L. and Parmar, B. S. 1975, J. Indian Soc Soil Sci. 23: 131-34

Saxena, P. K. and Singh, H. G. 1980. Effects of rates and time of application of castor cake on potato production on vertisols in Rajasthan, Indian J. Agron. 25:651

Shanmugam, M, Rathanaswami, R and Nadarajan, N. 1995. Madras Agric. J. 82:85

Schumutterer, H. 1988, Potential of azadirachtin containing pesticides for integrated pest control in developing and industrial countries. J. Insect Physiol 34: 713-19

Subbiah S, and Kothandaraman, G.V. 1980. Effect of neemcake blending of urea on ammonium and nitrate content of soils under field conditions. J. Indian Soc. Soil Sci. 28: 136-37.

Singh, U and Gurumurti 1984 oilcakes from oilseeds of forest origin- their potential as fertilizer Indian J. forestory 7: 12-18

Singh, M and Singh, T.A. 1989. Fertilizer News 34 (7): 39-43.

Subba Rao, B, 1990. 40th Ann Conv. DSTA, Pt.1, 79-93.

Surve, S. P. and Daftardar, S.Y. 1985. J. Indian Soc. Soil. Sci. 33: 182-186.

Thopate, A.M., Murkute, S. B. Phonde, D. B, More, B.B. and Hapase, D. G. 1991, Proc DSTA Convention part I,A: 312-15

Tiwari. S.C. 1989. Effect of neemcake blending of use on ammonicacal and nitrate content in vertisol. J. Indian Soc Soil Sci. 37:380

Vasconcellos, E. and Cabral, F. 1993. Use of environment implications of paper mill sludge as an oranic fertilizer. Environmental Pollution 80: 159-162

White, A.W., Giddens, J.E. and Mories, H. D. 1959. Proc. Soil Sci. Soc. Amer 23:365.

Yadav, D.V. 1992. Indian J. Sugarcane Technol. 7: 1-16

Yadav, D.V. 1995. Recycling Sugar factory press mud In: Recycling of Crop, Animal, Human and Industrial wastes in Agriculture (Ed. HLS Tandon) FDCO, New Delhi.

Yadav V.K. and Srivastava, L.L. 1987. Mineralization of urea coated with nitrification inhibitors and their effect on the population of Nitrosomonas and Nitrobacter. J. Indian Soc. Soil Sci. 35 :138-140.

5

Canning and Food Processing Wastes

India produces around 33 mt. of fruits and 50 mt. of vegetables annually (Chadha and Bhargava, 1997). It is estimated that 10 to 15 percent of total produce is available as organic waste for recycling in agriculture (Nand *et al.*, 1996). In addition to processing of fruits and vegetables results in production of around 5 mt. of solid wastes (Table 66). Most of these wastes contain organic and plant nutrients (Table 66). There are still large amounts of wastes that are not being completely utilized. In most cases, these wastes are pro-duced from isolated sources and not by large centralized.companies. In practically all cases, horticultural and vegetable products are processed in relatively small canning plants and distributed for use generally in the same geographical area where the produce is grown, this is true world over since fruit and veg-etable wastes cannot be usually transported long distances in view of their bulk, high moisture and low to moderate feeding value relative to conventional feed.

Table 66: Annual potential of Fruits and Vegetable processing

Fruit/ Vegetable	Nature of Waste	Potential Quantities of waste (000 tonnes)
Mango	peel and stone	3145
Banana	peel	832
Citrus	peel	415
Apple	peel pomaces and seeds	412
Tomato	skin, core and seeds	90
Pineapple	skin and core	25

Source: Nand *et al.*, 1996, 188

General aspects of waste handling

1. *Economics*: The amount of dry matter in the wastes produced by a processing plant may not be sufficient to justify the inconvenience and collection cost including transport and perhaps making changes in the method of feeding and feed storage. Before considering the use of certain wastes in the livestock feed, the consumer should determine the

feed value and economics in relation to present standard ration used and whether or not the waste product can be substituted partially or completely for standard and costly feed.

2. *Potential of available wastes and its nutrient composition:* Many cannaries produce known volume of wastes regularly and the composition is relatively uniform. However, such a waste material, not used previously as an animal feed, should be subjected to proximate analysis (A.O.A.C 1975) mineral analysis, pH and dry matter determinations. For consistent quality the various types of waste products may include their separation or the removal of excess waste from the waste material.
3. *Removal of metal, glass or similar contaminants*: In many canneries, organic wastes from the processing system may include nonusable wastes, such as hardware, glass, cans and nails. The canneries wastes are generally disposed off in landfils if it is not used as livestock feed or further processed. Naturally metal or glass materials must be removed if it has to be used as animal feed. The primary factor of the relatives of waste materials are their energy and protein content. The value of even standard feedstuffs will vary widely such as their price, nutrient content and other factors.

I. Wastes from Fruits

1. Apple Processing Waste

Apple pomace is generally fed fresh from the factory or ensiled. In most cases, rice hulls are added in an approximate 1:13 proportion of rice hulls to apple processing wastes (on a volume basis). The rice hulls are added to increase the amount of liquid pressed from the liquid waste before the waste leaves the processing plant. Addition of rice hulls does not appear to reduce the feeding value, although rice hulls are essentially an indigestible portion of the waste material. There were indications of digestive impaction in a feeding trial conducted by Wilson *et al.* (1970) in which 70% Handbook of Organic Farming & Biofertilizers of the total dry matter of the ration consisted of the rice hull/ apple waste product. However, this impaction decreased weight gain and feed efficiency only slightly compared to the gain and efficiency calculated from the energy values of the materials.

2. Pineapple Wastes

Pineapples wastes include the dried residues from the canning of pineapple, slips, suckers, immature fruits, stumps and dead leaves. It is generally a bulky material and as a result has been pelleted for improved handling. It is usually considered as a roughage source for ruminants. Pineapple bran was found to be highly digestible and palatable as feed in the Hawaiian studies. (Otagaki *et al.*, 1961).

3. Cassava

Cassava is widely grown in tropical areas, with the principal food value contained in the tuber. Cassava contains a cyanogenic glucoside and accompanying enzymes, which can liberate hydrocyanic acid, which is toxic to animals and humans. The concentration of these toxic substances is higher in bitter (0.27%) than in sweet cassava (0.007%). Cassava should be dried or ensiled prior to feeding to animals to reduce the toxicity. However, for both types of cassava, the concentration of cyanogenic glucoside varies with growing conditions, moisture, temperature and plant age (Castillo *et al.,* 1964)

4. Citrus Pulp

Citrus pulp is dehydrated and can be utilized in dairy rations. It is not only a high energy feed but also increases feed consumption thereby resulting in a higher performance rate and increased feed efficiency.

II. Wastes from Vegetables

1. Potato Processing Waste

Potato processing wastes are becoming more important as livestock feed for several reasons. Many potato processors are changing from the "Wet" caustic processing method which usually results in wastes of pH 9.0 to 13.0 when large amounts of tubers are used late in potato processing season to a dry peel process which requires less alkali and water. The dry peeling process has low sewage cost and if utilized, the potato wastes can be transported economically to agricultural land for use as feed or Canning and Food Processing Wastes fertilizer. The high pH of processing wastes makes it a desirable fertilizer particularly in areas of acidic soils.

2. Tomato Processing Wastes

Tomato waste (also called pomace) is a mixture of tomato skins, pulp and crushed seeds resulting from tomato juice, catch-up, puree or other products. Tomato pomace is high in thiamine, contains moderate levels of riboflavin and has considerable carotene content. Tomato pomace has rather high crude protein (23.9%) table 67. Tomato pomace is ordinarily not palatable when fed fresh but palatability increases with drying. Tomato contains very high amount of water, therefore it can also be used by ensiling with conventional feed stuffs.

Table 67: Chemical composition of some food processing wastes

Waste	Total solids (%)	N (%)	Ash (%)	C/N ratio
Tomato peel and seeds	29.5	4.20	4.3	13.0
Mango peel	26.4	0.56	3.5	71.0
Pineapple waste	93.4	0.85	7.8	45.0
Orange peel	88.3	1.09	5.8	39.0
Deoiled orange peel	88.0	2.22	6.4	19.5

References

AOAC 1975. Official Methods of Analysis. Association of official Analytical Chemists, Washington, D.C.

Castilo, L.S. Aglibut, F.B. Javier, T.A., Gerpaciv, G.V., Ruyaoan, R.B. and Ramin, B.B. 1964. Philippine Agric. 47: 460-74.

Chadha, K.L. and Bhargawa, B.S. 1997. In: Proc. Symp on Plant Nutrient Needs, Supply, Efficiency and Policy issues: 2000-2025. Nat. Acad. Agric. Sci., New Delhi pp 114-132.

Nand, K., Madhukar, K., Sudhakar, B. and Srinath, K. 1996 In: National Seminar on Industrial Waste Management.

Otagi, K.K., Loffgreen, G.P., Cobb, E and Dull, G.G. 1961, J. Dairy Sci 44: 491-497.

Reed, A.D., Wildman, W.E., Seyman, W.S., Ayers, R.S., Prato, J.D. and Rauschkolb, R.S. 1973. Soil recycling of cannery wastes 27 :6-9. Univ. of California, Div. Ag. Science.

Wilson, L.L., Kurtz, D.A., Ziegler, J.H., Rugh, M.C, Watkins, J.L., Long, T.A., Borger, M. L. and Sink, I.D. 1970. Accumulation of certain pesticides in adipose tissues and performance of Angus, Heerford and Holstein steers fed apple processing wastes J. Ani. Sci. 31: 112-117.

6

Organic Wastes from Fermentation Industries: Their Uses

Solid wastes, which are left after the fermentation process is complete, may be potential sources for use as fertilizers and feed stuffs.

I. Antibiotic Fermentation Wastes

Antibiotics are produced by microoganisms and have capacity, in dilute solution, to inhibit the growth or even destroy the other microorganisms. They are produced on a large scale using fermentation process except chloroamphenicol which has been produced synthetically since 1948. Solid wastes remaining after the liquid containing the antibiotic has been extracted from the fermentation broth need to utilised. This is usually done by filtration. The solids are a heterogeneous mixture of hyphae of the producing organism, unused media, defoaming oils and metabolic products of the micro-organisms, the most important of which is residual antibiotic. The levels of residual antibiotic can vary a great deal.

Fertilizer Value of Antibiotic Wastes

The analysis of tylosin, oxytetracycline and penicillin wastes are as follows:

Tylosin: water 83.6%, pH 6, organic matter (dry basis) 85.5%, nitrogen 6.34%, P 2.0%, K 0.59%, Ca 0.30%, Na 0.21%, Mg 500ppm, Fe 275 ppm, Mn 7ppm, Cu 245 ppm, Pb 15ppm, Cd 1.5ppm, antibiotic content 2000 ppm.

Oxytetracycline wastes: water 65.7%, pH 2.2, organic matter (dry basis) 50%, N 2.80%, P 0.43%, K 0.07%, Ca 0.71%, Na 0.35%, Mg 60ppm, Fe 460 ppm, Mn 11 ppm, Cu 3.5 ppm, Pb 5 ppm, Cd 1.5 ppm, antibiotic content 1500ppm

Penicillin wastes: Water 80.8%, pH N.A organic matter 90.8%, (dry basis) N 5.92%, P 1.36%, K 0.38%, Ca 0.72%, Na 0.25%, Mg 550 ppm, Fe 60ppm, Mn 6 ppm, Cu 5 ppm, Pb 5 ppm, Cd 1 ppm, Antibiotic content - not available

There have been very few reports in literature on the use of antibiotic fermentation wastes as fertilizer. Molof (1962) reported that in U.S.A. antibiotic wastes were mixed with sawdust and animal manure to produce compost which was marketed as organic fertilizer. But no further reports of the process have been seen in the literature. Uhliar and Bucko (1974) reported an increase in fresh and dry matter, and numbers of root nodules of lucern (grown in Czechoslovakia) when sprayed with penicillin and streptomycin wastes although no indication about their nutrient status, antibiotic content and levels at which they were applied were given. There has also been some newspaper reports in the investigations by Pfizer of Groton, Connecticut into the use of fermentation wastes as fertilizer for potatoes (Anon, 1973: 1974, Warner, 1973). The addition of antibiotics to the soil could affect the soil microorganism as well as plants growing in the soil as a result of one or more of these process :

1. The chemical instability of the molecule of antibiotic.
2. Adsorption on clay minerals and organic matter
3. Microbial degradation.

Effect of antibiotics on soil microbial processes

There have been several studies on the effect of pure antibiotics on microbial population. Several of these investigations were carried out on pure culture of soil microorganisms in laboratory media (Pramer and Starkey 1962). However, the activity of an antibiotic on soil may be very different from that in the laboratory media. This is supported by a study on the influence of streptomycin on nitrification (Pramer and Starkey, 1951). Nitrification was completely inhibited in solution medium by 1 ppm of streptomycin. In sand, a comparable inhibition required 1000 ppm, while in soil a concentration of 10,000 ppm was required. The result of such studies indicate that relatively few antibiotics influence the microorganisms in soil. There are reports that generally all concentrations of antibiotic wastes increased the microbial activity as measured by O_2 uptake, CO_2 release and N mineralization compared with untreated soil (Bewick 1978). The uptake of streptomycin and its effect on plant growth was studied by Mukhejee and Bag (1974) and Mukherjee *et al.*, (1975).

Indian Scenario: India has made rapid progress in the production of drugs and other chemicals using fermentation process. The left materials or residues contain large amount of organic matter and minerals in solid and liquid form. It is estimated that there is generation of 7000 to 8000 litres of waste water with solid content ranging from 40,000 to 50,000 mg/ I having BOD in the range of 30,000 to 60,000 mg/l for each kg of antibiotic produced (Bewick, 1980). This waste water also contains about 2 to 6% N on dry basis besides other essential plant nutrients. Presently this waste water is processed through water treatment plants. This material can be used for enrichment of compost produced from crop residues and city wastes to enhance the manurial value of compost.

II. Winery Wastes

It has been reported that 1 tonne of grapes gave rise to approximately 747 litres of new wine, 88 kg pomace and 33 kg stems (Amerine and Joslyn, 1951).

Stems: There are stems and leaves of vines which account between 2 to 6% of the total weight of the fruit. There have been few reports on the potential uses of this material, although Scheffler (1954) suggested that stems could ground up in water and the resulting liquor distilled for its alcohol content. In U.S.A the stems are usually disposed of by spreading on the land and were used as organic materials for soil improvement (Amerine and Cruess, 1960).

Grape Pomace: This is a mixture of skin, seeds and pulp of the grapes. After extracting the juice for wine production, about 10 to 20% of solids of the weight of the fruit containing 35 to 70% moisture in it are left out. The report suggests that grape pomace contains approximately 11% crude protein, 6% fat, 4% fibre and 8% ash (Amerine and Cruess, 1960). The feed value of grape pomace was considered and found to be as good as wheat straw as feed supplement for the animals. Before feeding to animals, it is dried, crushed and mixed with molasses to improve its palatability. Agostine (1964) reported that its feed value could be improved by addition of lime to raise the pH. However, its use as animal feed does not appear to be practised widely and reports suggest that the material may be of use as fuel.

It's potential as organic material for soil fertility was also investigated. The chemical analysis of the by product showed but it contained 1.5 to 2.5% N, 0.5% P and 1.5 to 2.5% K on dry weight basis (Jacob and Proebstring 1937). Its fertilizing value was equivalent to F.Y.M and was suggested to improve the hysical condition of the heavy soils. It was also noted that heavy application to soil gave rise to phytotoxic effects due to intermediate substances formed in the soil during the early stages of decomposition. After unitial period of decomposition, the toxic effects disappeared and the soil supported luxuriant plant growth. Perone (1958) suggested that this material should be composted before its application to field so as to over come the phytotoxic problem noticed by the earlier workers.

Lees: These are the spent yeast cells which are left after completion of the fermentation of the wine. Spent yeast cells contain potassium bitartrate in large amounts, wine making represents the largest single source of tartrates in the world. Lees are unsuitable for animal feed, as the potassium bitratrate contained in it is a potent laxative. Lees do find a market in distillery where they are mixed with water to distill off the residual alcohol. The wastes after the operation may be mixed with other wastes and converted into compost for use in agriculture or horticulture.

III. Wastes from Cider Production

The wastes from cider production are apple pomace, yeast and filter aid.

Pomace: Cider/ pomace consists of the remains of skin, pulp and seeds from which most of juice and soluble consituents have been extracted in the making of cider. Much of the pomace is further treated to extract the pectin which is subsequently sold. The analysis of dry pomace showed that it contained as follows:

Dry matter 87.5%, 6.2% crude protein, 20.9% crude fibre, 2.5 ether extract and 2% total ash.

The levels of pomace are difficult to estimate as there are no recently published figures.

Uses: In U.K. much of the pomace after pectin extraction is used as animal feed additive. Wet pomace after mixing with fodder can be used for preparation of silage on the farm before using as animal feed. Ensiling can be easily done on the farm (Smith, 1950). Many workers reported that cider pomace was readily palatable as animal feed.

Barker (1933) suggested that apple pomace could be used as manure. Apple pomace was composted with phosphate and used as manure for grasses. The manure contained 1.6 to 1.7% N, 0.4 to 07% P_2O, and 0.2 to 0.6 K_2O. In France, this material was mixed with bone meal or rock phosphate to produce organic manure. Phosphate is solubilized during the process of composting.

Yeast: Spent yeast is usually mixed with pomace and sold as animal feed. The yeast produced in fermentation of cider, in U.K is usually mixed with the pomace and sold as animal feed. In other developed countries, it is likely that yeast is utilized in the same manner.

Other fermentative wastes: The quantity of wastes produced by other industrial fermentations including enzymes, vitamins and steroids are virtually impossible to determine due to high degree of secrecy. All that can be discussed here is which compounds are being produced on industrial scale by fermentation, will produce some wastes which can be sold as feedstuff or manures, if the security problems could be overcome.

References

Agostine, A. 1964. Utilization of by products of wines and vines, Wynboer, 32: 13-16.

Amerine M. A and W.V. Cruess 1960. The Technology of wine making. The avi Pub. Co. Westport, Connecticut.

Amerine, M.A and Joslyn M. A. 1951, Table Wines, University of California Press.

Anon, 1973. Norwick Bull. Oct 18.
Anon, 1974 The Hartford Courant. Oct 16.
Barker, B.T.P. 1933. J. Min. Agric. 40: 710-715.
Bewick, M.W.M. 1978, Soil Biol, Biochem 10: 403-407.
Bewick, M.W.M. 1980, In: Handbook of organic waste conversion. Van Nostrand Reinhold C. New York pp 165-186.
Jacob, H.E. and Proebstring 1937, Wines and Vines, 18:2-23.
Molof A.H. 1962, Biotech and Bioeng, 4: 197-209.
Organic Wastes from.Fermentation Industries- their Uses
Mukerjee, S and Bag, A 1974. Ind. J. Exp. Biol. 12:360-62
Mukherjee, S. Bag, A. and Paul, A.K. 1975. Biol. Plant. 17: 60-66.
Olive, T.H. 1949, Chem. Eng. 56: 107-10
Perrone, A. F. 1958, Composting of Pomace Piles. Wine Inst. Tech. Advis.comm. May 26, 1958
Pramer, D. and Starkey, R.L. 1951, Science 13: 127
Pramer, D. and Starkey, R.L. 1962, Bact. Proc. 52: 15-30
Schreffler C. 1954 Recovery of sugar from grape stems. Wine Inst Tech. Advis. Comm. Dec. 3, 1954
Smith, E.L. 1950. Apple Pomace Silage. Agriculture 57: 328-332
Uhliar,T and Bucko, M. 1974. Rostlinna Vyroba 20: 923-930
Wamer, N.J. 1973, once dumped at sea, antibiotic by product fattens potato. The providence Journal, Oct. 22.

7

Tannery Sludges in Agriculture

There are about 3000 tanneries processing 6000 tonnes of hides and skin annually in India. Large amounts of solid and liquids are generated in tanneries containing N and heavy metals. The treatment of tannery waste waters generates sludge which contain a large variety of organic and mineral components. The amount of sludge from a single tannery may be obtained up to 10,000 tonnes/ Yr. Fresh Sludges undergo preliminary fermentation on liquid separating plots which are stored in piles where further decomposition takes place. As a result of decomposition process, partially biodegraded material resembling compost is obtained. It looks like deep black earthy material. The chemical composition indicates the possibility of using it in agriculture. The hides of infected animals many contain anthrax germs and be contaminated with heavy metals particularly chromium in the sludge which depends on the technology for hide tanning and the methods used in a sewage treatment plants. The sludge from tanneries using chromium in the treatment process are not suitable to be used as fertilizer as they contain high amount of chromium. Co-composting of these wastes with crop residues could provide good quality compost and the heavy metal content would get diluted and reduced per unit weight.

I. Properties of Tannery Sludges

Initially the material is made alkaline with milk of lime and then coagulated with ferrous sulphate or aluminum sulphate. The precipitated sludge has 95 to 98% water, water removal is carried out on liquid separating plots or vacuum filters. The sludge therefore contains about 17% dry matter after removal of some of its water. When the draining installations develop some problems, the sludge may contain much less dry matter. Then its use in agriculture becomes limited due to increase in the transport costs etc.

The colour of the drained sludge ranges from brown gray to black. Since the decomposition processes continue, unpleasant smell if any disappears after incorporating them in soil.

The specific gravity of tannery sludge ranges from 1.005 to 1.200 g/ cm3 and is dependant on the mineral content. The pH of the sludge varies from 6.9 to 9.7. Tannery sludges have high absorption capacity which is relevant in increasing fertility of light soils with poor absorption capacity.

The average chemical composition of tannery sludge on dry basis is as follows:

Dry matter 17.2%, organic matter 57.3%, ash 42.7% and fats 6.9%. Dry matter is affected with the consistency of sludges. However, sludges from vegetable tanneries were found to contain higher percentage of dry matter than sludges from vegetable-chromium tanneries. Tannery sludges are characterised by large variations in ash content (12.0-66%). The large variation in the content of ash and organic compounds are due to variability in nature of tannery sludges that are treated with various amounts of coagulants. There is also a large variation in fat content (1.1 to 26.2%) of sludges.

Fatty substances have no deleterious effects on soil properties because fat gets decomposed in soil quite fast. However, proper aerobic conditions must exist and sufficient calcium is required to neutralize the fatty acids.

The nitrogen and mineral content of tannery sludges are variable. The mean values in these sludges reported by Mazur and Koc (1976 abc) are as follows:

Total nitrogen 3.78%, NH-N 0.25%, Phosphorus 0.25%, Potassium 0.09%, calcium 3.83%, magnesium 0.27%, sulphur 2.86%, sodium 1.39%, iron 0.95% and chromium 1.29%.

The chemical composition of solid wastes of leather industry reported by AICRP on Microbiological Decomposition and Recycling of Farm and City wastes (Bhopal) are as follow:

pH 9-10, Dry matter 46.0 to 67.9%, organic carbon 27.4 to 36.4%, Fats 27.5 g kg-1 and Cr13 2.8 to 11.0g kg-1.

These sludge contain high quantity of N, Ca, S, Fe and Cr. but have small quality of phosphorus and very little potassium. The occurrence of ammoniacal nitrogen more than 7% of the total nitrogen speaks for the decomposition processes of nitrogenous compounds during storage of sludges. The ratio of nitrogen to phosphorus and potassium indicates the need for supplementation of these sludges with phosphorus and potassium, especially in soils which are poor in these element.

The high content of sodium is due to use of this element in the preservation and tanning of the hides. Sodium in small amounts has a positive effect on plant growth. However, too high amounts of this element may destroy the soil structure or may lead to some other undesirable effect.

In such cases, liming of soil is necessary. However, tannery sludge contain more calcium than sodium and thus negative effect of sodium can be counteracted by high level of calcium. A desirable constituent of these sludges is sulphur which may show up the positive influence on the yield and quality of crops.

Chromium-in tannery sludges causes the maximum problem. It is reported that if chromium content in sludge does not exceed 3% it has no detrimental effect on crops. (Andrzejewski, 1970, Schreiber, 1963). The toxic effect of chromium ion on crops was found at high rates of application of the element, amounting to 1000 kg chromium per hectare (Liska and Kozik, 1963, Andrzejewski, 1971). Low rates of chromium showed beneficial influence on the crops (Moulinier and Mazover, 1968, Andrzejewski, 1970, 1971). Bertrand and wolf (1968) reported that chromium application to soils that were poor in this element increased potato yield by 42%. In tannery sludges chromium occurs as trivalent element, a form in which it is less toxic than it is in the hexavalent form. The alkali reaction of tannery sludges ensures low solubility of chromium (Blood, 1963, Moulinier and Mazover, 1968, Rosyzk 1968, 1969, Andrzejewski 1970,1971).

The rate of mineralization of tannery sludge was found slower as compared to farmyard manure. About 21 to 36% of the total organic matter of tannery sludge was decomposed during 6 month period in inclubation, studies. The rate of decomposition under field conditions is generally slower mainly depending on the soil moisture and temperature.

Effect of Tannery Sludge on Crop Yield

It was reported that the sludge in pot experiments increased the yield of barley, buck wheat, oats and sunflower. Raw sludge composted with calcium carbonate were studied. The sludge fermented with calcium carbonate had some what better effect on crops than raw sludge. Tannery sludges with high chromium content had a deleterious effect on plant growth (Szafranek *et al.* 1973). Application of tannery sludges at the rate of 7.5 tonnes/ha on dry basis increased the yield of wheat crop under field conditions. Farmyard manure gave some what better result than tannery sludges with third and fourth application. Potato tuber yields were increased by 17 to 22% due to sludge treatment. Liska and kozik (1963) reported 10 to 17% higher yields of potato due to use of tannery sludge as compared to moderate use of chemical fertilizers. The combined application of sludge and chemical fertilizers increased the potato tubers yield by 42%.

III. Effect on Soil Properties

Tannery sludges were found to increase the amount of humus content in soil but was less as compared o soil amended with FYM. Total nitrogen was also increased by 14 to 25%. Its application had no positive effect on phosphorus, potassium and magnesium contents of the soil. It is therefore, necessary to apply supplemental rates of commercial fertilizers so as to counteract soil mining of these plant nutrients.

The build up of chromium in soil due to tannery sludge application is a serious problem. However, only 1 to 4% of total chromium was found soluble in acetic acid. It gets solubilized due to soil processes and becomes available to crops in due course.

Tannery sludges may be used as organic fertilizer in agriculture production. This should be used on responsive crops such as potato, beet roots and forage crops.

References

Andrzejewski, M. 1970. Polish Agric Annu. A. 96 Pt. 2: 315-328.

Andrzejewski, M. 1971. Polish Agric Annu. A. 6: 75-97.

Bertrand, D and Wolf, A. 1968. Compte Rendu hebd. Séance. Akad Sci Paris, D-226:1494-95.

Blood, J. W. 1963. N.A.A.S. Quart Rev. 14: 97-100.

Liska, J. and Kozik 1963. Za Vysoku Urodu 9: 347-349.

Mazur, T. and Koc, J. 1976 a. Soil Sci. Ann. 27: 113-122.

Mazur, T. and Koc, J. 1976 b. Soil Sci. Ann. 27: 123-135.

Mazur, T. and Koc, J. 1976 c. Soil Sci. Ann. 27: 137-147

Moulinier, H. and Mazover R. 1968. Ann. Agron 19: 553-67.

Roszyk, E. 1968. Soil Sci. Ann. 19: 223-246.

Roszyk, E. 1969. Soil Sci. Ann. 20: 135-154.

Schreiber, H. 1963. Przeglad Skorzany 2: 27-32.

Szafranek Cz, Mazur, T. Koter, M, Koc J. and Czapla J. 1973. Zesz Nauk Art olszt. Rolnictwo. 3: 179-198.

8

Rock Phosphate Basic Slag and Fly Ash

I. Rock Phosphate

Rock phosphate is natural source of phosphorus. It occurs as natural deposits of rock in Morocco, U.S.A., Egypt, Poland, Russia, Tunisia, Algeria, Brazil, Nauru and some Islands in the Pacific and the Indian ocean. It contains 20 to 35 percent phosphoric acid (P_2O_5) but this phosphorus is in insoluble form. In India, some of rock phosphates occur in Udaipur and Jaisalmer districts (Rajasthan), Purulia (West Bengal), Singhbhum (Zharkhand) and Mussorie (Uttrakhand). Most of these phosphates are classified as low grade phosphate since its phosphoric acid content is less than 20-25%. The total reserve of low grade rock phosphate is estimated to be around 165 million tonnes of which nearly 70 million are reported to occur in Rajasthan. The other economically viable deposits are in Mussorie (Uttarakhand), Jhabua district (Madhya Pradesh), Singhbhum (Zharkhand) Kasipatnam (Andhra Pradesh) and Purulia district (West Bengal). Rock phosphate is the basic raw material for manufacture of superphosphate and other kind of phosphatic fertilizers. It is reported above that India has mostly low grade rock phosphate. Hence India has to spent huge amounts of foreign exchange to import high grade rock phosphate containing around 32-35%, P_2O_5 for manufacture of phosphatic fertilizers.

Mineralogical Composition

Of the Indian phosphate rock deposits analysed at the Bhabha Atomic Research Centre (BARC) Bombay (Bhujbal and Mistry 1981), the one found in Rajasthan (Udaipur) contained large amounts of quartz (7.07%) and considerable amounts (6.37%) of calcite and dolomite. The sample from West Bengal (Purulia) contained un-usually large amounts of clay minerals.

Madhya Pradesh (Meghanagar) phosphate rock samples showed presence of large amounts of quartz (24.52%) and clay minerals (10.57%) and some quantity (4.88%) of goethite. Sample from Uttar Pradesh (Mussoorie) showed in qualitative manner, presence of hematite, calcite and dolomite (all together around 30%), and a fair amounts of iron pyrites (about 5%) in a free form.

The presence of high percentage of iron pyrites crystals which occur occluded in the phosphate pellets, in the mineralogical suite of Mussoorie phosphate rock may be fortuitous.

If Mussoorie phosphate rock is used for direct application, the sulphuric acid release, from the weathering of iron pyrites, can make the phosphate of the rock more readily available to plants.

Chemical Composition of Mussoorie Phosphate Rock

Chemical composition of Mussoorie phosphate rock is given in Table 69. The whole of this mineral was reported to occur as francolite. The high percentage of sulphide sulphur (4.4 to 5.8%) has been attributed to the large amount of pyrites plus pyrrhotite, occurring as crystalline particles (grain size 10 to 50, µm) within the apatite matrix. The organic carbon content (1.14% to 1.40%) is significant.

Table 69: Chemical anaysis of Mussoorie Phosphate Rock

Constituents	Weight per cent
Phosphate (P_2O_5)	18-24
Carbonate (CO_2)	13.8
Calcium (CaO)	38.5
Magnesium (MgO)	5.6
Potash (K_2O)	0.25
Alumina (AI_2O_3)	0.83
Silica (SiO2)	6.60
Iron Pyrite (FeS_2)	4.41
Organic carbon (C)	1.14
Copper	10-50ppm
Zinc	100-150 ppm
Molybdenum	5-7 ppm
Boron	10-50 ppm

The total P_2O_5 content of a phosphate rock (designated on rock basis) is of little consequence, if the rock is meant for direct application. For, it is not the total P_2O_3 content, but the degree of its substitution by CO_3 (or CO_3: P_2O_5 ratio) in the apatite lattice composition (as will be shown hereinafter) which determines its reactivity in soil.

Another unique feature of Mussoorie phosphate rock not found in other rock, is its organic carbon content (1.14 to 1.4%). This unusual feature is considered highly significant in so far as its use for direct application to soil is concerned. The presence of organic matter in Mussoorie rock is ascribed to geological settings. These can be of: (a) igneous, (b) metamorphic, or (c) sedimentary geologic origin. Apatites in deposits from igneous and metamorphic sources are quite stable and unreactive, as they are coarsely crystalline substances and have not been used for direct application purposes. To make their phosphorus available for use in agriculture they need acid or high temperature treatment. But the sedimentary rock deposits, which are submicro-crystalline, are of major commercial importance for direct application. They are not so unreactive or inert, as those from igneous or metamorphic sources, and comprise younger apatites, formed relatively in more recent geologic era mostly in marine environments. Such phosphate rocks of sedimentary origin are collectively known by the name 'Phosphorite'.

It is recognized that phosphate rocks of sedimentary origin (un-like those from igneous or metamorphic origin) have a wide range of useful characters, making them as a good source of phosphorus for direct application. Some of these characters are :

- They are micro-crystalline or crypto-crystalline in nature;
- Have a fairly open loosely consolidated aggregates of micro-crystals; and
- Have large internal surface area for reactivity.

The known Indian deposits like Kasi-patnam (Andhra Pradesh), Purulia (West Bengal), Singhbhum (Zharkhand) are of igneous origin, and Jharmarkotra, Maton (Rajasthan) and Jhabua (Madhya Pradesh) are of metamorphic-cum-sedimentary nature (Bhujbal and Mistry, 1981). Mussoorie deposits are exclusively of sedimentary origin and should, therefore have relatively higher reactivity amongst the known Indian deposits.

The marine phosphate cycle postulates that an enormous mass of phosphorus rich sea life (fishes etc.) dies naturally and sinks to the sea floor. The sea bottom mud thus contains, besides phosphorus, organic residues which gradually decompose. Phosphate rock deposits of sedimentary marine origin, containing organic mater, thus indicate a rock of comparatively much younger geological age with 'soft' and porous internal texture. Mussoorie rock has accordingly an inherent character of easy decomposability in soil. This is in sharp contrast to the rock from igneous and metamorphic origins, where the primeval molten magma or high temperature and pressure, render the rock 'hard' for dissolution in the soil.

The Apatite Group of Minerals

In phosphate deposits of economic or commercial interest, phosphorus is found associated with calcium, as tricalcium phosphate. Tricalcium phosphate, or its hydrated forms namely hydroxyapatite, are sparingly soluble in water and fairly stable. But nature, in its benevolence to preserve these deposits, through the ages, for the use of posterity, has made it still more resistant to weathering by combining it with calcium fluoride. Three molecules of tricalcium phosphate, are associated with one molecule of calcium fluoride to produce in nature, a mineral called fluorapatite 3Ca3(PO_4)2 CaF2 or C10(PO_4)6F2. This belongs to the mineral family called the Apatite Group.

Characterization of Indian Phosphate Rocks

Following the technique of appraising phosphate rock reactivity evolved at the IFDC (USA) Lehr and Mc Clellan (1972), scientists of the Biology and agriculture Division of the Bhabha Atomic Research Centre of the Indian Atomic Energy Commission Bhujbal and Mistry (1981), characterized some samples of major phosphate rocks found in India with a view to evaluate their relative value as phosphate fertilizer for direct application. The main conclusions drawn are:

- Andhra Pradesh rock samples indicated a pure form of fluorapatite-chlorapatite, with no carbonate substitution.
- Phosphate rocks from Rajasthan, West Bengal, and Madhya Pradesh have an apatite having chemical composition essentially of fluorapatite type with negligible degree of carbonate substitution.
- Mussoorie rock phosphate samples indicated that the type of apatite in it is francolite with relatively largest degree of carbonate substitution amongst the Indian samples analysed.

Table 70: Characterization of principal Indian phosphate rocks

Phosphate rocks	**Unit-cell dimension a-axis°**	**Moles of * $P0_4$**	**co_3**	**Mole + + ratio CO_3: PO_4****
Mussoorie (UP)	9.363	5.69	0.30	0.0527
Purulia (West Bengal)	9.371	5.92	0.08	0.0135
Jhabua (Madhya Pradesh)	9.373	5.98	0.02	0.0033
Jhanmar Kotra (Rajasthan)	9.378	5.99	0.01	0.0017
Kasipatnam (AP	9.435	6.00	Nil	Nil

Data calculated from BARC's Report No. 1114 (1981)

*From Table of the above report

**Calculated from moles of phosphate and carbonate in the empirical formula of the rocks.

The unit-cell dimensions of these phosphate rock samples, the molar values of phosphate and carbonate, the molar ratio of CO: PO, and the reactivity rating of typical rock samples, arranged in decreasing order of their reactivity are shown in Table 70.

On the basis of the CO_3: PO_4 mole ratio as the index of reactivity, the characterization studies made by the BARC suggest that the theoretical relative inherent reactivity of Mussoorie-phos is about (i) four times as much as that of the Purulia (West Bengal) sample and (ii) Sixteen times as much as that of the Jhabua (Madhya Pradesh) samples. Rajasthan (Jhamar Kotra) and Andhra Pradesh (Kasipatnam) have practically no reactivity.

It has been estimated that acid soils an India comprise about 33 per cent of the total area of the country and those with pH less than 6.0 occupy 20 per cent of total cultivated area. Acid soils in Himalayan region extend from Kashmir in the northwest to Arunachal Pradesh in the northeast. In Peninsular India, acid soils are found in almost all states except the western most states where rainfall in low. The states where acid soils occur predominantly are Jammu and Kashmir, Assam and other north-eastern states, West Bengal, Bihar, Orissa, Andhra Pradesh, Madhya Pradesh, Karnataka, Maharashtra, Tamil Nadu and Kerala.

Uses in Agriculture

Rock phosphate is recommended for direct application in acidic soils but only small amount of rock phosphate has been used as a direct fertilizer because most of our soils are near neutral to alkaline in reaction. In other countries pulverised rock phosphate has been found to give satisfactory results where soils are acidic and deficient in phosphorus. Even in these countries very little rock phosphate is used directly as fertilizer. Mostly rock phosphate is converted to superphosphate by treating ground phosphate rock with almost an equal quantity by weight of sulphuric acid. This treatment produces a brownish-grey mixture containing monocalcium phosphate and calcium sulphate (gypsum) in practically equal quantities which provides water soluble phosphoric acid, which can be taken up by plants readily.

The use of rock phosphate directly as fertilizer is feasible even in neutral to alkaline soils if rock phosphate solubilizing microorganisms are used to treat seeds or seedlings before sowing crops. Beneficial effects of rock phosphate application to soil alongwith phosphate solubilizing microorganisms as inoculants has been reported to increase the yield of several crops like, paddy, wheat, soybean, cowpea, lentil, chickpea and potato (Gaur, 1990)

The use of rock phosphate in composting of farm and city wastes has also been demonstrated in preparation of phospho-compost or organo- phos. Rock phosphate is converted mostly to citric acid and water soluble phosphate during the bioprocessing of organic wastes into compost (Gaur, 2001). Most of the rock phosphate deposits in India are low grade which can not be utilized for manufacture of phosphatic fertilizers. The strategy to employ phosphate solubilizing microorganisms and/or composting with rock phosphate would be rewarding in utilizing the natural source of efficient p-solubilizing microorganisms and indigenous phosphate deposits for increasing crop productivity and saving foreign exchange by reducing import of high grade rock phosphate from other countries to meet our phosphorus requirement of crops.

II Basic Slag

This is a by product of steel factories. Depending on the phosphorus content of iron ore, it contains 6 to 20 percent of P_2O_5. Slag is produced when the impurities of iron ores combine with lime in the converter at a certain temperature, being lighter than iron, rise to the surface of molten mass and are poured off. On cooling, this material appears as a hard, massive black cinder and is called Bassemet basic slag (Collings 1955). Basic slag is a double silicate and phosphate of lime [(Cao)5 P_2O_5 SiO_2] but some chemists consider it a tetracalcium phosphate (Ca_4 P_2O_9) or $(Cao)_4$ P_2O_5 while a few others consider it to be basic siliconoxy apatite.

Slag is generated @ 200-250 kg/ tonne of steel produced. Indian iron ores contain very little phosphorus and thus produce better quality steel. Indian slags (Steel Authority of India) contain small quantity of phosphorus than p-rich slags produced by European industry. The phosphorus content of SAIL basic slags varies from 1 to 4 percent and it is known as low grade basic slag. The high-grade European slag containing 15 to 18 percent P_2O_5 and is a popular phosphatic fertilizer in the central Europe.

It is estimated that over 10 million tonnes of solid wastes (fly ash, sludges, flue dusts and waste refactories) are generated every year by SAIL plants at Bokaro, Bumpur, Durgapura and Rourkela. Out of this about 8 m.t. are reported to be dumped (Mathur, 1995). At present, the use of basic slag as a fertilizer in agriculture is negligible because it is not marketed in ground form.

Use of Basic Slag as a Fertilizer

Basic slag is the cheapest source of phosphorus available to the farmers of Germany, France and England. The consumption of basic slag has gone up at present due to expansion of steel industry.

The American Official Association of Agricultural Chemists (AOAC) have characterised finely powdered slags for fertilizer use which contain 12% P_2O_5

(5.2% P) out of which at least 80% shall be 2% citric acid soluble and 70% will pass through 100 mesh sieve and 50% through a 50 mesh sieve (Collings, 1955). The basic slags which do not fulfil these specifications are termed as low grade basic slag. Basic slag is heavy dark brown in colour powder and solidifies under moist conditions. It is alkaline due to its lime content.

Basic slag can be used as a fertilizer cum-soil conditioner because it contains lime (CaO- 30-40%) and citric acid soluble phosphorus. Pant and Pathak (1975) increased the water soluble phosphorus content to 90 percent by mixing and grinding steel slag with urea in 1:2 ratio. The water soluble P when added to acid soils which are rich in Fe and Al, it gets chemically fixed in soil and P-use efficiency is very much reduced. The phosphorus of basic slage is rendered quickly available by carbon dioxide as follow:

$$(CaO)\ 5P_2O_5SIO2+8CO_2+6H_2O=CaH_4(PO_4)_2+4CaH_2(C0_3)_2+SIO_2$$

Basic slag can be composted with organic materials which would convert the insoluble phosphate into soluble form due to production of organic acids and carbon dioxide. Crop response to basic slag is usually greater than ground rock phosphate because of its lime content. The active calcium content of basic slag is as active and efficient as the calcium in calcium carbonate. Basic slag also contains varying amount of Mg, Mn and micronutrients. The superiority of basic slag over water soluble sources of phosphorus such as SS P or DAP decreases as soil reaction changes from acid to alkaline.

Indian basic slag is poor in phosphorus but contains high quantities of lime and micronutrients. Basic slag can benefit soil productivity because it contains on an average 35-45% Cao which is as efficient as $CaCO_3$ in neutralising acidity. Since over 98% district of India are low *or* medium in available form of soil phosphorus and *use* of some amount of phosphorus is essential for sustainable yield, its role in supplying phosphorus may also be considered. The beneficial effect of basic slag on the yield of different crops in different soils and agroclimatic conditions conducted by Indian scientists over 4 decades has been summarised by Mathur(1995). Mandal and Jha (1970) prepared compost with fresh dung, straw with and without basic slag and Singhbhum rock phosphates equivalent to 12 kg P_2O_5/ha and tested the effect on maize corp. Basic slag compost (9t/ha) was superior to rock phosphate compost in context of high nitrogen and phosphorus uptake of maize crop but the maize yield was statistically non-significant. In red sandy loam soil (pH 4.8) from Bangalore, pot studies revealed that basic slag was as good as SSP, rock-P and dicalcium phosphate in increasing the yield of finger millet at 60 kg P_2O_5/ha (Varadan *et al.* 1977).

Acid Soils in India

In India, 30% of land is covered by acid soils. Out of this 6.5 million hectares are strongly acidic ($pH<5.0$) 29.5 m.ha. have pH between 5.0 to 5.7 and 23.07 m. hectares have pH between 5.6 and 6.5. Liming of these soils is a routine management practice. Basic slag differs from other liming materials because it is a silicate of calcium and magnesium and not as oxides or carbonate. But it has been reported that basic slag is comparable to limestone on CaCO, equivalent basis. Basic slag has a liming value of 76 to 86%. In addition it is supplementary source of phosphorus. However, basic slag has not been used by farmers in India because of its non availability in ground form.

Several studies including field experiments conducted during the last 30-40 years in different agro-climatic condition of India indicate a good potential for using basic slag in acid soils. It could not find a place specifically as a phosphalic fertilizer because of its very low P content (2-4%). However, when it is used at normal liming rates, it can add substantial amount of P to the soil. The beneficial effect of basic slag on crop yields are comparable when used alone or in combination with organic matter.

III. Pyrites and Sulphur

Pyrite (FeS_2) deposits at Amjhore in Bihar are estimated at 350 million tonnes of 40% S and 1500 million tonnes of 10% S and occur in a thickness of about 0.6 m under shale deposits. The mining of sedimentary pyrites in Bihar and Rajasthan is done for acid production but large amounts of low grade materials are available for use as soil amendments. Owing to advantage of location, pyrites are now favourably competing with gypsum for the reclamation of alkali soils in Uttar Pradesh and Bihar. Sulphur in the pyrites is present in a highly reduced state. Upon exposure to atmosphere, it readily accepts oxygen and autooxidises to S compounds of higher oxidation state. The rate of reaction depends upon moisture content, oxygen availability, temperature, particle size, and accumulation of toxic materials. Biological oxidation of pyrites is largely carried out by *Thiobacillus ferrooxidans* but only at pH less than 4.0 which is not the case in alkali soils.

The oxidation of pyrites starts as soon as the material is brought out of the mine into open. The following reaction occurs in the presence of water and both the products directly or indirectly react with the soil.

$$2FeS_2+50_2+2H_2O=2FeSO_4+2H_2SO_2$$

According to the above reaction immediate effectiveness of amendment would depend upon the date of mining and the storage conditions. Many of the earlier conflicting reports on the effectiveness of pyrites vis-a vis gypsum emanated from the fact that the comparisons were done on total S basis and not on soluble S basis. The efficiency of pyrites is enhanced considerably with increase in its soluble S basis content. Singh *et al.* (1980) discussed oxidation problems, effect of particle size, dose, and method of application of pyrites in the reclamation of calcareous alkali soils. The acids produced upon oxidation of pyrites react with $CaCO_3$ of the soil to release Ca_2+ for exchange with Na on the clay. Due to these intermediary steps and problems in complete oxidation of the amendment, the efficiency of pyrites is between 70% and 80% of that of gypsum at the recommended dose of application. Further researches are needed to increase oxidation of the field applied material, and on storage of the amendment in the godown and farmer's fields without losing its soluble S content. Official apathy towards this amendment has so far prevented meaningful researches on some vital aspect of use of pyrites in agriculture. *Sulphur*: Sulphur is biologically oxidized in the soil to form sulphuric acid which reacts with $CaCO_3$ of the soil.

$$2S+30_2=2SO_3$$

$$SO_3+H_2O=H_2SO_4$$

As with pyrites, oxidation of S in alkali soils is likely to be very slow. Basu and Tagore (1943) reported the highest yield of sugarcane on a black clay-alkali soil by applying a combination of sulphur and F.Y.M. Gidnavar *et al.* (1972) observed that during a period of eight years annual application of sulphur at 300 kg ha' together with 5 cartloads of F.Y.M. reduced pH and salt content and increased infiltration rate in saline-alkali soil, where rainfed cotton and wheat were grown in alternate years. Sulphur is costly and is imported for production of sulphuric acid. Its efficiency and cost prohibits its use as amendment in alkali soils.

IV. Fly Ash

Fly ash, an amorphous ferro-alumino silicate is a waste by product of power stations and is one of the serious pollutants causing environmental hazards and disposal problem. One hundred thousand tonnes of fly ash is thrown out every day by thermal power stations in India which is expected to go up by 3 to 4 times with expansion of coal based power generating plants. India is at present annually producing 10 million tonnes of fly ash from thermal plants.

A large quantity of pulverised coal is used in the process for raising steam in the boilers resulting in fine powder of fly ash. It is then taken away from the boilers by gases and is extracted out with the help of multicyclone collectors or electrostatic precipitators or a combination of both. The ash is mixed with water and the slurry is piped into ash dykes. The ash settles down to a great extent but huge quantities of water containing dissolved and suspended impurities are discharged into nearby water bodies mostly rivers. The main drawbacks of this conventionally used methods are:

a) Requirement of large land area for disposal and storage
b) Degraded land area after the ash pond attains full capacity.
c) Air pollution problems due to dry ash getting air borne in dry months.
d) Breach in earthen dams, resulting in the total impairement of water quality in the receiving stream and loss of fish and aquatic life etc.

Use of Fly Ash as Building Material

The effective and economic disposal and reuse of fly ash has attracted attention of scientists and technocrats. The conversion of natural resources and reuse of waste products is of economic importance and the need of the day.

Extensive experiments and studies at Roorkee indicate that waste generated in the form of fly ash can be successfully utilised for manufacture of Portland Pozzolana cement, clay-bound fly ash bricks and lime-fly ash bricks, light weight rice husk-fly ash bricks. Fly ash can be used for manufacture of pre-cast building components and floor cleaning powders. It can be used in landscaping and in filling of road, rail and canal embakments. Besides, valuable elements such as vanadium, gallinium, cadmium, uranium etc can be extracted out of fly ash.

Use of Fly Ash in Agriculture

Fly ash is composed of relatively large quantities of components of silicon, aluminum, iron, calcium and smaller amounts of compounds of magenesium, titanium, sodium and potassium and traces of several other compounds. The pH of the fly ash varies from basic to acidic depending upon the source of coal (Table 71). Fly ash with acidic reaction can be used for soil amendment.

An article "coal ash as a boon for chinese agriculture" was published in daily M.P. chronicle, Bhopal on November 1, 1989 by Lu Guoyuan which showed the beneficial effect of the use of fly ash on agricultural crops from experiments conducted in China (Table 72).

Table 71: Ranges of elemental concentration in coal and flyash (ppm)

Element	Coal	Fly ash
Antimony	0.05- 8.9	1-15
Arsenic	0.5-93	2-500
Barium	20-1600	500-7000
Beryllium	0.2- 4.0	1-10

Boron	1.0 -216	—
Cadium	0.01-300	0.1-50
Chromium	2.0-400	50-300
Cobaft	1.0-34	5-100
Copper	3.0-180	50-650
Lead	1.0-218	5-1000
Manganese	5.0-240	50-500
Mercury	0.02-1.91	0.02-0.4
Molybdenum	1.0-30	5-40
Nickel	3.0-90	5-100
Selenium	0.04-7.7	1-20
Telluriumh	0.02-3.0	2-30
Thallium	0.2-36	15-70
Tin	0.1-51	30-30
Uranium	0.9-3.7	5-20
Vanadium	2.0-78	100-500
Zinc	1.0-1444	50-5000

Table 72: Increase in yield of crops in China due to application of fly ash

Crops	**Increase in yield (kg ha-^{1}) over control**
Wheat	4500
Soybean	1875
Com	3000
Sweet potatoes	37500

Encouraged by the report, M.P. Pollution Control Board, Bhopal undertook a pilot project study using mixture of normal soil with fly ash.

Their experiments showed that grain and straw yield of wheat crop was maximum when 50 percent fly ash was applied to soil.

The fly ash contains a range of elements (Table 71, 73) which has been reported by several workers (Adriano *et al.*, 1980 Chatterjee and Rattan 1988, Uni 1984-85).

There may be considerable variation in the composition of fly ash from different power plants and in different samples of fly ash from the same plant (El-Mogazi *et al.*, 1988). Therefore different samples of fly ashes are sometimes difficult to compare. The literature in context of their effect on growth of plants in soil covered with fly ash is insufficient. Nearly all the investigations have been conducted in soil-fly ash mixtures rather than where soil cap is involved. Moreover, the soil properties may vary considerably from one soil to another hence comparison between them are difficult to make. The properties of fly ash depends on the composition of parent coal conditions and other factors (Adriano *et al.*, 1980).

Maiti *et al.* (1990) evaluated the samples of fly ash obtained from three different power stations of West Bengal whose mechanical and chemical characteristics of the fly ash are presented in tables 73 and 74.

Table 73: Some physical characteristics of the fly ash.

	Location (IPS)*		
	Bandel	**Kolghat**	**Santaldih**
No. of samples Mechanical composition	10	10	10
Sand (%)	36-37	33-38	76-88
Silt (%)	25-42	54-56	6-8
Clay (%)	22-38	8-12	6-16
Bulk density (mg m^3)	0.81-0.86	0.93-0.97	1.15-1.16
Water holding capacity (kg kg-^{1})	72-80	49-66	56-58
Pore space (%)	54-56	50-52	51-53

*.TPS: Thermal power stations

The pH of the materials ranged from 7.8 to 8.6 (Table 74) indicating that the samples may be used to reclaim acid soils. Electrolyte conductivity of the fly ash is low and is not likely to pose problem of soil salinity.

Table 74: Physicochemical characteristics of the fly ash.

Characteristics	**Bandel**	**Kolaghat**	**Santaldih**
PH	7.9-8.6	8.1-8.3	7.8-8.3
EC (dsm-1)	0.026-0.158	0.144-0.151	0.013-0.025
Organic C (%)	1.53-1.68	1.23-1.45	0.66-0.75
Total N (%)	0.07-0.08	0.08-0.09	0.10-0.11
Total P (%)	0.22-0.23	0.28-0.29	0.30-0.32
Total K (%)	1.18-1.19	1.15-1.29	1.15-1.32
Avail N (ppm)	20.0-32.8	17.0-21.0	12.0-14.0
Avail P (ppm)	48.0-52.0	57.0-60.0	33.0-43.0
Avail K (ppm)	72.0-104.0	88.0-92.0	92.0-96.0
Exch. cations [c mol (P^+) kg-^{1}]	2.68-3.80	4.42-4.60	2.65-2.92
Exch. Ca^{+2}+ -do-	1.18-1.19	1.30-1.41	0.96-1 .00
Exch. Mg^{+2}+ -do-	0.62-1.65	1.52-1.58	1.06-1.52

Exch. K^+ -do-	0.21-0.32	0.21-0.23	0.23-0.29
Exch. Na^+ -do-	0.56-0.62	1.21-1.38	0.19-0.28
DTPA-extractable Cu (ppm)	1.48-1.61	1.09-1.74	1 .09-1 .29
Fe (ppm)	13.82-14.66	16.10-17.14	19.52-21.04
Zn (ppm)	1.96-2.12	2.36-2.64	2.46-2.72
Mn (ppm)	1.48-1.60	1.00-1.14	1.22-1.52
Water soluble S (ppm)	50.0-125.0	129.8-182.3	16.3-17.5
NaCI extrac. S (ppm)	52.0-70.0	105.0-109.3	33.7-35.0
Ca Cl^2 extrac. S (ppm)	55.0-137.5	146.8-195.7	13.7-26.2
Organic S (ppm)	52.6-59.1	47.0-49.2	27.5-32.1
Total Fe_2O_3 (%)	12.5-14.4	8.7-9.1	12.6-14.3
Total CaO (%)	3.41-3.43	1.18-1.92	2.03-2.36
Total MgO (%)	2.01-2.06	0.74-1.17	0.82-1 .22
Total SiO_2 (%)	58.0-60.0	68.0-77.1	47.1-49.2
Total K_2O (%)	1.48-1.50	1.39-1.56	1 .39-1 .59
Total AI_2O_3 (%)	1.82-1.92	2.26-2.45	2.94-3.36

Particle size distribution of different fly ash samples showed wide variations. The sample from Bandel consisted of finer fractions whereas in other cases the percentage of finer fractions was less. Kolaghat and Santaldih samples consisted of particles having diameter greater than 20 μm. Bulk density of fly ash from Bandel was higher than of Kolaghat and Santaldih samples. The water holding capacity of fly ash was very high. The higher values obtained at Bandel may be due to presence of higher amount of clay particles. The pH of the ash ranged from 7.8 to 8.6 (Table 74) indicating that the samples may be used for reclamation of acid soils. Electrolyte conductivity of fly ash is low and is not likely to pose a problem of soil salinity when applied to soil. Cation exchange capacity of the fly ash samples is generally low, mg^{+2} is the dominant cation in the exchange complex followed by Ca^{+2}. As expected organic carbon, total and available nitrogen are low. In all samples available phosphorus is quite high (72 to 104 ppm). The content of total phosphorus is very high ranging from 0.22 to 0.32%

The available content of micronutrients in ash samples is quite low and may be ascribed to higher pH of the ash samples. Water soluble sulphur iş found to vary among the samples. The Kolaghat ash sample contains appreciable amount of sulphur which may be used in sulphur deficient soils.

Total Fe_2O_3 ranges from 8.7 to 14.4 percent among the samples. Cao and Mgo contents vary from 2.03 to 2.36 percent and 0.82 to 1.22 percent respectively in Santaldih but the variations are somewhat lesser at Kolaghat and higher at Bandel. The silica (SiO_2) contents in the fly ash sample are quite high.

From he foregoing discussion, it is clear that the use of fly ash may be useful in acidic soils. In light textured soils, the application of fly ash to soils would boost the water holding capacity. Over and above fly ash seems to be a good source of phosphorus and potassium for plants.

Fly ash has a vast potential for its use in agriculture as an amendment to improve the physical condition and plant nutrient status of soils. The agronomic experiments have shown that by the use of fly ash the yield of crops can be increased by 15 to 25 percent over control. Even higher increases have been reported in some cases. However, the presence of heavy metals in fly ash has been a cause of concern to many users. This negative effect of fly ash may be looked into and investigated where large quantities are planned to be used for agricultural crops. The findings of work done by projects commissioned under Fly ash mission (DST) on different crops under different agroclimatic conditions including negative effects if any, on soils, plant and microbes have been reported as follows:

Benefits of Use of Fly Ash

1. **Soil Texture:** Addition of 70 tonnes ha1 of fly ash has been reported to modify the texture of sandy and clayey soils to loam (Fail and Wochock, 1977 and Capp, 1978). The changes in soil texture are expected both in agricultural and strip mined soils. It was reported that the application of fly ash changed the texture of black cotton soil from clay to silty clay and sandy to silty soil (RRL, Bhopal).
2. **Modification of bulk density:** The grain size distribution especially the silt size range of fly ash affects the bulk density of soil. Chang et *al.* (1977) observed that among five soil samples (Reyes silty clay) showed increase in bulk density from 0.89 to 1.01 when corresponding rates of fly ash increased from 0 to 100%. But soils with bulk densities varying between 1.25 to 1.60, a marked decrease in bulk density was observed by addition of fly ash. Page (1979,1980) also reported a decrease in soils' bulk density due to amendment of soil with fly ash. Optimum bulk density improves the soil porosity, root penetration and moisture retention capacity of soils.
3. **Water holding capacity:** The application of fly ash improved the available water content of a sandy loam soil by 120 percent and of a sandy soil by 67 percent (PAU, Ludhiana). RRL, Bhopal reported that amendment of soil with fly ash increased the porosity of black cotton

soil and decreased the porosity of sandy soil and thereby saved irrigation water by about 26 and 30 percent respectively. Chang *et al.* (1977) reported that addition of 8 percent fly ash increased the water holding capacity of soils.

4. **Soil pH:** In India most of the fly ash is alkaline in nature. This property can be used to ameliorate acidic soils (Elseewi *et al.* 1978, Phung *et al.* 1978). Page *et al.* (1979) observed that experiments with calcareous and acidic soils revealed that fly ash addition enhanced the pH of the calcareous soil from 8 to 10.8 and that of acid from pH 5.4 to pH 9.9. The use of excessive quantities of fly ash to alter soil pH can increase the soil salinity especially with unweathered fly ash (Sharma *et al.* 1989). Some fly ashes are acidic in reaction which may be used for reclamation of alkaline soils. In one of the project sites of fly ash mission at Phulpur, Allahabad, IFFCO has done some work on the reclamation of alkaline soils and observed that pH of these soils can be ameliorated to near neutrality using acidic fly ash.
5. **Effect on Crust:** Fly ash application helps in reducing surface encrustation which is a problem in red soils (RRS, Raichur). This effect in turn can improve soil aleration and help in increasing the germination percentage of seeds.
6. **Effect on crop yields:** The application of fly ash on agricultural land has been widely evaluated. The crop response may vary from beneficial to toxic depending on the concentration of various elements in it (Kalra *et al.*, 1997, 2000, Grewal *et al.*, 2001). Studies have shown that fly ash application may improve K, Ca, Mg and S status of deficient soil (Srivastava and Chhonkar, 2000, Deshmukh *et al.*, 2000). In order to improve the N and P status, the use of bioinoculants can be explored whether these organisms will be able to perform well in presence of fly ash. The positive effect of fly ash incorporation in soil has been reported by several workers.

1. Regional research laboratory (RRL) reported that on an average an increase in the yield of about 50-60% in brinjal, around 45% in potatoes and peas, around 40% in tomato and 29% in cabbage due to application of fly ash over control.
2. Punjab Agriculture University reported that application of 10 tonnes fly ash ha1 increased the yield of wheat and cotton by 12.1 and 15.9 percent respectively.
3. College of agriculture (Raichur) registered an increase in the yield of groundnut by 32.4 percent due to application of fly ash @ 20 tonnes ha^{-1}.

4. Under fly ash mission project, at Farakka location the wheat yield was increased by 40 percent with the application of fly ash @ 200 tonnes ha^{-1}.
5. At Dhanbad, the yield of paddy was increased by 35 per cent due to application of 200 tonnes of fly ash ha1. Paddy and wheat crops grown with fly ash showed tendency for their early maturity.
6. At IIT, Kharagpur, the yield of groundnut increased significantly showing the residual effect of fly ash applied to the preceding rice crop.
7. At Bakreshwar, the farmers have observed that crops grown in fly ash treated plots were resistant to pest attack as compared to untreated control plots.

Effect of Fly Ash on Dynamics of Soil Microbes

Investigations were carried out at IARI (Gaind and Gaur 2002, 2004)on the effect of fly ash obtained from NTPC, Badarpur near, Delhi. The fly ash was mixed with unsterilized sandy loam alluvial soil at the rate of 10, 25, 50, 75 and 100%. Soil and fly ash alone were kept as necessary checks. The experiment was conducted in plastic pots and the soil was kept moist in order to get a sample which was neither too sticky nor too dry. All the treatments were triplicated. The plastic pots were covered with polythene sheet to check any loss of moisture and incubated at 32+1°C for 45 days. The enumerations of bacteria, fungi, phosphate solubilizing microbes (both bacteria & fungi) and *Azotobacter* were carried at periodic intervals of 15 and 45 days by dilution plate method using nutrient agar, Czepek's dox agar, Pikovskaya (1948) and Jenson's (1952) media respectively.

Both soil and fly ash had an alkaline pH 8.0 (Table 75). The water holding capacity of fly ash (74.2%) was about two folds that of the soil (35.6%) and also more than their combination (1:1). Fly ash contained more nitrogen (0.20%) and less available phosphorus (11.03 ppm) than soil. The carbon content of fly ash (0.29%) did not differ much from soil and soil-fly ash combination (0.27%, 0.28% respectively).

Table 75: Physico-chemical properties of fly ash.

Treatment	pH	WHC (%)	EC (mMhos/ cm)	Carbon (%)	Nitrogen	Available P_2O_5 (ppm)
Fly ash	7.80	74.20	0.36	0.29	0.200	11.0
Soil	7.98	35.60	0.20	0.27	0.035	20.0
Soil +fly ash (50:50)	7.93	63.41	0.29	0.28	0.117	15.0

Effect of Fly Ash on Soil Microflora

Fly ash *per se* was found to be devoid of any bacterial population but a few fungal colonies were observed in the unsterilised material (Table 76). The population of native soil bacteria, fungi, PSB/PSF and *Azotobacter* was augmented when fly ash was mixed with soil in various proportions. The effect was most pronounced on fungal population which recorded an increase of 120, 127, 142, 124 and 65 times with 10, 25, 50, 75 and 100% fly ash respectively on 45th day (Table. 76). The population increase was least for *Azotobacter* that may be due to their low numbers in soil and other environmental conditions. None of the fly ash-soil combination showed adverse on the numbers of soil microflora. The use of fly ash showed beneficial effect on the soil microflora tested which included bacteria, fungi, nitrogen fixing and phosphate solubilizing microorganisms. However, dissimilar results were obtained by Pichtel and Heyes (1990) who reported a decrease in the total count of bacteria, fungi and actinomycetes when fly ash was mixed with soil. Fly ash used by them had a highly alkaline pH of 12.2 whereas, fly ash used in the present studies had a pH of 8.0. The pH of fly ash is influenced by concentrations of different ions in equilibrium extract. The compatibility of heterotrophic microflora with fly ash may be attributed to the favourable physiological environment such as slightly alkaline pH, better WHC etc. The presence of some fungal colonies in the unsterilized fly ash may be due to the air borne contamination.

Table 76: Effect of fly ash** on soil* microbial population (104) gr1 soil

Type of Microorganisms	Soil		Soil: fly ash mixture					
			4:1		4:3		1:1	
	Days of incubation		Days of incubation		Days of incubation		Days of incubation	
	15	45	15	45	15	45	15	45
Bacteria	113	345	117	373	121	355	14.3	372
Fungi	31	40	52	278	62	360	5.8	380
PSB	04	16	06	28	08	31	0.7	16
PSF	05	10	06	18	12	29	1.8	12
Azotobacter	02	06	03	09	05	10	0.2	9

*Unsterilized soil

**Fly ash (bacteria not detected, fungi 17x101g1)

PSB - phosphate solubilizing bacteria

PSF - phosphate solubilizing fungi

Fly Ash as a Carrier for Preparation of Bacterial Inoculants

The average viable count of different P-solubilizing and N2 fixing cultures viz. P. *striata,* B. *circulans* and A. *chroococcum* and A. *brasilense* added/ml was 7.9, 8.6, 16.8 and 24.8 x 10^4 respectively (Gaind and Gaur, 2004). The population of *P. striata* increased 100 times compared to 10 fold increase recorded by other three test culture after 21 days of incubation (Table 77). The population of the introduced bacteria exhibited an increasing trend up to 75 days followed by a steady decrease at 120 to 210 days of incubation period and the most affected organism was B. *circulans*. Soil- fly ash (1:1) mixture was the best carrier for the growth of *P. striata* as compared to soil and fly ash alone (Plates 10 & 11). However, A. *brasilense* and A. *chroococcum* and *B. circulans* proliferated best in fly ash alone (Plates 12-14).

Table 77: Enumeration of inoculated bacterial population (g-1) in sterilized soil, fly ash and soil-fly ash (1:1) mixture.

Treatment	Bacteria	Incubation period (days)			
		21 d ($x10^5$)	**75d ($x10^5$)**	**120d (x104)**	**210d ($x10^2$)**
Soil	*Pseudomonas striata*	80.2	46	52	58
Fly ash	*Pseudomonas striata*	75.4	47	48	50
Soil + fly ash (1:1)	*Pseudomonas striata*	67	97	120	78
So3	*Bacillus circulans*	02	10	1.25	-
Fly ash	*Bacillus circulars*	07	86	1.20	-
Soil + fly ash	*Bacillus circulans*	12	40	0.75	-
Soil	*Azospirillum brasilense*	64	200	168	158
Fly ash	*Azospirillum brasilense*	30	283	215	190
Soil+fly ash	*Azospirilium brasilense*	28	54	110	112
Soil	*Azotobacferdvoococcum*	10.9	80	56	76
Fly ash	*Azotobacterchroococcum*	6.3	104	132	116
Soil + fly ash	*Azotobacterchroococcum*	15.3	64	42	53

The criteria for selecting a carrier is based on its being rich in organic matter content, mineral nutrients, high WHC, neutral pH and inexpensiveness and easy availability. The fly ash used as a substrate in the present experiment though fulfilled some of the requirements (rich in nitrogen, better WHC) but was poor in organic carbon. Fly ash per se though favoured the growth of both nitrogen fixers (A. *chroococcum* and A. *brasilense*) and PSB *(B. circulans)* but their shelf life was only up to two and a half months and a downward trend in population was recorded.

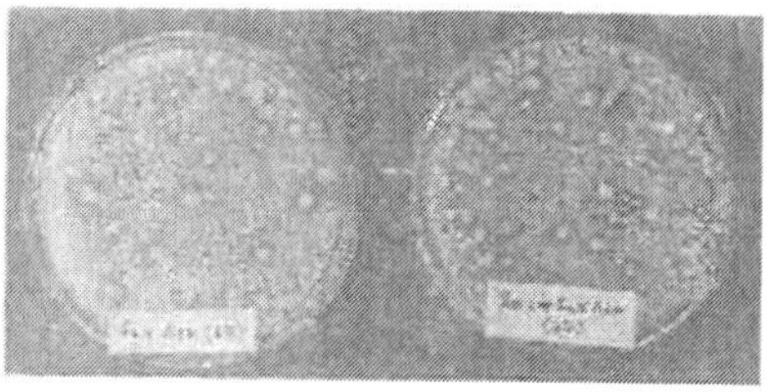

Plate 10: Number of *Psuedomonas striata* (strain 27) in soil-fly ash mixture compared with Fly ash alone

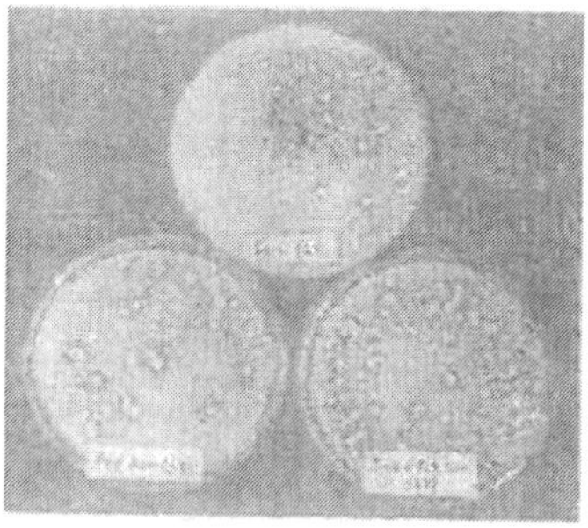

Plate 11: Colonies of *Pseudomonas striate* (strain 27) in different carriers

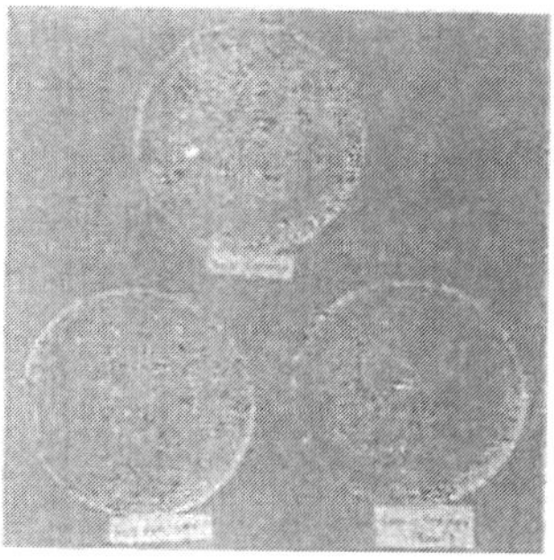

Plate 12: Colonies of *Azospirillum brasilense* in different carriers

Plate 13: Colonies of *Azotobacter chroococcum* in different carriers

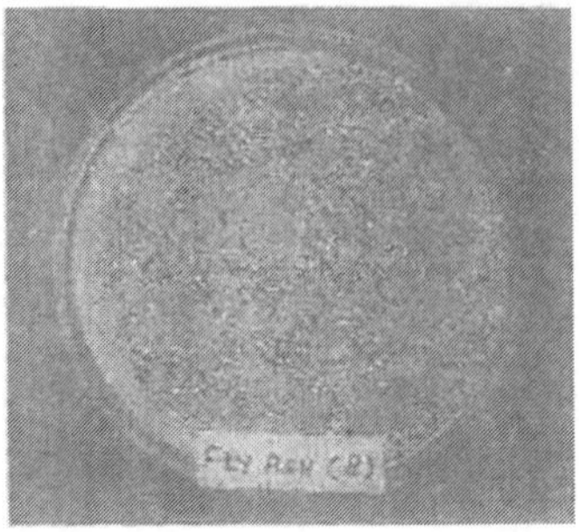

Plate 14: Colonies of *Bacillus circulans* in fly ash

The most affected organism was *B. circulans*. However, soil-fly ash combination was best for the proliferation of *P. striata.* Soil being a buffering agent must have fulfilled some of the requirements in which fly ash was lacking. The results showed that fly ash after amendment with organic matter may prove to be a good carrier material for different microbes. The better growth of nitrogen fixer in fly ash alone may be attributed to its being rich in mineral nutrients and having better WHC than soil.

Effect of Fly Ash on Soybean Crop

The application of fly ash to soil at different rates increased the availability of soil phosphorus in all the treatments as compared to control (without fly ash). Fly ash applied @40 and 60 t/ha showed the same status of available P_2O_5 of soil (26.8 ppm) under soybean (Table 78). The inoculation with *P. striata* further enhanced the soil P_2O_5. The inoculation effect on P- availability was more pronounced in treatments having 40 and 60 t/ha fly ash application (34.7 ppm) and both the levels were statistically at par with each other but were significant over control (Gaind and Gaur, 2002). The available phosphorus showed a declining trend towards the maturity of crop.

The introduced PSB (P. *striata*) was found to establish itself well in plant rhizosphere as is clear from the enumeration studies carried out at different intervals. The addition of fly ash to soil at varying rates showed significant favourable effect on the inoculated bacteria (Table 78). The population showed a slight increase towards the crop maturity though variations were non-significant.

Table 78: Rhizosphere population of phosphate solublizing bacteria and available P_2O_5 content of soil under soybean crop.

Treatments	Available P_2O_5 (ppm)		Population ($x10^4$ g^{-1})	
	30 days	90 days	30 days	90 days
Control (Fo)	14.3	10.2	-	-
Fo + P. *striata*	19.3	17.6	14	16
F_{20}	24.3	19.8	02	02
F_{20}+ P *striata*	26.8	20.2	18	21
F_{40}	26.8	20.4	02	02
F_{40}+ P *striata*	34.7	27.0	28	35
F_{60}	26.8	18.6	01	02
F_{60}+ R *striata*	34.7	24.8	29	31
F_{80}	19.2	15.6	01	01
F_{80}+ P *striata*	29.4	25.4	16	12
SE (m) ±	2.4	3.1	0.81	0.81
CD 5%	7.0	9.2	2.14	2.14

FO, F20, F60, and F80= rates of fly ash application to soil (t/ha)

Table 79: Grain yield and uptake of plant nutrients by soybean crop.

Treatments	Grain yield g/pot	N mg/pot	P g/pot	K ug/pot	Ca ug/pot
Control (Fo)	12.7	0.52	58.67	46.48	43.5
Fo + *P. striata*	21.0	1.01	114.45	66.21	46.3
F_{20}	18.2	0.86	88.55	58.46	45.5
F_{20}+ *P. striata*	30.4	1.49	176.32	101.45	47.3
F_{40}	20.8	1.20	109.82	56.58	52.3
F_{40}+ *P. striata*	35.0	2.41	205.80	106.73	57.0
F_{60}	27.0	1.70	135.81	80.20	47.0
F+ *P. striata*	26.7	1.76	153.79	87.15	49.0
F_{80}	19.5	1.18	113.10	45.72	42.0
F_{80}+ *P. striata*	18.2	1.16	105.00	49.57	50.0
SE (m) ±	1.82	0.09	0.01	2.75	1.1
CD 5%	5.40	0.24	0.03	8.05	3.2

F_0, F_{20}, F_{40}, F_{60}, and F_{80} = rates of fly ash application to soil (t/ha)

The addition of fly ash to soil under soybean showed a significant positive effect on crop growth and grain yield (Table 79). In all the fly ash treatments, grain yield increased significantly over control (Gaind and Gaur, 2002). However, the best results were recorded with fly ash applied @60 t/ ha and the increase over control was about 110%. The inoculation with *P. striata* also enhanced the grain yield and the maximum increase of 66% was recorded with F_{40} + P. *striata* over Fo+ *P. striata.* However, fly ash applied @ 80t/ha did not show proportionate positive effect on the grain yield. F_{40}+ P. *striata* treatment was recorded as the best treatment for getting significantly higher grain yield.

Table 80: Effect of fly ash on the uptake of trace elements by soybean crop.

Treatments	Zn ug/pot	Fe ug/pot	Mn ug/pot	Cu ug/pot
Control (Fo)	15.80	63.23	5.94	2.54
Fo + *P. striata*	29.33	138.30	15.29	3.89
F_{20}	22.27	115.73	19.28	5.30
F_{20} *P. striata*	42.82	261.28	27.83	7.65
F_{40}	26.83	122.75	16.22	4.46
F_{40}+ *P. striata*	53.52	258.89	20.99	6.46
F_{80}	84.86	177.03	17.70	5.45
F_{60}+ *P. striata*	34.80	170.57	17.59	5.41
F_{80}	26.18	123.11	17.72	3.42
F_{80}+ *P. striata*	22.96	123.93	17.49	5.10
SE (m) ±	2.2	12.81	2.60	0.52
CD 5%	6.0	38.40	7.80	1.50

F_0, F_{20}, F_{40}, F_{60}, and F_{80} rates of fly ash application to soil (t/ha)

The phosphorus uptake by grain (Table 79) was also increased in all the treatment having fly ash application varying from 20 to 80 t/ha. A maximum increase of 133% was obtained in F_{60} treatment. However, P. *striata* inoculation gave the best result with F_{40} dose and the increase over inoculated control (F_0+*P.stratia)* was about 80%. Higher rate of fly ash i.e F_{80} did not show any increased positive response on phosphorus uptake by grain. A similar trend was observed for nitrogen uptake (Table 79).

A favourable effect on the uptake of potassium and calcium was recorded with fly ash applied at a rate up to 60 t/ha and 40 t/ha respectively. However, the highest rate of 80 t/ha depressed the uptake of potassium substantially and that of calcium also. The use of P. *striata* increased the uptake of potassium appreciably and of calcium slightly. The combined effect of fly ash and P. *striata* was most pronounced in F_{40} + P. *striata* treatment where maximum uptake of potassium was recorded (106.73 ppm). Similarly F_{40} + P. *striata* treatment was also best for calcium uptake.

The application of fly ash and seed inoculation with *P. striata* exhibited positive effect on the uptake of trace elements viz. Zn, Fe, Mn and Cu by soybean grains (Table 80). The seed application with P. *striata* increased the uptake of Zn, Fe and Mn by 2-3 folds, except for Cu where it was slightly less. An appreciable increase in the uptake of trace elements was recorded with fly ash applied @ 20 t/ha along with *P. striata*. However, further additions of fly ash did not augment the uptake of trace element to that extent.

The integrated use of PSB and fly ash was effective in augmentation of trace element uptake up to F_{40} only. However, maximum Mn uptake was recorded in F_{20} treatment (with or without inoculation).

The favourable effect of fly ash on grain yield of soybean crop was also clearly demonstrated. Maximum yield of soybean grain was obtained with F_{40}+ P. *striata* inoculation. Inoculation of soybean crop with P. *striata* increased the grain yield by 66% even at a lower dose F_{20} of fly ash. However, no additional effect of inoculation was recorded at F level of fly ash. The positive response of soybean crop to applied fly ash was also observed in increased available P_2O_5 content of soil, total P and nitrogen content of grain which may be due to its compatibility with soil micorflora. In addition to the inoculated organism, native flora also helped in solubilizing the insoluble phosphorus and uptake of other macro and micro-nutrients upto certain level of fly ash. The favourable effect of fly ash used as a fertilizer on yield of soybean crop was in agreement with the finding of Guoyani (1989) who also reported an increase in the yield of rice by applying fly ash in soil. Field studies conducted by M.P. pollution control board (Bhopal) also reported increased spike length of wheat and profused growth of crop in plots with increased fly ash content mixed with soil as compared to check plots with no fly ash. It can be concluded that fly ash is more effective if applied @ 60 t/ha, whereas the inoculation with P. *striata* gives better results with lower dose of fly ash @ 40 t/ha. However, fly ash @ 80 t/ha did not show proportionate *increase* in grain yield and nutrient uptake by the crop.

Studies on Possible Negative Effects of Fly Ash Application

1. Ground water

Fly ashes contain a small amount of trace and heavy metals which may percolate down and pollute ground water. The solubility of these elements is <10% (Rohriman, 1971). Natusch (1975) observed in a laboratory experiment on leaching potential that 5 to 30% of toxic elements especially Cd, Cu and

Pb are leachable. Gralloway *et al.* (1976) observed that atleast 10% of total Cd would be solubilized in the acidic pH range of 3 to 5. It is unlikely that these will have any major effect on the quality of ground water. However, monitoring of this aspect may be advisable.

At Central Fuel Research Institute (CFRI), Dhanbad it was observed that the quality of ground water did not change with the application of fly ash and all the parameters including the trace and toxic metal contents were within the permissible limits. Some other research organisations also observed that fly ash has no significant polluting effect on ground water.

2. Uptake of heavy metals and toxic elements by plants

Fly ash has ppm level concentration of heavy metals, when applied to soil these elements may get absorbed by plants grown on it which may ultimately enter into food chain. However, the absolute quantities of these elements in fly ashes are low which may not result into negative effect. The data on trace element uptake and accumulation by plant are limiting despite fairly, intensive research over the last 25 years. Boron in fly ash is readily available to plant and investigators consider B to be limiting factor in unweathered fly Ash utilisation (Townsend and Gillham 1975, Elseewi *et al.,* 1978, Ciravolo and Adriano 1979). RRL, Bhopal conducted a study regarding the uptake of heavy and trace metals by some vegetable crops and it was observed that the uptake was quite low and remained within the normal range. Central Fuel Research Institute, Dhanbad observed that there was no significant differences in uptake of trace and heavy metals between control and fly ash treated plots. Although Fly Ash contains a moderate amount of trace and heavy metals, the uptake and accumulation of these by plants is very negligible.

3. Radionuclides

There have been several reports in the literature on the presence of radionuclides in fly ash but studies on their impact have been few (Gowiak and Pacynas, 1980). The radiochemical pollution of uranium and thorium series is always present in fly ash (Eisenbud and Petrow, 1964). The concentration of natural uranium varies from 14 to 100 ppm although in exceptional cases it may be as high as 1500 ppm whereas that of thorium is less than 10 ppm. The fly ash concentrates besides other gaseous and trace metal oxides, contain several redioactive contaminants like 222Ru & 220Ru (Sharma *et al.,* 1989). Bhaba Atomic Research Centre, Mumbai is of the opinion that most of the Indian coals have very low levels of radioactivity which is well below the hazardous limit. Hence radioactivity of fly ash may not be a limiting factor for its application for agriculture purpose. Central Fuel Research Institute, Dhanbad observed that there is no significant uptake of radioactive elements by plants and also that there was negligible cumulative build up of these contaminants in soil when fly ash applied for agriculture purposes.

References

Adriano, D.C., Page, A.T., Elseewi, A.A. Chang, A.C., and Straugan, 1. (1980). Utilisation and disposal of fly ash and other coal residues in terrestrial ecosystem. A review. J. environ. Qual. 9: 333-334.

Bhabha Atomic Research Centre (1981). Report No. 1119.

Basu, J. K. and Tagore, V. D. (1963). Indian J. Agric, Sci. 13: 157- 181.

Bhujbal, B.M. and Mistry, K. B. (1981). Bhabha Atomic Research Centre Report No. 1114.

Capp, J. P. (1978). Power plant fly ash for land reclamation in the eastern united states. In: Reclamation of Drastically Disturbed lands, Eds. Schaller, F. W. and Sutton, P. Soil Sci Soc., Am, Madison, Wisconsin 339.

Chatterjee, R.K. and R.K. Rattan (1988). Agricultural Utilisation of fly ash, In: National workshop on coal ash Utilisation in India. June 28, 237-252.

Central Fuel Research Institute, Dhanbad.

Cirovolo, T.G. and Adriano, D.C (1979) Utilisation of coal ash by crops under green house conditions, In Ecology and Coal Resources Development, Wali, M, Ed, Pergamon Press, New York, p 958.

Chang, A.C., Lund, L.J., Page, A.L. and Wemeke, J.E. (1977). Physical properties of flyash amended soils. J. Enviorn Qual. 6 :267-270.

Collings, C.H. (1955). Commercial Fertilizers-their sources and use. 5th ed. Mc. Graw, Hill Book Co. Inc N.Y.

Deshmukh, A., Matti, D.B. and Bharti, B. (2000). J. Soil Crops 10: 69-71.

Eisenbud, M. and Petrow, H.C. (1964). Radioactivity in the atmospheric effluents of power plants that use fossil fuel, Science 144:288.

Elseewi, AA, Binghman, F.T. and page, A.L. (1978). Growth and mineral composition of lettuce and swiss chard grown on flyash amended soils, In: Environmental Chemistry and Cycling Process, Conf. 760429, Adriano, D.C. and Brisbin, I. L. Eds. U. S. Department of Commerce, Springfield, VA 568.

EL-Mogazi, D., Donald, J. L. and Weinstein, L. H (1988). The Science of Total Environment 74 : 1-37.

Fail J.. and Wochok, Z. S. (1977). Soyabean growth on flyash amended strip mine soil. Plant Soil, 48: 473.

Gaind, S. and Gaur, A. C. (2002). Bioresource Technol. 85: 313- 315.

Gaind, S. and Gaur, A. C. (2004). Evaluation of Fly ash as a carrier for diazotrophs and phosphobacteria. Bioresource Technol. 95: 187- 190.

Gaur, A.C. (1990). Phosphate Solubilizing Microorganisms as Biofertilizer. Omega Scientific Publishers New Delhi pp 176.

Gaur, A. C. (2001). Organic Manure, a basic input in organic farming. Indian Farming 51 (3): 3.

Gidnavar, V. S. Gumaste, S. K. and Krishnamurthy, K. (1972). In Research series. Univ. Agric. Science, Hebbal, Bangalore pp 303-311.

Gowiak, B.J. and Pacyna. J.M. (1980). Radiation dose due to atmospheric release from coal-fired power stations. Int. J. Environ. Stud. 16:23.

Gralloway, J.N., Likens, G.E. and Edgeston, E. S. (1976). Acid rain precipitation in north eastern United States, pH and acidity Science 194: 722.

Grewal, K. S., Yadav, P. S., Mehta, S. C. and Ostwal, M. C. (2001). Crop Res. 21: 60-65.

Guoyari, Lu, (1989). "Coal ash a boon for chinese economy". Article appeared in the daily M. P. hronicle. Bhopal November 21.

Indian Institute of Technology, Kharagpur.

Jensen, H.L. (1953). Proc. Soc. Appl. Microbiol. 14: 89-94.

Kalra, N., Goshil, S. C. Chaudhury, A., Chaudhury, R. and Sharma, S.K. (1997). Bioresource echnology. 61: 39-41.

Kalra, N., Harit, R. C., Sharma, S.K. and Kumar, V. (2000). Bioresource Technol. 75: 91-93.

Lehr, J. R. and Mc Clellan, G. H. (1972). T. V. A. Bull. Y-93.

Maiti, S.S. Mukopadhyay, S. K., Gupta, S. K. and Banerjee, S. K. (1990). Evaluation of fly ash as a useful material in agriculture. Jour, Indian Soc. Soil Science. 38: 342-344.

Mandal, S.C. and Jha, K. K. (1970). Agricultural utilization of basic and other steel slags produced in India. Tech. Bull No. 5. Agricultural Research Institute, Kanke, Ranchi.

Mathur, B. S. (1995). Utilisation of Basic Slag in Agriculture, In: Recycling of Crop, Animal human and Industrial wastes in Agriculture (Ed. HLS Tandon) FDCO, New Delhi pp 125-142.

Natusch, D. F. S. (1975). Characteristics of pollutants from coal combustion and conversion process In: Toxic effect on the Aquatic Biota from coal and Oil Shale Development. Quarterly Progress Rep. Oct-Dec Natural Resource Ecology Laboratory. Colorado State University Fort Collins, 73.

Page, A. L. Elseewi A.A and Struaghan, I.R. (1979) physical and chemical properties of flyash from coal-fired plants with reference to environment impacts. Residue Rev. 71: 83-120.

Page, A. L., Elseewi, A. A., lund, L. J, Bradford, G.R., Mattigod, S., Chang, A. C. and Binghum, F. T. (1980). Consequences of Trace Element Enrichment of Soils and Vegetation from the Combustion of fuels Used in Power Generation. University of California, Riverside, 158.

Pant, G. N., and Pathak, D. N. (1975). J. Indian chem. Soc. 51: 101-103.

Phung, H.T., Lund, I. J. and Page, A. L. (1978). Potential use of flyash as a liming material in environment chemistry and cycling processes, Conf. 760429, Adriano, D. C. and Brisbin I. L. Eds. U.S. Department of Energy, p 504.

Picktel, J. R. and Hayes J. M. (1990). Influence of fly ash on soil microbial population. J. Environ. Qual. 19:593-597.

Pikoskaya, R.I. (1948). Mikrobiologiya 17: 362-370.

Punjab Agriculture University (1993). Utilisation of flyash in agriculture and revegetation of dumping sites. Annual Progress report.

Rohriman, F. A. (1971). Analysing the effect of flyash on water Pollution. Power, 115:76.

Regional Research Laboratory, Bhopal.

Sharma, S. *et al.* (1989). Flyash dynamic in Soil-Water systems, Critical Reviews in Environmental Control 19:251-275.

Singh, R., Singh, N. T. and Arora, Y. (1980). J. Ind. Soc. Soil. Sci. 28: 38-41.

Srivastava, A. and Chhonkar, P. K. (2000). J. Ind. Soc. Soil Sci. 48 :850-853.

Townsend, W. N. and Gillham, E.W.F. (1975). Pulverised Fuel ash as a medium for plant growth. In: The Ecology and Resource Degradation and Renewal, Chadwick, M. L. and Goodman, G. T. Eds. Blackwell Scientific Oxford, p 287.

University of Agricultural sciences, Raichur.

Unni, K. S. (1985). Impact of Sami and Koradsi thermal power stations on the aquatic ecosystem. Annual progress report submitted to Department of Environment. Government of India p 2-3.

Varadan, K. M., Satyanaryana, T. and Havanagi, G. V. 1977, J. Indian Soc. Soil Sci. 25: 380-390.

9

Organic Agriculture

The terms such as organic, natural, biological, biodynamic, altemative and sustainable farming have entered the agriculture vocabulary. Inspite of these differences in terminology, basically identical ideologies, ideas and technologies are reflected. The common objectives of all the system are to develop farming methods to restore, maintain or build-up soil productivity, to protect the environment instead of causing pollution, produce safe and healthy food, and preserve or help the small farmer.

In Japan, the farmers study the nature, learn from it and apply the knowledge on their farms. For instances growing plots should also be mulched with straw or plant residues, this imitates the conditions in natural forest where ground is never bare. Instead of using pesticides, strong herbs or plants can be grown around vegetable plots to ward off the insects. To keep the soil health natural, organic fertilizers are used. Soil microorganisms and earthworms take care of the processes involved in decomposition and mineralization of the organic substances, fix atmospheric nitrogen, solubilise insoluble forms of phosphate and produce natural growth promoting substances and antibiotics,

The concept of sustainable agriculture is gaining acceptance and momentum because of rapid degradation of natural resources, increase in the cost of production in conventional agriculture system and deterioration in the quality of rural life caused by environmental pollution. Moreover it is risky and uneconomical to practise conventional agriculture in over 70 percent of the cultivated area in India which is presently under rainfed farming. Much of the know how to practice organic farming models has rested on the individual farmers till now. They however do not have the benefit of appropriate technology and scientific information and also institutional support to ensure success of the new farming system.

Many times organic agriculture is either presented in negative terms because it involves nonuse of synthetic fertilizers, pesticides and antibiotics or in terms of alternative agriculture where use of organic manure, crop rotation, minimum tillage etc. are promoted. Such specifications neglect attitudes, goals, values and redesign component of the new farming system. Organic agriculture as a philosophy has its roots in a set of values that reflect a state of empowerment of awareness, of ecological and social realities and the ability of the individual to take effective action. It combines the design and management practices with natural processes to create an ecosystem which can achieve sustainable productivity without the use of external inputs such as chemical fertilizers and pesticides. Organic farming thus promotes recycling of resources through methods of crop rotation, rotational grazing, alley cropping, intercropping, recycling of plant and animal remains and by products, green manure, off - farm organic manures, vermicompost, organophos, biogas after detoxication, mineral bearing rocks and water tank silt for maintaining soil fertility and crop productivity and use of natural, biological and cultural methods for control of insects, pests and weeds.

I. Fundamentals and Aims of Organic Farming

Organic agriculture is a holistic production management system which promotes and enhances biodiversity and biological activity. It essentially promotes and enhances the environmental health, at the same time produces food. It is based on use of low external inputs and non-use of synthetic fertilizers and agrochemicals, it also takes into consideration the fact that regional conditions require locally adopted systems.

Organic agriculture develops a system that ensures that all forms of life from livestock to microbes are conserved and profitably utilized in production.

Organic farming as for as possible employs farm inputs which can be reused or recycled, are generated on site and create minimal pollution to local external environment. It specially excludes all products and processes of genetic engineering and related technologies.

It produces food diversity that is free from toxins and is of high nutritional value and has good shelf life and of a quality suitable for direct consumption and small scale processing.

Organic farming assures a basic minimum quality of life inclusive of nutritional availability, production and processing while providing them a safe working environment.

Basic Rules of Organic Agriculture

Organic agriculture is the practice of agriculture that develops a viable and sustainable agroecosystem. The basic requirements for organic farming are as under;

Soil: Soil is a living matter in which seed is nurtured and finds expression, which is central to organic farming therefore, techniques like hydroponics or acroponics do not find place in organic farming systems. Soil fertility is maintained and enhanced and the biological activity of soil enriched. Soil loss through erosion has to be prevented. Organic material is to be returned to soil to maintain and enhance the humus content of the soil. Such organic materials must be produced at the organic farms or obtained from other sources especially free of residues of chemical fertilizers, pesticides and agrochemicals. Farms either should become self sufficient in organic materials or manures or should buy, organic manure certified for use at organic farms from other source such as compost plants. However it may not be possible to stop contamination from external sources although in such cases the toxic residues are likely to remain well below the permissible limits. Cultivation has to guard against the occurrence of pesticide and herbicide contamination and carriage of inorganic chemical as fertilizers by irrigation and drainage. Manual and animal driven agriculture implements should be used. High-mechanized implements should be excluded. Soil and water conservation measure should be an integral part of organic farming practices.

Seed: Seed should preferably be of organic origin locally adopted indigenous varieties which are resistant to pests and disease. Seeds should be generated on farm or must be obtained from existing organic farm or may be got from other certified organic farms. Seed that has been modified by genetic engineering or related technologies will not be permitted.

Pest and diseases: The pest and diseases are accepted fact in organic farming system. Organic farming system should be managed in a way that ensures that losses from pests, diseases and weeds are minimised. Pest, disease and weeds should be controlled by prevention and cultural methods such as balanced programme of organic manuring including green manuring, crop rotation, companion planting, activating natural enemies, predator management and use of mechanical methods and mulching for elimination of pest and weeds. Plants extracts such as neem cake and cow urine can be used for pest control

Animal and livestock fodder: Livestock in an important and integral component of organic farming system because of the ability of animals to convert organic matter into a source of farmyard manure thereby ensuring close nutrients cycle. They also provide animal farm energy for farm operations. Livestock should be provided with good quality organic fodder and proper veterinary care.

Storage and preservation: Organic produce must be stored, transported and sold to the consumer in its most original and fresh state. Care should be taken that the quality of organic produce is not lost by association with other produce not cultivated organically. Transportation of organic produce must be proper and any damage to the product must be minimized.

Pest at the storage or processing stage must be controlled by means of physical barrier, sound and light. Temperature and atmospheric control processes are permissible. Botanical extracts and herbal pellets can be used for preservation.

Additives and substances that diminish or alter organic nature of organic produce are to be avoided. Processing methods should be based on mechanically, physical and biological processes. The use of synthetic chemical for preservation or methods like irradiation are not acceptable.

Packaging and labelling: Packaging must take care to prevent material contact to diminish the organic purity of the produce. Biodegradable material should be preferred medium of packing. Waste generating and pollution causing packaging are discouraged.

Labelling of organic produce must declare the fact of fully organic or under conversion period. Showing of ingredients is essential on the labels of processed items.

II. Standards of Organic Farming

1. Choice of Crops and Varieties

General

i) All seeds and plants material should be certified organic.

ii) Varieties cultivated should be adapted to soil and agroclimatic conditions and be resistant to pest and disease.

iii) In the choice of varieties, genetic diversity should be considered.

Standards

i) When certified organic seed and plant materials are not available, chemically untreated conventional materials shall be used.

ii) Where no other alternatives are available, chemically treated seed and plant materials may be used

iii) The certification programme shall define conditions for exemptions and set time limits for use of chemically treated seed or plant materials.

iv) The use of genetically engineered seed pollen, transgenic plants, or plant material is not allowed.

2. Crop Rotations

General

Rotations should be as varied as possible and aim to.

i) Maintain soil fertility.

ii) Reduce nitrate leaching.

iii) Reduce weeds, pest and disease problems.

iv) Legumes in rotations are recommended.

v) Permaculture, alley cropping and perennial plants (as border plant) must be promoted.

Standards

The certification program shall require the rotation of nonperennial crops in manner that takes into account pressure from insects pests, disease and weeds while maintaining or enhancing soil organic matter, microbial activity, soil health and soil fertility.

Organic Manuring

General

i) Sufficient quantities of organic matter should be added to soil to increase or at least maintain its biological activities and fertility.

ii) Organic biomass produced on organic farms should be the basis of manurial programme.

iii) Organic manure units like compost pits, biogas plants, biodynamic manure pits etc. should be set up on each farm.

iv) Use of improved manual and animal driven agriculture implements should be promoted.

Recommendations

i) Fertilization operations should minimise nutrient losses

ii) Accumulation of heavy metals and other pollutants should be avoided

iii) Biofertilizers of microbial origins should be used as supplementary and not as a replacement to organic nutrient recycling

iv) When human faeces is processed via composting or biogas plants, care has to be taken that it does not contain pathogenic microorganism and pollutants such as heavy metals.

Standards

i) Organic manuring shall form the basis of soil fertilization

ii) The certification programme shall set limitations to the total amount of organic material brought into the farm unit, taking into account the specific nature of crops.

iii) Mineral fertilizers shall be used in agriculture as supplementary role to organic materials.

iv) Mineral fertilizers shall be supplied in their natural composition and shall not be rendered more soluble by chemical treatment

v) The certification program shall lay down restrictions for use of inorganic inputs such as micronutrient mixtures but micronutrients may be used if required in the initial stages.

vi) Chilean nitrate and all synthetic nitrogenous fertilizers including urea are prohibited.

4. Pest, Disease and Weed Control

General

Organic farming system should be managed in such a way that losses from pests, diseases and weeds are minimized. Emphasis is placed on the crops and varieties well adopted to agroclimatic condition. A balanced manurial programme is followed to maintain high biological activity, soil fertility etc.

Recommendation

i) Pest, diseases and weeds should be controlled by use of preventive cultural techniques, which limit their development e.g. green manures, balanced manurial programme and suitable crop rotations.

ii) The natural enemies of pest and disease should be protected through proper habitat management such as hedges, nesting sites etc. be encouraged.

Standards

i) Products used for pest, diseases and weeds control prepared at the farm from local plants, animals and microorganisms are allowed.

ii) All equipment from conventional farming system should be properly cleaned and free from residues when used on organically managed areas.

iii) The use of synthetic agrochemicals such as herbicides and pesticides are prohibited.

iv) The use of genetically engineered organisms e.g. earthworms and microorganism or their products is not permitted.

5. Pollution Control

General

All measure should be taken to minimize pesticide contamination from outside and within the farm by wind drift, drainage and irrigation.

6. Soil and Water Conservation

i) Soil and water resources should be handled with a view to sustainability.

ii) Relevant measure should be take to prevent erosion, salinisation of soil, excessive and improper use of water and pollution of ground and surface water.

Standards

Clearing of lands through means of burning organic matter e.g. slash and burn, straw burning shall be restricted to minimum. Relevant measure shall be taken to prevent erosion. Excessive exploitation and depletion of water resources shall not be allowed. Relevant measures shall be taken to prevent salinisation of soil and water,

7. Labelling

Standards

When the full standards requirement have been fulfilled, product should be sold as "Produce of organic agriculture" or similar description. Mixed products where not all ingredients including additive are of organic origin may be labelled in the following way (raw material weight).

i) Where a maximum of 95% of the ingredients are of certified origin, product may be labelled "certified organic" or similar description and should carry the logo of Certified Programme.

ii) Where less than 95% but more than 70% of the ingredients are of certified organic origin, products may not be called organic. The word organic may used on the principal display in statements like "Made with organic ingredients" provided there is clear statement of the proportions of the organic ingredients

iii) Where less than 70% of ingredients are of certified organic origin the indication that an ingredients organic may appear in the ingredients list. Such products may not be called "organic"

III. Conversion to Organic Agriculture

Conversion requirement refers to the time between the start of organic management and certification of crops and/or animal husbandry is known as the conversion period.

The most salient aspect of organic agriculture is the emphasis on working with natural soil processes. The system has to make use of existing soil nutrients, water cycles, naturally occurring energy flows for production of nutritious foods which are free from harmful chemical residues. Organic, food movement promotes the health of humans and their environment by encouraging farmers to use agricultural methods that neither deplete the soil nor disturb the natural balance of the environmental systems or the farmers.

The shifting from conventional agricultural to organic sustainable agriculture can be approached by adopting principles of zero waste concept, increased efficiency substitution and redesigning of the farm.

Shifting from conventional agriculture to organic sustainable agriculture can possibly be attained by following the principles with increased efficiency substitutions and redesigning of the farm. For changeover to organic system from conventional farming, the use of organic manures and biofertilizers, regulation of natural processes of biological nitrogen fixation, insoluble phosphate solubilization and biocycles are promoted/ practised for maintenance of soil fertility and for meeting plant nutrient needs.

Pest control is affected by using biopesticides, allelochemicals, botanical pesticides, predators, parasites, mycotoxins, growth hormones and resistant local crop varieties and by timely agricultural operations. Efficiency of irrigation is increased by adopting drip and sprinkler and mulching with organic wastes to reduce evaporation losses by regulating the soil temperature and water use.

By Substitutions and Alternate Inputs

Resources dependent and environmentally impacting products can be substituted, replaced or complemented by those materials which are ecofriendly. Organic sources of plant nutrients such as compost, FYM, Gobargas plant spent slurry, green manure, crop residues, agro-industry byproducts should be used in the places of nonrenewable resources such as chemical fertilizers. Similarly the recent trends of the use of biopesticides, biological and cultural control of insect pests may replace the use of agrochemicals.

IV. Redesign and Management

The farming practices and operations should be well thought, redesigned and managed after identifying the existing problems. The farm should become self reliant in resources through on-farm diversity.

Perma Culture

Permaculture or natural approach where one relies on the productivity on permanent structure such as trees, perennial species or other undisturbed plant, soil system for food has been perceived to be the ultimate goal of sustainable farming system. The benefits from such a system is a long term goal but once the agro- ecological system is established, periodical *Alley cropping* attention is only required.

Alley cropping: This is essentially an agroforestory system in which shrub or trees are grown in alley formed by hedgerows of nitrogen fixing leguminous trees or shrubs planted on the contour over or across slope (Kang *et al.*

1981,1990) The hedgerows are cut back when the main crop is planted and kept prunned during the cropping seasons to reduce competition with the crop for water and plant nutrients. During the non- crop period the hedgerows are allowed to grow to protect the land from #erosion and to produce biomass. Kang and Wilson (1987) described the concept of alley cropping, its soil management and potentials.

Alley cropping has been referred to as "Controlled bush fallow" and retains the basic feature of bush fallow or shifting cultivation system. Trees and shrubs in the alley cropping provided green manure or mulch for companion food crops and facilitate nutrients uptake by recycling from deeper soil layers. It provides prunnings or loppings as mulch and shade during fallow period to suppress weeds. It provides contour over barriers on sloping land to reduce soil erosion and nutrient runoff losses.It provides biologically fixed nitrogen to the companion crop. It provides favourable environment for growth and multiplication of macro and microorganisms. It provides forage for ruminant animals, construction material and firewood. Increased yield with reduced inputs form a cardinal advantage of the system.

The major advantage of alley cropping over the traditional shifting cultivation or bush fallow system is that the crop and the fallow phase can take place concurrently in the same field. This allows the farmer to crop for an extended period without returning the land to bush fallow.

Mac Rae(1990), quoted that approximately 4000 farmers have converted to sustainable practices in Canada, at least 30000 in USA, several thousand in Europe and many more are in transition phase. Thampan (1995) has published case studies and Directory - network of those practicing organic sustainable agriculture in India. These farmers are testimony to the feasibility and success of organic farming.

In India most of the farmers are practicing LISA (low input sustainable agriculture) because they cannot afford expensive off-farm inputs. They however have not assimilated the concept of organic agriculture fully because of poor information and extension services. In developed countries the prime factor for the change from conventional agriculture practice to organic agriculture has been due to the fear of soil and water pollution and environmental degradation which has affected the human health. In developing countries, the depressed economy is forcing many more farmers to opt for alternative farming methods to reduce input cost and maintain financial stability. Although the yields during initial duration of change over may be lower in organic farming system, the studies have shown that the total input cost are substantially reduced and net profits are similar or even higher than in the conventional agriculture particularly due to premium prices of organically grown food products.

Decreased yields in organic agriculture are reported during initial years of around 3 to 4 years on conversion from conventional agriculture. This is possibly caused due to delay in reestablishment of certain biological processes due to toxic residues left by the components of chemical agriculture. It also may take time to build up the efficient useful microbial biomass including that of autochthonous and zymogenous microflora and fauna which are essential in biogeocycles and plant nutrition. Organic farming has been drawing attention from all over the world and demand for organic food is growing by leaps and bounds.

V. Certification of Organic Commodities

The organically grown food products may sell at premium price particularly vegetables, fruits, spices, some grains and beans. The premium prices are 10-15 percent over conventionally produced products. In organic farms certain certified inputs have been recommended so that the products may attain certification standards. These standards serve to differentiate the products and help to get premium prices.

An organic label indicates to the customer that agriculture product was produced using certain organic production methods. In other words, organic is process claim rather than a product claim. An apple produced by practices approved for organic production may be very well identical to an apple produced under other agricultural management regimes.

Several countries and a number of private, certification organizations have defined organic agriculture. In the past the differences in the these definitions were significant but the demand for consistency by leading multinational traders has lead to great uniformity.

The international federation of organic agriculture movements (IFOAM) a non government organization internationally networking and promoting organic agriculture has established guidelines that have been widely adopted for organic production and processing.

All India federation of farming (AIFOF) has formulated a certification programme to give quality symbol to the organic products which meet the criteria specified by the certification council of AIFOF. The council gives the license or permits for the use of quality symbol "Certified Organics" to the producers engaged in processing, packaging and distribution of raw materials or products. These standards are in accordance with the standards of international federation of organic agriculture movements (IFOAM) Germany.

Products for Use or Non- use for Manuring and Soil Improvement

Banned chemical fertilizers and pesticides are restricted in organic agriculture. The use of sewage and sludge are also banned. Permitted organic matter produced on the farm such as farmyard manure, slurry and urine, composted poultry manure, vermicompost, compost produced from organic residues, green leaf manure, leaf litter, biodynamic preparations, *Azolla*, crop residues such as straw, groundnut haulms, mulches from sugarcane trash, straw etc. kitchen wastes out of organic food, coir pith, plantation by products and composted tea and coffee wastes and ash from different organic source are permitted for use in organic farming.

Restricted Use: Farmyard manure, urine, compost, straw, coir pith, plantation products, oil cakes, mulches, and calcified seaweed procured from other organic farms.

Biofertilizers *(Rhizobium, Azotobacter, Azospirillum,* Blue-green algae, phosphate solubilizing microorganisms, cellulolytic and lignlolytic microorganism, Mycorrhiza) sawdust, organic byproducts of food and textile industries without any synthetic additives, gypsum, limestone, rock phosphate, basic slag, potassium mineral, trace elements.

VI. Products for Plant Protection Measures

There are 3 categories of pesticides. The use of some are banned, some are for restricted use and many are permitted.

Banned for Use

Synthetic and chemical pesticides.

Permitted for Use

Mechanical traps, pheromones (permitted in traps and dispensers only) chromatic traps, algal preparation, bacterial (*Bacillus thuringiensis*) beeswax, biodynamic preparations, chitin nematicides (natural), Ca(OH)2,CO_2, chloride of lime, clay (bentonite, perlite, vermiculite, zeolite), coffee grounds, corn gluten meals (weed control), dairy products (e.g. milk, casein) C_2H_5OH, sabadilla(unprocessed) silicates (sodium silicates, quartz,) Na_2CO_3, lecithin, lime, natural acid (vinegar) neem (*Azadirachta indica*) restricted plant based repellents, propolis, potassium bicarbonate, soft soap, soda, sea salt and salty water, gelatine, sulphur (Calcium polysulfide) mulches from natural sources, common unprocessed salt.

Restricted Use

Diatomaceous earth, neem (Azadirachta indica) plant and animals oils, release of predators of parasites, viral, fungal and bacterial preparations; sterilized insects, sulphur copper salt (cuso4) $KMNO_4$, Chloride of lime soda,light mineral oils, deris root (Rotenone) *Pyrethrum cinerafolium*, *Ryania spectosa*, tobacco, tea, Cassia amara, nematicides from natural sources e.g. Tayestus, mulches from natural sources, common unprocessed salts.

Global Demand for Organic Food

Under Indian condition attempt is required to be made to analyse the impact of organic farming. Thampan (1995) observed sporadic success by adopting organic farming by farmers of tree based cropping system involving the combination of trees, shrubs, seasonal field crops and livestock components. The various products thus obtained and increase on farm employment have strengthened socio-economic security of the farmers.

In tribal areas of states like Chhatisgarh, and hilly areas of Uttrakhand, the farmers belonging to small and marginal groups practice organic farming depending fully on local availability of organic resources but these farmers are deprived of premium prices on organic food due to operation of middle men. Efforts are required to develop a network of cooperative society in the entire tribal zones and system for facilitation or "Organic Certification". This will help in raising the standard of living of organic farmers of tribal, hilly and other areas. The prospects of organic farming is profound and constrains are limited and can be overcome. The research, extension and training wings of the departments of agriculture have a large role to play for promotion of organic agriculture for an overall impact.

During the last decades organic farming in India has attracted many farmers across the country and many of them have been growing successfully crops like coconut, spices, sapota, banana, grape etc. with organic manures and biofertilizers. Spice board has been implementing programmers to encourage growers to adopt organic farming methods and techniques. Organically grown spices such as black pepper, small cardamom, ginger, turmeric, cloves and maize are exported. There is an increasing demand for organic tea, coffee, cashew nuts, fruits and vegetables from developed countries. The quantum of export form India is small. There is scope for exporting organic fruits and vegetables on a large scale.

An Example of Sustainable Organic Agriculture

Warlis are Adivasis (tribals) inhabiting north Mumbai (Maharashtra state) and in Dang district and Dharampura taluka of Bulsar district in south Gujrat who have survived for Millenia in harmony with their environment. They consider themselves a part of living nature and hence are careful not to interfere unduly with the environment. It has developed over several Millenia by integrated study of plants, animals and human beings and is ecologically sound. Their priority is to grow enough paddy and pulses alongwith oil seeds, fiber plants and vegetables so that they can be self sufficient in food. They have developed an unique method of multiple cropping which is workable even under drought conditions.

Warlis usually cultivate a few of more than 15 major traditional varieties of paddy that are available to them. These have different requirement of water and soil, mature at staggered times having varied incidence of pest and diseases and possess varied flavours. Under normal monsoon, the main varieties give good yield, under abnormal rains, the other such as more hardy varieties still produce some food with lower yields.

High yielding varieties yield nothing or a little under abnormal rains conditions. The average yield over a number of years could, therefore, be higher with their traditional varieties than high yielding varieties. One black variety is harvested within 60 days of sowing. Most paddy varieties have to be stored for some time before use but this can be consumed immediately. Such varieties may be essential for other farmers as well if monsoon becomes more erratic because of global warming.

Warlis have found that the yield of HYV's with the use of chemical fertilizers is good in the first year but subsequently the yields get reduced unless higher dosages of chemical fertilizers are applied. The practice of adding more and more chemical fertilizers impoverishes the soil of fertility. If at all they use chemical fertilizers for a year or two, then they switch back to organic manures to restore the soil fertility. They use farmyard manures, leaves and twigs from their jungles which are available, paddy straw that were used in previous year's thatch and twigs from old fences are recycled. All theses are burnt together with some cowdung on seedling fields.

Some Warlis plant Dhedhar a species of *Sesbania* as green manure crop is grown in the main fields. just after the first rainfall, this is ploughed in after about 3 weeks when the field is puddled immediately before paddy transplanting. The seeds are obtained from wild plants or from those growing in hedges.

Warlis intergrate paddy with number of other crops such as *tur* (*Cajanus* cajan), Jowar (*Sorghum vulgare*) or chaoli (*Vigna unguiculata*) and paddy in the same seedling beds. On the border of these beds, they plant ambadi (*Hibiscus cannabinus*) lal ambadi (*Hibiscus sabdarifa*) or Bhindi (*Abelmoschus esculentus*). After paddy seedings are removed for transplanting, the rest thrive on residues of nutrients not removed by paddy seedlings.

Other monsoon crops are an oilseed, khorsani (*Guitotia abyssinica*) and a pulse blackgram (*Vigna mungo*). After harvest of paddy, chickpea (Cicer *arietinum*) val *(Dolichos lablab)*, rai *(Brassica juncea)* and other crops are grown in these fields since they survive on the residual moisture in soil.

These legumes help in maintaining soil fertility. Where low lying areas stay damp after the monsoon, a second crop of paddy is often sown.

Several other field crops are also grown including some vegetables such as *spinach (Spinacea oleracea)* and loni *(Portulaca oleracea).* In addition cucumber and pumpkin creepers are trailed over their huts while the backyard has a place for trees like shegva (*Maringa oleifera)*, ber (*Zizyphus mauritiana*) and mango.

Warlis rarely use chemical pesticides because of the harm they do to paddy ecosystem. They use their indigenous methods of pest control such as putting of leaves of Khair (*Acadia catechu)* in the water channel .They also plant branches of trees like shinde *(Phoenix sylvestris*) Kuda *(Hollarhena antedysentrica)* which might possess some pesticidal action. Because of organic method of farming, there seems to be little damage by pests. Pest resistant crop varieties are also chosen. The variety harbara grown has small pods unlike that usually cultivated as a cash crop by non-Adivasi farmers. It has been found that this latter variety is more susceptible to pests.

Many Warlis own land on hillsides, which are rocky and infertile. But even on these they grow crops like nagli *(Eleusine coracane)* and Kodra *(pospalum-scrobiculatum).* Some species of Kodra are toxic but they know the technique for detoxifying them.

Their agricultural practices have the aim of ensuring sustainability even at the cost of low output and immediate income. This is exactly the opposite of conventional agriculture, which promotes maximum income now without caring for the future.

References

AIFOF 1995. All India Federation Organic Farming. C. Satya Bhawana Raghunath Nagar, Thane (West), 1-54

Kang, B.T.Wilson, G.F. and Sipkens, L. 1981. Alley cropping maize (Zea mays) and leucaena (Leucaena leucocephala) in southern Nigeria. Pl. Soil 85:267-277

Kang, B.T. and Wilson, G.F. 1987. The development of alley cropping as a promising agroforestory technology pp 227-244. In: Agroforestory: a decade of development (Eds:H.A. Steppler and K. R. Nair)ICRAF. Nairobi, Kenya

Kang, B.T., Reynolds, L. and Atta-Krah, A.N. 1990. Alley Cropping. Advances in Agronomy.43: 315-59

MacRae, R.J., Hills, S.B., Mehuys, G.R. and Henning, J. 1990. Farm scale agronomic and economic conversion from conventional to sustainable agriculture. In: Advances in Agronomy, Brady N.C. (ed.) 43:155-198.

Thampan, P.K. 1995. Case studies, In; Organic Agriculture (Thampan, P. K. Ed.). Peekay foundation, Gandhi Nagar, Cochin- pp. 309-351.

10

Composting of Organic Wastes

Proper soil management without impairing soil health is the prerequisite for achieving higher productivity from agriculture land. The arable soils of tropics and subtopics are poor in organic matter due to high temperature and intense microbial activity. The importance of organic matter in improving soil fertility through maintenance of optimum levels of soil organic matter as well as plants nutrients is well recognized. There is need to encourage more ecofriendly and productive farming system.

A growing number of farmers, ecologists and scientists are coming to conclusion that conventional farming can not be sustained because in the end it does more harm than good and it is high time to think of sounder ways of farming which are healthy for the soil, environment, plants, animals and human beings.

The practices of economically sound, viable and culturally appropriate agriculture is termed as sustainable agriculture. The excessive and imbalanced use of chemical fertilizers and pesticides on soil and plant in not only harmful to soil microflora and fauna but also reduces the productivity potential of the land. The conventional intensive agriculture has depleted the soil organic matter content and plant nutrients besides the susceptibility of crops to pest and disease. Nitrates the ultimate product of N- fertilizers pollute the ground of water and food, has higher levels of non-protein nitrogen (NPN). Increasing levels of NPN in human bodies have been reported. Soil is a dynamic living system and if it is not manured to feed soil biota, soil health gets deteriorated which is essential for cycling of plants nutrients.

Organic farming is successful on farms where organic biomass are generated in and around farms and whatever rural organic materials become available should be mobilized for this purpose.

Composting

Composting will conserve and recycle organic materials generated in villages, farms and cities, safeguard rural public health and improve the quality of the rural environment all at low cost using ecofriendely microbial processes.

Composting is a key technology for preparation of organic manure. Organic manure is a basic and valuable input is sustainable agriculture.

Importance of Compost

There are varieties of benefits from compost but its primary use is to supply humus to soil and yet humus can not be artificially produced. There is continuous depletion of soil organic matter content particularly in tropics and therefore, there is an urgent need to supplement this fraction of the soil by a source from outside the system. It was found that a good compost could produce the result that are comparable to fertilizers in mineral deficient soils. The demand for a ready source of humus and plant nutrients may increase as the farming becomes slowly mechanized and livestock production and dairy become a specialized industry. Mechanisation of agriculture has been taking place in some parts of the country and the use of crop residues as cattle feed at farm level is decreasing. The surplus crops residues like straw etc. should not be wasted but composted and returned back to soil.

Benefits of Compost

1. Studies have shown that organic manure has a definite short term and long term advantages when applied to soil to raise crops.
2. The exact mechanism of benefits seems to be tied to the effects of humus substances like humic and fulvic acids.
3. It is a storehouse of energy for the growth and multiplication of microoganisms and their activities in soil which is vital in plant nutrient cycling of different elements.
4. It supplies both macro and micronutrients such as nitrogen, phosphorus, potassium, calcium, sulphur, iron, manganese, zinc, boron etc. for plant growth.
5. It provides growth-promoting substances, which stimulate early plant growth and maturity of crops.
6. It acts as a soil conditioner because it improves the physical condition of soil such as tilth, water holding capacity and cation exchange capacity.
7. It has been found to be very effective material for erosion control.
8. The end product is safe for humans and animals because of all pathogens including *Bacillus, anthrax, Salmonella enteridis* and several other pathogens are eliminated due to high temperatures (55-65°C) attained in the compost piles.
9. It increases the biological activity of soil and produces crops with less plant diseases.

10. Antibiotic substances produced in composting process control plant and animal pathogens.
11. The use of well-decomposed stabilized manure is similar to humus and causes minimum disturbances of soil microorganism and crops yields are improved with the application.
12. Pathogenic bacteria that are associated with sewage sludge and municipal refuses are destroyed when added to compost heaps.

I. Composting Process

Composting Process may be defined as a biochemical process in which microorganisms and earthworms as important agents, breakdown organic matter to relatively humus-like material. It is distinguished from digestion or fermentation because the decomposing material is primary discrete solids with open pore spaces rather than solids in a liquid or semi -liquid environment. Composting biomass vary from anaerobic to aerobic and from mesophilic to thermophilic conditions of decomposition. Aerobic thermophilic composting process generally provides the most rapid and thorough decomposition of readily oxidisable or putrefiable matter. The process has the capability of converting different types of farm and rural organic wastes to a by product which has a lower bulk volume than organic waste, is stable and has a potential of being recycled into ecological sphere without destruction of its high energy value. The technology makes it feasible to recycle organic waste back to the soil without causing pollution of land and water resources. At present composting is the only solid waste disposal process that salvages and reclaims organic solid wastes.

Historical Aspects

This is one of the best methods for preparation of organic manure and has developed into a key technology for hygienic disposal and recycling of farm and city organic wastes. In India, the use of dung as manure was known since Regvedic age (2500-1500 BC).The value of animal wastes was known in post vedic age between 500 BC and 500 AD (Kautiyla's Arthshatra). The civilizations that practiced good soil husbandry survived for thousand of years while other civilizations that ignored the basic fact have disappeared. The Mayas of central America were probably more advanced in mathematical science than any other, as the fantastic cities that still stand show, but they were on point of extinction long before Cortes and his Spaniards landed to plunder the Aztecs. The Maya civilization lived in forest countries and the people grew their basic crop of maize by burning off trees and growing crops after crops until the fertility was exhausted. The peasants from the villages that surrounded the cities brought their produce in but none of the wastes was returned to land. As domestic animals and wheeled transport were not known to these people, the inevitable happened and within a thousand years the civilization had burnt itself out with constant construction and movement involved due to poor agricultural policy followed. On the contrast, the history of Chinese for fifty centuries, showed that the Chinese had followed a policy of the conservation of soil fertility, as all wastes, animal, human and vegetables were returned to soil to build the humus content of their soils.

Scientific Aspects

The scientific aspect of composting was initiated at Rothamsted by Hutchinson and Richards (1921) and was adopted by Howard and wad (1931). Composting has been the basis for organic method of farming. Howard systematized the concept of compost making in India and showed that the best compost was prepared when crop residues and cattle dung were mixed together in the ratio 3:1.

Table 81: Chemical composition of some organic farm wastes (%) on dry weigh basis.

Organic Material	Cellulose	Hemicellulose	Lignin	Protein	Ether Soluble
Rice Straw	35.8	22.9	11.6	3.1	3.3
Wheat Straw	37.8	26.9	18.8	2.4	3.8
Barely Straw	40.4	25.3	17.8	3.9	6.1
Oat Straw	36.4	26.6	18.0	2.7	6.3
Saw Dust	53.6	9.8	28.2	1.8	3.0
Groundnut Shells	16.8	42.2	15.7	14.8	1.2
Cattle Dung	28.6	26.0	17.7	8.3	3.4
Sheep Dung	18.7	18.5	20.7	23.5	2.8

Principle of Composting

Compost can be prepared from organic materials from different sources such as crops residues, straw, paddy husk, groundnut shell, wastes of safflower, sunflower and mustard, sugar cane trash, press mud and habitation wastes like city wastes/garbage, kitchen wastes, vegetables and fruit market wastes, sewage and sludge etc. In general some of the organic raw materials contain 16.8 to 53.6% cellulose, 9.8% to 26.9% hemicellulose, 11.6 to 28.2% lignin, 1.8% to 23.5% protein, and 1.2% to 6.3% of ether soluble substances (Table 81) besides 2-30% of hot and cold water soluble fractions (sugars, starches, amino acid, urea, ammonium salts). The ash contains 5-30% minerals, (Gaur 1997, 1999)

The matter undergoes intensive decomposition under thermophilic and mesophilic condition in pits, heaps or windrows. The decomposition process is carried out by bacteria, fungi, actinomycetes, protozoa, earthworms, termites, ants etc. The role of active cellulolytic and lignolytic organism is of a paramount importance in composting process. When the organic matter is broken down in presence of air (oxygen), the process is aerobic. In the absence of oxygen the process is known as anaerobic composting. This type of fermentation takes place in biogas gas plant, waste stabilization ponds, in marshy soils and composting process in Acharya's method of composting.

Factors Affecting Composting and their Regulation

The most important factors which regulate the composting process, are as follows:

(1) C/N - ratio (2) Shredding of organic material (3) Blending or Proportioning of raw material (4) Moisture and Aeration (5) Temperature (6) pH (7) Microbial Inoculants (8) Amendment with rock phosphate (9) Destruction of pathogens and parasites

1. Carbon: Nitrogen Ratio

The carbon nitrogen ratio of organic materials is the most important factor in composting and so it is influenced by the proportion of carbonaceous and nitrogenous components present in the residues to be composted.

Microorganisms require carbon for growth and nitrogen for protein synthesis. A C/N ratio 30 to 40 is optimal for efficient composting and ratio between 25 to 40 have been found satisfactory. If the organic material is low in nitrogen, that is if C/N ratio is wide, biological activity decreases and several successions or cycles of microbial activities may be required to complete the process of composting. Thus finished compost takes longer time to be produced. High C/N ratio materials are cereal straw such as rice, wheat, sorghum, pearl millets, oats, barely, maize stalks, besides sawdust, coir dust and pith etc. Therefore, the C/N ratios of organic materials vary widely from 20 to 300. As the stock of available nitrogen is exhausted during composting, the activity of nitrogen fixing bacteria predominates and there is gain in nitrogen from the atmosphere. When C/N ratios of raw materials is less than 30: 1, the proportion of nitrogen is in the excess of the requirement of microorganism. In such conditions although the bioprocess goes on, the unassimiable nitrogen is lost by volatilization as ammonia or by denitrification. Low C/N ratios under unfavorable condition cause increasing loss of ammonia. About 20-40% nitrogen may be lost as ammonia during the preparation of compost from animal dung, night soil, water hyacinth and green immature organic materials. It is clear that simple nitrogenous compounds are more susceptible to losses and control of this for a greater retention of nitrogen in the final compost in of practical value is obtaining better quality compost.

The time of composting of high C/N ratio materials can be reduced by adding a mineral nitrogen sources or a nitrogen source such as activated sewage sludge, legumes residues, grass cutting, aquatic weeds, slaughter house wastes, green leaves and sugar industry wastes.

2. Shredding of Materials

The bioprocess can be accelerated if raw materials are shredded into smaller parts since then it becomes more susceptible to microbial action due to greater exposure of the surface area (Gaur and Sadasivam 1980). The most desirable particle size for composting may be in the range of 3-5 cm although large sizes can be composted satisfactorily. However the advantages of shredding on farms or by individuals is doubtful when one considers the additional cost involved. The effect of chopped and unchopped straw on the quality of compost has been reviewed (Gaur, 1999). The results showed chopping helped early composting of paddy straw and sugarcane trash. However, good mature compost was also produced from unchopped farm wastes like crop residues by increasing the period of composting by over 30 days which is required by chopped materials.

3. Mixing or Proportioning of Organic Materials

The C/N ratio and moisture percentages are two important parameters to be considered when mixing together different kinds of material for composting. There may not be any need of blending if the C/N ratio of the material is between 25 and 50 although a range of 30 to 40 would be a better range to aim for substances poor in nitrogen such as sawdust, coirpith, coirdust, paper,cereal crop residue and stalks and stubbles etc. which can be mixed with comparatively nitrogen rich materials such as leguminous wastes, vegetable and fruit wastes, water hyacinth., fish scraps, slaughter house wastes, biogas slurry and so on to obtain a near optimum ratio around 30 to 40. Similarly the wet and dry material can be mixed together to obtain a desirable percentage of moisture and to avoid anaerobic conditions.

Soil at the rate of five to ten percent can be added to moist material to reduce the moisture content and absorb ammonia while composting materials of low C/N ratios. Dry soil may be added if sufficient dry organic wastes are not available, soil may be added to high C/N ratio organic material to buffer acid conditions and to act as diluents for retarding anaerobic process, it can also be used for improving the appearance of the finished compost by giving it a more granular structure.

4. Moisture and Aeration

Moisture content and degree of aeration are important factors in composting. If the amount of moisture in the compost pile is below forty percent (w/w) decomposition will be aerobic but slow and degree of aeration is an important factor in composting. Aerobic decomposition will occur at any moisture content between thirty and one hundred percent if adequate turning is provided but higher moisture content should be avoided. The optimum moisture level for aerobic composting is 50-60%, however, a range of 40-80% will be quite satisfactory depending on the nature of material to be composted. Gaur *et al.* (1982) maintained moisture content of 80- 85% for aerobic composting of fibrous materials including straw. For anaerobic composting maximum moisture level is not important.

5. Temperature

With the multiplication of microorganism in the composting mass, the heat of exothermic reactions is retained due to large mass. When the temperature rises above 40°C, the mesophilic microorganisms are replaced by thermophilic microorganisms. High temperature is essential for destruction of pathogenic organisms and weed seeds. The thermophilic condition generally occurs within two to seven days of the start of the operations. The temperature in the middle of the pile goes up to 55-70°C after which it gradually cools to ambient temperature. Decomposition is fastest in the thermophilic range, the optimum temperature for organic matter decomposition into carbon dioxide and water has been found to be 60°C (Wiley and Pierce 1955) whereas Wiley (1957) and Schulz (1961) reported the maximum temperature of 71°C to be the best but should not be allowed to exceed this for long as decomposition will be slowed down due to thermal kill of microorganism. Only a few thermophilic organisms actively, carry on decomposition above 70°C. While composting high C/N ratio materials such as wheat and rice straw, sorghum slalks and jamun (*Syzquim cumine*) leaves maintaining C/N ratios 48 to 50, it was observed by Gaur *et* al. (1982) that the temperature of compost did not go beyond 52°C (Plate 15). This indicated that extent of temperature rise will depend on the type of organic material being composted as well as the size of the pile.

Plate 15: Composting of high C/N ratio materials in pits

6. pH

Decomposition will occur satisfactorily in neutral pH because most of microorganisms grow best under neutral pH conditions. Under aerobic conditions initially a neutral raw material will experience a decrease in pH. The drop is due to the activity of acid forming bacteria. It is followed by an increase in the pH due to proteolytic metabolism and liberation of ammonia from amino acids and account for subsequent rise in pH. Most materials decomposing aerobically will stay within pH range that is conducive for microbial growth thus eliminating any need for pH control (Eberhardt and pipes 1972).

7. Microorganisms and Microbial inoculants

Composting is dynamic process resulting in succession of a group of microbial populations with specific metabolic functions, all of which are interrelated in holistic process. Most of the organic wastes and crop residues carry indigenous populations of bacteria, actinomycetes and fungi, which contribute significantly in the composting process.

The fungi and acid producing bacteria appear during mesophilic stage (Chang and Hudson 1967 and Goleuk *et al.,* 1954). As the temperature increases above 40°C, these organisms are replaced by thermophilic bacteria, actinomycetes and fungi. Spore forming bacteria develop at still higher temperature and finally, mesophilic bacteria and fungi reappear as the temperature again falls.

Many aerobic bacteria are initially present and multiply but after their increased activity as the temperature rises their numbers decrease. A minimum is reached at 55-65°C but their numbers increase again as the temperature drops below 50°C.

The role of mesophilic bacteria may be involved in the raising of temperature of compost for the development of thermophilic bacteria which colonize the compost mass. Mesophilic bacteria, which flourish during a limited time consume most of the readily degradable carbohydrates and proteins. *Bacillus* spp. are especially involved in degradation of proteins, amino acid, peptones and blood meal. Actinomycetes degrade starch very actively and also water soluble fractions of raw materials. Thermophilic bacteria attack proteins, lipids and hemicelluloses.

Lipids are degraded extensively whereas degradation of cellulose and hemicellulose is comparatively slow. Lignin is most resistant compound to biodecomposition. Actinomycetes such as *Thermonosopra* curvata and many fungi are involved in cellulose decomposition. Mesophilic fungi are present in compost as it warms up below 40°C and they are quickly replaced by thermophilic fungi but reappear as the heaps cool to 40°C. They probably persist in the outer layers of the heap during thermophilic stage. hermophilic fungi occur in the temperature ranges between 40°C and 60°C.

Microbial Inoculants

It has been observed in preparation of sawdust compost that inoculation with spores of celluloytic fungus *Coprinus ephemerus* and amendment with plant nutrients (NPK) hastened the period of composting from the usual one or two years to three months and produced a compost that did not immobilize nitrogen as compared with the fresh sawdust (Davey, 1953. Wilde, 1958). The work done in India on microbial inoculants has shown their benefits in composting of farm wastes (Gaur, 1999).

8. Amendment with Rock Phosphate/ Fly Ash

The application of rock phosphate is important in composting when raw material have a wide C/P ratio (rice straw, dry leaves etc.). However, the quality of compost prepared from a mixture of rice straw and water hyacinth was improved when rock phosphate was added both with and without pyrites. The results showed that an application of rock phosphate at five percent (w/w) increased the nitrogen content of compost by 30 percent over control. The enrichment of compost with rock phosphate besides increasing phosphorus content increased the content of calcium and micronutrients, particularly zinc, iron, manganese and copper. The benefit of rock phosphates in composting has been reported by several workers in India (Gaur *et al.* 1995).

The amendment of paddy straw with 2-3 percent of Mussorie rock phosphate (MRP) alongwith combined inoculation of *Aspergillus niger* and *Azotobacter chroococum* accelerated the process of composting by reducing the organic carbon percent and simultaneous enrichment in nitrogen and phosphorus content up to 3 percent of rock phosphate (Gaind and Gaur 2000).

The progressive levels of MRP decreased the carbon percent of biomass significantly and caused significant effect on nitrogen enrichment of composted biomass to 1.7 percent due to application of 2 percent MRP (Table 82). The C/N ratio was brought down to 16.7:1 and the available to P_2O_5 content went upward (25.5 mg/g compost). The quality and maturity of compost was improved due to microbial inoculants used in the process.

Table 82: Carbon, Nitrogen, C/N ratio and available P_2O_5 contents as affected by rock phosphate levels and inoculants (Pooled statistical analysis)

Rock Phosphate (%)	Org. Carbon (%)	Nitrogen (%)	C/N Ratio	Available P_2O_5 mg/g compost
0	38.9	1.5	26.1	17.2
1.0	34.3	1.6	21.3	23.3
2.0	28.0	1.7	16.7	25.5
3.0	25.1	1.5	16.6	26.5
5.0	19.2	1.0	18.9	15.5
CD at 5%	0.9	0.05	0.08	0.8
Uninoculated	29.11	1.31	22.2	22.23
Inoculated *A. niger*	29.06	1.35	21.5	24.51
Inoculated *A. chroococcum*	27.91	1.52	21.5	22.72
Inoculated *A. niger* + *A. chroococcum*	27.83	1.50	18.5	25.36
CD at 5%	0.666	0.036	01.568	0.593

Fly ash was composted with wheat straw and 25 rock phosphate (w/ w) for 3 months. Fly ash addition at 20% level resulted in lowest C/N ratio of 16:1 and highest total and available phosphorus. The microbial population and phosphate solubilizers showed an upward trend with period of composting (Gaind and Gaur, 2003).

9. Destruction of Pathogens and Parasites

Composting temperature in windows can easily rise to 70-75°C and can be held for comparatively long periods. Upto ten days is normal, and if in windrowing plant, turning is done when the temperature has reached a plateau, the temperature will drop sharply during the turning and restore itself a slightly lower peak. This peak will get progressively lower with each turning till the compost is finally matured and the peak temperature is only slightly above the ambient.

The destruction of pathogens and parasites of animal and human should be taken into consideration while composting materials containing sewage and sludge wastes. The fact that fly larval breeding can effectively be controlled by composting solid wastes has been long known.

Acharya (1948) reported that pathogenic organisms in rural solid wastes which may lead to serious outbreaks of diseases, can be destroyed by composting. Pathogenic microorganisms from composted solid wastes could not be isolated (Van Vuren, 1950, Blair 1952). Gotaas (1956) recommended a temperature of 60°C for thermal kill of common pathogens. *Salmonella typhus* does not survive beyond 46°C and death of these organism occurs within an hour at 55-60°C and at 60°C. *Shigella sp.* which cause dysentery and Escherchia *coli* common in sewage are also killed. *Mycobacterium tuberculosis* var. *hominis* does not survive at 66°C or above and their death occurs within less than twenty minutes. *Entamoeba histolytica* cysts which cause amoebic dysentery are destroyed within a few minutes at 45°C and only within a few seconds at 55°C. Tapeworm and hookworm are quickly killed at 55°C. *Ascari lumbricoides* eggs which develop as round worms and suck blood die within an hour at temperature above 50°C. Other pathogens are also killed within a few minutes at 55°C and above.

Bhaskaran (1957) reported that pathogens present in human excreta can be effectively destroyed by thermophilic stage of composting. Morgan and McDonald (1969) found that *Mycobacterium tuberculosis* did not survive composting process.

Wiley and Westerberg (1969) reported that bacterium *Salmonella*, poliovirus type I and an intestinal helminthic or metazoan parasite *Ascaris* lumbricoides were destroyed within a few days during aerobic composting of sewage and sludge at temperature of 60°-70°C. Cooper and Goluek (1977) reported that the population of faecal coliforms, *streptococci* and *coliphage* were reduced drastically during the composting process.

Nandkishore (1980) reported that pile height and degree of aeration influenced significantly. The thermal death point of *Salmonella typhimurium* was found to be 65°C but an exposure at 50°C for twenty four hours was sufficient to kill the inoculated *S. typhimurium* in raw wastes or soil.

De Bartoldi *et al.* (1980) observed that coliform bacteria *Salmonella* sp; Streptococcus sp. and *Aspergillus fumigatus* were inactivated at varying temperature in the range of 58 to 75°C. *Streptococcus* and *A. fumigatus* were relatively more persistent. The maximum temperature of about 70°C was, however, obtained during the first 10 days of composting.

The extent and duration of high temperature in compost heaps as well as antibiosis as a result of mixed population of microorganisms provide reasons to believe that pathogens and parasites are not able to survive the composting process. However, to achieve this physically, at least turnings of the heap are necessary to develop high temperature.

Anaerobic composting at mesophilic temperature range does not satisfactorily destroy the pathogens/parasites in a short time. Anaerobic composting of sewage or night soil should be preceded by aerobic composting for at least a week. The natural death of parasites and pathogens will occur in the anaerobic environment and microbial antagonism will eventually eliminate them but this will generally take a longer period of time, about six months.

The composting process can serve useful purpose to obtain pathogen free manure for use on agricultural or horticultural land. The full benefit of the process can be realized if the variable factors such as moisture, aeration, temperature, carbon-nitrogen ratio and nutrient balance are properly controlled.

Heavy Metal in Compost

It has long been recognized that plants require not only nitrogen, phosphorus and potash for their growth but also small amounts of other metal such as iron, calcium, magnesium, copper, zinc, sodium, manganese, chlorine, cobalt, molybdenum and selenium.

The use of compost can increase the content of these metals in the soil. Although the total metal in soil is increased, but it may not be in available form for the uptake of these metals by plants. Their availability would depend on the soil organic carbon and pH, the higher is the humus content and pH, the lower is the availability of some metals to the plants.

The quantities of lead and cadmium present in municipal compost when it is applied to soil and mixed into the top 12-20 cm of soil gets diluted. Moreover the plants would remove some of the metals for their growth so that the actual build up would be slower.

The only difference between an essential trace element and a heavy toxic metal is one of the quantity and position .If zinc, lead and boron are naturally present in the soil, then they are referred to as essential trace elements and if they are present in sewage sludge, then anticompost or antisludge lobbies refer to them as toxic metals.

In practice, a vast amount of research has been carried out *to* investigate the complex relationship between municipal compost and plants by research workers of several universities and institutions and so for none have found any cause for concern in the use of municipal compost.

Microbiological Process and Possible Control

A biodegradable organic waste is converted into dark brown humus like substances by bacteria, fungi, actinomyces, protozoa etc which are widely distributed in nature. Special cultures such as cellulolytic fungi or cultured animal dung slurry for 7 to 10 days can be sprayed to hasten the process of composting and for sanitization of organic materials.

In composting, a few important factors (Table 83) such as C/N ratio, moisture, aeration, pH, temperature microbial inoculation etc. may be controlled to produce a good quality mature compost.

Table 83: Important factors affecting composting

C: N ratio of feed	**25 to 35**
Particle size	10 mm for agitated systems and forced aeration, 50 mm for long heaps and natural aeration
Moisture content	50 to 60 percent (higher Values possible when using bulking agents)
Air Flow	0.6 to 1.8 m air/day/kg volatile solids during thermophilic stage, or maintain oxygen level at 10 to 18%
Temperature	55 to 60°C held for 3 days
Agitation	No agitation to periodic turning in simple system and short bursts of vigorous agitation in mechanized system
pH control	Normally not necessary
Heap size	Any length, 1 .5 m high and 2.5 m wide for heaps using natural aeration, with forced aeration, heap size depends on need to avoid overheating.
Activators	Use of efficient cellulolytic fungi; and biofertilisers

When composting begins, the mesophilic microflora (25-45°C temperature range) are responsible for the initial bioprocess resulting in decomposition of water soluble and ready biodegradable components like sugars, proteins etc and release of exothermic energy. Thus temperature of the compost heap is increased to thermophilic temperature range (50°- 65°C) replacing mesophilic population by thermophilic species of microorganisms. The rise in temperature is influenced by the nature of organic matter and to a great extent by aeration (oxygen supply) of the heaps. Composting organic residues usually have a wide range of active microflora which participate in the process of composting but Indian research workers have developed celluloytic cultures for decomposition of high C/N ratio material such as straw, leaves, sugarcane trash, bagasse etc. Nitrogen and phosphorous content can be improved by use of nitrogen fixing bacteria, *Azotobacter chroccoccum* and phosphate solubilizing fungi, *Aspergillus awamori* or bacteria, *Bacillus polymya.* Rock phosphate or bone meal at the rate of 1 to 5 percent can also be applied in the composting process to produce nutritionally balanced organic manure.

Some specific earthworms also activate the process of composting Earthworms are helpful in grinding and pulverizing of organic material, making it more conducive to microbial action

II. Methods of Composting

Indore method of composting was developed by Howard (1930) and Bangalore method by Acharya (1938). The composting methods were further improved by Indian scientists during the last 2 decades. The brief description of different methods is outlined in table 84.

There are several options to select an appropriate method depending on the nature and quantity of wastes to be processed, requirement of plant nutrients to be applied including humus substances on the farm, availability of labour etc.

The low nitrogen and phosphorus content of organic manure and their bulkiness are hampering its large-scale production and widespread use although it is well known that organic manures improve physical, chemical and biological properties of soil (Gaur *et al.* 1984).

Researches were carried out in AICRP on microbial Decomposition and Recycling of farm and City Wastes (Bhardwaj and Gaur 1985) to evolve and to develop simple and low cost technology for rapid composting and preparation of enriched compost.

Table 84: Brief description of the composting methods

Method	General Description	Processing Time
Indore method (Pit Method)	Pit Size 1m deep, 1.5 m wide and a suitable length. Organic residues put in layers of 20 cm thickness followed by a layer of dung, urine earth, 15d old compost as inoculum in the ratio of 1:1:1 Pit filled layer by layer within a week	4-6 months
Indore method (Heap Method)	Heap method good during rainy season Heaps prepared overground, size of heap 2m wide at base, 1.5m high and 2 meter long. 20 cm organic matter layer followed by second layer of dung, sewage sludge, biogas slurry, weeds etc. till completion of heap, cover with soil/hay/ plastic sheet to retain heat	3-4 months
Bangalore Method	Basically to compost city garbage & night soil in trenches recommended in areas with less rainfall. 15-20 cm thick layer of city organic wastes followed by second layer of night soil and finally cover with soil	6 months
Improved Microbial Method	Composting in pit or heap using crop residues such as paddy or wheat straw, ragi wastes, sugarcane trash leaves etc. Rock phosphate applied at 1 -3% to narrow down C/P ratio. Celluloytlic fungi are used for inoculating the organic matter. Rapid digestion of cellulose / lignin. Quality of compost improved and process is hastened	6-8 weeks
Enriched compost (Microbial technology)	In addition to the above process compost biomass is inoculated with nitrogen fixing bacteria (Azotobactet) and PSM (Asperg-illus awamori) after 1 month of composting. Quality of compost improved in plant nutrient and humus contents.	10-12 weeks
Phospho-compost	Composting of organic wastes with 25% rock phosphate (5% P_2O_5) alongwith cattle dung @ 10% and fresh compost or FYM @ 1%. Phosphocompost applied at the same rate of superphosphate (P_2O_5) gives the same response in crop yield.	3 months
Vermicompost	Use of earthworms in composting such as Eisenia fetida, Eudritus eugianiae or Perionyx excavatus, vermicompost include degraded organic matter plus worm casts, earthworms, their cocoon and other microorganisms Pit Size : 1.25m length, 0.6m width, and Oi5m depth -Organic materials and animal dung etc. are used for composting	3 months

Rapid Composting

It generally takes about 6 months to get a finished compost. One of the objectives of research work conducted by ICAR project was to evolve a simple low cost technology for rapid composting. Rapid method will be more acceptable because there would be greater turnover in a lesser area of land with less effort (Plate 16).

Plate 16: Rapid composting of farm wastes in cemented compost pits at IARI, New Delhi

Cellulolytic Microorganisms as Inoculants

Fungal isolates such as *Aspergillus niger, Penicillium* sp. *Trichoderma* viride, T. *ressei, Trichurus spiralis, Chaetomium* sp and *Paeciliomyces fusisporus* were found efficient in hastening the decomposition of cereal crop residues/straw, banana leaves etc, (Gaur *et al.* 1982). Several experiments were conducted to study the role on effect of cellulolytic fungi in composting of crop residues richer in cellulose, hemicellulose and lignin.

The findings of several trials conducted with different type and combination of organic materials under different agroclimatic conditions with and without cellulolytic inoculant on the quality of finished have been summarized (Table 85).

Within 8 or 10 weeks a good quality compost from paddy straw, cereal residues or leaves was prepared which contained about 1.38 to 1.76 percent nitrogen with C/N ratio 18.9 to 12.3:1. The beneficial effect of cellulolytic fungi in the composting of dairy farm waste mixed crop residue, sugarcane trash has also been reported. The use of cellulolytic culture not only helped to prepare compost in 8-10 weeks but also reduced the bulkiness of manure by 5-10 percent. In presence of the inoculants, the enrichment in total nitrogen was quite pronounced.

Table 85: Enrichment and hastening of compost with celluloytic fungi

Location	Substrates	Total N%		C/N Ratio	
		Uninocu-lated	Inoculated	Uninocu-lated	Inoculated
New Delhi	Sorghum + wheat straw	1.38	1.82	20.6	12.0
New Delhi	Banana leaves	1.90	2.80	19.3	11.6
New Delhi	Jamun leaves	1.44	1.56	20.4	15.7
New Delhi	Paddy Straw	1.15	1.30	23.9	18.9
New Delhi	Paddy Straw	1.52	1.76	15.5	12.3
Ranchi	Wheat straw+Water hyadnth	1.00	1.37	28.4	18.6
Ranchi	Rice Straw	0.90	1.20	41.0	30.0
Hissar	Mixed Crop Residues	1.38	1.52	24.0	20.0
Pune	Sugarcane trash	0.98	1.54	35.0	24.0

Cattle Dung as an Alternative to Cellulolytic Culture

Since the use of cellulolytic inoculant for rapid composting is often constrained due to non-availability of cultures to the users, the effectiveness of animal dung as starter cultures for rapid composting was evaluated (Table 86).

Table 86: Effects of Cellulolytic fungi and Cattle dung as starter in composting of organic wastes. Starter Culture

Starter Culture	Rice Straw- Water Hyacinth Mixture (1:1)			Sugarcane Trash		
	Total M (%)	C/N Ratio	Wt. Loss (%)	Total N (%)	C/N Ratio	Wt. Loss (%)
Check	1.03	23.0	52	1.25	25.2	57
Trichurus spiralis	1.31	18.3	60	1:48	19.0	65
Pencillium digitatum	1.18	20.3	55	1.31	202	61
Paecilomyces	1.31	19.3	59	1.40	21.2	63
fusisporus						
Aspergillus niger	1.28	20.0	57	1.45	21.1	65
Cattle dung (10% on dry wt. basis)	1.30	19.4	56	1.46	18.5	60
Original material	0.88	58.4	-	0.36	124	-

It was confirmed by repeated experiments carried out at different locations that application of about 10% fresh cattle dung on dry basis to the composting of organic residue was as effective as the use of cellulolytic inoculants. This was clearly demonstrated by the average compiled results from such trails.

Enrichment of Compost with Microbial Inoculants

Enrichment of compost with nitrogen fixing bacteria (*Azotobacter chroococcum)* and phosphorous solubilizing fungi improved the nutrient content of the final product (Gaur 1987,1999)

Inoculation with *Azotobacter* and phosphate solubilising culture in presence of 1% rock phosphate is a beneficial input to obtain a good quality compost rich in N and available P_2O_5 (Table 87). The humus content was also significantly higher in material treated with microbial inoculants. The quality of compost produced from dairy farm waste was improved by its inoculation with *Azotobacter* and *Aspergilus* awamori. Similarly sugarcane trash compost containing 1.34% N with a C : N ratio 26 was obtained with bioinoculants. Using chopped paddy straw, compost with 1.78% N and C:N ratio of 13.3:1 was produced with the inoculants against Composting of Organic Wastes 1.30% N and C:N ratio 20:1 without the inoculants .Banana leaf compost prepared with similar technology, contained 2.8% N and a C : N ratio of 11.6 against 1.90% N and C:N ratio 19.3 in the control treatment.

Table 87: Enrichment of compost with bioinoculants

Location	Substrates	Nitrogen %		C: N ratio	
		Without inoculation	With inoculation	Without inoculation	With inoculation
New Delhi	Wheat Straw + Sorghum stalk	1.38	1.82	20.6	12.0
New Delhi	Paddy Straw	1.52	1.82	15.6	11.6
New Delhi	Paddy Straw + Leucaena (4:1)	1.16	1.85	22.2	12.2
New Delhi	Chopped Paddy straw	1.30	1.78	20.0	13.3
New Delhi	Unchopped Paddy straw	1.26	1.72	25.2	15.5
New Delhi	Banana leaf	1.90	2.80	19.3	11.6
Kanpur	Dairy farm waste	0.58	0.64	20.1	15.5
Pune	Sugarcane trash	1.05	1.34	26.0	33.0

Phosphocompost

The majority of Indian soils are low in available phosphorus. Phosphatic chemical fertilizers are expensive and in short supply. Moreover the fertilizer use efficiency of phosphatic fertilizers is quite low since it gets chemically fixed by soils. organic manure helps in making available the fixed soil phosphorous. In this context utilization of low grade rock phosphate which is less expensive can be used to prepare phosphocompost resulting in bioconversion of insoluble phosphorus into available form for ready utilization by crops. The researches were conducted by the scientists associated with ICAR project on recycling of farm and city waste on the production of phosphocompost and its effect on growth and yield of crop plants.

Powdered rock phosphate composted with organic material was found to be effective as a source of superphophate in augmenting yield of crops. Organic waste is composted with cattle dung, soil, fresh compost or FYM in the proportion of 8:1:0.5:0.5. This organic mixture is composted with low grade rock phosphate applied at 2 different rates to produce phosphocompost. The composition of phosphocompost prepared with 25% Mussorie rock phosphate (5%P_2O_5) contained high amount of citric acid soluble phosphorus (15.24 mg P_2O_5/g compost).

The use of phosphocompost as a source of phosphorus was investigated on crops under field condition. The yields of crops showed that phosphocompost supplying 60kg P_2O_5/ha was comparable to single superphosphate applied at the same rate (Table 88 & Plates 17a, b, c; 18 a, b; 19 a, b).

Table 88: Influence of phospho-compost on grain yield of crops

Source of Phosphorus	**Wheat (kg/ha)**			**Ragi (kg/ha)**
	Hissar	**Kanpur**	**Palampur**	**Bangalore**
Superphosphate (60 kg P205)	4760	2570	3045	2810
1 500 kg phosphocompost (5% PA) added as rock P	4610	2320	2890	2950

Plate 17a: Effect of NK without phosphorus on rice

Plate 17b: Effect of superphos- phate (60 kg P_20_5) on rice

Plate 17c: Effect of phosphocompost on rice

Plate 18a: Effect of NK + Phophocompost (1.5 t/ha) containing 25% MRP/ 5% P_2O_5 on wheat

Plate 18b: Effect of NK + Phophocompost (1.5 t/ha) containing 12.5% MRP/25% P_2O_5 on wheat crop

Plate 19a: Effect of superphos- phate (60 kg P_2O_5ha) on cowpea crop

Vermicomposting

It is a biological process to convert solid wastes of animal and plant origin into valuable organic manure under aerobic conditions through the activity of earthworms (Plates 20-22). Along with earthworms, bacteria, fungi and actinomycetes participate in the process of vermicomposting. The waste such as kitchen, city and swage waste can also be utilized in Vermicomposting. It is an appropriate method for disposal of nontoxic solid and liquid wastes. It is cost effective in recycling of poultry, horse, piggery, animal dung etc. using low energy. Earthworms feed on organic wastes, consume 2 to 5 times of their body weight and after using 5 to

Plate 19b: Effect of phosphocompost (1.5 tones/ha) prepared from 25% Mussorie rock phosphate (5% P_2O_5) on cowpea crop

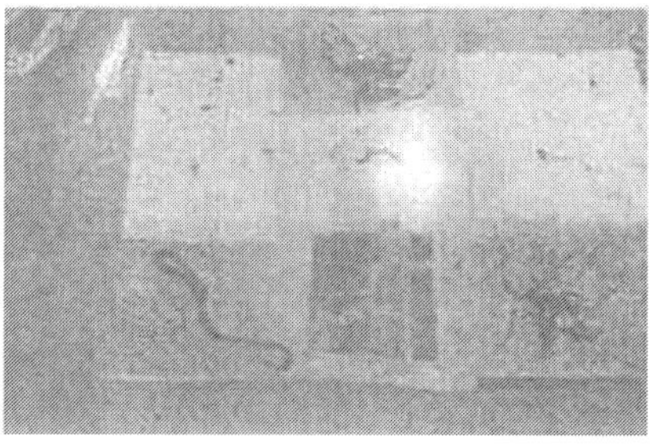

Plate 20: Earthworms convert organic wastes into Vermicom- post under aerobic conditions

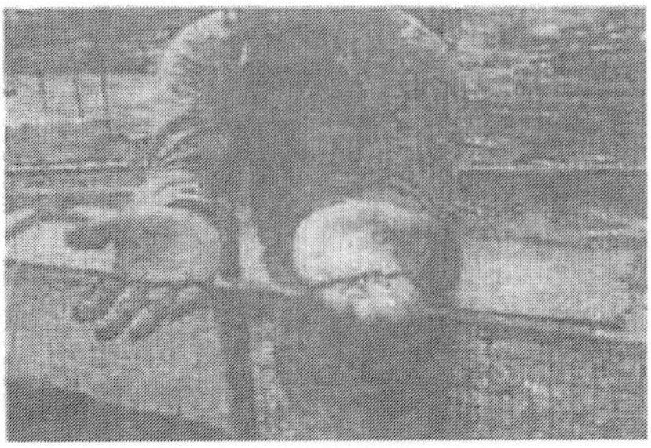

Plate 21: Earthworm compara- tively large size in vermicomposting also used

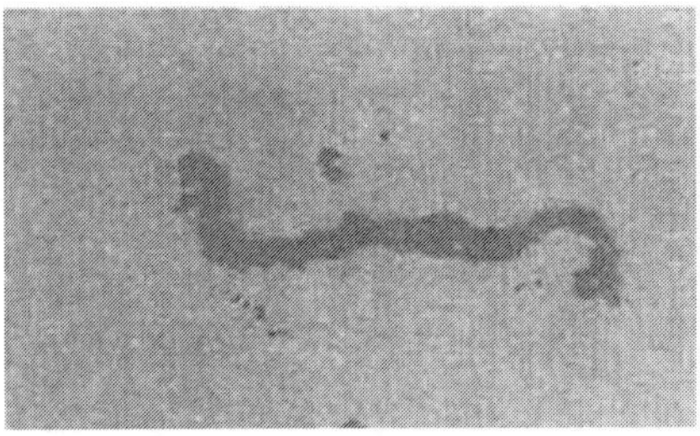

Plate 22: Commonly used earthworm Eisenia fetida in vermicomposting

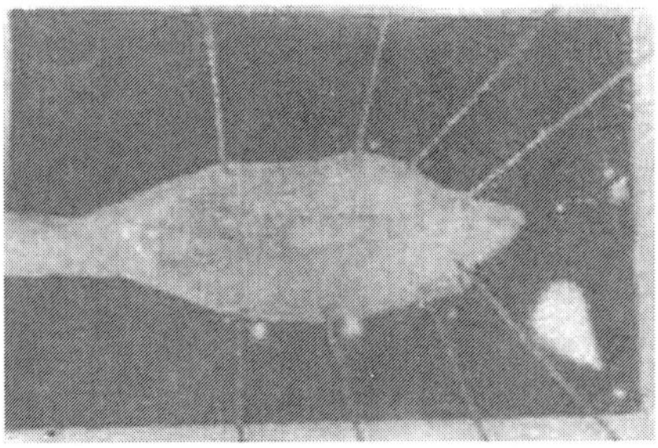

Plate 23: Gizzard in Drawida ghatensis (endogen)

10% of feedstock for their growth, excrete mucus coated undigested matter as worm casts. Worm casts consist of organic matter that have undergone physical and chemical breakdown through activity of muscular gizzard (Plate 25), which grinds the material to a particle size of 1-2 micron. Vermi- casts are good sources of plant nutrients, enzymes, growth hormones etc. Vermicompost is mixture of worm casting (faecal excretion) organic material including humus, earthworms, their cocoons and other microorganisms.

Earthworms and Food Quality

Food produced with the aid of earthworms, bacteria and plant interactions is considered of good quality and healthy product. It has no residual nitrates and pesticides and thus is safe to consume. The spoilage due to fungal attack and insect infestation is less. The keeping quality is improved. Food thus produced by use of organic methods could be certified based on residual nitrates levels which can be easily detected than pesticide residues. Organic produce gets dehydrated without getting spoiled.

Vermicomposting has attracted the attention of farmers and scientists alike world wide, since it is a natural organic product which is ecofriendly and leaves no adverse residues either in food or the environment.

About Earthworms

They are long and cylindrical in shape and varying in size having a large number of grooves on the body. Earthworms belong to phylum Annelida of animal kingdom. There are 2500-3000 species of earthworms in the world which are adopted to different range of environment. More than 350 species have been identified in India. The species of earthworms are differentiated mostly on the basis of external features such as length of earthworms, numbers of segment, position of clitellum and external apertures present on the body.

Life Cycle of Earthworms

Although hermaphrodite two mature earthworms are required to propagate. At the time of egg laying, the clitellum is transformed into hard gerdle like capsule called cocoon. Shedding of cocoon takes place about 10 days after fertilization (Plate 24).

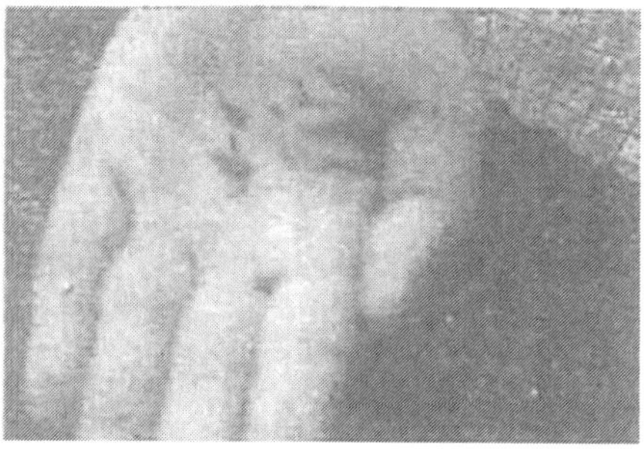

Plate 24: Cocoons of earthworm

The number of fertilized ova in each cocoon ranges from 1 to 20, only a few of them survive and hatch. The juvenile earthworms crawl out of cocoons, maturing of juvenile and again formation of cocoons takes a period of 50-60 days. Normally the average life span of an earthworm is about 1 year during which they may produce about 10 times.

Type of Earthworms in Vermicomposting

i) Epigeics (surface feeders) are important in vermicomposting. The epigeics such as *Eisenia fetida* and *Eudrilus eugianiae* are exotic worms and Perionyx excavatus is a native one which are being used for vermicomposting in India.

ii) Anecic are feeders on leaf litter and soil at upper layers of soil. This group such as *Lampito mauritti* is indigenous and active in situ decomposition of organic wastes and residues in soil.

iii) Endogenous, They live within soil and ingest large quantities of organically rich soil. They make generally horizontal furrows and deposit their casts within the furrows. Endogenous are pale in color, thick and larger in size. They have low reproductivity and low in decomposition activity.

Both epigeic and anecic groups of earthworms are slender, shorter in length and red to dark brown in color. They have high production activity and efficient in recycling of organic materials. Increased attention has been paid to Eisenia fetida which has been found to be a potential agent in vermicomposting of a wide range of agricultural wastes (Edward *et al.* 1985, Kaplan *et al.* 1980, Gaur, 1999) and can grow at a wide range of temperature varying from 0-40°C (Philip 1981)

Earthworms require safe handling and protection from predators such as rats. The maintenance of proper moisture and temperature are necessary for formation of good quality vermicompost.

Method of Vermicomposting

Selection of Enclosure

A size of vermibed depends on quantity and nature of organic wastes to be composted. The vermicompost can be prepared in wooden boxes, plastic buckets, tanks or pits made of brick or concrete or in heaps (Plates 25-29).

The bottom of wooden box should be lined with plastic sheets .For composting of kitchen wastes, buckets or wooden box of appropriates size may be used. A few holes in the bottom of container should be made for proper drainage of excess water. In outdoor vermicomposting, the site of windrows should be under shade on an elevated area so as to protect it from water stagnation particularly in high rainfall regions and rainy seasons.

Plate 25: Vermibed made of bricks under a tree

Plate 26: Special vermibed for composting animal dung waste with cover

Plate 27: Vermicomposting in different size containers

Plate 28: Vermicomposting in cement pots in a culture room

Setup of Vermicomposting

Vermibed :organic matter of different C/N ratio and maturity are arranged in layers. The bottom most layer of vermibed is formed by placing pebbles, stones etc. followed by coarse sand of about 3cm thickness to facilitate proper water drainage. This is followed by moist layer of loam soil of about 15cm. For epigeic species, soil is not required for preparation of vermibed, but when anecic species or a combination of epigeic and anecic species are used, layer of soil as mentioned above would be necessary.

Plate 29: Vermiculture room

Addition of Earthworms

One hundred epigeic and anecic earthworms are added to this soil, small lots of fresh or dry cattle, goat or pig dung are spread over the soil layer followed by a covering of 10 cm layer of organic matter. Water is sprinkled to maintain optimum moisture. Poultry manure should be handled carefully when it is fresh because it may cause toxicity. Kitchen and farm wastes (Plants 30 & 31) can be used as long as they are properly treated to give proper C/N ratio. The vermiheap is covered either by gunny bags or broad size plant leaves to protect the worms from birds. Moisture should be maintained at 50%. Appearance of juvenile earthworms in about one months is a good sign of the process of vermicomposting.

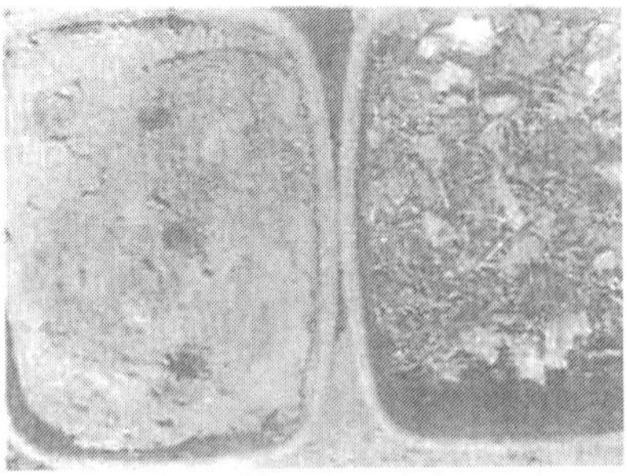

Plate 30: Kitchen wastes, good substrate for earthworms

Plate 31: Collection of organic wastes for vermicomposting

Application of Feed Materials

The renewal of application of feed should be started after a month of composting. After removing the cover, fresh organic matter about 5 cm thick should be spread. Application of thick layer in vermicomposting is not advised because it may kill the worms due to high temperature .The compost biomass should be checked both for temperature and moisture level. Once the vermibed is filled up it is allowed complete decomposition.

Finished Vermicompost and Harvesting

Organic matter is converted into amorphous odourless dark brown vermi compost which also contains vermicasts and cocoons. Further addition of water is stopped. Earthworms move downward and compost is separated and placed on a plastic sheet or ground in the form of cone under bright sunlight which comples earthworms to move in the bottom of compost biomass. The recovered earthworms can be used again for vermicomposting,

The compost may be sieved using 2-2.25 mm (Plate 32) sieve and stored in polythene bags to retain about 15% moisture (Plate 33).

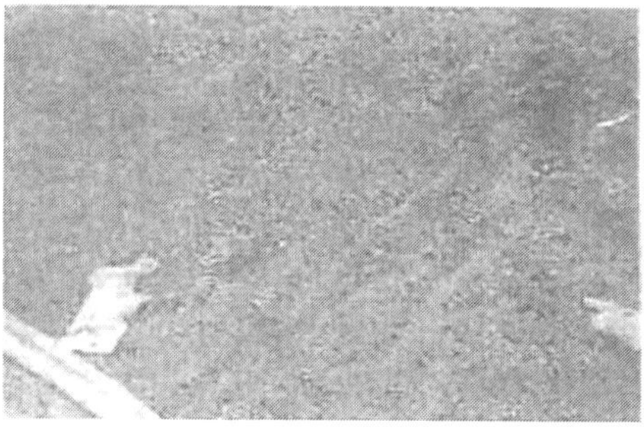

Plate 32: Sieved vermicompost

Plate 33: Baged vermicompost

Precautions

1. Don't cover vermicompost heaps with plastic sheets because it may trap heat and gases.
2. Do not overload the vermicompost heap to avoid thermophilic temperature which will adversely effect the population of earthworms.
3. Dry condition kill the worms and waterlogging drive them away. Thus optimum moisture level is necessary.
4. Addition of higher quantities of acid rich substances such as tomatoes and citrus wastes should be avoided.
5. Precautions should be taken against attacks by red ants who may also feed on cocoons by spraying a suitable decoction for their control.

Vermicompost from Farm Wastes

Influence of *Eisenia fetida* was studied in vermicompost of paddy straw, *leucaena* leaves and vegetable wastes (Saravanan, 2000). During the process samples were drawn at periodic intervals and analyzed for organic carbon, total nitrogen, humus, total and available P and some enzyme activities. Finally the effect of compost was evaluated on growth, yield and nutrient uptake by wheat crop.

There was a progressive reduction in organic carbon (43.6%) over the initial carbon of the substrates and 23.5% over uninoculated control (Table 89).

Table 89: Chemical characteristics of Vermicompost*.

Treatment	Carbon (%)	Nitrogen (%)	Humus (%)	C/N ratio	Total Phosphorus (%)
Paddy Straw (P.S)	34.02	1.24	7.24	27.44	0.073
P.S. + E. fetida	31.18	1.45	8.34	21.88	0.077
Leucaena leaves (L.L)	37.16	4.22	6.04	8.81	0.210
L.L. + *E. fetida*	30.24	4.40	6.98	6.87	0.220

Vegetable Waste (V.W)	39.96	2.94	3.28	13.59	0.450
V.W.+ *E. fetida*	30.56	3.20	4.82	9.55	0.460
P.S.+LL+V.W (1:1:1)	36.51	3.42	3.49	10.74	0.258
P.S.+LL+V.W (1:1:1) + *E. fetida*	28.42	3.65	5.06	7.79	0.289
C.D at 5%	2.41	0.18	0.53	1.19	0.02

*Vermicompost analyzed after 3 months of composting.

The amount of total N was appreciably increased in inoculated vermicompost .The highest value 4.40% was obtained with *Leucaena* leaves inoculated with *E. fetida* resulting in 15.8% increase over initial value 3.80%. The C/N ratio was also significantly decreased due to introduction and presence of earthworms. Highest amount of humus (8.34%) was obtained in paddy straw compost inoculated with *E. fetida*, a 15.2% increase in humus content over un inoculated compost.

Cellulose, hemicellulose and lignin contents decreased gradually. The maximum reduction of cellulose (43.72%) hemicellulose (36.99%) and lignin (6.70%) was recorded in vermicomost when compared to initial value of the substrates. Lignin was decomposed to a lesser extent due to earthworms inoculation. The enzymatic activities of cellulase (2.42 I.U. ml^{-1}) phosphatase (64 u moles g^{-1} h^{-1}) and dehydrogenase (0.48 u moles g1h^{-1}) was observed in vermicompost (Table 90). The variation in the enzymatic activity at different intervals was due to availability of substrates, ingested microorganism and intestinal mucus.

Table 90: Changes in Enzymatic activities during composting

Treatments	**Cellulase (I.U. ml^{-1})**	**Phosphatase (u moles g^{-1} %$^{-1}$)**	**Dehydrogenase (u moles g^{-1}h^{-1})**
Paddy Straw (P.S.)	1.50	28	0.08
P.S.+ E *fetida*	1.85	37	0.14
Leucaena Leaves (L.L)	1.58	26	0.09
L.L.+ *E. fetida*	1.74	34	0.18
Vegetables wastes (V.W)	1.43	23	0.04
V.W. + E *fetida*	1.66	29	0.08
P.S.+L.L.+V.W.	2.08	41	0.26
P.S.+LL.+V.W.+ *E.fetida*	2.42	64	0.48
C.D. at 5%	0.13	3.42	0.02

*Samples analyzed after 2 months of composting.

Vermicompost from Coir Wastes

The nutrient content of compost prepared from coir waste and topica peel prepared with and without earthworms were compared in respect of plant nutrients and C/N ratio (Table 91) The compost prepared with the use of earthworms was lower in C/N ratio (19-23:1) as compared to compost prepared without earthworms (C/N ratio 62:1). The contents of N,P,K Fe, Mn, Zn and Cu were higher in vermicompost as compared to nonvermicompost prepared from coir waste (Ramesh and Gunathilagraj, 1996).

Vermicompost from Farm Wastes Inoculated with Cellulolytic Fungi

Vermicompost was prepared from paddy straw, *Leucena leucocephala* and *Jamun* leaves mixed in proportion of 1:1:1 (w/w). These raw materials were amended with fresh cowdung 30% (w/w) and soil (10%w/ w) Moisture was maintained at 70% during pre- vermicomposting for a period of 2 weeks in pits. Partially decomposed organic biomass was then inoculated with *Trichoderma reesei* (cellulotytic fungi) and earthworms *(E.fetida)* in required numbers.

Table 91: Nutrient content of raw and vermicomposted coir waste

Treatment	N (%)	P (%)	K (%)	Fe ppm	Zn ppm	Mn ppm	Cu ppm	Organic Cprbon	C/N	Microbial Ratio activities (mg of Co2 evolved 100 g sample)
Air dried coir waste	0238	0.011	0.648	1.83	7.768	11.50	Z820	27.72	119.1	116.08
Coir Waste compost	0.416	0.018	0.764	2.21	10.124	14.69	3.60	25.72	61.9	13252
Coir waste vermicompost	1.018	0.040	0.976	2.59	13.402	22.18	5.20	23.14	232	173.60
Coir waste cow dung vermicompost	1.118	0.044	0.990	2.70	13.770	22.14	5.205	17.04	19.1	175.84
CD (0.05)	0.0221	0.032	0.022	0.09	0.1504	1.29	0.638	6.42	5.7	1.067

Dual inoculation with *E. fetida* (earthworm), and cellulolytic fungi (*T. reesel)* was effective in composting due to greater build up of microbial population compared to inoculation with either of them or unincouclated control. The changes in nitrogen, C/N ratio and humus content of compost made with and without earthworms are presented in table 92. The maximum reduction in organic carbon was observed when the material was inoculated with *Trichodema reesei* and *E. fetida.* Earthworms alone or in combination with T. *reesei* decreased organic carbon content by 30% and 40% respectively within 60 days of composting. However, in control the organic carbon was decreased by 2% only during the same period of composting. Nitrogen was increased and C/N ratio was decreased progressively to 18.4 from the initial value of 54.7 in the dual inoculation compared to 31.1 in control. Rock phosphate addition helped in hastening the process of decomposition and resulted in production of compost with higher amount of citrate soluble phosphorous without having any unfavourable effect on the earthworms and microorganisms. Humus content in compost with dual inoculation was 6% compared to 2.9% in uninoculated control and around 5% in other treatments when inoculation was done by either the fungi or the earthworms. Dual inoculation produced compost of good quality containing higher amount of nitrogen and humus with low C/N ratio.

Table 92: Characteristics of Vermicompost inoculated with cellulolytic Fungi

Treatments	30 Days			60 Days				90 Days				
	Org. C %	N %	C/N ratio	iumus %	Org. C%	N %	C/N ratio	Humus (%)	Org. C%	N %	C/N ratio	iumus (%)
Organic matter (O.M)	40.6	0.92	45.1	2.1	38.3	0.98	39.0	2.7	35.7	1.12	31.8	2.9
OM+ *Eisenia fetida (E.f)*	35.0	1.18	29.6	2.5	30.9	124	25.5	3.3	29.1	1.39	20.9	5.1
OM+5% Rock Phosphate *Tnchodemta* reesi (T.R.)	36.0	1.14	31.5	2.6	31.1	120	25.9	3.7	28.7	1.36	21.1	4.9
OM+5% RP+Tr+E.f	352	1.16	30.3	3.8	30.8	1.30	23.6	4.4	262	1.42	18.4	6.0
OM+12.5% RP	36.7	1.12	32.7	2.9	32.3	129	25.0	4.1	27.0	1.30	19.5	5.1
OM+12.5% RP+E.f	34.1	1.14	29.9	3.1	31.6	131	24.1	4.9	272	1.46	18.6	5.5
O.M, (Initial)	44.3	0.81	54.7	—	—	—	—	—	—	—	—	—

Source: I.A.R.I., New Delhi

Vermicomposting of Coconut Plantation Wastes

Studies conducted at Central Plantation Crop Research Institute (CPCRI) Kasargod have showed that one hectare of coconut plantation produces about 8 tonnes of dry organic matter per year. As the use of this biomass as fuel and thatching material has declined over the years, its proper disposal and use has become problem in many coconut gardens.

A survey of soil in the CPCRI at Kasargod has reported about the occurrences of wide diversity of earthworm population comprising, *Perionyx excavatus, Perionyx sansbaricus,* Octochetona spp., *Lampito mauritii, Polyheretima elongata, Eisenia foetida, Perionyx* spp (pink coloured), *Megascolex cochinensis, Megacolex spp.Eudrilus eugeniae* and *Dichogaster*. These worms are being maintained in pure culture in CPCRI.

Eudrilus eugeniae was located in a heap of coconut palm wastes which were found fully decomposed leaving behind only midribs of the leaves. Further surveys have confirmed their presence in cowdung heaps in the Agrostology unit of this institute. Among the local earthworms tested in vermicomposting of coconut palm wastes, this strain of *Eudrilus* was found most effective in composting and hence more studies were conducted.

Preliminary studies have shown that this local strain *Eudrilus* is different from the African night crawler *Eudrilus eugeniae.* This strain of *Eudrilus* located at Kasargod differs from the African strain in its colour, vigour and cocoon colour. The local isolate is more pinkish and vigorous and the cocoon case is thick and dark. It is also capable of forming furrows in soft feedstock materials and vermicasts are deposited around the burrow holes. Also, it is able to decompose organic wastes in absence of cowdung. Much faster multiplication of this worm on coconut palm trees was also observed as compared to the Africans strain. According to taxonomists, this local strain may have to be described as a new species of *Eudrilus.*

Prabhu (1998) have standardized the method of composting of biomass left exposed in the field for 3-4 months. This naturally weathered biomass is mixed with cowdung at the rate of 10 percent w/w and allowed to predecompose for 1-2 weeks before vermicomposting. Earthworms are introduced at the rate of 1kg per tonne of organic waste. The compost windrow/bed is mulched with grass or straw and has to protected from direct sunlight. The windrows are kept moist. The vermicomposting is completed within 2-3 months leaving behind only midrib of the coconut leaves. On an average, 70 percent recovery of vermicompost was obtained.

The composition of vermicompost was organic carbon 17.84%, nitrogen 1.8%, phosphorus (P) 0.216% potassium (K) 0.16%, having a C/ N ratio of about

10:1. Total microbial population and number of other beneficial microbes were also more in compost compared to the raw material.

This technology was also tested in large pits in coconut plantations in coastal sandy soils and was found to work. Efforts are being made to popularize this technology. Earthworm cultures are supplied to the farmers for the vermicomposting of their wastes.

Coir Pith

Coir pith is another organic material, available in huge quantities from Coir processing units and is resistant to decomposition due to presence of lignin. Local *Eudrilus* sp was tested for vermicomposting of coir pith. Coir pith alone was not able to support the growth and multiplication of earthworms. Encouraging results were obtained when coir pith was sieved to remove the fibre pieces and was mixed with about 30 percent cowdung by weight. The compost analyzed after 3 months of bioprocessing showed the following composition :-

Organic carbon 22.2%, Nitrogen 1.0%, C/N ratio 22.0, P 0.48%, K 0.7%. Coir pith with decaying organic matter has been found to undergo fast decomposition. Further studies are underway to use mixed culture of earthworms such as *Pheretima elongata* and *Lampito mauritii* alongwith the *Eudrilus* sp.

Vermicomposting of Pine Needles *(Pinus Roxburghii)*

Pine needles rich in resin are available in forest of Uttranchal hills in abundance lying unused which was utilized for vermicomposting (Rajio Singh *et al.* 2002.). Pine needles not only increase the soil acidity but also do not allow any other vegetation to grow under the trees. Being high resin material, pine needles are neither consumed by animals nor easily compostable for use as organic manure. Moreover, it is a biggest source of fire hazard in forest area of hill regions during summer months in particular.

The organic matter content of cultivated soils in these areas is gradually depleting due to heavy rainfall followed by soil erosion and loss of nutrients. For maintenance of soil fertility in these areas, the forest litter like pine needles must be converted into manure for its use.

The work carried out at Defence Agriculture Research Laboratory, Pithoragarh during 2000-2001 showed the process of vermicomposting of pine needles.

Treatment of Pine Needles

Pine needles are high C/N ratio material (70:1) contain polymers of volatile oils and thus take longer time in decomposition. Therefore a treatment with urea solution to pine needles was given to decrease the C/ N ratio of the pine needles. The chopped pine needles were submerged in varying urea solution of 2-10% for 72 hours and thereafter thoroughly washed with tap water to remove ammonia content. Material was then sundried for 48 hours. The result showed that 6% urea treatment on an average reduced the oleo resin content in pine needles by approximately 50 percent.

The treated sundried pine needles were mixed with cowdung in 1:1 proportion and filled in plastic trays (Size 1.5'x1.0x0.26')with one or two small hole in the bottom of trays to drain of the excess water if any. Adult *Eisenia foetida,* one hundred in number were introduced in each tray. It took about 6 months to convert the pine needle-cowdung mixture into fine vermicompost. Pine needle treated with 6% urea solution recorded highest earthworm population. The control has around 10.54% average resin content which was reduced to 4.90%. Pine needles treated with 6% urea recorded highest earthworm population and average body weight and thus succeeded in vermicompsoting of pine needles in a shorter period. It contained 1.0% nitrogen as compared to control (0.3%).

Method For Production of Vermiwash

Vermiwash is the liquid form of organic manure collected after washing vermicompost and body of earthworms with water. This vermiwash may provide the major and minor plant nutrients which may be present in the easily available form for plant nutrition. This also provides additional plant growth substances that are not found in chemicals. As the vermiwash contains vitamins, hormones, enzymes, antibiotics etc. application of vermiwash to the plants confers pest and disease resistance and also improves plant production and food quality.

Vermiwash production technique can be easily adopted by a farmer. The procedure for production of vermiwash is described below. (Vijayalalitha and Charumithran 2001). A plastic basin of 20-25 cm height can be used as a culture tank. A small hole of 0.5 cm diametrer is put at the centre of the basin and it is kept on a stand. A plastic bucket of 2 litre capacity is put at the centre of the basin to collect the vermiwash. A layer of nylon net is kept at the bottom of the basin to prevent the passage of earthworms alongwith vermiwash and above that is filled with a layer of small stones or pieces of tiles to a height of 5 cm and another layer of coconut fibre. The basin is filled with organic waste and cowdung in the ratio of 8:1. After partial decomposition of waste for 10 days, 100-200 number of *Eudrilus eugeniae* variety of earthworms are introduced in the basin for vermicomposting. The content are stirred once in a week and optimum moisture is kept. After compost is ready within 5-6 weeks, vermiwash can be collected.

For collection of vermiwash 2 litres of cowdung dissolved water is poured in the basin. The water will wash the compost and the body of the earthworms and vermiwash will be slowly collected through the hole in the bucket. This liquid is called vermiwash. This is kept for about 24 hours for draining of all the water. The vermiwash can be diluted 4-5 times and used as foliar spray on crop plants.

Rating of Substrates for Composting

As substrates differ a great deal in their chemical and physical characteristics and hence suitability for composting, a general chart rating them is given below.

Example of Substrates	C:N ratio	Suitability
Fish scrap, poultry manure, night soil activated sludge, pig manure, sheep dropping, meat scraps, cottonseed meal and other oilseed residues.	1-19	Highly suitable due to their high nitrogen content
Sea weed buttercup, amaranthus, lettuce cabbage and other vegetable wastes which are fresh green and succulent including wastes from food processing units	19-27	Moderately Suitable
Saw dust, flax waste straw, coir waste etc. including all crop residues with high lignocellulose content high carbon and low moisture and low nitrogen content.	27-208	High in carbon and low in nitrogen, less suitable but compost-able with amendments and use of microbial cultures

Precautions in Preparation and Application of Compost

1. Minimum time temperature requirement for killing pathogens in compost piles is 18-21 days with continued temperature of about 55°c. Lower temperature requires longer time of composting. The frequency of turning is important for aeration and maintaining high temperature for killing pathogens A heat range of 50-65°C is sufficient to eliminate faecal coliform bacteria to non detectable levels of most probable numbers (MPN) method. *Mycobacterium tuberculosis* was normally destroyed on 14th day of composting, if temperature averaged 65°C, Poliovirsues are destroyed within 30 mts. at temperature range of 60-70°C.

   There are no references to any sanitation workers who have been infected by fungi, as a result of handling and no restriction should be put on the use of compost by farmers.

2. Matured compost as soon as it is ready, should be applied in the soil to get the maximum benefit of components of compost such as humus, plants nutrients and biological activity.

Rural Composting as a Commercial Activity

Although the farming community is fully aware about the benefits of organic manures in crop production but it seems that they are not usually motivated to produce compost even from unutilized rural organic wastes by using appropriate technology probably due to labour cost, collection of organic wastes and social reasons. Therefore, to establish rural composting based on scientific lines, there is an urgent need to commission compost plants at the village, block and district levels as a commercial enterprise with suitable incentives, or subsidies from the government and non government agencies. This will provide employment opportunities to rural youth and small farmers on full time or part time basis. This enterprise may also be planned to run on cooperative lines involving several small and medium farmers including women in a village or a block level.

To run the compost plants efficiently, the village youths and farmers should be trained by establishing training centres so as to learn the special skills of composting, working knowledge for maintaining pure cultures of cellulolytic and lignolytic microorganisms and specific composting cultures of earthworms for vermicomposting.

Enriched and quality compost will be prepared, sieved and graded at these commercial or cooperative compost plants which will bagged and used or sold to local farmers at competitive price. For quality control, enriched and graded compost should be tested for organic carbon, total nitrogen, phosphorus, potassium sulphur, zinc, pH and humus content.

The transfer of appropriate technology at village, block and district level for production and use of compost by small and medium levels farmers should be carried out by state agriculture departments in collaboration with experienced specialists in the line of work.

The use of low grade available natural sources such as rock phosphate, bone meal pressmud, spent wash, non edible cakes, poultry manure, gypsum, fish meal, green leaves, grasses etc. should be extensively used in composting.

Microbial cultures of cellulolylic microorganisms, nitrogen fixing bacteria and phosphate solubilzing fungi or bacteria should be used to improve the quality and acceptance of the new product viz. Organo · biofertilizer or Organo-phos.

Vermicomposting technology should be transferred to rural areas to be practised by women farmers to produce vermicompost as a business activity. This would not only provide employment but also profits from the sale of vermicompost for use in kitchen gardens, vegetable production and in orchards.

III. Maturity of Compost

The criteria of maturity of compost is essential for achievement of the objectives of composting. This can be judged by studying several indices such as BOD (biological oxygen demand) VFA (volatile fatty acids) CEC (Cation exchange capacity), nitrate-N ratio of reducing sugars, C: N ratio and the germination test. (Haga, 1990). Compost prepared from good quality organic material in general contains 40 kg nutrients/tonne (1.5% N+ 1.0%P2O5 + 1.5% K2O). From utilization point of view of the final product, the information on nutrients contents as well as on presence of any heavy metals or phototoxic substances are equally important. Humus content in the compost can be important criteria to judge the maturity of compost.

The standards for characteristics of compost in most cases are not available though this aspect is receiving attention. As an example, Anthonis (1994) has suggested the following parameters (Table 93).

Table 93: Standards for good quality compost

Minimum Nutrient Content	1-3% N, 1.5-3%P2O5, 1.5%K2O
Moisture Content	Not Exceeding 15-25%
Organic Matter	At least 20% Carbon
C:N ratio	Between 10: 1 to 15 :1
PH	Around Neutral, 6.5-7.5

A wide range of methods (Table 94) to evaluate maturity of compost have been recommended. The method such as physical, chemical and biological may be selected to judge the maturity and quality of compost depending on the infrastructural facilities and expertise. For quality control, some basic criteria have been described below.

Quality Control

Compost marketed for agriculture use must indicate specifications regarding its origin, chemical composition and degree of stabilization. The raw materials used in composting and their origin should be specified. If blended or mixed substrates are used the ratios of the different substrates may be mentioned. The physical, chemical and biological properties of compost vary depending on the type of raw materials such as agricultural, municipal or agro-industries.

Table 94: Methods to evaluate maturity of compost

Physical	Chemical	Biological
Particle size*	pH*/**	Microbes*
Colour-	EOT*	Annelids

Moisture*	C/N ratio */**	Worms
Porosity	Nitrogen*	Collembola
Bulk density	Sulphur *	Thermopiles
Water holding capacity	Starch - iodine test*	Pathogens Microbes/**
Temperature *	Heavy metals	Verticillium test*
Total solids	Pesticide residues	Plant growth response test
Volatile solids	Plant nutrient content	
Total ash content	Cellulose	
	Lignin	
	Protein	
	CECT*/**	
	Humification Index	

*Regular/** Quality control test

Many methods have been suggested and evaluated to assess the maturity of compost as a medium for plant growth.

1. Chemical Method

The chemical methods are often more rapid and take less time.

Carbon/nitrogen Ratio

Carbon and nitrogen are the building blocks of plant and animal cells and they are important in microbial activity. Hence they are the most studied parameters during the composting process. The C/N ratio below 20:1 is considered to be indicative of mature and stable compost.

Nitrogen Forms

The changes in ammonia, nitrite and nitrate forms have also been suggested as indicators of maturity. Spohn (1978) reported that the presence of ammonia indicated that the compost was not ripe where as the presence of nitrate nitrogen indicated that it was well ripened. However, it was indicated by him that N test alone were not sufficient and that should be complimented with sulphide test and a cress seed germination test.

pH

The change in pH during composting period does occur and has been considered as possible indicator of the maturity of compost. Generally, the pH drops during the initial period composting and then increase to a range of 6.5 to 8. Jimenez and Garica (1989) noted that acid pH values indicate to lack of maturity due to short composting time or the occurrence of anaerobic conditions. Often with very acid food processing wastes (grape, pomace, cranberry) the initial pH is acidic and it remains acidic even under aerobic conditions for a long period of time.

Others

The other parameters such cation exchange capacity (CEC) changes in the major organic constituents of raw material such as cellulose etc., humification parameters, optical density, BOD etc. may also be considered for establishing the maturity of compost.

2. Physical Method

Temperature and Heat Output

The judgment of decomposition of compost can be made by recording the temperature during several stages of composting. It can only be guide and the comparison with plant growth requirement have to be made.

Colour, Odour and Structure

Mature compost should be dark brown to black in colour regardless of the raw material used. Mature compost should smell like forest soil.

Typical soil odour is caused by actinornycetes. The smell of soil is primarily the result of two gases such as geosmin and 2-methylisoboneol which are by products of fungi and actinornycetes. If these two gases are present in the compost, it is possible that they could be used to determine the maturity of compost. Odour and colour are too subjective to provide accurate assessment of maturity and structure is difficult to be described and measured accurately.

Plant Test

Plant test are conducted to test maturity (Gaur 1982). Only plant test can give the final answer whether or not compost is matured. A feasible plant test is using cress which gives result within five days from planting (See Chapter 21).

Microbiological Test

The microbiological - related tests are:

i) Carbon dioxide evolution

ii) Oxygen depletion/uptake

iii) Microbial changes in the population of fungi, actinornycetes, bacteria etc.

iv) Enzyme Activity

The following parameters, however, are suggested by the author for quality control of compost (Table 95).

Table 95: Parameters for Quick evaluation of the quality of compost

Parameters	**Good Quality**	**Poor Quality**
Colour	Brown to Black	Varies
Smell	Pleasant earthy smell	Bad odour
pH	6.5 - 8.0	<6,>8
C/N ratio	10-20	<10,>20
Temperatures	30-45° C	>45°C
Moisture	25-30%	>30%
Humus %	>4%	<4%
Plant Assay	Good Growth	Growth Inhibition
Nitrogen	>1.25%	<1%

A well humified and stabilized product is a valuable organic fertilizer for sustainable agriculture. It can be easily handled, stored or directly used on land without any negative effect on the environment. Thus the criteria for judging a good quality compost are dark brown to black in colour, crumbly structure, largely insoluble in water but dissolves to a large extent in dilute alkali, sodium pyrophosphate or ammonium oxalate, has a carbon: nitrogen ratio of 10 to 20:1 and has beneficial effect on the soil properties and the growing vegetation.

References

Acharya, C.N. 1948, Indian Fmg. 9 :125.

Athonis, G. 1994. Agro-Chemical News -in- brief 17(2): 12.

Bhaskaran, T. 1957. Indian J. Agri. Sci. 27: 91

Blair, M.R. 1952. Pub. hith., Johannesberg.15: 70

Chang Y and Hudson, H.J. 1967. J. Ecol. Studies. Trans. Br. Mycol. Soc. 50: 649.

Cooper, R.C. and Golueke, C.G. 1977. Compost Sci. 18:8

Davey, C.B. 1953. Soil. Sci. Soc. Am. Pro. 17: 59

De Bertoldi et a/1980. Compost Sci. and utilization 21: 32

Eberhardt, D.T. and. Pipes, W.O. 1974. Illinois composting applications In: Large Scale composting (Ed. M.j. Satraine) Noyes Data Corp. Park Ridge, New Jersey.

Gaind, S. and Gaur, A.C. 2000. GEOBIOS. 2721 -24

Gaind, S. and Gaur, A. C. 2003. Bioresource Technology 87(1): 125-127.

Gaur, A.C. 1982, A. Manual of Rural Composting. FA.O. Projects Field Document No 15, pp 102.

Gaur, A.C. 1999. Microbial Technology for Composting of Agriculture Residues by improved methods., I.C.A.R. New Delhi, pp 78

Gaur, A.C. Neelkanthan, S and Dagran, K.S. 1995. Organic Manures. ICAR, New Delhi pp.159

Gaur, A.C. and Sadasivam, K.V. 1980. Recycling of organic wastes in Agriculture, Research Bull. No. 21,1.A.R.I. New Delhi-12 pp50

Gaur, A.C. and Sadasivam, K.V.Magu, S.P. and Mathur, R.S. 1982. Agri. Wastes 4:453-468

Gotaas, H.B. 1956, Composting - Sanitary disposal and reclamation of organic wastes, W.H.O., Geneva

Haga, K. 1990. Norsk land brusk forsking 4: 245

Howard, A and Wad, Y.D. 1931. The Wastes Product of Agriculture. Their utilization as Humus. Oxford university Press, London

Hutchinson, H.B. and Richards, E.H. 1921. J. Min Agri 28: 398

Jimenz, E.J. and Garcia, V.P. 1989. Biol Wastes 27: 112-142

Morgan, M.T.. and Mc Donald, F.W. 1969, J. Environ Health 32: 101.

Nandkishore, T.G. 1980 Studies on Urban Solid Waste Composting M.Sc. thesis, Univer. Agric. Science, Bangalore pp166

Prabhu, S.R. 1998. Tree World 8 (4): 3-4

Rajio Singh, D.S. Bhakuni and Narendra Kumar 2001. A potential biowaste for preparation of vermicompost in central Himalayas. Defence Agriculture Research Lab. Pitthoragarh (Uttranchal) Ramesh, P.T. and Gunathilograj, K. 1996. Madras Agric. J. 583:26- 28

Saravan and, P.M. 2000 influence of Earthworm in Biodegradation of organic residue and microbial activities. Ph.D. Thesis, PG School IARI, New Delhi

Van Vuren, J.P.J. 1950. Compost Bull No. 310. Dept. Agric., Union of South Africa p. 39

Vijaylalitha, S.V. and Charumithran, K. 2001. CARD-Krishi Vigyan Kendra, Kolabhagom, Kerala

Wilde, S.A. 1958. Forest Products J. 8: 323

Wiley, J.S. 1957. Proc. 12th Indust. Wastes conference. Purdue University. Engr. Bull.17::596.

Wiley, J.S. and Pierce, G.A. 1955. Proc. Amer Soc. Civil Engr. J. Sanit. Engr. Division 81:846.

Wiley, B.B. and Westerberg, S.C. 1969 Appl. Microbiol. 18: 994.

11

Underground Biodiversity and Biofertilizer

The biosphere or part of the earth and its atmosphere where life is found is made of different size of ecosystem regulated by solar energy. Ecosystems have both biotic and abiotic components. The biotic component comprises closely interacting groups of flora and fauna having varying energy needs which comprise heterotrophic, chemoautotrophic and photoautotrophic microorganisms.

Biodiversity refers to the richness of life as manifested by their morphological, cultural and biochemical characteristics of microorganism and their interdependence in habitat. It is because of biodiversity that each ecosystem functions as the basic life support system of the planet earth. Biodiversity sustains the food and nutrition security of human beings and mediates the ecological process by resisting environmental perturbations in a given soil and climatic conditions.

Presently greater emphasis is being placed on the biodiversity of above ground biotic components belonging to plant and animal kingdom. The living organisms in soil and their diversity in relation to their contribution to sustainable agriculture often do not receive adequate attention. Without taking into considerations the magnitude and understanding of the global diversity will be fragmentary. However, management of below ground biodiversity is being focused presently by several national and international bodies.

I. Biodiversity in Soil

Soil is a natural habitat for diverse groups of macro and microorganisms related to the plant and animal kingdom. They vary in Underground Blodiversity and Biofertilizer shape, size and morphology and are found from microscopic to macroscopic bodies belonging to flora and fauna. There are many organisms which are beneficial in improving soil fertility and plant productivity but some pathogenic types damage the agricultural crops by causing damage and plant diseases such as root and stem rot.

The decomposition of plant and animal residues in soil, mineralization (conversion of organic nutrients to inorganic form) of plant nutrients, solubilization of insoluble forms of nutrients, biological nitrogen fixation, mediation in natural cycles etc. are processes in which soil biota actively participate. In the absences of soil organisms, the biological processes will stop, making life of plants and animals impossible on the earth.

The underground biodiversity comprises of several types and forms of fauna and flora

Soil Fauna

Their biomass is smaller as compared to soil flora. Soil fauna or soil animals comprise large varieties of organism with different shape and size and adaptive strategies. Based on their size and adaptive strategies, they are divided into microfauna, mesofauna and macrofauna

Microfauna

Organisms belonging to this group are very small with an average size of less then 0.2 mm. They are found in soil pores and water films that cover soil particles and feed on microflora and other microfauna. Many are parasitic on higher plants and some persist on plant and animal residues. Nematodes and protozoa are the main representatives of this group.

Protozoa are one celled organisms confined mostly to surface horizon of soil. They are slightly bigger than bacteria and usually the most varied groups such as flagellates, amoebae and ciliates. The availability of food materials, temperature, aeration and pH are the factors influencing their population Their population in normal soils varies between 10,000 and 1,00,000 per g of dry soil. Protozoa are important predators of bacteria resulting in partial sterilization of soil exerting a regulatory control on the bacterial population of soil. They may prove beneficial in control of pathogenic microorganisms in the root zone of crop plants.

Nematodes are found in all types of soil and are among the most abundant soil fauna. Their numbers are highest in cultivated soils and ranging from 10,000 to 10 million per square meter. Nematodes are composed of those that are predatory on bacteria, fungi and protozoa, parasite on higher plants and those that feed on decaying organic matter. Nematodes often cause serious damage to plants and over 500 species of plant parasitic nematodes are known to exists in soil. These round and spindle shaped microfauna have adaptable mouthparts and pointed form. As such, it is easy for them to penetrate plant tissues. Though a major crop pest, nematodes also have the potential to damage and suppress the population of fungi, bacteria and actinomycetes in the root zone.

Mesofauna

Organisms of this group range in length from 0.2 to 2 mm. Population densities in favourable soil conditions, can rise up to 10 million per meter of soil surface. Mites and collembola are of common occurrence in soil. These organisms degrade plant litter and are predators of other mesofauna and fungi. Although present in all soil layers, maximum population is recorded in the rhizophere of plants.

Macrofauna

This group comprises of small mammals, insects, mites, centipedes, millipedes, slugs, snails, spiders, earthworms etc. They exceed 10 mm in size and include predators and feeders of plant materials. They contribute to soil fertility by mixing organic matter in top and subsoil thereby improve physical condition and other properties of soil. On death, their bodies are added to soil organic pool witch are degraded in soil and release plant nutrients. The most important members of this group from point of view of soil fertility are termites, ants and earthworms.

Termites and ants are very common in tropical, subtropical and mediterranean soils. Through their activities, they modify the organic substance of soil either by digestion or translocation. Termites utilize undecompensed organic matter as food resource and thus play a role in their degradation in soil. Their biomass in soil is much less then earthworms. Termites may damage the growing plants when present in large numbers in soil, depleted of all organic matter on the surface soil making it vulnerable to erosion.

Earthworms are well known representative of soil macrofauna having the maximum influence on the physical and biological properties of the soil. They are found in large numbers in soil which are moist and rich in organic matter, exchangeable calcium and phosphates. In rich cultivated soil, their population may range between 0.6 to 2.5 million per hectare with live weight ranging between 300 to 1250 kg per hectare. In unfavourable soils, their population may drop to even less than 30,000 per ha with a biomass of about 15 kg per hectare.

Earthworms are classified into detrivores and geophages. The detrivores include epigeic and anecic forms and geophages comprises the endogenic earthworms. The epigeic bring the fragmentation of organic material restricted on the soil surface. This activity has no effect on soil structure. The anecics feed on litter mixed with soil and after digestion produce surface casts. The endogenic earthworm live within the soil deriving energy and nutrition from organically rich soil which they ingest. The common examples of detrivores which are involved in humus formation and verimcomposting are *Perionyx excavatus*, *Eisenia feotida*, *Eudrilus euginae*, *Lempito mauriti*, *Octachaetona* serrate and *Octachaetona thustoni*. Earthworms ingest organic debris alongwith sufficient soil particles. The ingested material is subjected to enzymatic and grinding action before excretion as 'casts'. In rich soil, the casts may work out to be more than 15 tonnes per ha. The organisms may pass through bodies annually around 40 tonnes soil plus organic matter per hectare. It is claimed that earthworms casts are richer in organic matter, available plant nutrients, and microbial population. The accelerated microbial population and their activities are helpful in improving the physical, chemical and biological properties of the soil. However, data substantiating these claims are fragmentary, hence systematic studies need to be taken up to confirm the claims of the promoters of vermicomposting.

Soil Flora

It is represented by diverse types of microorganism such as bacteria, actinomycetes, fungi and algae. The living root of higher plants are also important component of soil. The biomass of soil flora is higher than the biomass of soil fauna. This group performs important functions such as soil granulation, organic matter mineralization, fixation of atmospheric nitrogen, solubilization of phosphates, production of plant growth promoting substances etc. The organic transformation initiated by soil fauna are carried forward by microflora and cumulative effect of interaction such as synthesis of microbial polysaccharides, binding effect of fungal mycelia and plant roots bring about significant improvements in soil structure and physical condition of soils. Soil enzymes, Vitamins, amino acid and hormones etc. are produced in soil due to microbial processes.

Bacteria

Soil bacteria is the most abundant group of all microflora present in soil. Indian soils in normal arable condition have 10 to 25 million per gram dry soil. These are single celled simplest and smallest forms of life, present in the soil. Their growth rate is faster than other microflora of soils, Most of the soil bacteria are heterotrophic and depend on their needs of carbon and energy on native and added fresh or dead organic matter. Heterotrophic bacteria alongwith other microorganism, extensively participate in organic matter decomposition, release of plant nutrients and humus formation. The chemoautotrophic are of special significant because they participate in nitrification process and nutrient cycling such as sulphur, iron etc. Soil bacteria in general are involved in cycling of plant nutrients in different ecosystem.

Soil bacteria as a group are quite prominent in fixation of atmospheric nitrogen and is of great relevance in sustainable agriculture. Nitrogen fixing bacteria form symbiotic, associative symbiotic or asymbiotic association with higher plants. They form symbiotic associations with higher plants of the leguminous group. The belong to different species of Rhizobia. In some legumes such as *Sesbania* rostrata nodules are formed on stem as well. The inclusion of leguminous trees, pulses and oilseed crops in the farming system or crop rotation is of great importance in substantial gains of atmospheric nitrogen.

The associative and asymbiotic bacteria are categorized into aerobic, microaerophilic, facultative and anaerobic. The typical examples of nitrogen fixing bacteria in this area are *Azotobacter, Clostridium* and others such as *Bacillus* spp, *Klebsiela planticola, proteus vulgaris*, *Acetobacter, Alcaligens* and *Herbaspirillum* etc.

This soil health, namely the population of beneficial bacteria is influenced by soil conditions and farming system. Organic matter, moisture, temperature, pH and nutrient content affect the growth and activity of soil bacteria.

Actinomycetes

They are unicellular and filamentous organism. They are smaller in size and closer to bacteria. They occur in large number in well aerated moist soils, rich in organic matter. They are highly sensitive to acidic conditions and do not function in soil pH below 5. Their population.density in India is found to range between 10-30 lakhs per gram dry soil. These organisms are active as antibiotic producers and participate in decomposition of resistant plants material components like lignin and cellulose. *Frankia*, an actinomycete, is known to possess ability to fix atmospheric nitrogen in association with non-leguminous shrubs and trees of *Casuarina, Alnus, Myrica* etc. The quantity of nitrogen fixed has been found to vary from as low as 12 kg to as high as 320 kg per hectare per year.

Algae

Algae are present in diverse location such as land, fresh water and sea. The organisms present in upper layers of soil are chlorophyll bearing and derive their energy from photosynthesis. Algae are also found in subsoils as spores or cysts or as vegetative forms that do not depend on photosynthesis. These forms, however, represent only a small fraction of the total algae biomass.

These microorganisms are not as numerous as bacteria, actinornycetes or fungi. Their development on the surface of cultivated or virgin moist lands is frequently noted by naked eye but isolates can be obtained from lower depths as well. The counts of algae usually range from about 100 to 50000 per gram of soil samples taken just below, the surface of arable lands. Algae improves the soil structure by binding the primary and secondary soil particles. Blue green algae (Cynobacteria) fix atmospheric nitrogen both asymbiotically and symbiotically. Blue green algae found in the rice field comprise of unicellular, colonial and filament forms and fix nitrogen in the heterocysts which are large, thick walled empty cells, The blue green *Anabeana azollae* in symbiotic association with aquatic fern *Azolla* fixes atmospheric nitrogen in most soils.

Fungi

Different groups of fungi are present in soil. They are aerobic and active in presence of organic materials in general and in acidic soils in particular. The population density of fungi in normal cultivated soils are less compared to bacterial and actinornycetes. However, their biomass is more than that of bacteria. Fungi are active decomposers of cellulose, lignin, starch, gums etc in soil. They are efficient in carbon assimilation from decomposing organic residues into fungal hyphae, which serve as source of energy and nutrients for bacteria and other microflora to carryforward the biochemica! transformation. In unfavorable conditions such as drought weather they can survive by forming heat resistant spores. Their role is of much importance than bacteria especially in acidic forest soils and in mulched soils. In cultivated Indian soils, their population is found in ten thousand per gm dry soil.

Certain fungi such as spp. of *Glomus,* form symbiotic association with the roots of higher plants. Their association are known astnyeorrhizae, resulting in increase of nutrient absorption area of the root system. Mycorrhizae are beneficial for higher plants as they increase the availability of phosphorus, zinc, copper and molybdenum. Mycorrhizal associations improve greater water uptake and nitrogen fixation by *Rhizobium* and *Frankia* in leguminous and non-leguminous plants respectively and better resistance to disease by inhibiting root pathogens. Ectomycorrhiza and vesicular arbuscular mycorrhiza (VAM) have been found to have the maximum potential in forestry and associated only with tree roots with their mycelia developing on the surface of roots to form a sheath. The occurrence of VAM fungi have been found in wide range of cultivated plants and trees where the mycelia do not form a sheath on the root surface but penetrate into the root. Some fungi are pathogenic and cause root rot, damping off, wilt blight etc. in plants resulting in reduced yields and damage to the plants.

II. Rhizosphere and Nutrient Dynamics

Rhizosphere is central to microbial and nutrient dynamics and describes the zone of soil surrounding roots of legume in relation to symbiotic nitrogen fixing bacteria. It includes both, the area of influence and the physical location around the root. This definition in now extended to the zone of soil in which rnicrobial activity is affected by the roots of plant species which release organic substances (Dommergues, 1978) The rhizosphere has been further subdivided into (1) Ectorhizosphere (Outer rhizosphere) (2) Endorhizosphere (a zone where invasion and colonization of root cortical cells by soil microorganism occurs)

The distinct boundary of the root surface with the soil has been termed the "rhizosphere". The term was coined by Hiltner in 1904 based on the observations. Rhizoplane refers to the zone of attachment of microorganisms to the root surface. However, rnicrobial colonization of microorganisms to the root surface interface is often difficult to distinguish and, therefore, it is more reasonable to consider it as a continuum. Microbial colonization is minimal at the region around the root cap and is greatest in the zone of elongation behind the root tip (Lynch, 1990).

Rhizosphere Effect and R.S. Ratio

Microscopic examination reveals the presence of a vast microbial community surrounding on the surfaces of roots and root hairs. Bacteria are found to be localized in colonies and chains of individual cells. Filamentous fungi and actinomycetes are not observed as frequently. Protozoa particularly flagellates are situated in the water films on the root hairs and on the epidermal tissues.

Rhizosphere Effect

The root influence on soil bacteria as measured by plating techniques is often expressed as rhizosphere effect. This can be expressed based on R: S ratio. The rhizosphere effect is governed by root variation in plant species, its age and also influenced by cultural, soil and environmental condition.

R: S Ratio

Katznelson (1965) suggested Rhizosphere: Non rhizosphere ratio of the population to find out the degree of influence of plant roots on soil microorganism.

It is defined as the ratio of microbial numbers per unit weight rhizosphere soil (R) to the number in a unit weight of the adjacent nonrhizosphere soil (S). The rhizosphere effects are consistently greater for bacteria than for the other microorganism. Those responding in a significant way are short and Gram negative which generally make use of a larger percentage of rhizophere soil than the normal soil microflora. The percentage occurrence of short, Gram positive, coccoid rods and spore forming bacteria (*Bacillus* spp.) declines. *Arthrobacter* group, cocci or long nonspore forming rods are not generally influenced.

The genera of *Psuedomonas*, *Flavobacterium*, *Alcaligenes* and occasionally *Agrobacterium* are commonly reported. The other bacteria such as *Arthrobacter*, *Brevibacterium*, *Corynebacterium*, *Micrococcus*, *Bacillus* etc. show their presence in this zone but the environment is not so favourable. Anaerobic bacteria are also affected by the root in view of reduced oxygen tension due to enhanced root and microbial respiration.

Soil microflora is activated by growing plants and the nutrient cycling is more rapid in the rhizosphere than in the nonrhizosphere soil. The plant roots release number of low and high moleculare weight organic substances which are easily available to organisms. The most important contribution of plant is the excretion of root exudates and sloughed off tissue as source of energy and nutrients. The products excreted by plant under aseptic condition are amino acids, organic acids, carbohydrates, pentoses and hexoses and oligosaccharides, growth factors, enzymes and auxins. The plant begins to excrete organic compound in the environment from the beginning of seed swelling and germination. Generally the germinating seeds exude more nitrogen in the form of protein and peptides than in the form of amino acids where as the roots release more of free amino acids than in the form of protein and peptides. Other quantitative difference include that reducing compounds are released more by seedling than germinating seeds.

Rovira (1965) reported three possible ways by which the rhizosphere microorganisms can affect plant growth are the (1) production of metal chelating compounds, solubilizing some important plant nutrients (2) formation of certain growth promoting substances (3) protection against plant pathogens. Kransilnikov (1963) reported the production of biologically active substance by rhizosphere microorganism considered to be an important factor in soil fertility. There are evidence about uptake of organic compounds by plant roots by different workers (Winter 1961, Miller and Schmidt 1965).

Both positive and negative effects of bacteria on plant growth were observed by Hussain and Vancura (1970). The negative effect might be due to high concentration of B-indolyl acetic and which was produced by two strains (4 and 5) of *Pseudomonas fluorescens*. Favourable influence on plant growth was observed when the above strains were used as inoculant in nonsterile soils. The strains producing B- indolyl acetic acid and gibberalline-like-substances enhanced the growth of plants. *Pseudomonas fluorescence* (strains No 10) produced both these compounds in addition to biotin and pantothenic acid.

The influence of microbial metabolites on plant growth is most effective under condition when the effect of microorganisms on plants is continuous e.g. in case of microbial inoculants applied to seed survive and multiply in the rhizosphere during the vegetative and productive part of plants. *Pseudomonas* and probably other Gram negative bacilli dominated among microorganisms utilizing amino acids, organic acids and other early available substances. It is known that the compounds exuded by plant root were utilised preferentially by Gram negative bacilli from the root surface and rhizosphere soil.

In general, Gram(-); non spore forming rods with simple nutritional requirements which respond to amino acids, are stimulated to greater degree by roots than the coccoid and Gram(+) spore forming rods (Glick. 1995). The greater proportion of nonsporulating rods among the rhizopshere isolates and their ability to respond to nutrients such as amino acids reflects on the governing mechanism by which root and the rhizosphere are colonized.

Rhizosphere research has made dramatic changes in approach and methodologies during the last 90 years and has currently entered a phase where in *situ* detection of microbial activities is feasible through modern analytical techniques and tools. It is also being realized that single root often carries a multitude of symbioses such as mycorrhizal, actinorhizal and bacterial and therefore the net gain by plant is something more than the simple one to one relationship. This fact is receiving further support from the observation that common single molecules might be operating as attractants and that the host response to colonization by an alien may be manifested through expression of genes that produce proteins that exhibit serological crossreactivity amongst widely differing mutualistic partners. All of these statement are based on the data available widely in literature but is often not viewed as common facet of rhizospshere biology. The reason is not lack of apprehension but often, nonavailability of suitable tools to tackle a complex situation (Chanway *et al.* 1991).

In the context of introduced biofertilizers there is an increased emphasis on developing microbial consortia of nitrogen fixers, 'P' solubilizers, growth promoters and siderophore/ antibiotic producers so that growth promotion is more or less guaranteed and inter conversion and utilization of nutrients is not limited to one group with the resultant change in rhizosphere equilibrium. After all this zone is most critical in nutrient release, mobilization and competition.

Hormones are defined as a group of naturally occurring organic substances that influence the physiological processes at low concentrations. There is strong evidence that supports the hypothesis that plant growth and development in response to inoculation may be due to microbial production of plant growth regulatory substances. Pseudomonads in particular have been studied for phytohormones on account of their known growth promotory activity. Both, IAA and gibberellins have been reported from Pseudomonad fluorescens, especially GA3. Transformation of tryptophon to IAA has been observed for these bacteria and seed inoculation was found to improve plant growth and productivity. Some pseudomonad strains have also been found to secrete biotin and pantothenic acids, in addition to IAA (100 to 200 ug/g cell dry mass). Although cytokinin-like- substances too have been detected, their production is largely restricted to species of *Azotobacter, Agrobacterium* and *Rhizobium.*

The successful establishment of PGPR in an ecological niche such as rhizosphere is depend upon colonization, competition, antibiosis, HCN production and chemotaxis. Among these, colonization is an active process and involves survival of bacteria onto seeds, in soil, multiplication in response to seed exudates rich in carbohydrates and amino acids, attachment to the root surface and proliferation in soils containing indigenous microorganisms. However, an effective colonizer should be able to sustain under the stress such as water potent ion, temperature. pH, soil type, composition of root exudates and soil texture to be successful in the milieu of the rhizosphere. A good colonizer adheres strongly to the roots and possesses short doubling time thus achieving a selective advantage. For example, fluorescent pseudomonads show a generation time of 27 h on barley during first four days of plant growth but the duration increases to 103 h when the plants are 7 to 16 days old. Glycoproteins of root origin (agglutins) have also been shown to be involved in the colonization process in *Pseudomonas putida*. Recently, it has been possible to identify the genes responsible for root colonization but main colonization genes(s) are yet to be elucidated. Competitive advantage in case of fluorescent pseudomonad arises as a consequence of higher levels of iron chelating siderophores, secretion of antibiotics and release of hydrogen cyanide (HCN). Considerable data on each of these components exist which tend to suggest that each microbial strain can behave in a different manner and that growth promotion and biocontrol action represents two sides of coin.

The iron chelating siderophores are of increasing interest and can be important in the dynamics of iron utilization and competition between the plant and microbial community (Raaijmakers *et al.*; 1995; Marschner and Crowley, 1997). With the important role of iron in the development and proliferation of plant pathogens the rapid colonization of roots with fluorescent pseudomonas has been found to play a major role in increasing crop yields and decreasing the effects of some deleterious fungi and bacteria. The use of exudates to support colonization by these organism can be of direct benefit to plant.

III. Biofertilizers and Classification

The term biofertilizer includes soil microorganism which fix, nitrogen, mobilize or conserve plant nutrients. Biofertilizers are based on renewable sources available in soil and nature and are low cost inputs and eco- friendly. These inocula carrying the organisms when applied to soil promote specific biochemical activity in rhizosphere.

Definition

The term biofertilizer or microbial inoculant can be defined as the preparation containing strains of microorganisms which can augment the microbiological process such as nitrogen fixation, phosphate solubilization or mineralization, excretion of plant growth promoting substances or cellulose or lignin biodegradation in soil, compost or other environments.

Classification of Biofertilizer

Biofertilizer can be divided into nitrogen fixing bacteria, phosphate solubilizing and mobilizing microorganisms and organic matter decomposers etc. At present mainly four to five types of biofertilizers are commercially produced for use of farmers. Some of these microorganisms are also being reported to produce plant growth promoting substances which are known as plant growth promoting rhizobacteria (PGPR).

In recent years, biofertilizers or microbial inoculants containing microorganism such as nitrogen fixing bacterial, phosphate solubilizers, organic matter decomposers have been recognized as a part of strategy to supply plant nutrients. These biofertilizers, have an important role to play in rainfed agriculture, primarily because they are low cost inputs. In assured rainfall, they play a supplementary role alongwith fertilizing inputs. In India about 70% of the cultivated area is under rainfed condition. Of this, nearly 50% is under unassured rainfall conditions. Further crop like legumes are grown mostly under unassured rainfall conditions with little agronomic practices and without inorganic nutrient supply. Among the biofertilizers, *Rhizobium* inoculants for leguminous crops are the one known for the longest and also produced on a large scale. The other inoculants have been developed and recommended for the past few years but their use has not been significant due to several constraints such as their mass production, easy availability to farmers etc.

The chief characteristics and benefits of the biofertilizers listed above are narrated below:

Symbiotic System

Rhizobium: It belongs to Rhizobiaceae family and fixes 50-100 kg atmospheric nitrogen per hectare. Obviously, most interest among .biofertilizers surrounds *Rhizobium* inoculants because of their ability to fix nitroger. in association with leguminous plants which result, not only in meeting the nitrogen requirement of the plants but also the symbiotic system leaves behind sizeable amount of nitrogen for the succeeding. crop. The popularity of legume inoculants is related to cropping system including legumes and cereals practiced since time immemorial.

Air, we breath, is primarily a mixture of nitrogen, oxygen, carbondixoide and argon. A column of air over a hectare of land contains approximately 80,000 tonnes of inert N which is not available to plant and animals as such, but can be utilized by plants in association with bacteria having capability to fix atmosphere nitrogen.

Pulses, soybean and groundnuts occupy nearly 25 million hectares in India and the bacteria present in the nodules of these plant fix the atmospheric nitrogen to meet their needs. However, the *Rhizobium* bacteria present in the nodules of these crops are not always efficient. Therefore, the competitive, efficient bacteria are isolated, screened, selected, and produced as carrier based inoculant. These inoculants can be used for treating legume seeds so that when the seeds germinate and plant, grow, there are abundant efficient and infective strains of *Rhizobium* bacteria near the root of the plant to form efficient pink nodules. The method is inexpensive and is known as low input sustainable Agriculture (LISA) since the biofertilizer costs only Rs. 20 for treating seeds to be sown in a hectare of land.

In soils of high alkalinity or acidity, when seeds are pelleted with calcium carbonate or calcium sulphate after inoculation with *Rhizobium,* the harmful effects of alkalinity or acidity is minimized. In this way Rhizobium gets established in problem soils and nodulates plants effectively resulting in better crop productivity. The success of method of inoculation and pelleting has been proved in the case of soybean and mungbean crops.

Depending on the variety, plant protection measures, and agro- climatic conditions, legume inoculants increase the grain yields from 10- 50% as reported by ICAR coordinated project on Pulses. Field experiments have also shown the addition of residual nitrogen in soil by pulse crops. The grain yields of succeeding crops like wheat and rice are increased due to effect of residual nitrogen left back in soil.

The *Rhizobium* cultures for different legumes grown in *Kharif, Rabi* and *Zaid* such as black gram, green gram, pigeon pea, cowpea, lentil, chickpea, peas, soybean, groundnut, berseem, lucerne etc .can be used not only for improving the yield of the legume crop but also for sustaining the productivity on a long-term basis.

In recent years, there has been development in understanding the function of these bacteria. The domain of nodulating plants has widened since non-leguminous plants have been identified to be nodulating. Although the domain has been enlarged, this area has not been sufficiently exploited.

Biochemically, the enzymes involved in nitrogen fixation have been well explored. In addition to enzymes the role of bacteria in producing recognition factors and specificity factors have also been identified.

With regard to genetic development, the understanding of the structural organization of the nif system has been well advanced and the role of plasmids in nitrogen fixation in rhizobia has been well documented. It is now believed that the ability to fix nitrogen and to oxidise hydrogen can be plasmid borne. It is believed that some other factors relating to nitrogen fixation may also be plasmid borne.

There has been considerable progress in understanding the genetic control of nitrogen fixation. Inter-specific and intragenic transfer of nitrogen fixation properties in not difficult. However, progress in developing strains superior to those existing in nature has not been achieved. Further, some of the strains found superior in laboratory conditions, have difficulties in establishing under field conditions. The challenge, therefore, exists of developing rhizoidal strains that would suit to encounter the varying stress conditions that exist in tropics.

Frankia

Non-leguminous Symbiosis: Nearly 120 species, mainly trees and shrubs are now known to harbour nitrogen fixing organism in their root nodules which are not rhizobia. An actinomycete is known to develop association with nonlegume plants such as *Alnus, Casuarina. Ceanothus. Eleagnus, Myrica, Hippophae*, *Purshia* etc: and the organism has been named as Frankia.

The nonleguminous nitrogen fixing system are of modest agricultural importance, because their potential is not widely known. The extent of nitrogen gain by such Angiosperm varies enormously depending on soil type, climatic conditions and plant age. Nitrogen gains in the range of 12-200 kg/ha/year have been reported for *Alnus* species and 27-179 kga/ha/ year for *Hippophae* species. Tests of individual location suggest the nitrogen fixation of 58 kg/ha/ year for *Casuarina,* 60 kg/ha/year for *Ceanothus* but only about 10 kg/ha/per year for *Myrica*. Plants in the group have little agricultural potential as grain crops or forage but Alder could be available in Agroforestory. They may be used for upgrading poor soils. Its role as biofertilizer is insignificant.

Azolla

Azolla belong to family *Azolla*ceae, is a free floating water fern in symbiosis with *Anabaena azollae (Cynobacterium)* and fix appreciable amount of nitrogen in the rice fields. *Azolla* has been used as biofertilizer in some parts of India and south-east Asian countries. The association is utilizing the energy from photosynthesis to fix atmospheric nitrogen which may average from 40-

60 kg N/ha/rice crop depending on the soil, temperature, light intensity and phosphorus content since its survival is difficult at high temperature and soils low in available phosphorus. *Azolla microphylla* is used as biofertilizer in rice in Tamil Nadu.

Fresh biomass of *Azolla* such as *Azolla microphylla* at rate of 200kg /ha are broadcasted 7 to 10 days after rice plantation which covers the fields within 2-3 week period with 15-20 tonnes biomass production. Even inoculation of 50-100 kg biomass of *A. microphylla*, A. *piñata*, A. *caroliniana and* A. *filiculoides* per hectare will establish within 4-5 weeks in transplanted rice. It is also used as green manure because of its large biomass and high nitrogen content in flooded rice field within maximum temperature of 38°C. The increase in yield of rice crop due to *Azolla* is around 15-20 percent.

Non Symbiotic System

Azotobacter: It belongs to *Azotobacteriaceae,* is aerobic, chemo - heterotrophic and free living. The use of *Azotobacter* helps in saving 10 to 20 kg N/ha. It produces growth promoting substances which improve seed germination and growth of extended root system. It produces polysaccharides which improve soil aggregation. *Azotobacter* suppresses the growth of saprophytic and pathogenic microorganisms near the root system of crop plants. *Azotobacter* technology has not been received favourably in the western countries where farmers go for heavy application of chemical fertilizers. But in a country like India where application of chemical fertilizers in rainfed cultivated area is low and in irrigated area its application is much less as compared to many developed countries, its benefits have been reported. Thus saving of 10-20 kg nitrogen ha^{-1} as chemical fertilizers by small and marginal farmers will be of a great relief.

Azotobacter helps in maintaining better plant population, growth and yield of the crops such as cereals, sorghum, maize, mustard and cotton, besides vegetable crops like tomato, potato etc. In general in cereals, the yield increases are of the order of 15 to 30% and 10 to 20% in cash crops.

Azospirillium: The bacteria have been found to live within the roots.of sorghum, bajra (pearl millet) and ragi plants. They belong to family Spiriliaceae, are chemoheterotrophic and associative in nature. They fix atmospheric nitrogen at the rate of 15-30 kg N/ha and secrete growth regulating substances.

The use of Azospirillum inoculants help in increasing yield of millets. It is also recommended for rice, maize, wheat and sugarcane crops and as a co-inoculant for legumes. It significantly increased the growth, chlorophyll content and mycorrhizal infection in the roots. Increased growth and nutrient uptake by barley plants were observed when seeds were co- inoculated with A. *brasilense* and *Glomus versiforme.*

Bluegreen Algae or Cynobacteria

Bluegreen algae are also known as cynobacteria, forming nonfibrous slimy growth on the ground, and on the surface of soil and water. They have striking similarities with bacteria except they have oxygen evolving photosythesis. They are cosmopolitan in distribution and have ability to survive in extreme climatic conditions. They predominantly multiply vegetatively through fragmentation. A sexual reproduction is brought about by formation of exospores, endospores and akinetes. None of the BGA have sexual reproduction and form motile stages. Unlike, green algae, their food reserves is glycogen, which does not react with iodine to from violet colour. They belong to eight different families, phototrophic in nature and produce auxin, indole-acetic acid and gibberellic acid, fix 20-30 kg N/ ha in wet land rice fields. Application of BGA increases paddy yields by 15-20 percent.

The species of blue green algae which are known to fix atmospheric nitrogen and classified into three groups. (i) Heterocystous- aerobic forms (ii) Aerobic unicellular forms and (iii) Non heterocystous filamentous, microaerophilic forms. The composite inoculant of blue-green algae may consist of the cultures such as *Nostoc, Anabaena, Calothrix, Tolypothrix, Plecetonema, Gloeocapsa, Oscillatoria, Cylindrospermum, Aulosira* and *Scytonema.*

The agronomic potential of cynobacterial nitrogen fixation in rice field was first recognized in India by De (1939) who attributed the natural fertility of tropical rice fields to nitrogen fixing bluegreen algae. The rice field ecosystem provides a favorable ecosystem such as light, water, high temperature and available nutrients. Algal biofertilizer is a source of nitrogen and is environment friendly, does not contaminate ground water or deplete the resources. They also add organic matter to soil, excrete growth promoting substances and improve soil properties.

Phosphate Solubilising Biofertilizer

Phosphorus is one of the most important plant nutrients and may be critical nutrient for the optimum growth of plant. Most of our soils are low in available form of phosphorus and require phosphorus application. The proliferation of efficient strains of phosphate solubilizing microorganism in the rhizosphere of crops will render insoluble soil phosphate available to plants due to production and secretion of organic acids by them. The use of this biofertilizer will also increase the availability of phosphate from rock phosphate applied directly even to neutral to alkaline soils or when used for preparation of phospho-compost.

Rock phosphate is recommended in acid soils as phosphatic fertilizers. The amount of phosphatic fertilizers which gets chemically fixed in the soil or retrograded is also made available to crops due to their activities (Gaur, 1985, 1990).

Phosphate solubilizing microorganisms include efficient strains of bacteria, fungi yeast and actinomycetes in that order. However, species and genera of bacteria and fungi have mainly been reported as true phosphate solubilizers. There are few reports about phosphate solubilization in traces or small amounts by *Rhizobium,* algae and inefficient strains of phosphate solubilizers. The most important genera of bacteria include *Bacillus* and *Pseudomonas* and that of fungi are *Aspergillus* and *Penicillium.* Most efficient bacteria and fungi were reported as *Bacillus polymyxa, B. circulans, Pseudomonas striata, P.rathonis, Aspergillus awamori* and *Penicillium digitatum* (Gaur 2001) Microphos culture was developed by Gaur and coworkers (Gaur, 1990) and was made available to farmers by Division of microbiology,IARI at subsidized rates. The preparation of this microbial inoculant contains either *Pseudomonas striata, Bacillus polymyxa* or *Aspergillus awamori* which were isolated and identified as the most efficient cultures.

Multilocational trials have shown that PSB inoculation increased yield of rice by 10-20% wheat by 10-40% chickpea by 10-30% and potato by 30-50% with an additional average yields of 200-500 kg/ha alongwith a saving of 30-50 kg P_2O_5/ha equivalent of phosphatic fertilizer (Gaur 1985).

Aspergillius awamori a well known phosphate solubilizing fungi was found effective in preparation of compost as well as phosphocompost. The use of phosphocompost tested under field conditions in Haryana on the yield of pigeon pea, green gram and cluster bean showed the replacement of superphosphate by phosphocompost (Mishra and Bangar, 1986).

Mycorrhiza/ VA Mycorrrhiza

There are four main types of mycorrhiza such as ecto, vesicular arbsuclar (Va) ericoid and orchidacous. In ectomycorrhiza, the fungus forms a compact mantle or sheath over the root surface and the hyphae grow out into the soil. These are generally found in forest trees including many gymnosperms (e.g. pines, spruce, fir, larch) and angiosperms (e.g. oak, beech, eucalyptus, poplar). The mycorrhizal hyphae assume at least partly the function of root hairs. Detailed studies have been carried out with *Pinus* spp. symbiont. The ectomycorrhizal fungi help the plants in phosphorus uptake through increased surface area of absorption, offer protection from some soil bom plant pathogens and enhance rooting and survival of cuttings through production of growth hormones.

VA Mycorrhiza

These are obligate symbiont with a loose network of hyphae in soil and extensive growth within the cortex of the plant. In the host cells they produce highly branched hyphal structures called arbuscules and also vesicles. This occurs in most angiosperms as well as gymnosperms pteridophytes and bryophytes. Most plant species particularly leguminseae and graminae are normally mycorrhizal. Even here some species are prone to VAM infection than others.

Other crop where considerable VAM infections have been reported include sorghum barley, upland rice, cassava; grape vine, olive, citrus, cacao, tobacco, cotton, sugarcane, pineapple, lettuce, onions, cowpea, soybean, strawberries, apple, rubber, coffee, tea, papaya, oil palm and various ornamental bulbs.

Vam fungi can neither be cultured in the absence of living root nor isolated on agar plates by standard microbiological techniques. It is now well established that many plant roots can't grow adequately without VAM fungi, especially in phosphate deficient soils. In view of this, their impact in tropical agriculture will be greater than in temperate conditions.

VA mycorrhiza based on spore morphology were divided into four genera by Gerdemann and trappe (1974) such as Glomus, Gigaspora, Acaulospora and Scleroystis. These are grouped in family Endogonaceae, order Mucorales, class zygomycetes. A fifth genus Endogone, formerly the name used for all VAM fungi, is now restricted to species that form ecto or no mycorrhiza. It is the only genus believed to form true zygospores resulting from fusion of gametaniga and nuclei.

The starting point for practical studies is the isolation of local endophytes and their establishment in pot cultures. Most pot experiments have been done is sterilized soils in order to have nonmycorrhizal controls.

Vesicular-arbuscular mycorrhiza is a balanced mutualistic symbiosis in which both partners can benefit. There are three way interactions between soil, plant and fungus in which the movement of phosphate is of prime importance. Phosphate is first taken up by soil based hyphae, transferred into the root based mycelium via the connecting links (entry points) and then supply to the plant. The interactions are influenced by environmental conditions. Soil fertility is a key factor influencing the extent to which the plant can benefit from mycorrhiza.

It is well established by research work that VAM affects phosphate nutrition of plants particularly in P deficient soils. However, VAM can also enhance uptake of other nutrients such as zinc and sulphur as well as greater water uptake resulting in increased crop productivity.

The production of VAM fungi named "Nutrilink" as biofertilizer has been taken up by I.A.R.I., New Delhi on commercial scale. The cost of one packet of Nutrilink is Rs. 20/kg and 3-5 kg inoculants is sufficient for one acre of land.

Plant Growth Promoting Rhizobacteria (PGPR)

The rhizosphere effects are known since long but comprehensive studies into other rhizosphere bacteria (rhizobacteria) besides N2 fixers is of recent origin. PGPR belong to genera of *Agrobactericum, Arthrobacer*, *Bacillus, Cellulomonas*, *Pseudomonas* and *Xanthomonas*. Their mode of action has been mainly two ways, of colonization of root system and pre- empting the establishment of suppressing deleterious rhizosphere microorganisms on the roots. The presence and survival of enormous population of bacteria established on the planting material and roots play an important role in recycling of plant nutrients and thus reduces the energy source and nutrients for proliferation of pathogenic microorganism. Fluorescent Pseudomonads are known for their role as nutritional scavenger and for production of siderophore like compounds. The high proficient iron scavenging system through synthesis of siderophores is attributed to be mainly responsible for Fe depletion causing unfavourable environment for the pathogenic fungi. These organisms when used as biofertilizer were reported to increase crop yields.

Organic Matter Decomposing Cultures

Crops remove substantial quantities of plant nutrients from soil. The utilization of farm, rural and urban organic wastes is essential for maintenance of soil health and recuperation of plant nutrients. Organic matter can be recycled by in situ incorporation and mulching. Composting is a key technology for disposal and recycling of organic wastes to maintain hygienic conditions in the environment.

The bulk of organic wastes in made up of cellulose followed by hemicellulose and lignin which are slowly decomposed as compared to other components such carbohydrates, proteins, amino acids, fats/wax etc. The efficient cellulolytic and lignolytic cultures if used as compost inoculants can improve and enhance the composting process. The cellulolytic fungal cultures such as *Trichurus spiralis, Aspergillus niger, Paecilomyces fusisporus, Trichoderma viride* etc. have been identified and selected for rapid composting of cereal residues, sugarcane trash, banana leaves, water hyacinth, leucaena etc. mixed with crop residues and dairy farm wastes. The utility of cellulölytic culture in composting of different organic wastes have been demonstrated in a several multilocational trials (Gaur 1987, 1992). Within 8 to 10 weeks good quality compost from paddy straw could be prepared which contained around 1.7% N and C/N ratio of 12.3. Similarly other wastes were composted which contained more nitrogen and attained lower C: N ratios with compost inoculants as compared to control. The bulkiness of compost was also reduced by 10-15%.

The bacterial genera such as *Cellulomonas* and *Cytophaga* can also be utilized for hastening of the Composting process. However, systematic trials are required if it has to be practiced on a large scale.

Carrier based cellulolytic fungal cultures are being produced in some institutions, for example, Agricultural, college, Pune. Cellulolytic fungi are cultured on sorghum seeds and supplied to farmers for use as inoculants in composting of organic wastes. The compost material can be further enriched by fortifying with nitrogens fixing and phosphate solubilizing bacteria. Compost can be prepared using indigenous technology through a diversified small scale industries planned at village, block and district levels.

References

Chanway, C.P., Turkington, R. and Holl, F.B. 1991. Adv. Ecol. Res. 21:121-169.

Cook, R.J. Thomashow, L.S. Weller. D.M. Fujimoto, D. Mazzola. M. Bangera, G. and Kim, D.S. 1995. Proc. Natl. Acad. Sci. USA. 92: 4197-4201.

Dommergues, Y.R. 1978. In: Interaction Between Nonpathogenic Soil Microorganisms and Plants. Y.R. Dommergues and S.V. Krupa(eds.) Elsevier Scientific Publishing. New York. Pp. 1-37.

Gaur, A.C. 1984. Response of rice to organic matter. Indian experience. Proc. Internatl. conference on organic matter and rice .IRRI, Manila Pp 503-514.

Gaur, A.C. 1985. Phosphate solubilizing microorganisms and their role in plant growth and crop yields. Proc. Natl. Symp. Soil Biol. Hissar Pp. 125-138.

Gaur,A.C. 1987. Resources and Conservation 13:154-174.

Gaur, A.C. 1990. Phoshphate Solubilizing Microorganisms as Biofertilizer. Omega Scientific Pub. New Delhi Pp. 176.

Gaur,A.C. 1992 Bulky organic manures and crop residues. In: Fertilizers, Manures. Recyclable wastes and Biofertilizer (ed HLS Tandon) FDCO, New delhi Pp 36-51.

Glick, B.R. 1995. Can. J. Microbiol 41: 109-117.

Hussain, A. and Vancura, V. 1970 Folia Microbiologica 15: 468-478.

Johri, B.N. Rao, Ch. V.S. and Goel, R. 1997. In: Biotechnological Approaches in Soil Microorganisms for Sustainable Crop Production. K.R. Dadarwal (ed.) Scientific Publ. Jodhpur. Pp. 193-221.

Katznelson, H. 1965. Nature and Importance of rhizosphere. In: Ecology of Soil Borne Pathogens. Univ. of Califomia Press, Berkeley, USA.

Krasilinkov, N.A. 1963. The role of microorganism in plant life. In Recent Progress in Microbiology, University of Toronto Press, Toronto.

Lynch, J.M. 1990. The Rhizosphere. John Wiley and Sons. New York.

Marschner and Crowley, 1997.

Miller, R.H. and Schmidt, E.L. 1965. Soil Sci. 100: 323.

Mishra, M.M. and Bangar K.C. 1986. Biological Agri. and Hort. 3: 331-340.

Raaijmakers,J.M., Van der Sluis, I., Koster, M., Bakker,P.A.H.M., Wesisbeek, P.J. and Schipprs, B. 1995. Can. J. Microbiol. 41: 126- 135.

Rovira, A.D. 1965. Ann. Rev. Microbiol. 19:241.

Winter, A. G. 1961. Symposia Soc. Exptl. Biol. 15: 229.

12

Symbiotic Nitrogen Fixation in Legumes

Nitrogen is the most important plant nutrient for increasing crop productivity. A few prokaryotic microorganisms are able to fix N_2 directly mediated by enzyme nitrogenase. Annual biological nitrogen fixation (Table 96) is estimated to be around 175 million metric tonnes per year of which about 139 million tonnes (79%) are accounted for by terrestrial fixation (Burns and Hardy 1975).

Table 96: Estimate of annual biological fixed nitrogen in various terrestrial systems

Type of land	Million metric tonnes/year	Distribution (%)
Permanent Grasslands	45	32.3
Forest and Woodlands	40	28.8
Legumes	35	25.2
Unused Land	10	07.2
Non-legumes	09	06.5
Total	139	100.0

The nitrogen fixers involved are metabolically diverse but the process is similar and depends upon (1) nitrogenase (Nase) enzyme complex (ii) a high energy requirement (ATP) (iii) anaerobic conditions for nitrogenase activity (iv) source of strong reductant.

The symbiotic process is known to be one of the important contributors to the nitrogen cycle both from agricultural and ecological points of view. The soil bacteria of the genus *Rhizobium* colonizes and forms nodules on the root of leguminous plants such as clover, lucerne (alfalfa), pea, chickpea, pigeonpea, beans, groundnut, *Vigna* species etc.

The nitrogen uptake in legumes by some unknown process was demonstrated through agronomic experiments by J.B. Boussigault in 1838 which was opposed by Liebig and others. However, Hellriegel and Wilfarth's experiments (1886-88) clearly showed that the growth of non-leguminous plants, barley, oats etc. was directly proportional to the amount of nitrate supplied while in case of leguminous no such relationship existed and growth of the leguminous plants picked up in some cases after the seedling stage. Beijerinck (1888) isolated the bacterium from the root nodules in pure culture which lead to study their morphological characteristics and physiology.

The legume-*Rhizobium* symbiosis is estimated to account for 40 percent of the world's combined nitrogen. Pulses are the basic sources of protein in Indian diet. They play an important role in farming systems and sustainable agriculture due to their contribution in meeting about 80-90% of the nitrogen requirement of leguminous crops and show the residual effect on the following crop adding nitrogen to soil organic pool. Such effects are observed either by *in situ* incorporation of legume residues or via cattle dung and urine.

I. Present Status of Pulses and Oilseeds and Bacterial Inoculants

Pulses and oilseeds such as groundnut and soybean occupy about 30 million hectares in India. Pulses include green gram, black gram, red gram, horsegram, cowpea, chickpea, peas etc. Leguminous fodder crops like berseem and lucerne are grown for cattle feed. The cultivation of grain legumes provides cheapest source of plant protein to Indian population which are usually used as pulses in regular diet. Although the production of pulses has increased to some extent since independence, the per capita availability has declined. It implies that increase in production has not kept pace with demand of the growing population. The per capita availability of pulses has gone down from 60-70 gm in 1950-51 to 42 gm 1996-97. However, the per capita availability of total food grains has gone up from 350 gm in 1950-51 to 450 gm in 1996-97. The total availability of food grain production has gone up mainly due to increased production of rice and wheat. This has resulted in cereal-pulse imbalance in the diet and thus unfavourably influencing the nutrient security of the Indian population in general. The low productivity of pulses has been caused by several negative factors such as unremunerative and unstable prices to farmer, lack of adequate market support, non-availability of quality seeds, losses in pigeonpea due to mycopalsma inducted sterility in which floral structure changes to vegetative parts under unfavourable conditions, pod borer problem and laythrism due to consumption of grass pea *(khesari)* which used to be grown in UP, Bihar and Assam earlier.

Use of Appropriate Bacterial Cultures

The average pulse production which was 506 kg/ha during 1986-87 rose to 650 kg/ha in the year 1996-97. However, this was not even comparable to that of oilseeds in the same production year. The expectations were not translated into ground reality because the improved varieties as well as production technology could hardly reach 20 percent farmers.

To meet the projected demand of pulses at end of ninth and tenth- five years plans would be 18.5 to 21.0 million tonnes respectively. It would require a growth rate of about 4.5% to achieve the target. However, the pulse production has been fluctuating between 12 to 15 million tonnes per year. Therefore, future strategies should incur both short term and long term goals for increasing the pulse production.

Rhizobium is most effective bacterial inoculant known so far as it can supply nitrogen up to 300 kg ha-1 and thus supplement N need to a great extent (Biswas *et al.*, 2001). There is ample potential for increasing the average yields by way of using appropriate *Rhizobium* and phosphate solubilizing cultures. Legumes requires only a starter of 10-20 kg N/ha at sowing time. These cultures not only increase the yield but also save on nitrogenous and phosphatic fertilizers as well as upgrade fertility status of soils. At present hardly 15-20% of the pulse growers are using bacterial cultures. Research has shown that due to use of efficient cultures the yield of crops can be increased by 10-20% resulting in a net profit of rupees 1000-1800 ha^{-1} over control.

The Legumious Plants

The family is a group of flowering plants found both in tropical and temperate climate. They range from small widespread plants such as clovers to flowering plants such as Lupins, shrubs and trees such as *Acacia*.

Leguminoseae has been divided into 3 sub-families, the Papilonacease, the Mimosaceae and the Caesalpiniaceae. There are 700 genera and 14,000 specie of the leguminoseae of which 500 genera and about 10000 species belong to sub-family Papilonaceae. About 80 to 90% specie of Papilonaceae form nodules whereas only about 10% of Mimosaceae and 65% of Casesalpinceae bear nodules. The papilonaceae includes leguminous plants of agricultural importance such as berseem *(Trifolium alexandrinum)* white clover *(T. repens)* lucerne (alfalfa) lentil, peas, broad beans, French bean or kidney bean, soyabean, chickpea, groundnut, black gram, green gram, sannhemp and *Sesbania*. The mimosaceae consists of babul (Acacia) ironwood of India and sensitive mimosa where as Caesalpinaceae include *Cassia fistula* (Senna) Tamarind and Judas tree.

Rhizobia are capable of nitrogen fixing symbiosis with more than 1130 species of leguminous plants and parasponia among Ulmaceae (Trinick, 1973, Vincent, 1974) Most rhizobia are specific in their association with legumes. *Rhizobia* have been taxonomically grouped and designated to a particular species based on cross inoculation group concept proposed by Fred, Baldwin and McCoy (1932).

II. Classification of Rhizobia

Certain types of plants are colonized by certain types of rhizobia showing a degree of specificity. Each of these groups of legumes and their specific Rhizobium constitutes a cross inoculation group. A cross inoculation group refers to a collection of leguminous species that develop nodules when exposed to bacteria obtained from the nodules of any member of that particular legume group. The genus *Rhizobium* consists of six distinct species based largely on cross inoculation group concept. *Rhizobium* genus contains members that are gram negative rods, aerobic chemoheterotrophs having 59 to 65.5% (G+C).

More than 20 cross inoculation groups have been established but only seven have achieved prominence (Table 97).

Table 97: Cross inoculation groups of Rhizobium

Rhizobium species	Cross inoculation group	Genera of host legumes
Fast growing types Group I: Petrichous flagella (2-6) GC content of DNA (59.1-63.1%)		
R. meliloti	Alfalfa	*Medicago, Melilotus, Trigonella*
R. trifolii	Clover	Trifolium Clovers
R. leguminosarum	*Pea*	Pisum, *Lathyrus,* Lens, Vicia
R. phaseoli	Phaseolus	Phaseolus vulgaris
Slow growing Types Group II: Polar or Subpolar flagellum, G C content (61.6-65.5%)		
R *lupini*	Lupin	Lupinus, Omithopus
Bradyrhizobium Japonicum	Soybean	Glycine max
Rhizobium sp.	Cowpea	*Cowpea, Arachis, Cajanus,* Cicer, Wgna, *Phaseolus* (Lima bean), *Crotalaria, Sesbania*

Rhizobia thus can be categorized into two broad groups on the basis of metabolism i.e. fast growers and slow growers (Elkan, 1981). *Rhizobium* isolated from legumes such as *R. trifoli, R.* meliloti, *R. leguminosarum* and *R. phaseoli* are designated as fast growers, having generation time less than six hours whereas slow growers such as *B. japonicum, R. lupini* and *R.* spp. (Cowpea miscellany) have generations time exceeding six hours (Stowers, 1985).

Rhizobium Species and Their Characteristics

The updating of *Rhizobium* systematic has been done by the insurgence of sequence comparison data of the small subunit r-RNA (SSU r-RNA). The approach provided new insight of *Rhizobium* classification as it used the phylogenetic relationship among bacteria, *Rhizobium* leguminosarum remained the type strain that defined the genus (Young, 1996). *R. phaseoli* and *R. trifolii* only reference in host range from the above. *R. leguminosarum* has 3 biovars viz., viciae, trifolii and phaseoli.

The validity of cross inoculation has been questioned as many legumes get nodulated by rhizobia of other cross inoculation group. With repeated evidence of anomalous cross infection among different plant groups, a new classification has been proposed (Jordan, 1984). The slow growing rhizobia are grouped under the genus *Bradyrhizobium* and the fast growers under the genus *Rhizobium*. The new classification is more complex as the same host may be nodulated by either of them. However, the agricultural significance of cross inoculation groups still remains key feature of the established taxonomic system.

Two additional genera (*Azorhizobium* and *Sinorhizobium*) and five additional species *R. galegae*, *Sinorhizobium fredii* (formerly *R. fredii), S. xinjiangensis, R. tropici* and *A. caulinodans* of stem and root nodulating bacteria have been proposed and listed (Moore and Moore, 1992).

It was discovered in 1973 that a cowpea *Rhizobium* nodulated *Trema aspera,* a non-leguminous tropical plant. In mid 1970's, the dogma that rihzobia never fix nitrogen except within a nodule of plant was discovered because slow growing types such as Cowpea strains were grown in media entirely free of plant (Pagan *et al.,* 1975). The ability to fix nitrogen ex- planta, as it is termed is mainly restricted to slow growing types e.g. cowpea group rhizobia. Nevertheless, it seems probable that rhizobia alone are .unable to carry out the process for the benefit of agricultural production to a significant degree.

Presence of Rhizobium in Soils

Rhizobium is known to survive in soils unaffected by continuous use of nitrogenous fertilizer. They can survive at low temperature as well as tolerate warm temperatures up to 50°C for more than few hours. The bacterium survived for several years in dry stored soil although the mechanisms of survival were not reported (Sen and Sen, 1958). It is sensitive to agricultural chemicals used for plant protection measure (Nutman 1965, Gaur 1980). Soil protozoa are known predators of rhizobia (Alexander, 1977). *Rhizobium* is more tolerant towards salts than its hosts (Subba Rao et *al.,* 1972, 1974).

III. Process of Nodulation

The formation of nitrogen fixing nodules is a multistep process requiring interaction between the plant and the bacterium.

Nearly all soils contain rhizobia, normally not fixing nitrogen until these colonize in appropriate leguminous plants at their growth phase when leaves begin to appear. Plant excretes into the root zone growth promoting chemical which stimulates the growth of soil microorganisms including rhizobia. The rhizobia may then aggregate at distinct site adjacent to roots via chemotaxis (Schmidt 1979). Recent evidences suggest that polysaccharides on the surface of invasive bacteria are involved in or cause binding of these cells to the constituents on the surface of plants, they can potentially nodulate. Little or no adhesion is evident between rhizobia and plants of heterologous cross inoculation group (Bohlool and Schmidt, 1974, Dazzo and Hubbell, 1975).

In legumes, generally the invasion occurs though root hairs, which in presence of appropriate bacteria undergoes a curling under the influence of some chemical. Indoleacteic acid was earlier believed to cause curling of root hairs but its role in the process is not substantiated or the compound is only one of the contributors to the morphological changes. Moreover, *Rhizobium* excretes one or more compounds probably including a nucleic acid and a polysaccharide or protein which may be involved in deformation of root hairs (Holheim and Raa, 1973). The deformed hairs are then penetrated in the initial phase of the actual infection, resulting in root hair wall invagination which continues to develop into tube-like structure. Usually less than 5% of the infection finally results in nodules. A hypha-like infection thread is formed due to rhizobial penetration into the root hair. Due to production of polygalacturonase by rhizobia invagination of root hairs takes place in formation of infection thread through which rhizobia enter the host. Infection thread bursts upon to liberate bacteria in the host cytoplasm, there in to multiply. Shortly prior to or immediately following the release, a period of rapid cell division takes places in the host's cell. The structure finally consists of central core containing the rhizobia and a surrounding cortical area in which is found the plant vascular system and thus the bacterium establishes contact with host bundles.

A curious characteristic of the plant cell in the central portion of the nodule system is their possession of twice the chromosome number characteristics of the host. The doubling of chromosome number takes place in the nodule of polyploid as well as diploid legumes. The cells become emerged with bacteria which then cease to grow and often assume unusual shapes which are called 'bacteroids' rich in nitrogenase. Fixation of nitrogen then continues until seed formation when it stops and the nodules become senescent. With maturity of plant, nodules become brown, the bacteroids disappear and nitrogen released for used by plant. In the end nodule is cut off from the rest by means of cork layer. Some of the bacteroids are incapable of further multiplication and some remain viable as the nodule or the plant (annuals) ages and dies. These survive in soil for the next season's growth and activity.

The colonization process differs in character and timing from plant to plant. There are reports that certain plant proteins called lectins can interact with the specific rhizobia which colonize the plant and their property enables the plant to recognize and admit the type of *Rhizobium.* Recognition involved cell-cell communication via signals as proposed by lectin hypothsis (Albersheim and Anderson-prouty, 1975) and later supported by other scientists with certain improvements (Graham, 1980) Lectins are glycoprotiens, occur on legume root surface and bind to lipopolysaccharide/ extracellular polysaccharide in bacterial cell wall/capsule at certain specific determination of saccharide units. The relative capacity of plant bacterial association once established to assimilate molecular nitrogen refers to its effectiveness. A strain that does not permit fixation at a rate sufficient to meet the demands of the host is either partially effective or totally ineffective. An effective strain forms healthy nodules with a distinct pink color, the interior is distinctly red if they are cut open. Ineffective strains of *Rhizobium,* form pallid sometimes-greenish nodules. Once a plant is colonized by a certain strain of *Rhizobium,* others get to be excluded, and therefore it is important in agriculture to ensure both effective and infective strains are used.

The red colour in effective nodules is due to hemoglobin, a protein which is produced by the plant and supplies oxygen to bacteroids but avoids oxygen damage to their nitrogenase. Nitrogenase is produced by bacteroids. It consists of 2 iron containing proteins, one large containing molybdenum in addition. It is oxygen sensitive and consumes ATP. It is similar to nitrogenase of free living bacteria.

Events Leading to Nodulation and N_2 fixation in Legumes

- Attachment of rhizobia to root begins within one minute of inoculation.
- Number of attached rhizobia increases with time upto several hours.
- Root hair curling begins within 5 hours.
- Infection threads become visible within 3 days of inoculation.
- Nodules become visible within 5-12 days.
- N_2 fixation is often evident in 15 day old plant

IV. Factors Affecting Symbiosis and Nitrogen Fixation

The efficiency of the symbiotic nitrogen fixation depends on host plant, genetic makeup of the *Rhizobium* strain, and the environmental factor.

Competitiveness with *native Rhizobia*: The ability of *Rhizobium* strain to successfully compete with the native rhizobia for nodule site is an important feature. A number of abiotic conditions influence the competence of the *Rhizobium* inoculant strains and its establishment in the soil and the environment. Water stress affects adversely a variety of physiological processes in nodule (Rathore *et al.*, 1981), while soil temperature affects root exudation, growth and survival of rhizobia, root hair formation and infection process by microsymbiont. (Venkateswalu and Rao, 1987).

Thies *et al.* (1992) showed the influence of environmental variables on inter-strain competition between the applied and indigenous rhizobia. Establishment and occurrence of rhizobia for different legumes such as soybean, (Rangarajan and Hariharan 1988) chickpea, pigeon pea (Sharma and Khurana, 1988) mung bean (Ranagrajan and Hariharan (1988) and groundnut (Narula and Gupta 1986) have been extensively studied. Soil type, origin of isolate and host variety specificity influence the establishment of the inoculants.

The level of native *Rhizobium* population seems to be the major factor in determining the response to inoculation of legumes. Significant increases in chickpea yield have always been associated with low native population of chickpea rhizobia, less than 100 rhizobia g^{-1} soil (ICRISAT, the 1984). When *Rhizobium* population was greater than 1000 g^{-1} soil, response to chickpea inoculant was nominal.

Low native population is not the only criterion for the success of inoculation because native population adopted to soil and environmental condition are found to be highly competitive compared to the inoculated strains (Azad *et al.* 1991). The rhizospheric microbes also influence the competition ability besides the indigenous rhizobia. Impact of rhizophages (lytic and lysogenic) on the *Rhizobium* is well known (Joseph *et al.* 1985, Lesley, 1982).

Plant Host Specificity

The genetic characters of the host plant have a marked effect on the symbiosis process (Nutman, 1959; Parker 1957). The plant having higher C/N ratio are sensitive to nodulation (Hubbel and Elkan, 1967). Legume plant species not only differ in nodulation but cultivars within a specie differ significantly showing that host genetic characters are important determinant of nodulation. Besides the specificity of root infections, the development of root nodules is also influenced by both bacterial and plant genes. Genetic changes of any one of the partners may check the development of nodules and the process of nitrogen fixation. In chickpea, genetic studies with three non-nodulating mutants (PM 233, 665, 674) and two ineffective mutants (PM 405 and 769) indicated that the phenotypic expression of each of the five mutants was controlled by non -allelic recessive genes (Davis, 1985).

Host preferences exert influence on the competition of the nodulating inoculant strain (Hynes and O'Connell, 1990). EMS treated seeds of two varieties of soybean when inoculated with rhizobia having different degree of competitiveness showed that competitiveness of strains varied with host's original behavior. Considerable variability was observed with 49 lines of green gram inoculated with two *Rhizobium* strains and 17 varieties of guar. The studies suggest that selection of an appropriate strain of *Rhizobium* for a given host type is required, however, the specific reasons for such interactions are poorly understood. Specific strain and host cultivar combinations have potential to increase yield even in prevalence of high native rhizobia (PAU, 1994). Improvement of host cultivar through pulse breeding programme would be desirable to identify such strains and cultivars combinations keeping in view that environmental factors which greatly influence the symbiosis.

Chemotaxis and motility play a definite role in competition ability of rhizobia. Non-motile mutant of *R. leguminosrum,* R. *meliloti* and *B. japonicum* were competitive than their wild counterparts (Malak, 1992).

Bacteriocin production in soil is important in controlling inefficient strains present in the soil. Bacteriocin production has been reported in *R. trifolii.* Rhizobial toxins such as bacteriocin etc. are known to regulate the competition ability of Rhizobia (Triplett, 1990).

Ecological Factors

The effect of soil organic matter, temperature, moisture and nutrients can be observed on the symbiotic process by improving the growth and development of the macrosymbiont.

Organic Matter

Several workers have reported that organic matter exerts a favourable influence on the rhizobial numbers. Profuse nodulation of certain pulses in soil enriched with charcoal was reported by Iswaran *et al.* (1979). Gaur *et* al. (1975) reported better nodulation and yield of chickpea by application of farm yard manure. Response of groundnut to FYM was also recorded by Dongal and Zende (1976). Likwise Tripathi and Edward (1978a) reported that *Rhizobium* strains isolated from nodules of groundnut crop grown on FYM enriched soil were more efficient. Gaur and Mukherjee (1978) reported that nodulation and pod yield of groundnut crop in field condition was significantly augmented due to application of 2 and 5 tonnes of straw ha-1 in an alluvial soil. The residual effect of straw incorporation in soil was noticed on succeeding wheat crop resulting in significant yield increase. The favourable effect of straw incorporation on the yield of lentil and green gram was also reported by Gaur (1979).

Shivshankar and Shantaram (1980) recorded that *Rhizobium* inoculation alongwith application of 3 tonnes straw ha-1 with and without slurry significantly increased the seed yield of soybean under rainfed condition. Under irrigated conditions, 3 tonnes straw ha-1 increased the seed yield after one inoculation. Inoculation of seed with rhizobia and organic mulching increased the yield of soybean (Kaul and Sekhon, 1977) and green gram (Gaur and Mukherjee, 1980).

Humic Substances and Rhizobium Inoculants

The dry matter yield and nitrogen uptake by berseem (*Trifolium alexandrinum)* and yield, nitrogen uptake, nodulation and leghaemoglobin content of *dhaincha* (*Sesbania aculeata)* inoculated with specific rhizobia was favourably influenced by applied sodium humate extracted from FYM to soil (Gaur and Bhardwaj, 1971). Calcium humate was found to effect significantly the nodule development and yield of pea crop (Bhardwaj and Gaur 1968). Spraying of humic acids in small dose (10 ppm) 2 or 3 times increased the yield of soybean and green gram (Varshney and Gaur, 1974). Mathur and Gaur (1977) reported significant effect of humic plus fulvic fractions on the nodulation and yield of chickpea with and without inoculation with *Rhizobium* culture.

Temperature

It influences the population of *Rhizobium* and state of nitrogen fixation in legumes. Root temperature of 30°C and above are known to adversely effect the infection and nitrogen fixation process. *Rhizobium meliloti* was more tolerant than other rhiobia at higher temperature ranging from 30.5 to 42.3°C (Bowen and Kennedy 1959). There is a marked effect of root temperature on the efficiency of rhizobia (Gibson, 1967). Effective symbiosis is also hampered at high temperature (Burton, 1964). Hernandez Armenta *et al.* (1989) reported that at root temperature of 38°C, N_2 fixation was inhibited regardless of shoot temperature. However, at root temperature of 32°C, nodulation and nitrogen fixation were not influenced even when the shoot temperature was 38°C. It appears that high soil temperature adversely affected N_2 fixation in beans, soybeans and Lupin (Rawsthrone *et al.,* 1985, Legros and Smith, 1994).

Moisture

The soil moisture that is adequate for seed germination is also adequate for bacterial movement and nodule formation. The reduced moisture can retard nodule development, even after successful infection, higher senescence resulting in lower rates of nitrogen fixation. Nitrogen fixation in nodulated *Vicia faba* was maximum having soil moisture at field capacity and the activity was retarted at soil moisture above or below this level (Sprent, 1972). Singh *et al.* (1988) observed that flooding the soil decreased nodulation and grain yield of mungbean. Plants at 3 weeks of growth were more susceptible to excessive moisture than at the flowering stage. Water stress adversely affected the yield parameters in mungbean (Khurana *et al.*, 1988).

Plant Nutrients

Apart from, phosphorus, sulphur, potassium and calcium, micronutrients are required as well. A soil deficient in plant nutrients not only affects the growth of plant but also hampers the biological nitrogen fixation.

Nitrogen

The application of starter dosages (10-20 kg ha^{-1}) has been found to increase the growth, nitrogen fixation and yield of legumes and pulses. In the initial stages, the growth of young plants can be retarded in soils having inadequate supply of nitrogen. On the other hand several studies showed adverse effect of higher rates of application of N fertilizers on nitrogen fixation. Increased nitrate concentration significantly reduced their nodulation and dry weight in pea and chickpea (Cowie *et al.* 1990). Nelson (1983) suggested that effect of combined nitrogen on nitrogen fixation in legume depends on host species, *Rhizobium* strain, the form and quantity of N added and its time of application. Organic fertilizers such as soybean meal and farmyard manure did not inhibit nitrogen fixation while providing the plant with complementary nitrogen required for their growth (Hardy *et al.*, 1973 Gaur *et al.*, 1975).

Phosphorus

This is essential plant nutrient but its availability in most soils is very low. Legumes require higher phosphorus amounts to meet their energy needs for the process of nitrogen fixation. Approximately 21 mol of ATP are converted to ADP per mol of N_2 reduced. Phosphorus not only affects plant growth but also nodulation and nitrogen fixation (Lanka and Gautam, 1972, Sattar and Gaur 1989, Othman *et al.*, 1991, Gupta and Sharma, 1989, Sarkar and Banik 1991, Kumar *et al.*, 1993).

It is well established that in acid soils rock phosphate can be used as a source of phosphorus for crops. The work done in India has shown that rock phosphate can also be applied in neutral to alkaline soils when rock phosphate dissolving bacteria are used as inoculant (Gaur, 1983).

Dommergues (1982) in his review, referred to the role of endomycorrhizal infection in relation to P availability to *Vigna unguiculata*. It seems that endomycorrhizae can not replace phosphatic fertilizers but increase their efficiency thus reduce their rate of application. However, in most cases, rock phosphate is not readily absorbed even when they are infected with endomycorrhizae since endomycorrhizae do not play a role in solubilization of rock phosphate. This solubilization of rock phosphate by the addition of elemental sulphur (22 ppm) and inoculation with *Thiobacillus thiooxidans* was suggested by Swaby (1975). However, sulphur is an expensive item and imported in many tropical countries. Moreover, the performance of this treatment was not found superior to rock phosphate plus phosphate dissolving fungi where additional sulphur or pyrite as substrate is not required (Patil *et al.,* 1974-78).

Other Plant Nutrients

There are not many reports on the effect of potash on pulse crops. Ramamurthy *et al.* (1990) reported significant increase both in number and dry weight of nodules/plant in Cowpea, but Singh and Bajpai (1990) did not record any effect of potassium on Soybean yield raised under rainfed conditions. The moisture stress in the soil may be cause of no response to potassium application.

Molydenum is a part of nitrate reductase enzyme and thus improved nodulation and biological nitrogen fixation in chickpea and *Phaseolus vulgaris* (Rai *et al.*, 1984, Dhingra *et al.*, 1981, Sukhdeo 1985). The leghaemoglobin content of nodules in green gram and black gram was maximum due to treatment of seed with sodium molybdate followed by foliar spray (Pandher *et al.,* 1976).

Legumes in general require as much as sulphur as they need phosphorus. Sulphur acts as cofactor in oxidation of pyruvic acid and chlorophyll synthesis and also helps in nodulation and nitrogen fixation. Aulakh and Pasricha (1986) reported a maximum of 22 kg S/ha for Soybean crop followed by other pulses such as chickpea, green gram, pigeonpea, lentil and black gram. Elements like iron, and cobalt have been involved in synthesis of leghaemoglobin (Burris, 1966).

Effect of Pesticides

Pesticides such as insecticides, herbicides and fungicides are applied to soil to control plant pests and disease for increasing crop yields per unit areas per unit time. Numerous research reports have indicated that number of non target organisms of importance to soil fertility processes are adversely affected. Soil respiration is inhibited by insecticides at higher doses (Bardiya and Gaur 1968, Bartha *et al.,* 1967, Radwan, 1965). Nitrification has also been reported to be hampered with higher dosage of insecticides and other pesticides (Bardiya and Gaur 1970, Eno and Everette 1958, Radwan, 1965). Among the nitrifiers, *Nitrobacter* was found more susceptible to pesticides than *Nitrosomonas* (Gaur and Mishra, 1977).

Legume-Rhizobium symbiosis is also sensitive to certain pesticides resulting in lowering the efficiency of nitrogen fixation. Both the bacteria and plants may be affected due to persistent or accumulated residues of pesticides. The effect of different types of pesticides used and their effect in symbiotic nitrogen is discussed in the following pages.

Simazine at the rate of 3 kg ha applied annually for three years decreased the growth and nodulation of peas when it was grown after maize (Avorov, 1946). However, various concentrations of triazines were not toxic to *Rhizobium leguminosarum, R. japonicum* and *R. trifolii* under cultural conditions (Avorov, 1966; Kaszubiak, 1968a, 1968b). Hauk Pacewiczowa (1969, 1970) conducted pot culture experiments to study the effect of simazine on clover, lucerne, serradela and peas. Simzine at 0.1 ppm reduced nodulation in the legumes and yields were markedly decreased. Higher rates and the residues of the herbicide decreased the amino acid spectrum and modified the levels of free amino acids. Simazine at 0.5 ppm completely inhibited plant development of the legumes immediately following germination. Simazine (0.1, 0.05 or 0.01 ppm) reduced the yield of fresh plant material, decreased the content of nitrogen fixed and nodulation rate and changed the quantities and composition of plant proteins and free amino acids in nodules. Application of simazine (0.1-1.0 ppm) not only decreased nodulation but morphological changes in the nodules were also observed (Thomas and Hammond, 1971; Dunigan *et* al., 1970).

Misra and Gaur (1975) reported that application of simazine to soil adversely affected root nodulation, growth, yield and nitrogen uptake by groundnut crop (Tables 98 and 99).

Table 98: Effect of simazine, lindane and ceresan on nodulation and growth of groundnut crop

Treatment	Fresh wt. of nodules (mg)		Dry wt. of plants (g)	
	First uprooting (4 weeks)	Second uprooting (8 weeks)	First uprooting (4 weeks)	Second uprooting (8 weeks)
Control	195	501	2.64	4.95
Simazine 2 ppm	5	31	0.54	0.82
Lindane 1 ppm	55	62	1.93	2.50
Lindane 10 ppm	Nil	36	1.06	1.69
Ceresan (0.2%)	184	1028	2.72	4.55
Ceresan (2.0%)	120	889	2.77	4.39
Ceresan (20.0%)	41	281	2.53	2.99

Table 99: Effect of simazine, lindane and ceresan on yield and nitrogen uptake by groundnut crop.

Treatment	Yield		Nitrogen uptake	
	Dry wt. of pods(g)	Drywt. of leaves (g)	Pod (mg)	Leaves (mg)
Control	24.63 (2793)	20.63 (24.00)	1531.3 (1798.6)	332.1 (391.2)
Simazine 2 ppm	Nil	1.00 (1.00)	Nil	21.0 (21.0)
Lindane 1 ppm	9.48 (10.10)	7.63 (9.13)	594.4 (633.2)	105.2 (125.0)
Lindane 10 ppm	9.65 (9.63)	8.13 (8.63)	605.0 (614.3)	105.6 (109.6)
Ceresan (0.2%)	23.53 (25.00)	15.88 (18.73)	1475.3 (1570.0)	256.6 (301.5)
Ceresan (2.0%)	23.88 (24.03)	15.38 (18.38)	1490.1 (15015)	216.8 (284.8)
Ceresan (20.0%)	21.00 (22.10)	15.25 (15.73)	1247.0 (1312.7)	176.9 (119.7)
CD at 5%	1.72	1.70	359.4	33.9

Figures in Parenthesis's are with 1% legume Straw

Root nodulation was drastically reduced even at normal concentration (2 ppm) of the herbicide. AT higher concentration, complete inhibition was observed. The normal dosage of simazine caused stunted growth and yellowing of plants which died after 8 weeks. Simazine checked the pod formation completely and nitrogen uptake underwent a serious set back. The adverse effect of prometryne (20-5 ppm) on nodulation and other parameters of green gram and other legume was not detected by some workers (Appleman *et al.*, 1946, Avorov, 1966, Rethinam *et al.*, 1974).

The sodium and ammonium salts of 2, 4-D or MCPA in the concentration ranging from 0.01 to 10 ppm depressed nodulation when applied immediately after inoculation (Nilson 1957). Garcia and Jordan (1969) reported that 2, 4-D alone or in combination with dalapon reduced the nitrogen fixing capacity of nodules, their number and dry weight of *Lotus corniculates*. Dalapon had little effect on nodule number but the dry weigh per plant decreased. 2-4-DB directly acted on host plant rather than on the nodule bacteria, although it altered the metabolism of rhizobia in artificial media. Kust and Struckmeyer (1971) observed harmful effect of trifluralin on nodulation in soybean in green house experiment.

It was observed that under cultural conditions very higher concentration of 2, 4-D (50 ppm - 8000 ppm) was required to inhibit growth of *R. leguminosarum*

(Daitloff, 1970 a,b, Jenson 1969). The toxicity of 2, 4-D to R. meliloti and *R. Japonicum* was observed only at 250 ppm (Gaur and Misra 1972). Nilson (1957) observed the tolerance of *R. trifolii* to 2, 4-D, MCPA, 24 DB and MCPB upto 25 ppm and tolerance of *R. meliloti* to 2,4-D and MCPA. Rhizobia were found tolerant to herbicides and even at higher concentrations under cultural conditions.

Halogenated aliphatic acid such as TCA and Dalapon were not toxic to pure cultures of rhizobia and in some cases were found stimulatory to growth of rhizobia and nodulation (Avorov 1966, Jensen 1970, Garcia and Jodan 1969, Kaszubiak 1968 a,b Lakshmi Kumari *et al.,* 1973, Namedo and Dube 1973, Nutman 1948, Sud *et al.,* 1973).

Treflan applied at normal rates of application reduced nodulation in soybean and chickpea (Brock 1972, Dunigan *et al.*, 1970 Misra and Gaur 1975). The importance of time of application of herbicide to soils cropped with legumes was demonstrated by Hamdi and Twefik (1969). They reported that nodulation and growth of cowpea was inhibited by trifluralin at recommended field rates when applied on the day of planting but were stimulated when the herbicide was applied 27 days before planting.

Effect of Insecticides

Appleman and Sears (1946) observed that DDT did not reduce nodulation in soybean, pea, clover and lespedzea in sand culture experiment until concentration reached to 1000 kg per hectare. The results of Wilson and Choudhri (1946) confirmed their findings.

Colorade-9, an insecticide similar to DDT did not depress nodulation in common bean even at relatively high concentration. It was also recorded that DDT at the rate of 183 kg/ha significantly depressed number of nodules per plants. Application of 30 ppm of toxic material in the form of crude BHC was injurious to legumes but no such effect was observed where the same amounts of pure alpha, beta and epsilon isomers were used (Wilson and Choudhari 1946).

Braithwaite *et al.* (1958) showed that DDT and BHC at the rate of 1.5 and 10 kg/ha had a drastic effect on root nodulation in subclover but aldrin, diealdrin and chlordane did not show effect on root nodulation. Raghu and Mac-Rae (1967) found that 6 kg/ha of gamma BHC increased nitrogen fixation by blue green algae by checking the population of algae predators. Starks and Lilly (1965) studied the effect of seed treatment with BHC on 5 widely grown varieties of soybean and found no important difference in response to higher doses of BHC. Neither significant difference in seedling emergence nor in average weight of green matter was found, only the highest rate of BHC caused the serious damage to germination of seeds.

The favourable effect of DDT upto 40 ppm was observed on nodulation, growth, yield and nitrogen uptake of *Pisum sativum* inoculated with *Rhizobium leguminosarum* but higher levels were injurious to plant (Pareek and Gaur 1969). In another experiment (Pareek and Gaur 1970) reported that the normal dose and even ten times of the normal rate of application of DDT were not deleterious to plant growth, nodulation, yield, leghaemoglobin content and nitrogen fixation by *Vigna radiata* but 40 ppm and higher levels were significantly harmful. Addition of 1 percent organic manure appreciably reduced the toxic effect of the insecticide in both the experiments.

Lindane reduced nodulation and decreased the efficiency of nodules (Daitloff 1970 a, Misra and Gaur 1974, 1975) and other insecticides (Carbamates, cyclodenies, phenyls and phosphonates) did not effect nodulation (Daitloff 1970b, Salim *et al.*, 1970 Swamiapan and Chandy 1975). Taha, *et.al.* (1972) recorded that application of diptrex or endrin at normal field rates were found to stimulate a higher number of effective nodules and high rate of nitrogen fixation especially under non-sterile condition.

Gaur *et al.* (1975) showed that the toxic effect of the insecticide (lindane) was reduced due to application of farmyard manure (Table 100). Since the inhibition of nodulation with 25 ppm of Lindane was complete, the highest dose of farmyard manure applied at the rate of 0.6% carbon alleviated the toxicity and the lower doses were not effective.

Table 100: Effect of lindande and FYM on nodulation and dry matter of Chickpea

Treatment	Dry wt. of nodules/pot (mg)		Wt. of grain/pot (g)	Wt. of straw/pot (g)
	7 weeks	10 weeks		
Control	38.0	72.5	0.38	3.86
Lindane 25 ppm	0.0	-	0.00	0.35
Lndane 25 ppm + FYM (0.2% C)	17.7	17.0	0.29	1.26
Lindane 25 ppm + FYM (0.4% C)	18.2	27.0	0.27	1.26
Lindane 25 PPM + FYM (0.6% C)	30.0	121.0	0.90	2.57

FYM = Farm yard manure

Effect of Fungicides and Fumigants

Satisfactory nodulation was produced on *Rhizobium* inoculated soybean plants treated with semsan but cuprocide and ceresan prevented nodulation of soybean and canning pea plants (Appleman 1942). The experimental results of some workers like Allison and Torris (1944), Allington (1945), Miller (1948) and Vlitos and Preston (1949) showed that nodulation was not effected; on the other hand it was enhanced as a result of seed treatment with arason, phygon or Spergon. McNew and Hofer (1942) found that peas treated with spergon could be inoculated without significant reduction of nodulation provided the seeds were planted within half an hour after bacterial inoculation.

Ceresan proved to be very injurious to all of the legume seeds that were treated with this volatile chemical. However, nodulation was not affected by chemical seed treatment of alfalfa, mungbean, clover, Chinese red cowpea and Australian winter pea when chemical seed treatment was followed by *Rhizobium inoculation* of seed (Vlitos and Preston, 1949). Application of phygon, spergon, vancide 51, arasan SFX, cerenox and captan caused reduction in yield of subterranean clover forage in the situation where seed applied rhizobial inoculation was attempted using the recommended moist method immediately prior to planting (Vlitos and Preston, 1949). Phyzon was almost completely toxic to seed applied commercial inocula whereas others were not injurious to that extent (William *et al.*, 1960).

Rogor applied as a seed treatment prior to or at the same time of inoculation and lime pelleting is very damaging to the applied rhizobia. Poor establishment was obtained when the roger treated seeds were held for one month before inoculation and lime pelleting was carried out. It has very drastic effect on nodulation (Goss and Shipton 1965).

Ebbles (1967) reported that four soil fumigants, chlorpicrin, dazomet, farmaldehyde and D-D decreased nodulation but no adverse effect on plant growth was detected. With dazomet, formaldehyde and D-D the decrease in nodulation was largely attributed to extra soil nitrogen mineralized but almost complete suppression of nodulation by chloropicrin probably reflected its lethal effect on the *Rhizobium* bacteria.

The findings of several other workers and Mukewar and Bhide (1970) did not show any harmful effect of seed treatment with many fungicides viz thiram, zindeb, captan, agronol, brassicol, agrosol on the emergence and nodulation of soybean and groundnut. Ostwal and Gaur (1971) on the other hand found an increase in the yield of soybean due to seed dressing fungicide and antibiotic aureofungin both in recommended dose and ten times higher than recommended doses. Fungicides even ten time of normal rate of application showed no adverse effect on nodulation. Similarly, Balraman and Prasad (1973) reported that groundnut plants raised from seeds treated with both wet ceresan and *Rhizobium* sp., (alone or in combination) had the highest dry matter and nitrogen content as compared to plant raised from untreated seeds, Nodulation was also not suppressed. Daitloff (1970b) suggested that toxic effect of fungicides may be prevented by coating legume seeds with a layer of polyvinyl acetate after fungicide treatment but before inoculation with rhizobia. The data from the available literature suggests that fungicides have variable effect on rhizobia, nodulation and nitrogen fixing ability of legume.

V. Nitrogen Fixation

Out of 17000 species of legumes distributed the world over, only about 200 are cultivated for food, forage, oil fuel or organic matter. The estimates of quantum of nitrogen fixed by various legumes are presented in table 101. However, the amount of nitrogen fixed is variable because it is dependent upon environmental condition of plant, nutrient and the biotic factors.

Table 101: Estimate of nitrogen fixed by some cultivated legumes

Crop	Nitrogen fixed (kg/ha)	Crop	Nitrogen fixed (kg/ha)
Alfalfa	100-300	Cowpea	80-125
Clover	100-150	Pigeonpea	168-200
Cluster bean	37-196	Black gram	50-55
Peas	46	Chickpea	85-110
Lentil	35-100	Common bean	3-57
Fenugreek	44	Groundnut	50-206
Soybean	49-450	Green Gram	50-66

Source: Wani and Lee (1992) Peoples and Crasswell (1992), Subba Rao *et al.* (1993)

Alfalfa, clovers, soybean, pigeonpea and groundnut are vigourous nitrogen fixers. Moreover, valid estimate of N2 fixed by cultivated legumes are not available as in most cases, plant root, nodules and fallen leaves are not taken into account. In the tropics where plant residues are also removed from the field to feed animal, most grain legumes with an exception of permanent pasture deplete the soil nitrogen but the grain legumes only slow down the declining trend in soil nitrogen status.

The experiments conducted have shown based on the nitrogen fixed by the leguminous crops that about 50% of the nitrogenous fertilizer can be saved by inoculation with efficient rhizobia (Subba Rao *et al.*, 1993).

Residual Effect of Legumes

When grain legumes are grown in rotation or as mixed or intercrops, the succeeding or associated cereal crops do get benefited particularly in nitrogen from their secretions and residues (Table 102).

Table 102: Residual effect (fertilizer nitrogen equivalence) on

Legume	Succeeding cereal crop	Nitrogen (kg/ha)
Berseem/sweet clover/winged beans	Maize	123,83,70
Chickpea/cowpea/lathyrus	Maize	65,60,42
Pigeonpea/pea/ lentil	Maize	20-67.20-32,18-30
Blackgram/green gram	Sorghum	40,40
Greengram/Pigeonpea/cowpea	Wheat	16-68,40,38
Groundnut/intercrop cowpea/ groundnut intercrop	Wheat	28,13,12
Cowpea/chickpea/lentil	Pearl millet	60,40,40
Peas/Pigeonpea/Green gram	Peal millet	40,30,30

Source: Wani, Rupela and Lee (1995)

The findings presented show that pearl millet and sorghum crops grown after either kharif or Rabi grain legumes were benefited by receiving residual nitrogen ranging from 30 to 60 kg/ha. In case of maize, the benefit varied. Maximum residual effect of 123 and 83 kg N/ha was registered with berseem (*Trifolium alexandrinum)* and sweet clover respectively. Chickpea/cowpeal *Lathyrus* on maize showed the benefit of 65, 60 and 42 kg N/ha. The benefit of pigeonpea/peas/lentil varied in case of maize crop in the range of 20-67, 20-32 and 18-30 kg/N/ha respectively.

The increase in yield was of the order of 0.5 to 3 t ha-1 representing a 30 to 350% increase over yield in cereal cropping sequences (Peoples and Crasswell, 1992). Grain yield of maize crop succeeding medium duration pigeonpea grown at Hyderabad was 57% higher than the grain yield of maize following a fallow treatment (Kumar Rao, Dart and Sastry, 1983). In a long term rotation experiment probably only a small amount of nitrogen are released by legume roots and thus only a small portion of the nitrogen needed by adjacent plant can be provided by this mechanism. On the other hand when plants gets senescent, are killed or the above ground portion are cut, considerable nitrogen may be released by plant unable to utilize N, (Whittney and Kanehiro 1967). The increased availability of soil nitrogen contributed significantly to soil fertility in the legume based system. However, other mechanisms enhance nutrient availability, improve soil structure, reduce incidence of plant diseases improved mycorrhizal colonization etc. Increased mycorrhizal colonization other than nitrogen were also responsible for the yield increases (Cook 1988, Danso and Papstylianou, 1992, Wani and Lee 1995, Wani *et al.*, 1991).

The problem of inadequate availability of *Rhizobium* culture can be solved by expanding the Bacterial Fertilizer production. The carrier based Rhizobium inoculants requirement is about 15 million kg annually to cover the entire area under leguminous crops.

VI. Crop Responses to Rhizobium Inoculation

Most of the cultivated tropical soils have good population of rhizobia which nodulate legumes. The presence of nodules alone should not be taken as an index of establishment of an efficient nitrogen fixing system, competitive and efficient strains of Rhizobium used as inoculm will ensure maximum nitrogen fixation. Poor nodulation in certain areas is due to absence of effective Rhizobium strains. The occurrence of infective strains of rhizobia is fairly widespread. Out of 87 groundnut Rhizobial strains, only five were found to be efficient (Kulkarni and Joshi 1988). Absence of appropriate rhizobia and different environmental factors discussed in the preceding pages indirectly influence the efficiency of Rhizobium strains.

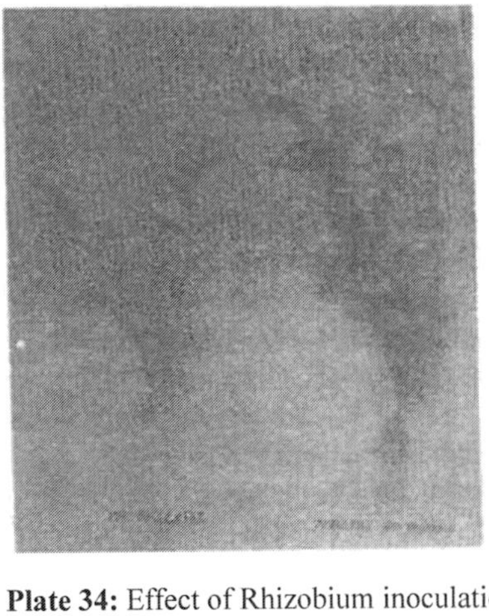

Plate 34: Effect of Rhizobium inoculation in chickpea in Aligarh soil

Rhizobia present in the soil and nodulating the roots of these plants are not always of the desired type. Therefore, the isolation of rhizobia from the nodules of different leguminous plants and screening them for efficiency are carried by research institutes and universities and the efficient cultures are tested in field and pot trials to find out the better strains for use as biofertilizer. Rhizobium inoculants in difference locations and soil types were reported to significantly increase the grain yield of chickpea (Bajpai *et al.*, 1974; Lehiri *et al.*, 1974; Patil and Medhane 1974; Chundwat et al 1976; Dube 1976; Subba Rao 1976; Rai *et al.*, 1977 and Singh 1978) (Plates 36-38) green gram (Sahu and Behera 1972; Gupta *et al.*, 1976; Rai and Singh, 1978) lentil (Dube 1976, Subba Rao 1976; Dhillon *et al.*, 1977, Ojha *et al.*, 1977, Lehri *et al.*, 1974; Bajpai and Gupta, 1977); Vigna radiata (Sahu 1973, Lehri *et al.* 1974, Reddy *et al.* 1978); pigeon pea (Singh *et al.*, 1976; Subba Rao 1976) and groundnut (Sahu *et al.*, 1972; Bajpai *et al.*, 1974; Rao 1974).

Plate 35: Effect of Rhizobium inoculation on chickpea

In a few cases the leguminous crops did not respond to Rhizobium inoculants; pea (Nandpur *et al.*, 1973); chickpea (Dhillon *et al.*, 1977; Panwar *et al.*,1977); Vigna radiata (Dhillon *et al.*, 1977) and lentil (Dube, 1976).

Plate 36: Rhizobium inocula- tion on chickpea in Delhi soil

The possibility of growing soybean as an economic crop in India was demonstrated only around 1963, although black seeded late maturing and poor yielding varieties had been under cultivation for long time in some parts of the country. The success of soybean crop depends on the availability of "efficient" culture of Rhizobium japonicum. There have been several reports on the favourable effect of inoculation of soybean seeds with R. japonicum on the grain yield and other parameters cowpea Rhizobium isolates. B&C of nitrogen fixation (Chhonkar and were better than isolate No. A. Negi, 1971; Thakur and Hasan, 1972; Singh and Saxena, 1973; Bajpai *et al.*, 1974; Sharma and Tilak, 1974; Sundra Rao, 1974; Tilak and Saxena, 1974; Dube 1976; Kumar *et al.*, 1976; Subba Rao 1976; Singh and Tilak 1977; Kumar Rao and Patil 1977; Tripathi and Edward 1978).

Plate 37: Performance of cowpea Rhizobium isolates. B$C were better than isolate No. A.

The effect of Rhizobium on grain legumes such as cowpea, (Plate 39) mungbean, urdbean, pigeonpea, chickpea, groundnut and lentil was examined under field conditions as reported by Subba Rao *et al.* (1993). The result showed that yield of the above crops was increased due to Rhizobium inoculation ranging from low responses to high responses with a few exceptions (Table 103).

Table 103: Increase in yield of grain legumes due to Rhizobium inoculants

Crop	Location	Increase in yield (%) over control
Cowpea	Krishi Nagar (Gujrat) Coimbatore (Tamilnadu)	1-36 No response
Mungbean	Lam (AP) Dholi (Bihar) Delhi (UT) Pantngar(UP)	1-17 &49 12-33 4-15
Urdbean	Podukottai (TN) Dholi (Bihar) Pantnagar (UP)	4-21 11-29 17-21
Pigeonpea	Varanasi (UP) Dholi (Bihar) Delhi (UT) Hisar (Mariana) Dohad (Gujrat) Sehore (MP) Hyderabad (AP)	4-19 12-25 18-28 24-43 33-67 20-41 No response
Chickpea	Varanasi (UP) Dholi (Bihar) Delhi (UT) Hisar (Mariana) Dohad (Gujrat) Sehore(MP) Hyderabad (AP) Bandnapur (Maharastra)	4-19 12-25 18-28 24-43 39-67 20-41 No response 8-12
Groundnut	Not given	21
Lentil	Pantnagar (UP) Ludhiana (Punjab) Dholi (Bihar)	4-26 No responese 5-17

Source: Subba Rao *et al.* (1993)

Table 104: Response of Rhizobium inoculant on grain (yield average response of several strains of Rhizobium of chickpea, lentil and rajmash)

Location	Crop	Grain yield (kg/ha)		Percent	
		Uninocu-lated	Inoculated	30/60 kg Nha^{-1}	Increase in yield due to *Rhizobium*
Coimbatore	Chickpea	735	854	788a	16.2
Bangalore	Chickpea	763	882	841 a	31.0
Delhi	Chickpea	1288	1330	1347a	3.2
Durgapur	Chickpea	1250	1662	-	33.0
Ludhiana	Lentil	749	884	-	12.&
Pantngar	Lentil	1584	1832	-	15.6
Varanasi	Rajmash	1569	1697	1665b	8.2
Vamban	Rajmash	612	690	664 b	12.7

Source: Khurana (1994)

a - compared with fertilizer N @ 30 kg N/ha.

b - compared with fertilizer N @ 60 kg N/ha

The yield response varied with the *Rhizobium* strains, location and crop. The average yield obtained due to different strains was calculated and compared with control yield obtained with fertilizers N applied @30 or 60 kg N/ha (Table 104).

The grain yields of chickpea due to *Rhizobium* inoculant was increased by 3.2, 16.2, 31.0 and 33% over respective control at Delhi, Coimbatore, Bangalore and Durgapur, respectively. The grain yield of lentil due to Rhizobium inoculation was increased by 15.6 and 12.8 percent at Pantnagar and Ludhiana respectively. The increase in yield of Rajmash due to *Rhizobium* inoculation was increased by 8.2 and 12.7 percent at Varanasi and Vamban respectively.

Indian soils usually did not have efficient strains of Bradyrhizobium inocula capable of nodulating the introduced cultivars of soybean. It was therefore, warranted to inoculate the seeds before sowing with the specific culture. In the initial years of soybean introduction in India during 1965 to 1969, soybean seeds were treated with imported carrier based *Rhizobium* cultures obtained from USA. However with the efforts of Indian Institutions, indigenous inoculants were developed and mass produced for use in soybean cultivation.

The proper selection of the strains for promising cultivars grown under different agro-climatic conditions, identification of suitable carrier material, production of *Rhizobium* inoculants and their testing in research stations and farmer's fields were carried out by All India Coordinated Research Project on Soybean at a few centers in the country.

The performance of soybean incoulants developed at IARI Centre was tested and found to be trustworthy over a long period of time (Table 105). The benefit of inoculating soybean seeds has been demonstrated time and again at different locations in the country and on average it has been found to increase the soybean yield by about 30 per cent over the uninoculated control (plates 40 a, b, c).

Table 105: Effect of Bradryhizobium japonicum on soybean yield

Year	Number of locations	Mean total bean yield (Kg/ha)		Percent Increase in yield over control
		Uninoculated (control)	Inoculated	
1968	2	2383	2756	16
1969	6	2106	3080	46
1970	10	1605	2384	49
1971	10	1408	1773	26
1972	10	1410	1739	23
1973	10	1659	2104	27
1974	10	1181	1393	18
1975	11	1197	1535	28
1976	7	1353	1670	24
1977	8	1267	1352	7
1978	1	1813	1958	8
1979	6	1051	1076	3
1984	3	1422	1595	12
1985	7	1742	2235	18

Source: Dr. V. Balasundram IARI, New Delhi

From a meager 0.622 million hectares of cultivated area in the country during 1981-82, it increased to 5.004 million hectare during 1996-97. The major cultivated area of 3.740 million hectares falls in Madhya Pradesh followed by Maharashtra (0.673 million ha) Rajasthan (0.429 million ha) Uttar Pradesh (0.072 million ha) and other states. Thus soybean production has increased.

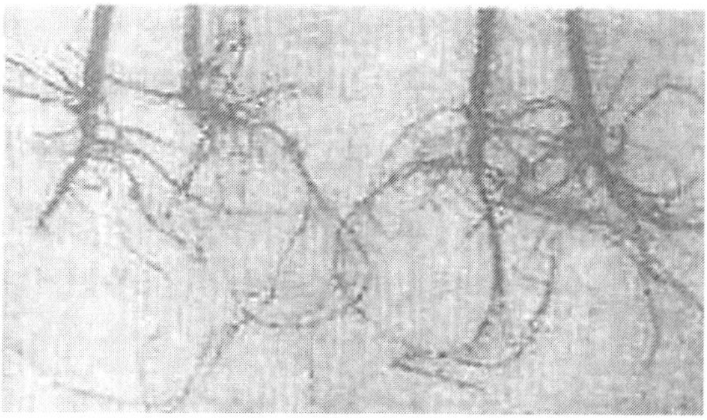

Plate 38 a: Poor nodulation in uninoculated soybean crop in aceptisol soil of Varanasi

from 0.466 million metric tonnes during 1981-82 to 4.028 million metric tonnes during 1996-97, which is a significant development. Soybean has emerged as the third largest oilseeds in India and the single largest foreign exchange earner among all oilseeds with an annual export of Rs. 1803 crores reached in 1995-96. The protein rich Indian soymeal has been in great demand in South East Asian and Far Eastern markets although facing competition from USA, Brazil and Argentina. However, our soybean yield of 800 kg/ha is the lowest in the world, because the nutrient requirement of essential plant nutrients are not being met at the optimum levels. Concomitant with the increase in the areas, the availability of good quality bacterial inoculants, to meet their nitrogen and phosphorus requirements has not been taken care of. Hardly 10% of seeds are being treated with legume inoculants at present by the farmers in Madhya Pradesh, a major soybean growing area.

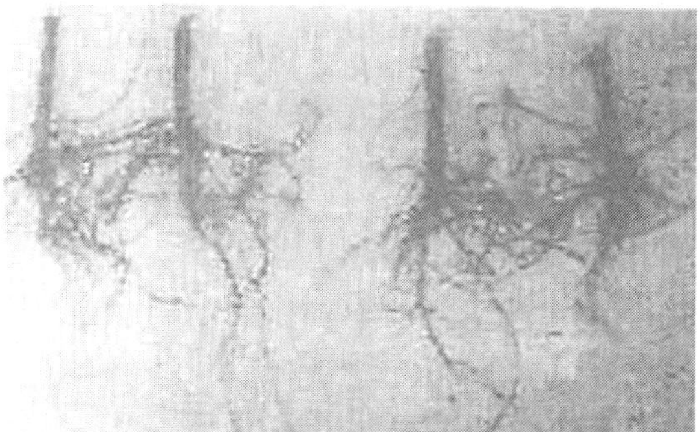

Plate 38 b: Good nodulation in soybean due to inoculation with USDA strain 110 (T3) and BAU stain (T4)

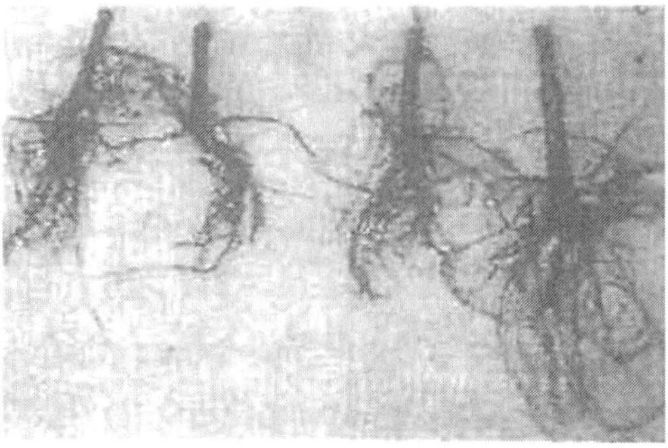

Plate 38c: Good nodulation in Soybean due to combined inoculation of USDA-110+PSB (T5) and BAU-118+PSB (T6)

Rhizobium and Pelleting

Pelleting of inoculated seeds with calcium sulphate, calcium carbonate or rock phosphate under adverse soil conditions as in saline, alkaline and acidic soils, was found helpful in improving the yield of berseem (Bajpai and Gupta, 1977), *Cajanus cajan* (Subba Rao, 1970), *Vigna radiata* (Gupta et *al.*, 1976), soybean (Chhonkar *et al.*, 1971; Kumar Rao and Patil, 1977), and groundnut (Sharma and Rao, 1978). In acidic soils, lime (calcium carbonate) pelleting is usually practised when rhizobium inoculated seeds are to be sown in acidic soils. Lime @ 10% (w/w) is coated on the inoculated seed forming a pellet around the seeds. This favours increased nodulation in problem soils.

VII. Technology for Mass Production of Rhizobium Inoculants

Isolation of Rhizobium from nodules

Rhizobium can be isolated from the root nodules of host plants. Rhizobia grow readily in culture media containing carbon source such as mannitol or glucose, require nitrogen in the form of ammonium or nitrate and several inorganic salts. In addition, one or several B vitamins are often needed for its growth. The vitamins which are required include biotin, thiamin, pantothenic (Grahm, 1963) and sometimes riboflavin.

Rhizobium are rod shaped, gram negative non-spore forming aerobic, typically motile bacteria measuring 0.5 to 0.9 um wide and 1.2 to 3.0 um in length.

The family Rhizobiaceae consists of *Rhizobium, Agrobacterium* and *Chromobacterium* (Breed *et al.,* 1957). Several carbohydrates are utilized by rhizobia, with the accumulation of acid but never produce gas. The rhizobia are quite similar to *Agrobacterium* but differ in minor cultural properties and its inability to form root nodules in legumes.

Isolation of pure culture of Rhizobium

The nodulation pattern and size and shape of nodules differ in legumes. A young leguminous plants (about 6-8 weeks) is carefully uprooted so that the nodules on the root remain intact and roots are washed in running water to remove adhering soil particles. Healthy unbroken, firm and preferably pink nodules are selected and washed in water. It is immersed in 0.1% acidified HgCl2 for 4-5 minutes. Surface sterilized nodules with $HgCl_2$ are washed repeatedly with sterile water and dipped in 70% ethyl alcohol followed by more washings with sterile water.

The nodule is crushed with the help of a sterile glass rod in a small aliquot of sterile water. Isolations are carried out by either serial dilutions prepared from the nodule extract and the aliquots of proper dilutions are spread on the yeast extract mannitol agar (YEMA) or the fluid from the dissected crushed nodules is spread on the surface of YEMA medium in plates with a smooth glass rod (Fred *et al.*, 1932). The plates are incubated up to 10 days at 28°C. Within 4-5 days, large raised slimy colonies will be formed (Plate No. 39) and later, it is possible that smaller colonies may also develop. Not all colonies of bacteria are rhizobia since bacteria of related genus *Agrobacterium* may also develop on the YEMA plates.

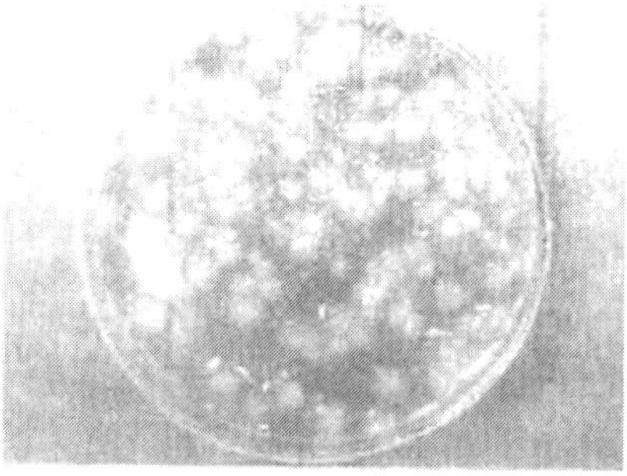

Plate 39: Colonies of *Rhizobium trifolii* on YEMA

Apart from the non-ineffective rhizobia, host-legume reaction still considered to be relevant test for the identification of rhizobia and differentiation from *Agrobacterium* spp.

Many of the biochemical and physiological characteristic such as catalase, oxidase, urease, liquefaction of gelatin, indole production, reduction of methylene-blue, methyl red, production of levan, vogespraskauer reaction, and fermentation of carbohydrate were not found suitable for identification of *Rhizobium* (Vincent, 1970). However, certain other cultural and biochemical characteristics are listed below which have been used from time to time.

Growth and Colony Characteristic on YEMA

The rhizobial colonies appear gummy, shining wet and raised, looking somewhat translucent with smooth edges. The consistency of the colonies may change depending on the source and *Rhizobium* sp. The following routine cultural tests are carried out to identify *Rhizobium*. sp. (Vincent, 1970).

Microscopy

The gram staining helps to classify bacteria into Gram (+) and Gram (-) cells. Rhizobia are Gram-negative rods.

Congo Red Test

This test is designed to differentiate rhizobia from agrobacteria. An aliquot of 25 ml 7 per cent Congo red solution in water is added to one litre of YEMA. On the this medium, rhizobia appears white, translucent, glistening, elevated comparatively small colonies with entire margin in contrast to stained colonies of Agrobacteria (Vincent, 1970).

Change in pH of Growth Medium

Norris and Date (1976) showed that when different strains were grown on YEMA medium containing bromothymol blue, strains of R. *meliloti*, *R. trifolii*, R. *leguminosarum* and *R. phaseoli* produced acid in this medium whereas the strains of *R. lupini, R. japonicum* and *Rhizobium* sp. produced alkalinity.

Hofer's Alkaline Test

Agrobacteria grow at higher alkaline pH 11 while rhizobia are unable to grow in Hofer's alkaline medium (Allen and Allen, 1958).

Growth on Peptone-glucose Agar

Agrobacteria grow very well in glucose-peptone medium whereas rhizobia grow very poorly in this medium (Klaczkowska et *al.*, 1968). However, Okon *et al.* (1972) reported that 2 strains of the four tested strains of *Cicer rhizobia* grew well in the glucose-peptone medium.

Lactose Agar

Agrobacteria utilize lactose by the enzyme ketolactase whereas rhizobia fail to utilize this sugar. This can be detected on agar medium containing lactose @ 10g/l. Rhizobia are grown on lactose agar plates for 4-10 days depending on the growth rate and the plates are flooded with Benedict reagent. The formation of yellow colouration due to Cu20 indicates the presence of agrobacteria.

Susceptible of High Salt Concentration

Agrobacterium radiobacter and A. *tumefaciens* and a few strains of medic rhizobia were able to grow in YEMA medium containing 2% NaCl, whereas other rhizobia's growth was hampered (Grahm and Parker 1964). However, salt tolerance of *Dolichos lalab* rhizobia from Indian soil was reported by Pillai and Sen (1979).

Plant tests for Screening of Efficient Rhizobia

The conventional method to screen or select the most efficient strains of *Rhizobium* from a large number of strains/isolates is to examine its potential for nitrogen fixing ability on its homologous legume. This is mainly done by growing the plants under aseptic conditions, inoculating the *Rhizobium* strain to test plant followed by examining nodulation and measurement of total N content or the dry matter yield of the plant after 4- 6 weeks of plants' growth. The total nitrogen is proportional to dry matter yield of plant, nitrogen percent remaining almost constant.

For Small Seeded Legumes

Thornton (1930) and Jensen (1942) used different media for seedling agar. In nodulation studies, various sizes of tubes plugged with cotton based on the need have been used. The volume of agar to be poured into the tube depends on the requirement of study.

Gibson Method (1963) modified the seedling agar by adding several trace elements (see Chapter 21). Nitrogen controls are usually maintained by adding KNO3 solution. This method allows the shoot system to grow freely in air while root system is still enclosed in the tube. Nitrogen-free agar (30 ml) is poured into tubes (200 x 25 mm). Cover the mouth of each tube with aluminium foil and tighten each cap with a rubber band. Then sterilise the glass tubes at 15 psi and at 121°C for 20 minutes, slanted and allowed to solidify in such way that the slope of the agar reaches the top of the tube. A small aperture is made on the foil and fitted with cotton wool which not only affords aeration but also acts as an inlet for adding nutrient solution in tube. Sterile conditions has to be maintained to prevent contaminants.

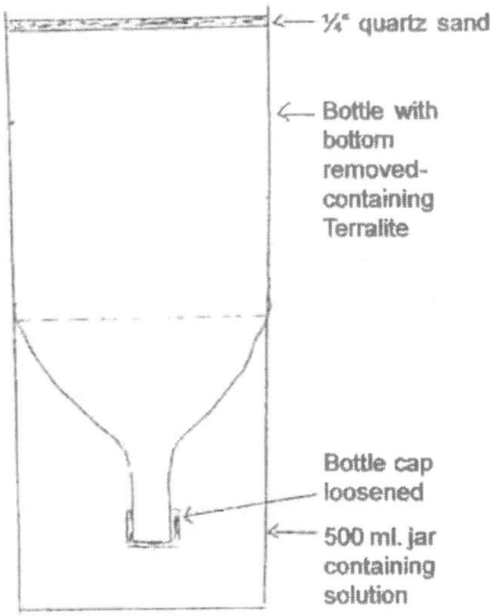

Fig. 2: Growth pot

Sterilization and Germination of Seeds

Seeds are sterilised, with 95 % ethyl alcohol, immersed for 3 minutes in 0.2% $HgCl_2$ acidified with 5 mg/l conc. HCI. Seeds are washed thoroughly (5 times) with sterile water. The seeds then are germinated in the folds of moist filter paper or on 3% plain agar (with water) in petri plates incubated at 28°C for germination which are inverted later to provide seedling with uniform straight roots.

Placement

Asceptically grown seedlings are then carefully placed in the second hole of Gibson's tube in such a way so that the emerging root just touches the surface of the agar slope in the tube. This is done by using a pair of sterile farceps. The lower portion of the tube is shielded with black paper to avoid light falling on the root system and the tubes are put in the racks for convenient handling.

One fourth strength of sterile seedlings solution is added through another hole made in the cap within 12 mm of the top of the each tube. Inoculum of specific *Rhizobium* grown for 2-4 day old is added through the ventilation hote on the aluminium cap. The tubes are incubated at 30°C under artificial light (2000 Lux) illuminated from above. After 4-5 weeks, quarter strength seedling solution should be added at 2-3 days intervals. After 7-10 days nodulation in the roots of the plants may be examined. After 4-5 weeks, dry matter yield of plants may be recorded and the nitrogen fixing efficiency is calculated in terms of increase in dry weight of plants over control.

Leonard Jar Method (Leonard, 1944, Vincent, 1970)

This method can be used for both small and large seeded legumes. A wide bottle of 750 ml capacity with its bottom cut off so as to provide a flat ground finish is inverted in a glass jar of suitable dimensions so as the neck of the bottle properly fits (Fig. 2). A suitable wick (cotton wool or ordinary cotton lamp wick) is passed through the inverted bottle neck in such a manner that it is not only dropped in the nitrate-free nutrient solution in the jar but also touches upper most region of sand medium in which the seed/seedlings are growing.

The cotton wick is meant for steady supply of plant nutrients to the growing plant by the capillary action of the sand. The sand substrate is wetted with 4 strength of the nutrient solution (Chapter 21) until the solution begins to drip through the wick at the bottom. The jar at the bottom is filled with same nutrient solution and the top portion of the assembly is covered with a petri dish. The assembly may be covered with water proof paper bag with a help of rubber bands so that the roots of the plants during growth are protected from light as well as contamination by other micro- organisms through the interface between the 2 glass assemblies. The assembly is sterilized in an autoclave at 20 lbs pressure (126°C) for 2 hours and used after it has cooled.

Surface sterilized seeds and washed with sterile water are sown on the sand, on germination, the petri dish is removed, seedlings inoculated with the culture suspension and covered with sterilized gravel to safeguard from aerial contamination. The proper controls are kept for evaluation.

Pot Culture Tests

Further testing of the efficiency of Rhizobial strains which have shown high efficiency in aseptic conditions are carried out by growing the specific host legumes in sterilized sand or soil. Fine grade river sand is washed well in tap water and dipped in acid for 2 days. The acid is removed and sand washed several times with tap and distilled water to get rid off the last traces of acid. Neutral sand allowed to dry, autoclaved (20 lbs pressure for 20 mts) and put in pots. *Rhizobium* inoculated seeds are sown and irrigated with nitrogen free nutrient solution at required intervals through container placed in bottom of the pot.

Further testing of the efficiency of a few efficient rhizobial strains are carried out by growing host legumes in sterilized soil and /or in unsterilized soil. In these cases the efficiency of inoculated strains of *Rhizobium* is measured from the nodulation taking into account number of nodules and/or dry weight of nodules and dry weight of vegetative part such as shoots.

By following the above plant tests, efficient nitrogen fixing strains of Rhizobia for each legume crop can be selected. Selection at this stage is done depending on the performance under pot culture conditions using unsterile soil which have to be tested under field conditions.

Field Tests

Finally, the rhizobial strains selected as above are further subjected to the field inoculation test. In this, the few efficient strains of rhizobia (obtained as above) are used to inoculate seeds of legumes growing in the field in randamized plots. Plots in which uninoculated seeds are sown serve as a 'control' and nitrogen treated (40 kg/ha) plots may serve as another control. Each rhizobial strain used for inoculation serve as one treatment.

Sufficient number of replications for each treatment are kept so that after 4-6 weeks growth (or at the pre-flowering stage) plants from 2-3 plots from each treatment are uprooted for nodulation and plants in rest of the plots are allowed to grow till maturity to compare the yield of grains obtained due to inoculation with different rhizobial strains with those of uninoculated and fertilizer nitrogen controls. From the results of such field experiments, the rhizobial strains with inherent nitrogen fixing efficiency (as tested under laboratory aseptic conditions or pot culture conditions) are further screened for their efficiency, leading to increased yield of grain of host legume under field conditions, where the inoculated rhizobial strains have to compete with the indigenous *Rhizobium* of the same species. It has been seen that the strains of *Rhizobium* which proved to be very efficient by the aseptic culture or pot culture conditions did not perform well under field conditions. This is mainly because efficient rhizobial strains (used as inoculum) may not be always more competitive than the indigenous rhizobia (which are inefficient) and therefore, are not able to occupy more infection sites on the roots of the host legume. Hence large number of nodules formed in the roots of the host legume will be due to the inefficient indigenous rhizobia and only a few will be due to the efficient inoculated strain. Under such conditions inoculation with efficient strains of rhizobia will not result in significant nodulation and increase in nitrogen fixation and grain yield. This means that rhizobial strains selected should not only be efficient nitrogen fixers but also be more "competitive" than the indigenous rhizobia of the same species.

Determination of the Competitiveness of Inoculated Rhizobial Strains

This mainly involves determination of percentage of nodules formed in a host legume due to the specific, efficient, inoculated strains and those formed due to be indigenous strains. This can be determined by two methods (i) serologically (ii) using antibiotic resistance as a marker for inoculated strains.

Identification of Inoculated Strains in Nodules by Serological Methods

.Heavy suspension rhizobial strains used as inoculum in physiological saline solutions (Antigen or ag) is injected in rabbits by standard procedures (Dadarwal and Sen, 1974) and ultimately the blood of the injected rabbit is collected and the red blood cells are separated by centrifugation. The liquid portion of the blood (generally called serum but when the serum is collected after injection of antigens it is called antiserum due to the formation of antibodies), is separately preserved. This is antiserum (As 1) prepared against the antigen (Ag 1) that is the *Rhizobium* used for inoculation of the legume. This antiserum will react (in agar diffusion tests) positively only where the *Rhizobium* used as inoculum or with the nodule (suspension) formed due to the inoculated *Rhizobium* is the same. No other *Rhizobium* strains or nodules formed by other rhizobia will act positively with the above antiserum As1. Hence, to determine the percentage of nodules formed due to inoculated *Rhizobium* (R1), all the nodules (or a representative numbers of nodules) formed on the legume host are collected, suspensions prepared and tested against the antiserum (As1) prepared against the inoculated *Rhizobium* (Dadarwai and Sen 1974 and Gaur and Sen, 1974). The nodule that will react positively will be considered to have formed due to the inoculated *Rhizobium* strain. Thus from the total number of nodules examined and number of the nodules giving positive reactions, percentage of nodule formed due to inoculated strain could be calculated. In the same manner, the antiserum of different rhizobial strains (used as inoculum) could be prepared and percentage of nodules formed due to the inoculated strain could be determined. The strain of *Rhizobium* bio-fertilizer which produces maximum number of nodules can be considered as the most competitive.

Determination of Competitiveness of Rhizobial Strain (Percentage of nodules Formed due to Inoculated Strain) by use of Antibiotic Resistance as a Marker

Rhizobium strains of the same species differ in their antibiotic resistance pattern. Different antibiotics like penicillin, streptomycin, terramycin, chloramphenicol etc. can be used for the determination of the resistance of different strains of *Rhizobium* used as inoculum. Nodules formed on the legume due to the inoculation of different strains are detached and their suspensions (in saline) are examined for their antibiotic resistance and compared with those of the inoculated strain. Nodules suspension showing similar antibiotic resistance pattern as that of the inoculated strain can be considered to be formed by the latter. In this way percentage of nodules formed by the inoculated strain can be determined (Jesey *et al.*, 1979, Stein *et al.*, 1982).

All the efficient strains (as selected by aseptic cultures and pot cultures test) are therefore examined for their ability to form nodules even in presence of indigenous rhizobia. Of all the efficient (with high inherent nitrogen fixing ability) strains of *Rhizobium,* the one which forms maximum numbers of nodules (say more than 60-80% nodules are then finally selected as efficient as well as competitive strains of rhizobia suitable for being used as an inoculum.

Screening of Efficient *Rhizobium* Strains Possessing Hydrogen Uptake System (Hup+)

It had been observed that nitrogen fixing nodules (or nitrogenous system) often evolve hydrogen in presence of ATP. However, certain *Rhizobium* strain formed nodules but resulted in very little or no hydrogen evolution. In case of the former the *Rhizobium* strain called Hup+ and in case of later Hup- strains. Hup+ strains possess the hydrogenase enzyme which enable the *Rhizobium* to utilize the hydrogen for its energy requirement. Hup+ strains are also said to possess hydrogen uptake system which has been also found to confer greater advantage in nitrogen process than those of Hup- strains.

Pahwa and Dogra (1981) examined 30 isolates of mungbean (*Vigna radiata) Rhizobium* and screened them for their hydrogen recycling ability under free living culture conditions using tri-phenyl tetrzolium chloride (TTC) reduction and screening procedure. The isolates which reduced the dye at earlier stages of growth were found to recycle hydrogen *in* vivo as well as *in vitro* (i.e. either in nodules or in free living condition). Nitrogen fixing efficiency of Hup+ and Hup- strains of rhizobia of mungbean were compared in green house conditions and it was found that there was 13-15% increase in dry matter of plants and 21-45% increase in total nitrogen content of mungbean plants inoculated with Hup+ strains of *Rhizobium* over the plants inoculated with wild type strain.

The efficient rhizobial strains of some species can also be obtained by taking into consideration the presence of hydrogen uptake system in those strains.

VIII. Measurement of Biological Nitrogen Fixation

The measurement of biological nitrogen fixation is of considerable importance which can be measured by various methods. Total nitrogen of the culture and samples of soils and plants etc. have been traditionally analysed by the Kjeldahl method and many studies of biological nitrogen fixation have been based upon such measurements. Although this method is useful but is not sensitive enough to detect small nitrogen gains. Gasometric measurements of N2 disappearance have been suggested but method of analysis is laborious.

Two techniques are especially suitable for detections of even slight nitrogen gains (Burris *et al.,* 1943). One involves the use of stable isotope $^{15}N_2$. If an organism is capable of nitrogen fixation it will incorporate 15N into protoplasm i.e. components when exposed to ^{15}N-labelled N_2. $^{15}N_2$ is a stable isotope of N_2 occurring naturally in atmospheric N_2 at an abundance of 0.3663%. This method is based on the fact that an organism growing at the expense of the substrate labeled with $^{15}N_2$ (combined nitrogen) in a system where no N_2 fixation occurs accumulates more $^{15}N^2$ than a similar system where N_2 fixation occurs. The validity of the estimation of this technique depends on the non N_2-fixing control. This method has a direct demonstration of N_2 fixation. However, this procedure is not widely adopted because ^{15}N tagged N_2 is expensive and measurement of this stable isotope usually requires a mass spectrometer, an instrument which is both expensive and needs proper maintenance.

Acetylene Reduction Assay

This technique involves determination of the microbial formation of ethylene (C_2H_4) from acetylene (C_2H_2). The reliability of C_2H_2 reduction as an indicator of nitrogenase activity has been established for a wide range of biological systems and this technique has been widely used for N_2 fixation studies (Hardy *et al.*, 1971, 1973). This method is based on the findings, that microbes that reduce N, which has triple bound in the molecule, can also reduce acetylene, a molecule with a triple bond. However, the reduction of N_2 by bacteria yields ammonia as considered below but the reduction of acetylene yields ethylene.

$$N = N \rightarrow 2\,NH_3$$

$$HC = CH \rightarrow H_2C = CH_2$$

The N_2 fixing systems are exposed to an atmosphere containing C_2H_2 in gas-tight containers. The atmosphere is sampled after suitable incubation period (0.5 to 24 hrs, depending upon the system) and analysed for C_2H_4 reduction and nitrogen fixation shows that 3 moles of C_2H_2 are reduced for each mole of N_2 fixed. Although some claim good correlations, but wide variations are found and this necessitates establishment of quantitative relationship between C_2H_2 reduction and N_2 fixed for a particular N_2 fixed for a particular N_2 fixing system.

This method has the advantage of being both sensitive and requiring an inexpensive substrate, acetylene. The gas chromatograph is more widely used instrument than the mass spectrometer but the reaction can also be measured by other methods (Hardy *et al.*, 1973).

The confirmation of the activity in many suspected species could be made by the method now available and at the same time certain organisms grow in the nitrogen free medium could be separated merely as scavengers or ghosts which utilize the element from the contamination of medium ingredients or uptake of ammonia or nitrogen oxides present in the laboratory air.

Maintenance of Culture Collection

A collection of microbial culture in a laboratory is essential for their timely access and availability to the prospective workers. The proper tagging and regular maintenance of the cultures is necessary in any permanent culture collection. Besides checking their purity, the most relevant feature of a good collection of rhizobia is the constant check in verifying the nodulation ability of the rhizobia on their respective host plants.

Agar Culture

The periodic sub-culturing at 2-4 weeks interval on YEMA (Chapter 21) has been a convenient practice for retaining the cultures in general work. However, the duplicates should be lypholized and kept for timely availability and use. The shelf life of the culture may be enhanced by preserving agar cultures at low temperatures (4°C) after their proper growth. In tropical countries, humified incubators may be used to avoid premature desiccation of cultures.

Agar Cultures Maintained in Paraffin Oil

The pure culture grown on YEMA agar slants are fully covered by pouring sterile liquid paraffin (medicinal grade). Oil is sterilized in moist tubes to ensure its complete sterilization. The paraffin stored cultures can be kept at low temperature (4°C) for proper maintenance.

Porcelain Bead Method

Airtight screw cap small bottles, silica gel and unglazed porcelain beads are needed (Norris 1963). The glass bottle and screw caps are sterilized separately in an autoclave, dried and assembled after cooling. Silica gel (3-4 g) is packed in a tough gauze or translucent cloth piece. The porcelain beads are washed well, air dried and both gel and the beads sterilized in a hot air oven. The silica gel pack is inserted into the bottle and stoppered. The sterilized beads are dipped in an actively growing Rhizobium culture broth removing the excess broth by drawing over sterilized cotton wool or suitable absorbent and inserted into the bottle referred above under aseptic conditions. Blue colored (dry) silica turns pinkish when it absorbs moisture indicating non-satisfactory status of the preservant. The cultures can be preserved and easily transported in this way. For reisolation, one bead can be transferred to fresh YEMA as and when required.

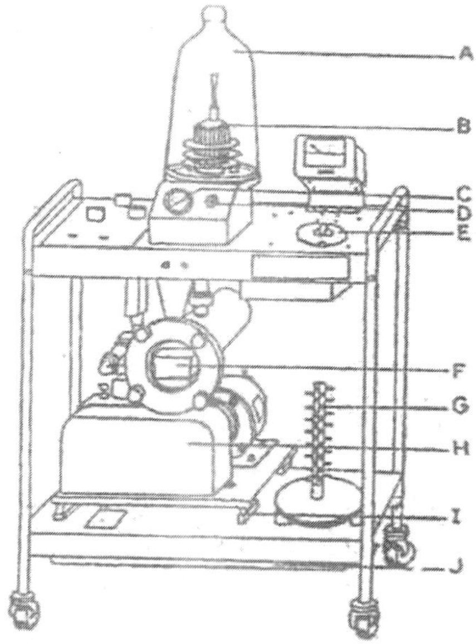

Fig. 3: A diagram of 'Speedvac' model SPS centri- fuge freeze dryer (not to scale) - **(A)** Bell-jar, **(B)** Am- poule carrier, **(C)** Vacuum dial gauze, **(D)** Air admittance valve, **(E)** Selector switch, **(F)** Moisture trap (P205), **(G)** sealing header, **(H)** Rotary vacuum pump, **(1)** Anti-vibra- tion mounts, and **(J)** Storage tray

Lyophilization

For this purpose, small ampoules are used. The tapering side is opened, plugged lightly with cotton wool and sterilized. The typed identification mark is introduced in the ampoule. A bacterial suspension from an actively growing agar slant is made from the bacterial growth. The suspended medium is an aqueous solution of 10% sucrose and 5% peptone which is sterilized prior to use. The cotton wool from each ampoule is carefully removed, 0.1 ml aliquots of rhizobial suspension is introduced and plugged with cotton wool. Ampoules are centrifuged in a centrifuge carrier which consist of glass bell-jar which is placed to a circular base plate by a rubber gasket (Fig. 3). Centrifugation increases the surface area by dispersion of the bacterial suspension on the wall of the ampoule which accelerates the freezing of the culture. Simultaneously, high vacuum is applied in the presence of desiccant such as phosphorus pentaoxide (P2Og) to remove all the moisture in the ampoule within few hours. Now the cotton plug is pushed half way into the ampoule and the stem of the ampoule constricted to facilitate sealing at a later stage. Drying in vacuum is continued for another 12 hours. About 10 g of loosely packed P2O, powder is required for every ml of water to be absorbed. The precaution in handling P2O, is required. The ampoule is now sealed, without burning the inside cotton which can be stored at room temperature. About 10-50% of the original population will survive during the lyophilization process.

Sterilization Methods

The sterilization of apparatuses and raw material is generally done in an autoclave designed to use steam under regulated pressure and temperature. It is double-jacketed steam chamber with inlet and outlet provision for steam along with measuring gauzes and safety valve. Microorganisms are killed by high temperature developed in the autoclave. The desired temperature is attained by increasing the steam pressure from 0 lb/square inch (100°C) to 15lb/sq.inch (121.5°C) or 20 lbs per square inch (1.26°C) and held at that pressure for certain period depending on the volume and nature of the materials to be sterilized.

Laboratory glasswares, powders, oils etc. are sterilized by dry heat or hot air in oven at a temperature of 160°C. This procedure is mostly used for sterilization of petridishes and pipettes etc. of common use.

Bacteriological filters made of asbestos pad (seitz), diatomaceous earth (Berkefield) porcelain (Chamberlain) and sintered glass filters are used for sterilization of liquids without denaturing the solutes. Laminar airflow system is commonly used for keeping dust free environment in the inoculation rooms and microbiological laboratories.

Inoculation rooms are equipped with ultraviolet lamps with wave length around 2650 A0 which has maximum bactericidal effect. These are used to sterilize inoculation rooms *or* hoods prior to handling of microbial work. The ultraviolet rays acts on the nucleic and prevents DNA replication in micro-organism. Gamma radiations penetrate microbial cells and kill them. Gamma rays are being now used extensively for sterilization of carrier material such as peat and lignite for mass production of microbial inoculants.

IX. Mass Production of Carrier Based Rhizobium Inoculants

Mass culture production of rhizobia has its beginning in 1895 when Nobbe and Hiltner introduced a laboratory grown culture of *Rhizobium* spp. under the name 'Nitragin'. Prior to 1923 cultures of rhizobia were mostly prepared in small bottle or cans. There are several forms of inoculants such as agar based, broth cultures, frozen concentrate, carrier based- granular soil inoculum, porous gypsum granules, natural peat granules and freeze-dried which can be used for inoculating seed or soil. The choice of the type of inoculants to be made available depends on several factors such as volume of demand, organizational considerations viz. Production for immediate use in nearby locations or R and D purposes, economics and its effectively for distribution over long distances and storage.

The work in India on rhizobial strains with a few legumes was started. In India commercial production of legume inoculants started in late 1965's and carried out during 1911-1922 (Joshi 1920, Hutchinson, 1922) when soybean was introduced for large scale production in the country. However, legume inoculants for fodder crops like berseem/Lucerne were prepared in laboratories of some Agricultural Colleges for use at Agricultural Farms and supply to some progressive farmers.

The carrier based inoculant is the most acceptable form which is produced today since the use in this form results in better survival of rhizobia on seed than do agar, broth or reconstituted lyophilized cultures. Additional advantages are long, shelf life, ease of manufacture and distribution. Technology involved in the production of *Rhizobium* inoculants has been improved and perfected for use at farm level.

Due to the diversity of grain and fodder leguminous crops grown in India and the season bound nature of individual crops, the inoculants are produced for each crop during the season. This practice also avoids storage problems in most of the places which have extreme summer season with temperature often exceeding 40-42°C. The quantum of production is obviously governed by the size of the potential market, availability of funds, technical expertise etc.

Inoculant production can be divided into four phases: (i) the selection and testing of strains of *Rhizobium* (ii) preparation of broth culture (iii) preparation of carrier based inoculants, (iv) packing and storage.

Selection of Rhizobium Strains

The basic criteria for strain selection are the ability of the rhizobia to form effective nodule with all the hosts for which they are recommended and their capacity to do their best under a wide range of field conditions* (Vincent, 1956). Effective nodulation is still the main criterion used in strain selection and forms the basis for host grouping, However, various ecological characteristics, such as their ability to form nodules in competition with native rhizobial population and to survive and multiply well in soil, have also been proposed as additional criteria. Prompt and effective nodulation over a range of root temperatures, ability to grow well both in broth and peat have also been identified as essential characteristics (Brockwell *et al.*, 1968).

Maintenance of Selected Culture

Maintenance of selected rhizobial strains requires methods that provides a maximum storage period (years) with minimal chance for variation (mutation) and contamination. Ideally this should be combined with tests for checks on serological type, host specificity and effectiveness.

Freeze-drying with additives such as sucrose, glucose, peptone, dextran or mixture of these has been successful with other organisms. Drying with a 10 percent sucrose or preferably 10 percent sucrose plus 5 percent peptone solution is generally used.

Without freeze-drying equipment, satisfactory storage may be achieved by drying cultures on porcelain beads over a desiccant (silica gel) in a screw-cap bottle. This technique can be used with minimal facilities and bottle may be sampled as required (1 bead at a time). However, the disadvantage is that the stock bead culture must be renewed every 2-3 years to maintain a maximum number of viable cultures. Survival can be improved by using 1 per cent maltose solution for preparation of stock based culture. Cultures may be stored for short terms on agar medium in screw capped tube or bottles in preference to liquid paraffin over agar. Subculturing of parent stocks should be minimized to reduce chances of variation by mutation, but frequently enough to maintain viability.

Preparation of Broth Culture

Rhizobia are heterotrophic and generally not highly demanding in their nutrient requirements. For mass production of inoculants, cheap sources of raw materials are used. Most of the rhizobia readily utilize mono and disaccharides. Slow growing strains prefer pentoses (arabinose) while fast growing ones utilize both mono and disaccharide. For mass production, sucrose is economical and thus preferred. Efficient utilization of these carbon sources, however, depend on aeration and the method of sterilization. Small quantities of yeast extract is used as source of organic nitrogen and growth factors. The pH of the broth is adjusted to 6.5-7.0 with KOH or H_2SO_4, Rhizobia grow best in the temperature range of 30-32°C (Burton, 1967).

Broth cultures with high population of rhizobia are desired to prepare high quality carrier based inoculant. Three to seven days old culture is examined microscopically and gram-staining test carried out to make sure it is free from contamination. About 3-6 ml of sterile water is transferred aseptically into the tube. The bacterial growth is scrapped with the help of an inoculation needle and the resulting bacterial suspension is transferred into a 250 ml conical flask containing 100 ml yeast extract mannitol broth which was plugged with non-absorbent cotton and sterilized at 15 lb-pressure for 20 mts and subsequently cooled. The flask are incubated at 28-30°C on a rotary shaker for 2 to 7 days depending on the nature or *Rhizobium*.

A short fermentation time is desired to reduce contamination. Increased number of viable cells may be achieved by varying the source and concentration of the yeast extract and mannitol taking into consideration the type of strain used.

Fermentors

Submerged culture technique is generally used for mass production of rhizobia. The size of small fermentors can vary from few litres to a working capacity of approximately 80 litres, capable of containing 50 litres broth. With an adequate autoclaving facility, these can be sterilized completely with the medium and all accessories. It is necessary to autoclave for at least one hour. The essentials of small fermentors' unit for inoculant production have been well illustrated by Van Schreven (1958), Date (1975) and Jauhri and Subba Rao (1984). In contrast to many commercially available units which have mechanical stirrers, these units are dependent on a source of air for aeration and mixing. However, use of a porous sparger and/or impeller ensure better diffusion of oxygen into the medium in large fermentation. For rapid growth and higher cell yield of rhizobia, vigours aeration is recommended. Air flow rates range from as little as 0.5 to 120 I air per litre medium per hour. The advantages with small fermentors are that only a small amount of broth is lost when it gets contaminated. Apart from this, 6-10 flask of 20 litres capacity may be connected with a single source of air if larger volumes of broth culture are required.

The inoculum should be suspended in log phase of growth and be sufficiently well grown to provide inoculum levels in broth (1-10%) depending on the organism but generally 10^6-10^7 rhizobia/ml of culture medium are required to initiate fermentation. Large inoculum rate reduces the time required to reach maximum viable number and, therefore, reduces the chance of contamination. Fast growers (4 h m.g.t.) reach maximum viable number in 36-72 h as compared with 108-192 h for slow growers (6-12 h m.g.t.). Maximum number of rhizobia may be maintained for another 24-36 h, depending on the growth conditions, although the proportion of viable cells will decrease. For example, Brockwell *et. al.* (1968) observed that clover *Rhizobium* strain TA 1 yielded a total of 5.2×10^9 cells per ml after 24-36 hr, growth, with 65% of cells viable, at 48 hr, total numbers had increased to 140×10^9/ml, with only 3% viable cells. Yeast extract source and concentration have marked influence on the total to viable cell ratio.

When the number of rhizobia (108-109 cells/ml) in the broth have been attained, it should be mixed with the carrier materials because storage in the fermentor may result in death of the viable cells.

Carrier Materials

The term 'carrier' refers to most solid medium such as peat, lignite, charcoal, soil, compost/FYM, pressmud, vermiculite, coirdust, polymers and their mixture which is impregnated with viable *Rhizobium* culture for preparation of specific Inoculants. The growth of bacterial cells when mixed with any of these carriers is known as carrier based Inoculant. They have completely replaced agar based or liquid cultures used earlier.

Peat soil rich in organic matter content is the most suitable and are widely used as carriers for preparation of legume inoculants. High quality peat-based inoculants are considered to be dependable in terms of quality. Good quality peat is not available in India although medium quality peat soils amounting to 5.5 million tonnes are found in Nilgiri valley in Ootacamund which are being used as carriers for biofertilizer in small quantities (Iswaran *et* al., 1969). It contains 21-56% organic matter and water holding capacity of 52-120%. Peat containing 50% organic matter was procured by IARI from Ooty during 1969-70 and was found to be a good carrier for *Rhizobium* production (Iswaran *et al.*, 1971). A count of 30 x 10^8/gram peat, was usually recorded after one week of incubation. The low grade peat retards the keeping quality, Due to non-availability of good quality peat, other alternate carriers were tested (Brockwell *et al.*, 1995).

The other carrier material widely used in India is lignite (Kandasamy and Prasad, 1971). Lignite, 3 million tonnes annually is available at Neyveli (Tamil Nadu) which contains about 28% organic matter and has 92% water holding capacity. It can also be obtained from Palana (Rajasthan) and Singrauli (MP). Charcoal is another carrier available in plenty which contains 77% organic matter and 90% water holding capacity (Jauhri *et al.*, 1979). Compost/FYM is available in large quantities and usually contains about 30% organic matter and 50% water holding capacity. Press mud, a byproduct of sugar factory is generated @ 5.45 million tonnes a year and readily available in almost all the states of India (Jauhri and Koshi, 1984). Recently water extending polymer crystals have proved to be very effective synthetic carrier for *Rhizobium* inoculants. It is non-toxic and holds moisture 30 times of its own weight. According to availability and cost at production site, choice is either lignite or charcoal.

Although charcoal is used by production units, quality criteria like fineness, extent of pyrolysis, water holding capacity, organic carbon, active surface area etc. are not properly followed resulting in production of inoculant containing inadequate number of rhizobia and low shelf life. Carrier materials for inoculants should meet the following requirements: (i) It should be finely ground to be thoroughly mixed with other components/ingredients, (ii) must have a good moisture holding capacity, (ii) be free from toxic substances (iv) have readily adjustable pH to 6.5-7.0. Interactions between rhizobia, carrier, method of treatment and storage are common and even peats from the same area can vary in suitability (Roughley and Vincent, 1967).

Preparation of Carriers

The carrier materials are sundried to five per cent moisture and pulverised to fineness of 100-200 mesh. Coarser materials form balls on mixing with broth and seeds are not properly coated with the inoculant. Lignite powder being acidic (pH 5.5) usually 1 kg of lime ($CaCO_3$) is mixed with nine kg of lignite to bring its pH to about 7. Charcoal powder being neutral in pH does not require lime. To enhance its effectivity for multiplication of rhizobia, 1% CaCO3, 0.5% KH_2PO_4 and 25% soil are thoroughly mixed with charcoal. Finally the carrier is mixed with 10% water before sterilization.

The carrier material must be sterilized before blending with broth culture to suppress the proliferation of contaminants. Better growth of *Rhizobium* is obtained-in sterile carrier. Best method is Gamma irradiation (5 megarods) for carrier sterilization. It is capable of penetrating the carrier material killing microorganisms. For bulk preparation, using autoclave, lime mixed lignite or charcoal powder is filled up to two third capacity of steel trays (size depending on the size of the horizontal autoclave) and sterilized at 15 PSI for 4 hours continuously or one hour each for three successive days. The carrier materials may be packaged and sealed in autoclavable polypropylene bags which can stand a temperature of 121°C and sterilized.

Blending of Broth Culture with Carrier

For preparation of carrier based inoculants, the rhizobial number of broth must attain a minimal population of 109 cells/ml which is mixed with sterilized carrier. The broth is added to 1/3 of the water holding capacity of carrier. Blending is done by thorough mixing of broth culture with sterilized carrier in trays manually. In case of bulk operation, mechanical mixtures can be used.

Packing and Labelling

The blended carrier is kept for 2-4 hours for curing. After curing, the inoculant is packed in polypyropylene bags, made of flexible sheet of pouch type of 50-75 micron milky opaque to prevent excess of light, low density, pouch type and 15 cm x 25 cm in size. It provides an excellent barrier to water loss. Dispensing of 200 gm inoculant in each packet is done either manually or by using automatic dispenser. The bags are sealed by an electric sealer leaving about 2/3rd vacant space to give proper aeration to the inoculant.

Low density polypropylen (LDPE) packets are used universally, as they can be readily sealed, allow high gas exchange with relative low moisture loss (Thompson, 1991). If production technology is based on sterilization of carrier material through autoclave, polypropylene or high density polypropylen (HDPE) pouches should be used. LDPE cannot withstand the sterilization temperature of 121°C. In case of sterilization through gamma irradiation, LDPE bags should be used as polypropylene is unstable when exposed to irradition.

The inoculant packet is labelled with relevant information regarding culture such as name of the culture, crop for which intended, expiry date, directions for storage/ use and precautions, certification (ISI) mark if any.

Incubation and Storage

Packets (Inoculants) must be incubated for a week in the room at 30°C so that the bacteria multiplies and attains the required count. The packets may then be stored in a cold room (4-15°C).

Distribution/Supply of Cultures

The inoculant should not be subjected to excessive temperatures, ideally it should be held at 4°C throughout transport and storage. Inocula in bulk should be transported in strong cartons or boxes. Distribution of cultures during summer months and rainy seasons requires suitable arrangements and strict quality control measures.

Suitable Adhesive Material

Gum Arabic (gum acacia powder) 5%, wall paper glue (5%) gum Arabic (4%) Carboxy methyl Cellulose (4%) milk powder (10%) sugar (10%) etc. are used as sticker and partly as nutrient source until infection process is completed by rhizobia. Adhesives should be supplied along with the culture to make it more convenient and logical. Unavailability of genuine gum arabic is a problem and usually it is adulterated with *Azadirachta indica* (Neem) which does not withstand autoclaving.

Methods of Application

Seed application: The coating of seeds with the *Rhizobium* inoculant or drenching inoculant suspension on crop rows are based on the principle to place the organisms on seed coat or at closest proximity of roots in rhizosphere. Cultures are mixed and coated on seeds so as to load each seed with 10^3-10^4 viable cells. The adhesives like gum Arabic 5-10% solution is prepared to serve as a sticker of *Rhizobium* cells on seeds. Usually 200 g *Rhizobium* inoculant are used for treating 10-12 kg pulse seeds. On an average 500g carrier based inoculant is sufficient to treat seeds required to be sown in one hectare. Under normal soils conditions, 10^3-10^4 viable cells seed1 bring the desired yield responses but it should be raised to 10^5-10^6 when seeds are sown in problem soils and adverse conditions. Treated seeds must be sown immediately and should not be mixed with acid forming chemical fertilizers.

Method of Seed Pelleting

The technique of coating of inoculated seeds with inert materials to protect them from problem soils and stresses is known as pelleting. This provides prolong survival to rhizobia on seed. The coating materials are calcium carbonate (lime) for acidic soils and gypsum for saline and alkali soils. Use of rock phosphate has also shown good results. The technique is as followed:

i) *Adhesive Inoculation Mixture* : One hundred grams pure gum Arabic (light coloured, non-toxic/preservative free) is dissolved in 225 ml hot water with mixing. The solution is neutralized with calcium carbonate. The carrier based inoculum (100 gm) is thoroughly mixed with the gum solution which swells to about 350 ml. This total quantity of slurry is sufficient to inoculate 7 kg of small, 14 kg of medium or 28 kg of large seeds (Roughley, 1970).

ii) *Pelleting* of Seed: The adhesive and seed should be mixed together in a mixer so that the seeds are coated uniformly. Three and a half kg of Calcium carbonate is required to pellet the inoculated quantity of different sizes of seeds mentioned above. Calcium Carbonate and *Rhizobium* inoculated seeds are quickly mixed usually within 1- 2 minutes until all seeds are evenly pelleted.

Precautions should be taken to avoid soft or powdery pellets which may be either due to addition of too much of lime or uneven mixing. Pasty- looking pellets with some exposed seed surfaces indicate use of too much of adhesive which can be rectified by addition of more lime. Clumping of small seeds may be caused due to addition of too much of adhesive which cannot be rectified even by adding more of lime.

X. Quality Control and Shelf Life

The quality of any product is the responsibility of the manufacturers. His product may in addition be subjected to official regulations/audit. It is essential for the producer to adopt rigid quality control of the product, but in the absence of any official regulation or control order the manufacturers may produce substandard products.

The desirability of control measures to test legume inoculants was recognized early in the History of the U.S. Industry. The US Federal Government was responsible for testing cultures and issuing licenses in 1930s. However, by the late 1940s regulatory control by Federal authorities was considered no longer necessary but was reimplemented in 1960s. Now in USA there is no state control and each manufacturer is free to control his own product and the industry is free to select its own strains and official control ensures that the product can form nodules on the legume for which it is recommended.

The technology is as fallows

1. Selection of strains of rhizobia that are compatible and effective N- fixer with particular legume.
2. Multiplying selected strains to high population densities 108-109/ml in bulk cultures i.e. liquid rhizobial culture or broth culture.
3. Incorporating liquid rhizobial culture into a carrier material usually finely milled peat or a substitute for packaging and distribution.
4. Finally coating the seeds of legumes with the carrier, inoculating the soil with the culture directly with the help of a seed drill or preparing granular inoculation and placing it in soil with seed drill.
5. An inoculum strain of *Rhizobium* recommended for a particular host must be able to form effective N-fixing nodule with that host under wide range of field conditions.

In a number of countries, Government agencies, universities and research Institutes supply inocula for commercial purposes. In à private enterprise it is often found that fully trained personnel cannot be employed and hence good quality inoculants are not produced. It was this situation which led to the formation of the Australian testing authority UDALS (University Department of Agriculture Laboratory Service) in 1956 and its current successor is AIRCS (Australian Inoculant Research and Control Service) in 1971. The Australian body developed out of co-operation between manufacturers, scientist and state department of Agriculture and functions without legislative backing.

The history of the development of AIRCS is worth noting in relation to the possible development of inoculant industries in other countries. Government laboratories vacated the commercial inoculant production field in 1954 in favour of private companies. Vincent (1956) initiated positive moves to secure improve duality and advocated setting up a licensing authority and quantitative and qualitative standards which should be the definite responsibility of the manufacturers. Poor quality inoculants continued to appear and an informal meeting attended by all active manufacturers and interested scientists, to establish mutually acceptable standards of quality and to ensure that only mother culture from a central collection should be used. Manufacturers provided financial support to the control laboratory, which tested the final product. Quality of the final product was still often poor so a further meeting of the same group of people resulted in the introduction of a system of progressive control in which the initial broth, the freshly manufactured inoculants and the product on sale were all tested. This marked the beginning of an effective quality control programme in Australia (Date 1969).

Standards set up in New Zealand and Canada are similar. The control body of South Africa functions in very similar fashion to that of Australia. In UK, Rothamstead Experimental Station gives the advice on inoculum production.

In Czechoslovakia, the minimum requirement is 300 x 10^6 rhizobia/ g whereas in Holland it is 4000 x 10^6/g. In New Zealand the standard reaches up to 10^8 cells/g whereas in Russia it ranges from 500-1000 x 10^6 rhizobia/g of carrier.

India has set explicit standards of both production and quality testing. Attainment of the standard allows the use of the ISI certification mark. However, the manufacturer is free to choose his own strains of rhizobia.

In practice the following procedure is adopted while making carrier based *Rhizobium* inoculants at the Indian Agricultural Research Institute, New Delhi.

The final decision to determine whether a culture is a *Rhizobium* or not is based on the following tests.

Test Characteristics of Rhizobium

Qualitative	**Characteristics**
Morphology	Short medium rods 1-3 urn motile when young
Staining	Gram negative
Growth on yeast mannitol agar26-28°C (medium 1)	Very little growth if any in 24 hr. Moderate to abundant growth in 3-5 days, white colourless, *(Rhizbobium trifolii, R. leguminosarum, R. phaseoli, R. meliloti)* or little growth after 5 days, slight to moderate growth in 1 0 days, *(R. Lupini, Bradyrhizcbium Japonicum* and cowpea rhizobia)
Growth on congo red yeast mannitol agar at 26-28°C	Colonies of *Rhizobium* usually absorb very little of the dye and so remain practically colourless.
Growth and change of pH on glucose peptone agar, 20 days at 30°C	*Rhizobium* grow poorly on this medium and causes little change of pH
Invasiveness	Nodulation of a legume under controlled conditions
Serology	Agglutination with *Rhizobium aritisera,* at 1/200 dilution or below

Quantative

a) *Total count.* This is performed by using a 1:100 dilution with a Petroff-Hausser bacterial counting chamber using phase contrast illumination. This gives a first assessment of the count and can also reveal gross contamination such as cocoid forms or 'chains' of cells of various bacilli or actinomycetes. The total count will include also non-viable cells and therefore can be used as a guide. Experience shows that a culture that fails to provide 109 rhizobia cells/ml in the total count is likely to provide less than 500 x 106 viable cells per ml in the plate count.

b) *Plate count* at 26-28°C: The normal plate count method is the best estimate of the number of viable rhizobia. But requires 4-5 days for fast growers and 10 days for the slow-growers. Plates are also checked for contaminants. Broths having 10°/ml viable rhizobia (minimum 5 x 10°/ml) and free of contaminants are accepted for the preparation of legume inoculants.

Test on Peat or Lignite Based Cultures

Plate counts give a reliable estimate for higher peat cultures. A peat culture is graded as follows:

- Below standard if the count is 10 x 10^6 rhizobia/g; carrier
- Doubtful if it is 100 x 10^6/g; carrier
- Satisfactory having 100 to 1000 x 10^6/g; carrier and
- Highly satisfactory if it is 1000 x 10^6 and above/g carrier.

When more accuracy is needed plant infection technique based on the ability of aliquots from successive dilutions to produce nodules on aseptically growing plants should be followed.

Shelf Life

The inoculant should be prepared and introduced in soil in such a way that it remains viable until the host seedling is able to accept an infection, competes with native rhizobia for infection site, nodulates host promptly and effectively over a range of environment conditions and persists well in soil for longer duration. Unless the inoculation meets these requirements of rhizobial strains, the inoculant will fail to meet the objective of its use, regardless of its quality at manufacture.

Bureau of Indian standards has prescribed that the inoculant shall not have any contamination with other microorganism and shall have a maximum period of six month expiry from the date of its manufacture. The inoculant shall contain a minimum 10^8 viable cells of *Rhizobium*/g of carrier on dry-mass basis within 15 days of manufacture and 107 within 15 days before the expiry date marked on the packet when the inoculant is stored at 25°-30°C.

Some progress has been made to improve the shelf life beyond 6 months by BAIF Laboratory Pune in collaboration with IARI, New Delhi. A technology is available to provide the culture at farm level whose shelf life can be extended to 2 years or more by freeze-drying (Jauhari *et* al., 1988). Such inoculants can be transported easily even in extreme climate without much loss in the viability of rhizobia. These can be reconstituted in an aliquot of sterilized broth and mixed with the sterile carrier supplied alongwith the culture. This inoculant can be mixed with seed and sown promptly.

References

Agarwal, R and Tilak, KVBR 1989. Ind. J. Microbiol 29: 229-232.

Albershiem, P. and Anderson Prouty, A.J. 1975. Ann. Rev. Plant Physiol 26: 31-52.

Alexender, M. 1977. Introduction to Soil Microbiology - John Wiley & Sons Inc. N.Y.

Allen, E.K. and Allen O.N. 1958. In Handbuch der Pfazenphysiologie. W. Rubland (Ed.) Vol B. Springer Verlage, Berlin pp. 48-118. Allington, W.B. 1945. Plant Dis. Rept. Supplement 159: 219-224.

Allision, J. L. and Torris, J. J. 1944. Phytopathology 34: 799-804.

Anonymous 1970. Report on ad-hoc Expert Committee for Evoloving a Mechinary for Quality Control of Rhizobial Cultures, ICAR, New Delhi pp. 1-42.

Appleman, M.D. 1942. Soil Sci. Soc. Amer Proc., 6:200-203.

Appleman, M.D. and Sears, O.H. 1946. J. Amer Soc. Agron. 38: 545-550.

Aulakh, M.S. and Pasricha, N. S. 1986. Fert. News 31 (9): 31-35.

Azad, S. Malik, K. A. and Hafeez, F. Y. 1991. Biol Fertil Soils 12: 107-111.

Avorov, O.E. 1966. Dokl. Vese, Akad, sci. Khoz Nauk 3: 16-19.

Bagyaraj, D.J. and Hedge, S. V. 1978. Crurr. Sci. 47: 548-549. Bajpai, P. D. Lehri, L. K. and Pathak, A. N. 1974. Proc. Ind. Natl. Sci. Acad. 40 B: 571-575

Bajpai, P. D. and Gupta, B. R. 1977. J. Ind. Soc. Soil. Sci. 25: 62- 68.

Balaraman, K and Prasad, N. N. 1973. Pesticides 7:11.

Bardiya M. C. and Gaur A. C. 1968. Ind. J. Microbiol 8: 233-238.

Bardiya M. C. and Gaur, A.C. 1970 Zbl. Fur Bacteriology 124:552- 556.

Bartha, R. Lanzilotta, R. P. and Pramer D. 1967. Appl. Microbiol 15 :67-75.

Bhardwaj, K, K. R. and Gaur, A. C. 1968. Agrochemica 12: 100- 108.

Bhargawa, J. K. Saxena, M. C. and Tilak, K. V. B. R. 1974. Proc. Indian Nat. Sci. Acad. 40B : 471-475.

Biswas, B. C. Das, S. and Kawle, S. P. 2001. Crop responses to Biofertilizers. Fet. News 46 (2): 15-18 and 21-24.

Bohool, B. B. and Schmidt, E. L. 1974. Science 185: 269-271. Bowen, G. D. and Kennedy, M. M. 1959. J. Agric. Sci. 16: 177-197. Braithwaite B.M., Jane, A and Swain F. G. 1958. J. Aust. Inst. Agri.

Sci. 24 : 155-157.

Breed, R. S. Murray, E. G. D. and Smith 1957. Bergey's Manual of Determinative Bacteriology 7th edition. Williams & Wilkins, Baltimore, USA.

Brock, J. L. 1972., Weed Res. 122: 150-154.

Brockwell. J. L. Dudman. W.G. Gibson, A. H. Heley, F. W. and Robinson. A.C. 1968. Trans. Int. cong. Soil Sci. 9th 2: 103-114.. Brockwell. J. L. Bottomely, P. J. and Thies. T. E. 1995. Pl. soil 174 : 143-180.

Burns, R.C. and Hardy R. W. F. 1975. Nitrogen Fixation in Bacteria and Higher Plants Springer Verlage, New York.

Burris, R. H. 1966. Biological Nitrogen Fixation Ann. Rev. Plant Physiol. 17: 155-179.

Burton, J. C. 1964. The Rhizobium-Legume Association. Microbiology & Soil Fertility. Proc. of 194 Biology, Oregon State Press, Corvallis. Poppler J. (ed.) Reinhold publishers Corp. New York.

Chhonkar, P. K. and Negi, A. S. 1971. Ind. J. Agri. Sci. 41: 741. Chhonkar, P. K. Iswaran, V and Jauhri, K. S. 1971. Plant Soil 35 : 449.

Chundawat, G. S., Sharma, R. G. and Shekhawat G. S. 1976. Ind. J. Agron. 21: 127.

Cook, R.J. 1988. Amer. J. Altern. Agric. 3:51-62. Cowie, A. L. Jessop, R. S. and Macleod, D. A. 1990. Aust. J. Expt. Agric. 30: 651-654.

Dadarwal, K. R. and Sen, A. N. 1974. Proc. Indian Natn. Sci. Acad. 40B: 548-555.

Daitloff, A. 1970a. Aust. J. Exp. Agri. Animal Husb. 10: 562-567. Daitloff, A. 1970b. J. Aust. Inst. Agric., Sci. 36: 293-294. Danso, S. K. A. and Papaslylianou, I. 1992., J. Agric. Sci. 199: 13- 18.

Date, R. A. 1969. J. Aust. Inst. Agric. Sci. 35: 27-37.

Date, R. A. 1975. Principles of Rhizobium strain Selection. In: Symbiotic Nitrogen fixation in Plant. Nutman, P. S. (Ed.) pp. 137- 150.

Davis, T. M. 1985. In: Nitrogen Fixation Research Progress. Evans, H.J., Bottomley, P. J. & Newton, W. E. (Eds). Martinus Nijhoff Publishers Dordrecht, The Netherlands. Pp 40.

Dazzo, F. B. and Hubbel, D. H. 1975. Appl. Microbiol 30: 1017- 1033.

Dhillon G. S., Kler, D. S. Walia, A. S., Chahal V. P.S. and Gupta, R. P. 1977. Ind. J. Agron. 22: 170-171.

Dhingara, K. K., Khurana, A. S., Phutela, R. P. and Tripathi, H. P. 1981. All India Symposium on Soil Biology and Ecology, held at Bagalore pp. 1-4.

Dommergues, Y. R. 1982. Whither Soil research, Panel Discussion, Trans 12th Internatl. Cong. Soil Sci., New Delhi pp. 138-149. Dongle, J. H. and Zende, G. K. 1976. Ind. J. Agron. 21:321-326. Dube G. N. 1976. In: Symbiotic Nitrogen Fixation in plant. Nutrman, P. S. (Ed.) Internatl. Biological Programme No. 7, 203-207. Dunigran, E. P, Allen, L. D. and Frey, J. P. 1970. La Agric. 13: 6-7.

Ebbels, D. L. 1967. Ann. Appl,. Biol. 60: 391-398.

Elkan, G. H. 1981. Intl. Review. Cytol. Suppl. 13, A. Atherley (Ed.) Pp. 1-14, Academic press, New. York.

Eno. C. F. and Everette, P. H. 1954. Soil Sci. Soc. Amer. Proc. 22: 235-238.

Fred, E. B. Baldwin, I. L. and Mc Coy, E. 1932. Root Nodule Bacteria and Leguminous Plant. University of Wisconsin Press, Wisconsin, USA.

Garcia, M. M. and Jordan, D. C. 1969. Pl. Soil 30: 317-334.

Gaur, A. C., 1979. Organic Recycleing- Prospects in Indian Agriculture. Fertil. News. 24: 49.

Gaur, A. C. 1980. Ind. J. Microbiol. 20: 362-370.

Gaur, A. C. 1983. In: Advances in Soil Sci. Paliwal K. V. (Ed.) Publishers Books & Periodicals, New Delhi. Pp 127-178.

Gaur, A. C. and Bhardwaj, K.K.R. 1971. Pl. Soil 35: 613-621.

Gaur, A. C. and Misra, K.C. 1972. Ind. J. Microbiol. 12: 45-46.

Gaur, A. C. and Misra, K.C. 1977. Plant Soil 46: 5-15.

Gaur, A. C. and Mukherjee, D. 1978. Ind. J. Agric. Sci., 49: 984- 987.

Gaur, A., C. and Mukherjee, D. 1980. Pl. Soil 56: 273-281.

Gaur, A. C., Magu, S. P., Sadsivam, K.V., Mathur, R. S. Kavimandan, S.K, 1975. Ind. J. Agric. Sci. 45: 329-332.

Gaur, Y. D. and Sen, A. N. 1974. Zbl. Bakt. Abt. 11, 129: 369-372.

Gibson, A. H. 1963. Aust. J. Biol Sci. 6: 28

Gibson, A. H. 1967, Aust. J. Biol Sci. 20: 1087-1104.

Goss, O. M. and Shipton, W.A 1965. J. Agr. (Western Australia) 6: 659-661.

Graham, P. H. and Parker, C. A. 1964. Pl. soil 20: 383-396.

Grahm, T. L. 1980. Recognition in Rhizobium-legume in Biology of Rhizobiaceae (Ed. A. Atherley, K. Giles) New York Academics Press.

Gupta, B. R. and Sharma, A. K. 1989. Farm Sci. J. 4 : 47-53. Gupta, R. P., Chahal V. P.S. Chahal, D. S. and Kalra, M. S. 1976. Jour. Res. Punjab Agri. Univ. 13:395-397.

Hamdi, Y. A. and Twefik, M. S. 1969. Farm Sci. J. 4: 47-53.

Hamdi, Y. and Twefik, M.S. 1969. Acta Microbiol Pol. 18: 53-67.

Hardy, RW. F., Bums, R. C. and Parchal 1971. Advances Chemistry Series 100, Amer. Chem. Soc. (USA). pp. 219-247.
Hardy, R. W.F., Burns, R.C. and Holsten 1973. Soil Boil. Biochem, 5 : 47-81.
Haszubiak, H. 1968. Acta Microbiol Pol. 17: 41-49.
Hauk Paciwiczowa, T., 1969. Pam Pulawski 37: 241-259.
Hauke-Paciwiczowa, T. 1970. Meded Rijksfac landb. Wet. Gent. 35 : 497-503.
Hernandez-Armenta, R., Wien, H.C. and Eaglesham, A.R.J. 1989. Crop.Sci. 29: 1260-1265.
Hutchinson, C.M. 1922. Report of Imperial Agricultural Bacteriologist. Scientific Reports for the Agric. Res. Inst. Pusa (1921-22) pp 38-42. Hynes, M.F. and O'Connelll, M. P. 1990. Can. J. Microbiol. 36: 864-869.
ICRISAT 1984. In: ICRISAT Annual Report 1983, Patancheru, India pp 219-223.
Iswaran, V. Sundara Rao, W. B., Magu, S. P. and Jauhri, K. S. 1971. Curr Sci, 48: 468-469.
Iswaran, V., Bhatnagar, R. S. and Sen, A. 1979. Ind. J. Microbiol. 19: 83-84.
Jauhri, K. S., Bhatnagar, R. S. and Iswaran, V. 1979. Pl. Soil 15: 105.
Jauhri, K. S. and Koshy, P. 1984. Zbl. Microbiol. 139: 93-103.
Jauhri, K. S. and Subba Rao, N.S.. 1984. Curr. Sci. 53: 94.
Jensen, H. L. 1942. Proc. Linn. Soc. N.S.W. 57: 205-212.
Jensen, H. L. 1969. Tidsskr. Pl. Avl. 73: 309-317.
Jordan, C. F. 1984. Rhizobiaceae In: Bergey's Manual of Systematic Bacteriology. Kreig, N.R. & Holt, J. A. (Eds.) Williams & Wilkinson Publisher Baltimore.
Jordan, C. F. 1985. Nutrient Cycling in tropical Ecosystems. John Wiley & Sons, London.
Joseph, M. V., Desai, J. D. and Desai, A. J. 1985. Appl. Environ. Microbiol. 49: 459.
Josey, D. P., Beynon, J. L., Johnston, A. W.W. and Beringer, J. E.
1979. J. Appl. Bacteriol 46: 343-350.
Joshi, N.V. 1920. Mem. Dept. Agric. India (Bact) Ser 1:246
Kandaswamy, R. and Prasad, N. N. 1971. Curr. Sci. 40: 496.
Kaszubiak, H. 1968a. Acta Microbiol. Pol. 17: 41-49.
Kaszubiak, H. 1968b. Acta Microboil PoL. 17: 51-57.
Kaul, J. N. and Sekhon, H. S. 1977. World Corp. and Livestock 29 : 248-249.
Khurana, A. L. 1994. Consolidated Report on Rabi pulses. All India Coordinated Pulse Improvement Programme, ICAR, New Delhi.
Khurana, A. S., Bhandari, S. C., Gupta, R. P. and Arora B. S. 1988. J. Res. PAU Ludhiana 25 (2): 250-256.
Kleżkowska, J., Nutman, P. S., Shinner, F. A. and Vincent, J. M. 1968. The identification and Characterization of Rhizobium. In : Identification Methods for Microbiologists. Part B. Gibbs. B. K. and Shepton, D. A. (Eds) Academic Press, London and New York. Pp 51-65.
Kulkarni, J. M. and Joshi, P. K. 1988. In Biofertilizers, Potentialities and Problems. (eds. Sen, S. P. and Patil, P.,) Naya Prakshan, Calcutta pp 51-56.
Kumar Rao, J. V. D.K. and Patil, R. B. 1977. ZBI. Bakt 11, 32: 628.
Kumar Rao, J. V. D.K., Dart, A. J. and Sastry, P V. S. 1983. Expl. Agric. 19:131-141.
Kumar, P., Agarwal, J. P. and Chandra P. 1993. Len Newsletter 20 (1):57-59.
Kust, C. A. and Struckmeyer, B. E. 1971, Weeds Sci. 19: 147-162. Lakshmi Kumari, M. and Subba Rao, N. S. 1973. Arch. Microbiol 93: 175-178.
Lanka D. and Gautam, O.P. 1972. Ind. J. Agric. Sci. 42: 676-680.
Legros T. and Smith D. L. 1994. Enviro Expl. Bot. 34: 117-127.
Lehri, L. L. Gangawar, B. R. and Mehrotra, C. L. 1971. Ind. Soc. Soil Sci. 22 : 66-69.
Leonoard, L. T. 1944. Method of testing bacterial culture and results of test of commercial inoculants, USDA Circ. No. 703, Washington, D.C.

Lesley, S.M. 1982. Can. J. Microbiol. 28: 180-189.
Malek, W. 1992. Arch. Microbiol 158: 26-28.
Mayer, R. J. Cambell, N. E. R., Hannuas, F. J. Simpson, F. B. Russel, S. A. and Evans, H.T. 1978. Proc. Nat. Acad. Sci. USA. 26 : 3258-3264.
Mathur, R. S. and Gaur, A. C. 1977. Microbiologiya 46: 146-154.
Mc New, G.L., and Hofer, A. W. 1942. Cancer 94: 11-12,24.
Miller, L. I. 1948. Phytopathology (abst.) 38: 18.
Misra, K. C. and Gaur, A. C. 1975. Zbl. Bakt. Abt.11 130: 598-602.
More, W. E. C. and More, L. V. H. 1992. Index of the bacterial and yeast nomenclature changes. Amer Soc. Microbiology, Washington D. C.
Mukewar, P M. and Bhide, V. P. 1970. Hindustan Antibiotic Bull 12: 75-80.
Namdeo, K. N. and Dube, J. N. 1973. Ind. J. Explt. Biol. 11: 114- 115.
Nandpuri, K. S. Singh, H and Kumar J. C. 1973. J. Res. Punjab. Agri. Univ. 10: 141.
Narula, N. and Gupta, K. G. 1986. Pl. Soil 93: 205.
Nelson, L. M. 1983. Can. J. Microbiol 29: 1626-1633.
Nilson, P. E. 1957. K. Lantbrucke-Hogksol Ann. 23: 219-253.
Norris, D. C. 1963. J. Exp. Agric. 31: 255-258.
Norris, D. C. and Date. R. A. 1976. Legume Bacteriology. Commonwealth Bureau of Pasture & Field Crops. Bull No. 57 Commonwealth agricultural Bureau, England.
Nutman, P. S. 1948. Ann. Bot. 12: 81-96.
Nutman, P. S. 1965. The relation between nodule bacterial and the legume host in the rhizosphere and the process of infection. Pp 231-246. In: Ecology of Soil borne Plant Pathogens. Baker, K. F. & Synder, W. C. (Eds.) Univ.of California Press, Berkely & Los Angeles.
Nutman, P. S. 1959. Jour. Exptl. Bot. 10: 250-263.
Nutman, P.S. 1965. Origin and development of root nodules. Handuch Pflanzen Physiologie 12: 1355-1379.
Ojha, S. N. Roy, R. K. and Jha, J. N. 1977. Ind. J. Agron 22: 250.
Ostwal, K. P. and Gaur, A. C. 1971, Hindustan Antibiotic Bull. 13: 73-74.
Othoman, N, M. M., Lie, T. A. Mennetie, L T. and G.Y. Wassink, 1991. PI Soil 135: 67-75.
Pagan, J. D. Child, J. J. Snowcroft, W. R. and Gibson A., H., 1975. Nature, London 256: 406-407.
Pahwa, K. and Dogra, R. C. 1981. Arch. Mikrobiol 129: 380-383.
Pandher, M, S, Chahal.V.P.S. and Gupta, R. P. 1976. Ind. J. Crop. Ecol. 3: 189-199.
Panwar, K. S., Singh, U. P. Singh, U. V. and Mishra, A. S. 1977. Ind. J. Agron 22: 145-148.
Pareek, R. P. and Gaur, A.C. 1969. Ind. J. Micobiol. 9:93-100.
Pareek. R. P. and Gaur, A. C. 1970. Pl. Soil 33: 297-304.
Parker, C. A. 1957. Nature (Lond) 179: 593-594.
Patil, R. B., Blasubramaniam, A. and Rajshkeharpa, B. J. 1974-78. Termincal Report of Atomic energy Project Role of Microorganisms in the Release of Soil Phosphorus to Plant, USA, Banglore pp 1-32.
Patil, P. L. and Medhane, N. S. 1974. Pl. Soil 40: 221-223.
PAU, 1994. Ann. Progress Report, Punjab Agri. Univ. Ludhiana.
Paul, S. and Verma, O. P. 1999. Ind. J. Microbiol 39: 249-251.
Peoples M. and Crasswell, E. T. 1992. Pl. Soil 14: 13-29.
Pillai, R. N. and Sen, A. 1979. Zbl. Fur Bakt. 128: 538.
Radwan, M. A. 1965. Forest Sci. 11: 152-159.
Raghu, K. and Mac. Rae, IC. 1967. Can. J. Microbiol. 13: 621-627.
Rai, R. and Singh S. N. 1978. Sci Cult. 44: 555-556.
Rai, R., Singh S. N. and Murtaza, M. 1977. Curr Sci. 46: 572-573.

Rai, R. Shukla, T. G. and Prasad, V. 1984. Ind. J. Chem. 17: 59 - 73.
Ramamurthy, V. Havanagi, G. V., Pinjappra and Shivshankar, K. 1990. Curr. Res. (UAS, Bangalore) 19 (8): 127-128.
Rangarajan, M. and Hariharan, K. 1988. Ind. J. Microbiol 28: 79.
Rao, H. S. N. 1974. Ind. Natn. Sci. Acad. B40: 650-665.
Rawthorn, S., Hadley, P. Summerfield, R. J. and Roberts, R. H. 1985. Pl. Soil 83: 279-293.
Reddy, O. R. Zahada, A A. and Rao, L., M., 1978. Ind. J. Plant Pysiol 21: 210-212.
Rathinam, P. Keservan, R. and Purshothaman 1974. Ind. J. Microbiol. 14:149-151.
Rathore, T. R., Chhnonkar, P. K. Sachan, R. S. and Ghildyal, B. P. 1981. Pl. Soil 60: 445-450.
Rougley, R. J. 1970. Pl. Soil 32: 675-701.
Roughley, R. J. and Vincent, J. M. 1967. J. Appl Bacteriol 30: 362- 376.
Sahu, S. K. 1973. Madras Agri. J. 60: 989-994.
Sahu, S. K. and Behera, B. 1972. Ind. J. Agron. 17: 359-360.
Salem, S. H. Szegi. J. and Gulays, F. 1971. Agro Chem. Talajt. 20 : 581-582.
Salim K. G. Mahmoud, S. A. Z., and E. L. Mokadem, M. T. 1970. Pl. Soil 33: 325.-329.
Sarkar R. K. and Banik, P. 1991. Ind. J. Agron 36: 91-94.
Sattar, M.A. and Gaur A. C. 1989. Thai J. Agri. Sci. 22: 129-136.
Sen A. N. and Sen, A. 1958. J. Ind. Soc. Soil Sci. 4: 215.
Sharma, P. K. and Khurana, A. I. 1988. Ind. J. Microbiol 28: 136.
Sharma, R. L. and Rao, A. V. 1978. Fert. Technol. 15:50.
Sharma, D. S. and Tiak, KVBR, 1974. Ind. J. Agri. Res. 8: 223.
Shiv Shankar, K. and Shantaram, M. V. 1980. Zeitschrift Fur Flazenerahrung und BodenKunde 143: 68-73.
Singh H. P. and Tilak, K. V. B. R. 1977. Ind. J. Agron 22: 57-59.
Singh N. P. and Saxena, M. C. 1973. Ind. J. Agric. Sci. 43: 62-66.
Singh, N. P. and Tagat, V. and Subedar Singh 1988. Annals Agric. Res. 9(1):65-71.
Singh, V. K. and Bajpai, R. P. 1990. Ind. J. Agron. 35: 310-311.
Singh, V. K. and Bajpai, R. P. and Sharma, B. D. 1967. Indian J. Agron. 21: 95-97.
Solheim, B. and Raa J. 1973. J. Gen. Microbiol 77: 241-247.
Sprent, J. I. 1972. New Phytol. 71: 603-611.
Starks K. J. and Lilly, J. H. 1965. J. Eco. Entomol 48: 538-543.
Stein, M, Brumfield, E.P.S. and Dye M. 1982. Ann. Appl. Biol 101: 261-267.
Stowers, M. D. 1985. Carbon Metabolism in Rhizobium sp. Ann. Rev. Microbiol. 39: 89-108.
Subba Rao, N. S. lakshmi Kumari, M. Singh C. S. and Magu, S. P. 1972. Ind. J. Agric. Res. 42: 386-388.
Subba Rao, N. S. laskhmi Kumari, M. Singh, C. S. an Biswas, A. 1974. Proc. Ind. Nat. Sci. Acad. 40B : 544-547.
Subba Rao, N. S. 1976. In: Symbiotic Nitrogen Fixation, Nutman, P. S. (Ed.) Internal. Biol. Programme No. 7 pp 255-268.
Subba Rao, N. S. Venkatraman, G. S. and Kannaiyan, S 1993. Biological Nitrogen Fixation, ICAR, New Delhi.
Sud, S. K. Kumar P. and Gupta, K. G. 1973. Zbl. Bakt. Abt. II 128; 423-425.
Sukhdeo, Mohandas 1985. Pl. Soil 86: 283-285.
Sundara Rao, W. V. B. 1974. Proc. Indian Nat. Sci. Acad 40B: 604.
Swamiappan, M and Chandy K.C. 1975. Curr. Sci. 44: 558.
Taha, S. M. Mohmoud, S.A.Z. and Salam, S. H. 1972. Symp. Biol Hung 11:423-429.
Thakur, B. K. and Hasan W. 1972. Ind. J. Agron, 17: 303-305.
Thies, J. E. Boholool, B. B. and Singhleton R. W. 1992. Can. J. Microbiol 38:393-500.
Thies, J. E. Singleton, P. W. and Bohlool, B. B. 1991. Appl. Env. Microbiol, 57:29-37.

Thomos, M. A. and Hammond, H. D. 1971. Ohio J. Sci. 71:21-29.
Thornton, H. C. 1930. Ann. Bot. Lond. 44: 385-392.
Tilak, KVBR and Saxena, M C. 1974. Proc. Ind. Natn. Sci. Acad. 40B: 476.
Tilak, KVBR and Singh G. 1974. Fertilizer News 39 (4): 11-17.
Tosh, B. N. and Sonoria, C. L. 1978. Fertil. Technol. 15: 35.
Trinick, M. J. 1973, Symbiosis between Rhizobium and the nonlegume. Trema aspera Nature (Lond) 244: 459-460.
Tripathi, S. K. and Edward, J. C. 1978 a. Allahabad Farmer 49 : 145-148.
Tripathi, S. K. and Edward, J. C. 1978 b. Indian J. Physiol 21: 248- 252.
Triplett, E. W. 1990. Appl. Environ Microbiol. 56: 98-103.
Van Schrevan, D.A. 1958. In. Nutrition of Legumes, Hallsworth, E, G (Ed.) Butterworth & Co. Ltd. Lond. pp 328-337.
Venkateswarlu, B. and Rao A. V. 1987. Biol. Fertil Soils 5: 18.
Varshney, T. N. and Gaur, A. C. 1974. Curr. Sci. 43: 95-96.
Vincent, J. M. 1956. J. Aust. Inst. Agric. Sci. 20:147-249.
Vincent, J. M. 1965. In: Soil Nitrogen, Bartholomew, W. V & Clark, E. C. (Eds.) Amer. Soc. Agron, Madison, wisconsin pp 384-435.
Vincent, J. M. 1970. A. Manual for practical Study of Root Nodule Bacteria. IBP. Handbook No. 15 pp 384-435. Blackwell Scientific, Publication, Oxford.
Vincent, J. M. 1974. In: The Biology of Nitrogen fixation, Quispel, Q (Ed.) pp 265-341. North Holland Publishers, Amsterdam.
Vlitos, A. J. and Preston, D. A. 1949. Phytopathology 39: 706-714.
Wani, S. P. Mc Gill, W. B. and Roberston, J. A. 1991.Biol Ferti Soils 12: 10-18.
Wani, S. P. and Lee K. K. 1992. In: Fertilizers, Organic Mnnure, Recyclabe Wastes and Biofertilizers, H. L. S. Tandon (Ed.) F. D. C. O. New Delhi.
Wani, S. P. and Lee K. K. 1995. Theory and Practical of Organic Agriculture, P. K Thampan (Ed.) Peekay Tree Crops Development Foundation Cochin.
Wani, S.P., Rupela, O. P. and Lee K. K. 1995, Pl. Soil 175: 229- 249.
Whiteny, A. S. and Kanhrio, Y. 1967. Agron. J. 59: 585-588. Williams, W. A., Harwood, L. H. and Hills, F. J. 1960. Agron, J. 362- 365.
Wilson, J. K. and Chaudhuri, R. S. 1946. J. Econ. Ent. 39: 537- 538.
Young, JPW 1996. Phyologeny and Taxonomy of Rhizobia Pl. Soil. 186: 45-52.

13

Frankia Non Legume Symbiotic Association

Frankia is potent nitrogenous biofertilizer for forest trees since it establishes symbiotic association with non-leguminous plants. Nitrogen fixing non-leguminous trees thrive in temperate condition. The species belong to Casuarinaceae are an exception and grow successfully in tropical areas. *Alnus* for temperate regions and the genus *Casuarina* for tropical and sub-tropical regions stand out examples of the benefits they provide to ecosystem by the way of nitrogen fixation. They have an excellent potential to adapt themselves to grow under diverse geographical and environmental conditions providing enough fuel, fodder, pulp and timber. The microsymbiont within these root nodules of non-leguminous plants has been identified and designated as *Frankia* (Becking, 1970).

Historical Developments

First of all Boussingault, a French farmer during 1838 drew attention that non-legumes were capable of fixing atmospheric nitrogen. The roots of *Alnus* were described as warty brownish biomass as large as man's fist (Woronin, 1866). Frank and coworkers (1887a, b, 1891) studied the nodulated nitrogen fixing trees regardless of whether they were leguminous or non-leguminous (actinorhizal). Frank believed that the bacteroids were protein deposits but subsequently he named the causative organism as *Frankia subtilis*. Moller (1885, 1890) reported the causative organism in the nodules as plasmodia and Frank also revised his opinion and pointed out that the causative organism was a fungus. Such conflicting results were due to imperfect techniques to differentiate between vesicles and sporangia. Later the organism was renamed as *Frankia bruchorstii* to honour Frank and his coworker, Brunchorst (Moller, 1890).

Symbiotic Association of Frankia and Non-leguminous Plants

In recent years interest in nitrogen fixation in forest ecosystem has been widely noticed. Around 1970 species mainly of shrubs and trees were reported to bear root nodules harbouring organism other than *rhizobia*. *Elegnus* was one of the first genus to be identified in the late nineteenth century but studies regarding nitrogen fixation were a relatively slow process due to reasons, namely the plants are tough, woody and often slow growing perennials inhabiting poor infertile soils.

I. Genus Frankia

Isolation of pure culture of nitrogen fixing actinomycetes despite several efforts was not achieved until 1978 when Callaham *et. al.* (1978) finally isolated the organism. The symbiosis is actinorrhizal caused by an actinomycete of genus *Frankia* belonging to Frankiaceae (Bier 1964, Bier and Rawat 1963). The microaerophilic nature of endophyte was indeed proved to be true.

The organisms show cross inoculation groups. For example the organism which nodulate *Alnus* shall not nodulate *Elaegnus* or *Ceonothus* though they may infect *Myrica*. But Myrica organism does not infect *Alnus*. Cross inoculation has been reported in three genera of Elaegnaceae suggesting similarity among themselves. More than 280 species of woody plants from 8 different families (Table 106) have been reported (Baker and Schwitzer, 1990).

Table 106: Actinorhizal families, genera and species

Orders	**Family**	**Tribe**	**Genus**	**No. of species**
Casuarinales	Casuarinaceae	-	Allocasuarina	54
			Casuarina	16
			Ceuthostoma	2
			Gymnostoma	18
CoriariaJes	Coriariaceae	-	Coriaria	16
Cucurbitales	Datiscaceae	-	Datisca	2
Fagales	Betulaceae	Betuleae	Alnus	47
Myricales	Myricaceae	-	Comptonia	1
			Myrica	28
	Elaegnaceae	-	Elaegnus	38
			Hippophae	2
			Shepherdia	2
Rhamnales		Rhamneae	Ceanothus	31
			Talguenea	1
	Rhamnaceae		Colletia	4
			Discaria	5
		Colletieae	Kentrothamnus	1
			Retanilla	2
			Trevoa	2

Rosales		Rubieae	Rubus	1
			Dry as	3
	Rosaceae	Drydeae	Purshia	2
		Cerocarp	Cerocarpus	4
		Eae	Charmaebatia	1
			Cowania	1

Source: Willis (1973); Becking (1982); Werner (1992)

Identification and Characterization

Frankia are primarily identified on the basis of size and morphology, physiology, growth characteristics and biochemical properties.

Morphology

Frankia are filamentous prokaryotes composed of branching septate hyphae ranging in diameter form 0.5 -1.0 μm depending on their age. Although *Frankia* are basically gram positive but this reaction may vary with the age of culture as in other actinomycetes. Thinner hyphae usually show as gram-negative while thicker hyphae show as gram positive.

Hyphae

Fine septate hyphae in the cells of the host are found especially in the initial period of infection (Becking *et al.,* 1964, Silver 1964, Gardner 1965). According to Gardner, the hyphae are branched and surrounded by a 'capsular' substance. At the ends of the hyphae round, bottle or club- shaped inflation's usually form vesicles. The ends of the hyphae in *Myrica cerifera* however, are thickened liked a club during the whole of the development of the nodule (Silver, 1964). As the vesicles mature the hyphae are destroyed. As a result, the link between the vesicles is lost. The diameter of this club shaped ending is 1.5-2.0 μm.

Sporangia

The reproductive structure which generally develops at the end of hyphae and some times in intercalary position on hyphae filament are called sporangia. These sporangia may be in different sizes and shapes. While globular shaped may be as big as 35μm the club shaped may be as big as 13-19pm in diameter. Mature sporangia are compartment bearing polyhedral sporangiospore 1.5-2.5μm in diameter. Hawker and Fraymouth (1951) consider that the germination of sporangia is connected with the release of motile spores. The sporangia of Frankia are not aerial, i.e. sporangia are borne not above the surface of the growth substrate, but within it.

Both Hawker and Fraymouth (1951) and Gardner (1965) consider that the vesicles are analogous to sporangia. Gardner considers vesicles as sporangia because they contain subunits and growth structures and they predominate in the winter. These vesicles are deformed and dissolved by the host cell in senescent nodule.

Vesicle

Vesicle is the term applied, for historical reasons, to terminal swellings on the tips of Frankial hyphae in symbiosis or in cultures starved for nitrogen. Vesicles take many forms in symbiosis. Some Frankia strains, in most media produce specialized spherical structures called vesicles, which are 2.5-5.0 μm in diameter. Other strains however, produce vesicles in nitrogen free media only. The shape of these vesicles varies from globular to club-shaped. The formation of spherical vesicles is not a characteristic of all Frankia strains since in many genera (*Casuarina, Allocasuarina, Coriaria, Myrica, Comptonia* and *Datisca*) no spherical vesicles are formed under nitrogen-fixing conditions. Four general types of vesicle morphologies have been reported (Baker and Seling, 1984) and these, combined with differences observed in vesicle anatomy and/or position within the host cell, result in characteristic structures within each host plant. The vesicles are considered to be the probable site of nitrogen fixation owing to the close correlation between vesicle occurrence and nitrogenase formation (Tjepkema *et al.*, 1980, Murry *et al.,* 1984).

Gardner (1965) described vesicles of shape with very dense contents. They characteristically had narrow projection like germ tube. Each vesicle is surrounded by a capsular substance, which approaches the membrane only at certain points.

Bacteroids

In the nodules of plants of this group, the endophyte is known to occur as hyphae, vesicles and bacteria-like cells or bacteroids as Gardner (1965) and some others called them. According to Becking (1966) the bacteria-like cells should not be called bacteroids because this term has a special meaning in relation to nodule bacteria.

Vesicles and bacteria-like cells are not usually encountered in the same host cell. The host cells containing vesicles are always alive and have a normal cytoplasmic structure including nuclei and mitochondria. Plant cells containing the bacteria-like cells are usually dead and have disorganized cytoplasmic structures. On the other hand, most bacteroid cells, retain their structure when the nodules age. Some of them, however, are dissolved (Becking *et al.,* 1964). The bacteroides may be rod-shaped or bent. Their dimensions vary within the limits 0.8 x 8 to 10μm.

The internal structure of the bacteroides is close to that of the subunits of the vesicles (Gardner, 1965). There is a well marked central nuclear region surrounded by granulated cytopiasm, plasma membranes, external membrane and capsular substance (Gardner, 1965). According to Becking (1966) the mucilage of the capsule is of a polysaccharide nature.

Here, as in the vesicles, the formation of plasmalemmasomes has been observed (Becking *et al.*, 1964; Gardner, 1965).

Frankia remain filamentous in the nodules without forming vesicles in two genera of actinorhizal plants, namely *Casuarina* and *Allocasuarina*. Under aerobic conditions the vesicles possess a multilaminate envelope that provides protection to the N-fixing enzyme nitrogenase from access of molecular oxygen (Torrey and Callaham, 1982). Naturally question arises as to how N-fixation commences in these two genera lacking vesicles to provide oxygen protection? Berg and McDowell (1987) answered this by providing correlation of absence of vesicles with presence of *Casuariana* infected by Frankia are modified by cell wall thickening by lignin like materials not found in non-infected cells. This lignin like material has been reported to be hydrophobic in nature. These lignin-like substances in the cell walls are believed to present a physical barrier to gaseous diffusion into the infected cells and the site of nitrogenase (Zeng *et al.,* 1985). The gas diffusion barrier represented by the lignin-like hydrophobic wall in infected cells of *Casuarina* would explain the lower oxygen tension within the infected cell without the vesicle.

Reproductive Torulose Hypha

Some strains of Frankia produce an additional structure as a result of conversion of a vegetative hyphae into a wide torulose hyphae called reproductive torulose hypha (RTH). These RTH get disrupted into spore like cells serving as surviving and regenerating structure particularly when vegetative hyphae get lysed as in *Casuarina*. Since RTH cells are capable of developing into new hyphae forming new colonies much faster than colonies produced by sporangiospores, they are considered to play an important role in reproduction.

II. Taxonomy and Systematics

The criteria used in classifying an actinomycete as a member of the genus Frankia include: (1) Cell morphology, sporangia and vesicle formation (in submerged liquid cultues), (2) Chemistry (Cell wall of type III, phospholipid of type PI), (3) serology, (4) DNA homology and (5) demonstration of infectivity and/or effectivity for a host plant (Lechevalier, 1984). Thus the new taxonomy in Bergey's manual of systematic Bacteriology Vol. IV tries to reflect the biochemical and cell physiological properties of the bacteria as well specific plant reaction and symbiotic specificity (Kreig and Holt, 1984).

All Frankiae isolated to date are Gram positive, filamentous, sporulating, branching prokaryotes. Therefore, Frankiae have been grouped with filamentous soil bacteria in the order Actinomycetales. A new family, Frankiaceae, has been created having *Frankia* as the sole genus (Becking, 1970).

Although an organism can be identified as belonging to the genus Frankia, but the occurrence of considerable physiological and morphological variation within the genus, makes species characterization a difficult problem. In view of this it has been decided that no species names should be used until a more complete understanding is reached as to what constitutes a species of Frankia. Strains have rather been designated by a combination of numbers and letters (Lechevalier and Lechevalier, 1986).

Type 'A' strains are genetically heterogeneous, relatively aerobic, use carbohydrate, and are usually non-infective on their own host plants. Type 'B' strains are serologically, genetically and chemically related. They are relatively less tolerant of O2, use carbohydrate poorly, and are generally infective to form nodules on the host plant of origin (Lechavalier and Lechavalier, 1986). The 'B' type (also known as serologic group I) have genomic homology levels between 67.4 to 94.1%. Group 'B' isolates have less than 30% homology with Ar 14, a serogroup, I strain, and less than 33% homology with Eull, a member of the 'A' assemblage. The 'A' group is therefore, genetically as well as physiologically heterogeneous.

III. Infection and Nodule Development

The host plant

- Host plant plays important role by intervening in potentialities of Frankia to express its symbiotic traits
- It supplies necessary energy to reduce molecular N through photosynthate translocation.
- A better efficiency of the system will be obtained if host plant characteristics are improved.
- Host plant contribution to symbiosis involves action on recognition, root hair invasion, formation of infection threads, nodule differentiation and Frankia morphology in tissue.

The Microsymbiont

- Early observation showed a hyphal morphology similar to that of actinomycetes.
- The organism grows very slowly in culture, usually taking 4-8 weeks for visible colonies to be formed.

- Species of Frankia have been named after the host on which they form nodules, but this practice was abandoned when it became clear that some isolates could infect a variety of hosts (through crushed nodules).
- Currently *Frankia* strains discovered are listed according the laboratory, which first isolated them and the host from which they were obtained.
- Morphology varies slightly with Rhizobia in number and position of flagella.
- *Frankia* may produce vesicles and sporangia. Vesicles are formed at the ends of hyphae.
- Vesicles are electron dense and are thickened by monolayers of lipid, which protects nitrogenase formed inside the vesicles, from O_2 inactivation.

Root Colonisation

- Chemotaxis does not play any role in root colonization as *Frankia* is non-motile.
- The roots of nodulated plants are negatively geotropic, but that of the unaffected plants are positively geotropic.
- The presence of endogenous auxin in uninoculated plants may also result in positive geotropism.
- Pieces of adhering root cap material forms excellent substrate for formation of *Frankia* colonies
- *Frankia* remains restricted to small area as dense colonies and does not grow all over rhizosphere.
- Further *Frankia* infection occurs just distal to root tip. Most infection however, occurs on root hair (as with *Rhizobium*).
- Like leguminous roots, the actionorhizal plant may also be infected by passing root hairs, perhaps through damaged epiderm cells.
- *Alnus cripsa* which lacks root hairs is associated with the absence of nodule formation.
- Some Actinorrhizal plants are infected through intercellular penetration.

Root Hair Infection

Adhesion by Roots

- Adhesion of *Frankia* to root hair tips in which the bacteria is bound by polar ends.
- Plants produce sugar binding protein called lectins enabling such binding

- Root hair surface as well as host-specific *Frankia* possess common antigens which are connected together by lectins binding the bacteria to the root hairs.

Deformation of Root Hairs

- Infection of actinorrhizal root hairs are mostly preceded by root-hair deformation, the first visible symptom following inoculation.
- Root hair deformation includes apical swollen region, flexuous and hooked root hairs and multiple swellings of root hair tip.
- These deformations are produced after a direct contact between the hair surface and the adhering *Frankia,* which in turn is related to subsequent infection.
- Root hair infection usually occur 1mm above the shortest developing root-hair within 6 hrs of inoculation.
- The cell walls of deformed root hairs show a distinct yellow-orange fluorescence after staining with acridine orange instead of green characteristic root-hairs mainly because of formation of polypectate-like polymer, deformation which occur 4-5 days after infection can be seen by naked eye.

The Infection Process

- *Frankia* enters as a hyphal structure, growing by cell division while still connected to original microcolony at the root hair surface.
- This enables Frankia to exert some pressure while penetrating the root hairs cell wall.
- As the root hair become increasingly deformed, *Frankia* hyphae are entrapped between segments of root hair.
- In these zones of entrapment, *Frankia* hyphal branches initiate the digestion of the primary cell wall of the root hair.
- In response to partial wall digestion the host plant cell begins to build additional wall material around the site of digestion.
- A second penetration in host plant, create a tubular growth in the root hair called encapsulation.
- Continued digestion by the bacterium and concurrent wall building by the same region becomes impossible due to induced host defense mechanisms (Tannin disposition).
- Prenodule initiation occurs after actual infection of root hairs. A proliferation of cells in hypodermis occurs near the base of infected root hair and N_2 - fixation begins.

Intercellular Penetration

- Majority of species are infected through root hairs. Some however, do *no* produce root hairs (Elaeagnaceae, Hippophae, Shepherdia). *Frankia* hyphae have been observed to grow between adjacent epidermal cells of the roots.
- *Frankia* filament apparently digest away a portion of the middle lamellar region to gain entry.
- The host plant, in response to the presence of bacteria, produces a dark staining substance, which is exuded in the inter-cellular spaces to surround the invading bacterium.
- No pre-nodule is formed, but a nodule primordia is initiated in the pericycle, differentiated vesicles may be observed indicative of N2-fixation potential.
- While remaining intercellular, the *Frankia* hyphae grow toward the primordia and enters the newly forming nodule.
- Hyphae within the cortical zone of the nodule, branch and penetrate individual cell in a way parallel to that observed in root hair infection.

Initiation of Nodule

- Actinorhizal nodules are modified lateral root. Each nodule lobe has a central vascular cylinder and the endophyte is confined to the cortical tissue.
- The meristematic activities by penetrating hyphae constitute the beginning of root nodule formation.
- The infected part increases radially producing visible starch granules called primary nodule.
- After a dormant period (few days to several weeks) the nodule-lobe meristem begins to grow and form an elongated nodule root which grows to 3-4cm length.
- New nodule lobes are initiated at the base of the existing nodule lobe forming a cluster of nodule roots.
- Time taken from infection to the formation of nodule varies in species from 11 to 70 days and in *Alder* sp. 3 to 15 weeks.
- The root nodules are pinkish red, but with age they acquire a brownish tint, but are white in some species. Presence of anthocyanin is thought to provide colour.
- The occurrence of growth hormone such as cytokinin and auxin in root nodule and their balance controls the growth of mature nodule.

Factor Affecting Actinorrhizal Frankia Symbiosis

- Inspite of potential, N_2-fixation in soils is much below the maximum level possible.
- Genetic limitations are imposed by the use of inferior plant and/or microbial genotypes.
- Abiotic constraints which limit the full expression of the genetic potential for growth and N_2-fixation.
- It is difficult to assess the extent and severity of a particular factor to tree growth and N_2-fixation
- Efforts to identify the occurrence of these constraints and to develop suitable management strategies will go a long way in improving tree productivity and N_2-fixation.
- Pattern of nodulation and N_2-fixation are difficult to study because of perennial growth habit and long life of actinorrhizal plants.
- Variability in nodule frequency per plants more so with age further aggravates the problem of sampling of nodules. Some nodules survive up to 8 Yrs, while others for several years on old roots. Some are borne on young roots.

Soil Temperature

- Plants and microorganisms from tropical climates have higher temperature optima than those from temperate.
- Symbiotic system is more sensitive to temperature stress than does plant growth.
- Low temperature delays infection and reduces nodule mass.
- Effect of low temperature on N_2-fixation in established tree is less marked as perennial nature of nodules continues N_2-fixation.
- A low soil temperatures of 15°C inhibited nodule formation but increased nodule number and nodule weight.

Soil Acidity

- Acid soils are characterized by high concentration of H^+, Mn^2+ and Al^3+ and deficiency of $Ca^{2+}PO_4^{3}$ and Mo.
- The situation may be corrected by liming but it may lead to problem of P, Zn, Cu and Fe deficiency. It is, however, a costly affair.
- Selection of tolerant plant and bacterial genotype offer appropriate management strategy.

- High Al beside being toxic to nodule initiation, adversely affects root growth.
- Excess Al interfere with multiplication of invading cells of microsymbiont by restricting root growth, but does not affect nodule development directly.
- It is the deficiency of Mo required by nitrogenase, affects nodule development and N_2-fixation directly.
- Reduced organic matter decomposition and lower energy substances (amino acids) influence survival, growth and colonising ability of *Frankia*
- Adequate supply of Ca is essential for nodule initiation process.

Soil pH

- About 50% of inoculated *Myrica gale* and *Alnus glutinosa* have been found to be nodulated at pH 4.2 but Hippophae failed to nodulate. All these 3 sp. grew well even at pH3 a situation quite different than in some legumes, where plant is more adversely affected compared to microsymbiont.
- pH optima for *Myrica gale*, *Alnus glutinosa,* and *Hippophae ramnoides,* are 5.4, 4.2 to 5.4 and 6.3 respectively
- pH optima are related to acidity of typical habitats.
- pH mainly affects forms of nitrogen besides availability of several nutrients.

Soil Salinity

- High concentration of salts may cause withdrawal of water osmoticaly from the nodules.
- There is wide variation in salt tolerance of host plants and associated *Frankia*
- Growth of microsymbiont and nodule functioning are more tolerant to salinity compared to nodulation or host plant growth.
- Many *Casuarina* grow well in saline environment.
- Treatment of *Casuarina* root nodule with high salt concentration resulted in high concentration of salts in non-infected cells but not in the infected cells, due to existence of a semi-permeable layer in cell walls of infected cells.
- Increasing salt concentration to 250 mM NaCl however inhibited *Frankia* -growth and nodulation.

- Variation in salt tolerance among *Frankia* strains have also been reported.
- Proper choice of actinorrhizal plant and associated *Frankia* is important while introducing actinorrhizal trees to saline lands.
- Some actinorrhizal plants and *Frankia* are able to tolerate high alkali but the mechanism is not yet known.

Soil Moisture

- Both deficiency as well as excess water availability in root are harmful to N_2-fixation.
- Water deficiency affects metabolism of host plant, besides affecting the process of infection, nodule development, nodule function and ultimately N_2-fixation.
- The size of nodule on *Casurina* decreased with increasing severity of moisture stress.
- Reduction in moisture below 25% markedly reduced acytelene reduction activity.
- However, as actinomrhizal trees have deeper root system as compared to agricultural plants, they are capable of extracting moisture from deeper layers and nodulate well.
- Highly lignified nature of actinorrhizal nodules facilitates N_2-fixation to function longer even when surrounding soil is very dry.
- Production of active lobes from existing nodules following first rain is an adaptation the actinorrhizal plants have developed to offset limitation of moisture stress.
- Waterlogging adversely affects N_2-fixation by restricting O_2 supply (anoxia) and reducing rates of N-diffusion to the nodules.
- Some actinorrhizal plants have developed adaptation to waterlogging through development of enlarged lenticells, pneumatophores and nodule roots.

Oxygen Supply

- *Frankia* can fix N_2 at atmospheric pressure of oxygen (PO_2).
- When there is low content of atmospheric O_2, N_2-fixation is suppressed.
- N_2-fixation is at maximum intensity in the presence of 12-15% atmospheric O_2 and stops if the content increase to 30-50%.
- The nitrogenase of *Frankia* is as sensitive to O_2 inactivation as other nitrogenases but vesicles protect *Frankia* nitrogenase from 0^2.

- The vesicles, from nodules have a laminated envelope that impedes diffusion of O_2 into vesicle containing nitrogenase.
- Presence of superoxide dismutase and catalase may also protect nitrogenase from oxidation via superoxide and hydroxyl free radicals.

Nutrient Availability

Combined Nitrogen

- Frankia-nodulated N- fixing plants can assimilate mineral N too through a functional nitrate reductase present in leaves, nodules and roots.
- Actinorrhizal plants are thus well equipped to utilize mineral and biologically fixed N.
- A high mineral N content of soil may result in partial or total elimination of N_2-fixation.
- Combined nitrogen in ammoniacal form inhibited N_2-fixation.
- Combined nitrogen in ammoniacal form inhibited N_2-fixation not by limiting distribution of photoassimilated to nodule but by damaging vesicles in the nodules or by initiation of a morphological block preventing formation of N_2-fixing vesicles.
- NH^+_4 may also affect the existing vesicles and nitrogenase activity.

Phosphorus

- P. deficiency limited the growth of host plants more severely than the N_2 fixation process.
- P deficiency by affecting host plant growth affects nodulation and N_2-fixation indirectly.

Calcium

- High requirement of Ca at early stages of nodule formation, adversely affects nodulation in acid soils.
- Ca deficiency restricts development of infection threads and other early POS infection.

Copper

- Host plant plays important role by intervening in potentialities of *Frankia* to express its symbiotic traits.
- Liming to neutralize soil acidity may result in Cu deficiency.
- Deficiency of Cu reduces N_2-fixation by adversely affecting nodules functioning.

Cobalt

- The role of cobalt in N_2-fixation is mediated through high requirement of Co for *Frankia.*
- Co is required for synthesis of B_{12} coenzyme, through involvement in nucleotide reductase and methyl-malonyl mutase.

Molybdenum

- The requirement of Mo for nodule function in both leguminous and actinorrhizal species far exceeds its requirement elsewhere in the plant.
- Mo is an essential constituent of the nitrate reductase required for N_2-fixing process.
- Nitrogenase and nitrate reductase share a common molybdo-protein subunit.
- N_2-fixation is directly related with nitrate reductase activity.
- A deficiency of Mo has been found to be associated with reduced size of root nodules.
- Mo deficiency in acid soils could be corrected by liming.
- The requirement of Mo for nodule functioning is far greater than its requirement for host plant.

Iron

- It is essential in respiration as well as in photosynthesis process as a constituent of cytochromes.
- Nitrogenase contains non-heme iron. As such all N_2-fixing plants have greater Fe requirements.
- A deficiency of Fe inhibits nodule initiation and functioning.
- The activity of enzyme nitrogenase are lower in nodules having Fe stress.

Nickel

- *Frankia* grown in Ni deficient medium lack hydrogenase activity, and released H_2 in atmospheric air.
- Addition of Ni to medium resulted in active hydrogenase with associated H2 in atmospheric air.
- Host plants often have efficient ways of Nickel uptake so as to provide enough of Ni to microbe to have enough active hydrogenase.
- Nevertheless an adequate supply of Ni is important for the expression of uptake hydrogenase activity.

Season

- Seasonal pattern of N_2-fixation has been observed in all actinorrhizal plants, which is a product of interactions between the local climate, phenology of host plant and *Frankia* endophyte.
- Nitrogenase activity is absent during winter, appears in mid-may when leaves unfold, reach at peak in July-august when shoots are fully expanded followed by gradual decline until October when all leaves have fallen in Myrica.
- Nitrogenase activity also depends on the amount of current season nodule tissue, which increases as nodule mass increases.
- Over wintering are known to have high levels of abscisic acid in nodules, which inhibit nodule growth. Elevated levels of cytokinins are effective in overcoming the growth inhibiting effect of abscisic acid.

Plant Age

- Stand of 0 to 3 and 13 to 16 year old *Hippophae rhamnoides* showed nitrogen gains of 27 and 139 kg/ha/yr.
- Nitrogenase activity has been found to decrease rapidly as the nodules age.
- However, total N accretion improves with the age of the stand because of greater nodulation of plants.

Frankia Strain and Host Genotype

- The strain of *Frankia* infecting the host plant as well as the genotypic variability of host plant species have been found to effect nodulation and N_2-fixation rates.
- A large variation has been observed in the number of nodules formed/ plant and also 2-fold variation in the nitrogenase activity of nodules depending upon the origin of inoculum and host genotypes.
- On equal weight basis the crushed nodule inoculum from spore positive nodule was 1000 times more infective than spore negative nodules.
- However, isolates from spore positive nodules were only 70% as effective in N_2-fixation as spore negative types.

Interaction with Other Soil Micro-Organisms

- Many strains of *Frankia* may fail to compete successfully with indigenous soil microorganisms

- However, increased nodulation of actinorrhizal plants has been observed in the presence of bacteria like *Pseudomonas cepacia* which is known to cause root-hair deformation thereby facilitating infection of actinorhizal plants by *Frankia.*
- The root system of actinorhizal plants generally form mycorrhizal association directly benefiting actinorrhizal trees by enhancing nutrient uptake particularly P availability, Increasing the longevity of feeder roots, alleviating water stress and protecting against root pathogens.
- Dual inoculation of *Frankia* and A.M. fungi resulted in higher concentration of N and P besides higher plant dry weight, number and nodule weight and increased nitrogenase activity.

Azospirillum spp. have been isolated from the roots of several actinorrhizal trees. However, the biomass Improvement resulting from *Azospirillum* inoculation is mainly attributed to the secretion of phytohormones by the bacteria rather than to minimal amounts of N2-fixation so often observed.

Overcoming Environmental Limitations

- Limitation imposed by unfavourable factors such as pH, salinity, available water and environmental factors may reduce long-term benefits of inoculation.
- The adverse effects of soil constraints could be alleviated through mulching to reduce soil temperature ad improve moisture availability, liming and application of gypsum to reclaim acidity and alkali problems, fertilizer application to correct nutrient deficiencies/and deep tillage to break compacted layers.
- Genetic selections, breeding or use of other genetic tools that may increase the photosynthetic capacity will enhance N_2-fixation.
- Building of desirable qualities like disease resistance, stress tolerance, and photosynthate allocation to nodules will accentuate the symbiotic process.
- Symbiotic N_2-fixation can also be improved through cultural practices that eliminates competition with weeds or which overcomes the stresses.

Areas of Potential Application

- Sericulture.
- Symbiotically fixed N is translocated to rest of the plant. It goes back to soil during decomposition of plant material and through root exudates made available for growth of associated plant communities.

Long Culture

- Timber production constitutes an important part of forest management.
- Benefits of inter-planting Douglas Fir with red alder has been demonstrated. Douglas Fir gave more yield when mixed with red alder.
- Mixed leaf fall not only provide high biomass but mineralise faster (1-4yr) compared to pine needle (7-10 yrs).

Monoculture

- Breeding for useful traits (shape, wood quality) has led to development of clones suitable for monoculture (Alder).
- The acceptable density and texture of Alder wood holds paint very well, flexible to facilitate easy production of furniture.

IV. Nodulation and Nitrogen Fixation

Several species of non- leguminous higher plants have been listed to fix atmospheric nitrogen (Werner 1992). Many species of the genera Casuarina, Myrica, Alnus, Elaegnus, Ceanothus and Coriaria are predominant nitrogen fixers. The most predominant belong to *Alnus* (47 spp.) *Casuarinaceae* (90 spp.) *Elaegenaceae* (38 spp Elaegnus) 2 each of *Hippophae* and Shepherdia, 38 species of *Myrica* and 31 species of *Ceanothus* of Rhamnaceae family. A low number of Rosaceae has also been reported to fix atmospheric nitrogen (Table 106).

Casuarina strains differed in their ability to from effective symbiosis with highly efficient strain of *Frankia* (Reddell, 1990). Girges *et al.* (1990). showed difference in their nitrogen fixing efficiencies. Preferential association between *Alnus* and *Frankia* strains was reported (Weber 1986). Studies on *Casuarina* spp. have shown favourable influence on nitrogen fixation due to phosphorus application (Reddell *et al.* 1988, Sanginga *et al.*, 1989).

The group of Alder tree (*Alnus* sp,) which fixes adequate amount of nitrogen shows its rapid growth and development in nitrogen deficient soils which has been extensively investigated. The trees grown under field conditions typically possess nodules that may attain large size comparable to the size of tennis balls. The soils under Alder tree are enriched with nitrogen than the surrounding soils.

The *Alnus*-type of root nodules; a coralloid root nodules by thickened dichotomous branching is formed. The nodule lobes are in the axil of lateral roots whose apical meristem grows slowly or is completely inhibited. In *Myrica/Casuarina* type of root nodules, the apex of each nodule lobe produces, a normal but negative geotropic root. Thus the root nodule mass gets covered with upward growing rootlets which may aid aeration in marshy on waterlogged situations.

Casuarians are native of Timor Island of Indonesia growing extensively from lower to hilly lands. The leaves of *Casuarinas* are not eaten by cattle and would be useful as mulch on soil for improving soil fertility (Parera, 1985). *Casuriana equisetifolia* is an important reforestation tree specie in Phillippines in duneland and at shoreline area (Baggayan, 1982) *Casuarina cuninghamiana* is an evergreen native tree of Australia which is cultivated overseas due to its rapid growth and to tolerance of harsh conditions such as drought, wind and salinity besides its use for windbreak, fuel wood, erosion control etc. (Merwin 1987).

Frankia is widely distributed in soils and can survive for long periods. Hence, actinorrhizal plants are usually well nodulated in field conditions (Smolander and Sundman, 1987).

Field grown plant nodules (e. g. *Alder)* are crushed and aqueous extract can be used as inoculum. The greenhouse plants develop nodules within two to three weeks. The nodules first appear as small lateral swellings on the root which develop into lobes at their apices resulting in a clustered coralloid structure. In host cells, the endophyte is seen in the from of hyphae, vesicles and bacteria-like cells, the nitrogen fixation in the species of nodulated *Alnus, Hippophae, Casuarina, Shepherdia* and *Ceanothus* were confirmed using N^{15} labeled gas.

Inoculation of *Casuarina equisetifolia* with *Frankia* could meet 53-65% of total nitrogen needs (Sougoufara *et al.,* 1990). *Casuarina equisetifolia* fixed nitrogen from 100 to 320 kg ha-1 Year-1. This nodulating non-leguminous tree has high calcium and phosphorus requirement for nitrogen fixation. Balasubramanian (1992) has suggested that the seedlings of *Casuarina* could be treated with *Frankia* especially in sandy soils for better results.

The extent of nitrogen gains by plant species varies widely with soil type, climatic conditions and plants age. The optimum pH for nitrogen fixation is near pH 5.5-6.0 although growth in nitrogen -free solution is rapid from pH4.2 to 7.0.

Myrica genus is known to be comprised of about 35 species in the world (Hutchinson, 1917, Shen 1983) and reported to nodulate (Shen 1983, Lechvalier, 1983).

M. *rubra* is distributed in temperate and sub-tropical areas of the far east including Japan. *M. rubra* is a tall dioecious evergreen tree and grows well even under shade and infertile clayey soil. Some varieties of this free provide edible fruits and the bark is used for tannin. Their outer colour is usually brown, inside tissue is white indicating absence of leghaemoglobin (Hiyoshi, Sasakawa and Yatazawa, 1985).

Nodulation and nitrogen fixation (acetylene reduction) was observed in two representatives of genus Datisca, *(D. cannabina)* showing a Mediterranean distribution ranging from indo-Arab region to Himalayas and central Asia and *D. glamorata* found in California and N. W. Mexico. In both these species root nodulation and nitrogen fixation was confirmed by Chaudhary (1979).

Root nodulation in Rosaceae was already known in three *Cerocarpus* species *(C. betuloides, C. montanus* and *C. paucidentatus*), nodulation was also observed in *C. ledifolius* growing as pioneer species in *Pinus flexilis* forests in California at high altitude. Root nodulation and nitrogen fixation (acetylene reduction) was confirmed in Rubus *ellipticus* from Java, Indonesia (Becking 1979). In all non-leguminous nodulated plants investigated so far, nodulation is a generic character and root nodulation habit in one specie out of 250 species of the genus *Rubus* (tribe Rubieae) is quite remarkable. Perhaps other nodulating species in the genus *Rubus* may be found by extensive surveys. *Purshia tridentata* is an important browsy plant in arid region of Western U.S.A and root nodulation and nitrogen fixation was reported by Bond (1976a).

Root nodulation in *Colletia* of Rhamnaceae was first observed by Bond (1976b) in *C. paradoxa*. Subsequently Medan and Tortosa (1976) reported nodulation in C. *paradoxa, C. spinosissima* and *Discaria* species such as *D. americana, D. serratifolia,* D. *trinervis* and *D. nana,* occurring locally in natural vegetations.

The species *Trevoa trinervis* is closely related to *Colletia* and bears root nodule and fixes atmospheric nitrogen in natural habitat (Rundel and Neel 1978) which is comparable to nitrogen fixing capacity of *Ceanothus* and *Cerocarpus* when measured under controlled conditions (Kummerow *et al.,* 1978). The leaves of *Trevoa trinervis* are drought deciduous.

Onset of drought conditions/stress, *Trevoa* leaves fall rapidly but photosynthesis is maintained by young stems and spines. *Trevoa* sp. closely resembles *Colletia* species which behaves in the same way such as shedding of leaves under unfavourable conditions and photosynthetic activity is carried out by phyllocladioid stems and spines.

Frankia as *an Inoculant*: A mixture of *Frankia* strains has been found to be more effective than a single strain (Prat 1989). *Frankia* Inoculants were prepared as liquid homogenate or air-dried. powder. Carrier based Inoculants with a long shelf life were developed (Burleigh *et al.* 1989).

Field Estimates of Nitrogen Fixed

Nitrogen fixing potential of actionorrhizal plants varies widely depending several factors (Table 107). It is as low as 18 kg N/ha/year in Myrica *faya* to modest 40-60 kg/ha/year in *Casuarina equisetifolia* and an impressive 320 kg/ha/year in *Alnus rubra*. Casuarinas are the most important plant species for tropical tree farming. Dommergues *et al.* (1990) reported that *Casuarina equisetifolia* fixes about 12-50 kg N/ha/year under semi-arid condition but under irrigated or rainfed conditions the quantity of nitrogen fixation can go up to around 80kg N/ha/year.

The non-leguminous nitrogen systems are of modest agricultural importance at present as their potential as nitrogen source is not widely taped. However, their nitrogen input globally may be enormous. Alnus alone can add 100kg N/ha/year to soil from leaf fall alone so that the overall contribution could be valuable in Agroforestory. Sometimes, soil gets enriched with nitrogen due to growth of these trees. It was reported that soil under *Alnus* rugosa registered 0.46 per cent nitrogen while the control soil contained 0.24 per cent (Voigt and Steucek 1969). *Alnus glutinosa* benefited non-nodulating tree (Delver and post 1968) by supplying fixed nitrogen.

Table 107: Nitrogen gains by some non-leguminous trees

Plant type	Nitrogen fixed kg/ha/year	Reference
Alnus spp.	20-179	-
Alnusrubra	320	Newton *et al.*, 1968
Hippophae sp.	27-179	Alexander, 1 985
Hippophae rhamnoides	9-15	Silver, 1977
Casuarina equisetifolia	40-60	Gauthier *et al.*, 1985
Casuarina equisetifolia	60-100	Dommergues 1987
Casuarina equisetifolia	12-60	Dommergues *et al.*, 1990
Ceonothussp.	50-60	Silvester, 1977
Coriaria arborea	130-190	Torrey, 1978
Myrica spp.	10-18	Alexander 1985, V'rtousek and Walker 1989

V. Role in Agroforesty System

Casuarina sp. have been successfully intercropped with cash crops like coffee, peanut, sesame and grain legumes.

Colonisation of Acid peat Land

- Extremely acid soils (pH 3.1-3.5) need to be limed before planting alders.

- *Myrica gale* is the main actinorrhizal plant naturally growing vigorously.
- Abundant nodulation of hosts show the presence of *Frankia* in these soils. It also indicates existence of *Frankia* strains able to tolerate acidity.
- However, as *Frankia* is more susceptible to acidity than its host, it is advisable to use inoculum for reforestation of acid lands.

Coastal Dunes

- Coastal dunes could be quickly colonized by introducing N2-fixing actinorrhizal trees.
- High salinity is an additional characteristic of coastal dunes necessitating introduction of salt tolerant plant e.g. *Elaegnus angustifolia, Hippophae rhamnoides, Myrica asplenifolia, M. faya.*
- *Casuarina* plantation has been successfully used in India.

Volcanic Sites

- The biological invasion of *Myrica faya* at volcanic site in Hawaii is fixing 18.5 Kg N/ha/yr.
- The success of *Myrica* was attributed to its N_2-fixing capacity, prolific seed production, effective seed dispersal by exotic birds, and its rapid growth rates.

Saline Lands

- Salt inundated soil can be planted with tolerant trees such as *Myrica* faya, *M. asplenifolia* and *Hippophae rhamnoides*.
- These species can withstand high salinity.

Eroded Soils

- About 0.5 t soil is lost/capita/year through erosion.
- The diversity of actinorrhizal plants available makes it possible to choose a particular species suitable for such soils, *Alnus*, *Myrica* and *Elaegnus* have been widely planted with success.

Summary and Conclusion

- The N-contribution by *Actinorrhizal Frankia* symbiosis is of great significance in present context. There is room for a greater understanding of the role of symbiotic and non-symbotic microorganisms in forest management and of interactions of mycorrhizae with these N,-fixers. The whole process of describing nodule initiation and development in actionorrhizal plants has been hampered by technical problems. The fact that the endophyte has now been obtained in pure culture, made the work under sterile conditions easier. Many of the plants studied produce quantities of tannins which obscure the finer structural elements in cell under light microscope. The presence of suberized host cells and capsulated endophte makes fixation for electron microscopy difficult and has led to conflicting reports. Generally superior varieties of N_2-fixing plants may be evolved in order to make biological N_2-fixation a potent tool.

- Management of quick growing and high yielding actionorrhizal trees 'for fuel, fodder on infertile land is the prime goal in several developing countries where fertilizer N could be spared for cash and staple crops.
- *Alnus nepalensis* is used as fuelwood and A. *rubra* for paper pulp in U.S.A. Casuarinas are used for fuel wood, timber pole and coffee shade in china, India and Thailand. *Elaegnus* spp. are used in Nursery in USA and *Hippophae rhamnoides* are used for fruit juice in China.
- The symbiotic association imparts actinorrhizal plants to be readily established on coastal lands, eroded soils, glacial tills, sand dunes, saline lands, mine soils and volcanic ash deposits.
- Actinorrhizal-Frankia follow steps similar to legume rhizobial infection and nodulation.
- The N_2-fixation potential of actinorrhizal plants varies from 55-300 kg/ha/yr and the plant species are capable of establishing quickly on degraded low nutrient status soils. These plants have great future in agroforestry and land reclamation.
- The development of plant transformation system, together with identification and transfer of genes necessary for actinorrhizal symbiosis would open the door for extending to other plant species the capacity to support a N_2-fixing symbiosis.

References

Alexander, M. 1985. Introduction to soil Microbiology. Wiley Eastern Ltd., New Delhi.

Allen, E.N. and Allen, O.N. 1981. The Leguminoseae: A Gource Book of Characteristics, uses and Nodulation. The University of Wisconsin Press, Madison, Wisc.

Baggayan, R.B. 1982. Habitat Philippines 3(1):2-5.

Baker, D.D.1990. Actinorhizal Plants: underexploited trees and shrubs for forestry. Nitrogen fixing Tree Research Reports 8: 3-7.

Baker, D. D. and Schwitzer, C.R.1990. In: The biology of Frankia and actinorrhizal Plants (Eds., Schwitzer, C.R. and Tjepkema, T.D.) Academic Press, San Diego. P3.

Baker, D. and Seling E. 1984. Frankia: "New light on an actinomycete symbiont" Biological, Biochemical and Biomedical Aspects of Actinomycetes. Oritiz, L.L., Bojalil, L.F., and Yakoleff, V., eds, New York: Academic Press. 563-574.

Balasubramanian, A. 1992. Role of microorganisms in Agroforestory In: Agroforestory, Extension and Training. Directorate of Extension. Education, TNAU, Coimbatore, India. p 43.

Becking, J. H. 1966. Interactions nutritionells plants-actiomycetes. Rapport General, Annles de l' Institute Pasteur. Suppl. 111: 211- 246.

Becking, J. H. 1970. Frankiaceae fam (19+0). Nov. (Actinomycetales) with one new combination and six new species of the genus Frankia Brunchorst 1886, 174". Int, J. Syst. Bacteriol. 20:201-220.

Becking, J. H. 1982. Nitrogen fixation in nodulated plants other than legumes. In "Advances in Agricultural Microbiology, Ed. N. S. Subba Rao. Oxford and IBH Publishing Co., New Delhi and Butterworths, London. pp 89-110.

Becking, J. H., De Boer, W. F. and Houwing, A. L. 1964. Electron microscopy of the endophyte of Alnus glutinosa, Antonie van leeuwenhoek: Journal of Microbiology and Serology, 30: 343-376.

Becking, J.H. 1979. Plant Soil 53: 541-545.

Berg. H. and McDowell, L. 1987. Endophyte differentiation in Casurarina actionrhizae, Protoplasma 136: 104-117.

Bier, J. E. 1964. Phytopathology, 54: 250-253.

Bier, J. E. and Rowat, M. H. 1963. Can. Jour. Bot. 41: 1585-1596.

Bond, G. 1976a. Observations on the root nodule of Purshia tridentata. Proc. Royal Society London B.193: 127-135.

Bond, G. 1976b. The results of the IBP survey of root nodule formation in non-leguminous angiosperms. In: Symbiotic Nitrogen Fixation in Plants (Ed. Nutman, P. S.) Cambridge Univ. Press. Cambridge, England. pp. 443-474

Burleigh, S. Baker, D. D. and Torrey, J. G. 1988. Mass Inoculum Protocol for Casuarina. Abstract, 7th Inter. Meeting, Frankia and Actinorrhizal Plants, Univ. of Connecticut, storrs, 7-10 August, 1988.

Callaham, D., Del. Tredici, P. and Torrey, J. G. 1978. Science 199:899-902.

Caudhury, A. H. 1979. Plant Soil 51: 163-165.

Delver, P. and Post A. 1968. Plant Soil 28: 325-336.

Diem, H. G. Duhouse, E., Simonet, P. and Dommergues, Y. R. 1989. Actinorhizal Symbiosis Biotechnology Symposium. The present and the future. Eighth Intern. Biotech. Sym., Paris, France, July, 1988.

Dommergues, Y. R., Diem, A. G. and Sougufara, B. 1990. In: Advances in Casuarina Research and Development (Eds. M. H. El- Lakany, J.W. Turnbull and J.L. Brewbaker). Desert Development Centre, A.U.C. Cairo, Eggypt.

Dommergues, Y. R. 1987. In: Agroforestory, A Decade of Development (Eds., Steppler, H. A. and Nair, P. K. R.), ICRAF, Nairobi, Kenya. pp. 45-271

Frank, A. B. 1887a. Sind die Wurzelansch wellungen der Erien und Elaegnaceen pillgallen. Ber. Deut. Bot. Geell. 5:50-58.

Frank, A. B. 1891. Ueber die aut verdauung von pilzen abzielende symbioss der mit endotrophen Mycorrhizen begabten pflanzen sowie der Legumensosen und Erien. Ber Deut. Bot. Ges. 9:244-253.

Frank, A. B. 1887. Ueber neue Mykorrhiza formen. Ber. Deut Bot Ges. 5: 395-406.

Gardner, I. C. 1965. Observations on the fine structure of endophyte of the root nodules of Alnus glutinosa. Gaerten. Arch. Mikrobiol. 51 : 365-383.

Gauthier, D., Diem, H. G. Dommergues, Y.R. and Canry, F. 1985. Soil Biol. Biochem 17: 375-379.

Girgis, M.G.Z., Ishac, Y. Z, El-hadda, M. Saleh, E. A. Diem. H. G. ad Dommergues, Y. R. 1990. In: Advances in Casuarina Research and utilization (Eds. EL-Lanky, M. H., Turnbull, J. and Brewbaker, J.L.) Desert Development Centre, AUC, Cairo, Egypt.

Hawker, L. E. and Fraymouth, J. 1951. A re-investigation of the root-nodules of morphology and life histories of the causative organisms. J. Gen. Microbiol. 5: 369.

Hiyoshi, T., Sasakawa, H. and Yatazawa, M. 1985. Nitrogen Fixing Tree Research Reports 3:33-34.

Kreig, N, R, and Holt, J. G. 1984. In: Bergy's Manual of Systemic Bacteriology IV, William and Wilkins, Baltimore

Kummero W. J., Alexander, J. V., Neel, J. W. and Fishback, K. 1978 Am. J. Bot. 65: 63-69.

Lechevalier, M. P. and Lechevalier, H. A. 1986. Genus Frankia Brunchorst 1886, 174. In: Bergey's Manual Systematic Bacteriology. Vol. E ed S. T. Williams, Baltimore: Williams & Wilkins.

Lechvalier, M.P. 1983. Plant Soil 78: 1-6.

Medan, D. and Tortosa, R. D. 1976. Boletin de Socieded Argentina de Botanica. 17: 323-336.

Merwin, Miles* 1987. Casuarina Field Trials in California. Nitrogen fixing Tree Research Reports 5 (5): 31-32.

Moller, H. 1885. Plasmodiphora alni. Ber. Dtsch. Bot. Ges 3: 102- 105.

Moller, H. 1890. Beitrage Zur Kenntnis der Frankia subtilis Brunchorst, Ber. Dtsch. Bot. Ges. 8: 215-224.

Murry, M.A. Fotaine, M. S. and Torrey, J. G. 1984. Growth kinetics and nitrogenase induction in Frankia sp. HFPAR 13 grown in batch culture Pl. Soil 78: 61-78.

Newton, M., E.L Hassan, B. A. and Zavitkovski, J. 1968. Role of red alder in western Oregon forest succession. In: Biology of Alder (Eds: J. M. Trappe, J. F. Franklin, F. G. Tarrant, and G. M. Hansen) USDA Forest Service, Portland, Oregon pp. 73-84.

Parera, Viator 1985. Nitrogen Fixing Tree Reports 3:29-38.

Prat, D. 1989. Plant Soil, 113: 31-38.

Reddel, P. 1990. Increasing productivity in plantings of Casuarina by inoculation with Frankia. In: Advance in Casuarina Research and utilization (Eds. M. H. El. Lankey, J. W. Turnbull and J. L. Brewbaker) pp 130-140. Desert Development Centre, A. U.C., Cairo, Egypt.

Reddel, P., Rosbrook, P.A., Bowen, G. D. and Gwaze, D. 1988. Pl. Soil 108: 79-86.

Rundel, P. W. and Neel, J. W. 1978. Flora 167: 123-132.

Sanginga, N., Danso, S. K. and Bowen, G. D. 1989. Pl. Soil 118: 125-132.

Schwitzer, C.R. and Tjepkema, J. D. 1990. In: The Biology of Frankia and Actinorrhizal Plants (Eds Schwitzer, C. R. and Tjepkema, J. P.) Academic Press, San Diego., p 14.

Silver, W. S. 1964. Root nodules symbiosis. 1. Endophyte of Myrica cerfera L. J. Bacteriol; 87: 207-218.

Silvestor, W. B. 1977. In: A treatise on Dinitrogen Fixation (Eds. Hardy, R.W. F. and Gibson, A. H.) pp 141-190. Wily & Sons, N. Y.

Smolander, A. and Sundman, V. 1987. Physiol Plant 70: 297-303.

Sougoufara, B., Danso, S. K., Diem, H. G. and Dommergues, 1990. Soil Biol. Biochem. 22: 695-701. Tjepkema, J. D. Ormerod, W. and Torrey, J. G. 1980. Vesicle formation and acetylene reduction activity in Frankia Cpll cultured in defined nutrient media. Nature, London, 287:633-635.

Tjepkema, J. D., Schwintzer, C. R. and Benson, D. R. 1986. Physiology of actinorrhizal nodules. Ann. Rev. Plant Physiol. 37 : 209-232.

Torrey, J. G. 1978. Bioscience 28: 586-592.

Torrey, J. G. and Callaham, D. 1982. Structural features of the vesicle of *Frankia* sp. Cpll in culture. Can. J.Microbiol. 28: 749-757.

Vitousek, P. M. and Walker, L. R. 1989. Biological invasion by Myrica faya in Hawii: plant demography, nitrogen Tixation and ecosystem effects. Ecol. Monogr. 59: 247-265.

Voight, G. K. and Steucek, G. L. 1969. Soil Sci. Soc. Amer Proc. 33 :946-949.

Weber, A. 1986. PI Soil 96 : 205-213.

Werner, D. 1992. In: Symbiosis of Plants and Microbes. Chapman and Hall, London p.171.

Willis, J. C. 1973. A Dictionary of the Flowering Plants and Ferns (8th edition), (Revised by H. K. Airy Shaw), Cambridge University Press, Cambridge, England (1971).

Zeng, S. Tjepkema, J.D. and Berg, R.H. 1985. Gas diffusion pathway in nodules of Casuarina cunninghaminana. Plant Soil 118: 119-123.

14

Production Technology for *Azolla:* A Biofertilizer for Rice

Azolla, a fresh water fem is a rich source of nitrogen especially for low land rice farming. It is found in rice paddies in many tropical countries like China, India, Korea, Philippines and Vietnam. The agronomic importance of *Azolla* lies in its ability to fix atmospheric nitrogen through the symbiotic association with a blue-green alga, Anabaena *azolae.* The symbiotic association can fix nitrogen proportional to its biomass produced. When *Azolla* is grown in paddy fields and subsequently incorporated as green manure, the nitrogen contained within *Azolla* biomass is mineralized during the process of decomposition.

Azolla is a genus of small aquatic fern comprising 7 living and 25 fossil species, which is globally distributed. It occurs on still water ponds, lakes, swamps, lagoons and paddy fields in both tropical and temperate regions. Because of its rapid growth, high nitrogen percent and its ability to grow in stagnated waters, it has been used as organic manure in Vietnam and China for centuries. Its use in India and other countries is relatively recent. *Azollla pinnata* is native to India and widely distributed.

Nature of Symbiosis and Symbiotic Partners

The association between *Azolla* and *Anabaena azolae* is permanent contrary to *Rhizobium*-legume symbiosis. *Anabaena* is present at all stages of its growth and development and is naturally transmitted from one generation to another regardless of the mode of its multiplication.

The main.relation between *Anabaena* and *Azolla* are the supply to the fem of nitrogenous compounds fixed by the bacteria from the atmosphere which in turn are using carbon compounds photosynthesized by the fern.

The symbiotic association can grow and develop without any source of nitrogenous compounds. This is possible by the fact that *Anabaena azollae* has enzymatic complex, nitrogenase, capable of converting N_2 to NH_3. The fixation of nitrogen into ammonia occurs in the heterocysts which also support vegetative cells and growth of *Azolla.* The heterocysts excrete a significant amount of fixed ammonia into the leaf cavity from where it is taken up by the fern.

Azolla and vegetative cells of *Anabaena* are photosynthetic. With non-symbiotic *Anabaena* species such as A. *cylindrica*, the organism should be self sufficient in carbon compounds and can not transform under these conditions more than 3 to 5% of its cells into heterocysts. It has been observed that in mature *Azolla* leaves, the proportion of heterocysts exceeds 30%. This is obvious that the part of the carbon requirements of the microsymbiont is met from *Azolla*.

Some recent observations show that the bacteria of the genus *Arthrobacter* may also be involved in the symbiotic process of nitrogen fixation which needs to be examined fully.

I. Morphology of *Azolla*

An *Azolla* plant, often called a frond consists of a main rhizome. rarely exceeding 3 or 4 cm in length. It floats on water surface and is covered by small alternate leaves closely overlapping the rhizome (Plate 40). From the main rhizome emerges at regular intervals, secondary rhizomes having the same general characteristics as the first one, followed by third which subsequently produce the fourth. This growth pattern gives the plant a triangular or circular shape depending on the species. Unbranched and adventitious roots are formed on the lower part of the rhizome towards the water where they grow vertically.

Plate 40: *Azolla* frond seen from above. The most devel- oped branch is separated from the main plant as a result of the formation of an abscission layer at its base (vegetative multiplication).

Each leaf is made of two lobes i.e. dorsal and ventral lobe. The dorsal lobe is thicker, chlorophyllous develops on the upper side of the stem and a central cavity shelters *Anabaena azollae*.

Anabaena azollae a procaryotic organism occurs as unbranched filaments formed by two types of cells. The one more numerous, measuring 6 µm x 10 µm is called vegetative cells and the other called heterocysts are slightly larger and have thicker cell walls. The heterocysts of the *Anabaena* is site of nitrogen fixation.

In very young leaves, filaments are made up of vegetative cells only and the number of heterocysts progressively increases during the leaf development until it accounts 30 to 40 per cent of all cells. Finally in some cases, some of the cells, change to akinetes, the presence of these conservation form in the leafy cavity is not well understood.

Reproduction in *Azolla*

***Vegetative reproduction*:** *Azolla* multiplies sexually by spores and vegetatively by fragmentation. The vegetative multiplication is an extremely efficient process. As the plant matures, the lateral branch separates itself from its main stem by the formation of abscission layer giving rise to new plants. After separating from the parent plant the new plant behaves independently. Under favorable conditions, the young plant doubles in biomass after every 3-5 days. *Azolla* can also be propagated through spores.

Sexual reproduction: *Azolla* produces both male and female spores. The spore bearing organs or sporocarps are distributed over the branches and are present only in older plants. The male reproductive organ is larger and bali shaped while the female reproductive organ is smaller and egg- shaped. The female organ matures in about one week. Both male and female organs fall down at maturity and sink at the bottom of water. Fertilization occurs under water (sporulating *Azolla* show megasporocarp and microsporocarp).

Each sporangium produces 32 or 64 spores which because of their small size are called microspores, consequently the sporangia and sporocarps, which protect them, are microsporangia and microsporocarps.

In the second case, a lone sporangium produces a single larger spore, hence the name megaspore, megasporangium and megasporocarp even though the latter is smaller (±0.5 mm) than the microsporocarps. At maturity stage, the spherical megaspore at this stage is topped by three or nine spongy structures called floats which in turn are covered by conical hood made of upper part of the indusium which is highly lignified and protects at its summit a small colony of *Anabaena*.

Plate 41: *Azolla* frond sporulated seen from bellow:

At maturit, microsporocarps liberate their microsporangia into water. Each of of the these will disintegrate into 3 or 4 small spongy masses called massulae which contain the microspores. The massulae are attached to the megaspore through glochidia at the bottom of the water. The microspores in the massulae germinate and develop into a small prothalus producing male gametes, ciliated and mobile. The megaspore (megasporocarp) germinates and becomes a prothalus producing female gametes. The antheroziods cross with oosphere through an unknown process and fertilization is achieved. The embryo star's growth and as it grows upwards its apex pushes away lignified indusium and freeing itself from the sporocarp, forms a new plantlet which floats on the surface of water.

Megaspores and microspores are resting stages of *Azolla.* By harvesting them by special methods, one can maintain the plant for several months. However, *Azolla* rarely produces sporocarps and the environmental conditions favourable for sexual reproduction are poorly understood. Moreover the harvest of large quantities of sporacarp is difficult and mass production of *Azolla* biomass from sporocarp is a slow process.

II. Taxonomy of *Azolla*

Azolla can be easily distinguished from various floating aquatic plants like *Lemna* or *Salvinia*, but specific differentiation within *Azolla* may be difficult. In absence of sporcarps, whose morphology is best identifying criterion may be problematic. The genus of *Azolla* has been divided into 2 sections having a total of seven species.

Rhizosperma: The megasporocarps have nine floats and the glochidia are absent or unbarbed.

Azolla **nilotica:** It is found from Sudan to Mozambigue and is a native of in East Africa. In favourable conditions, it grows profusely reaching several decimeters. Glochidia are absent.

Azolla **pinnata:** It is the only specie which is divided into 2 cultivars. *A. pinnata* var. pinnata and *A. pinnata* var. imbricata.

Glochidia Unbarbed

Azolla pinnata var. *pinnata:* It is found throughout Africa, south of Sahel, Madagascar and Australia. It is fairly easy to identify by the morphology of its vegetative structure, shaped like Christmas tree showing main axis and regularly arranged secondary branches.

Azolla ***pinnata var. imbricata:*** It is native of subtropical and tropical Asia. It is widely distributed in India and this variety has been used in China and Vietnam for many centuries. It resembles the var. pinnata but its shape is not clearly triangular.

Section *Azolla*: The megasporocarps have three floats and the glochidia are barbed.

A. caroliniana, A. mexicana and A. microphylla: All the three originate from temperate, subtropical and tropical regions of North and South America. These species differ from each other mainly by build up of their megasporal wall and thus are often difficult to identify at the vegetative phase.

A. filiculoides and A. rubra: They are native of America and the Far East (Korea, Japan and New Zealand) respectively. The two species differ from the Caroliniana group by the presence of unicelluar trichomes on the abaxial surface of the upper lobe of their leaves.

III. Productivity of *Azolla*

Under favourable conditions, a new *Azolla* plant doubles its weight after every 3-5 days. The plant is almost fully grown in 15-20 days. The value of *Azolla* depends on its biomass which is determined by several environmental factors. The growth of *Azolla* at the water surface under optimum conditions described earlier could be represented by a sigmoid (S. Shaped) curve, slightly asymmetrical, with the inflexion point (1) generally situated before the biomass reaches half the maximum value (Fig. 4).

The exponential phase (phase 1) depends on the species until the biomass covers the available water surface which represents fresh matter yield of about 1kg.m^{-2}. Later, the growth becomes practically linear (phase II) and continues until the biomass yields about 2kg.m^{-2}in some favourable cases, and even up to 8kg m-2 in some exceptional cases. This is followed by stationery (phase III).

The maximum biomass is reached (phase IV). Afterwards the degeneration of the biomass starts (phase V).

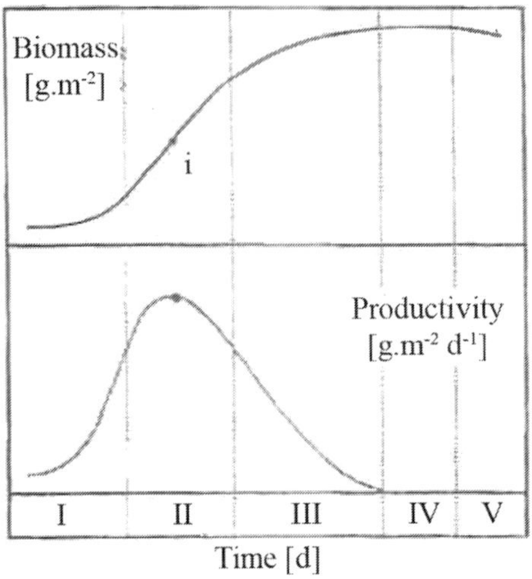

Fig. 4: Growth of *Azolla*

The growth curve depends on the environment conditions and *Azolla* ecotypes. It is generally useful to plant mixture of two or three different ecotypes, each of them characterized by high productivity during different seasons of the year.

Biomass Produced

One hundred to two hundred kilograms of fresh matter are produced daily from area of 1000 m2 or 1000 to 2000 kg fresh matter /ha/day.

The high nitrogen fixing ability, rapid growth, high biomass yields and nitrogen content indicate its potential as a biofertilizer for increasing rice yields with low cost and ecofriendly inputs. Watanabe *et al.* (1980) established nitrogen fixation by *Azolla* at the rate of about 1.1 kg N/ha/ day. Growth rates were summarised by Becking (1979) which showed that doubling time varies between 2 and 10 d for most species. Maximum biomass ranged from 0.8 to 5.2 t dry matter/ha and averaged 2.1 t dry matter/ha (Kikuche *et al.,* 1984). Nitrogen content ranged from 20-146 kg/ ha with an average of 70 kg N/ha

Chemical Composition of *Azolla*

Another criterion for selection of *Azolla* is the chemical composition which varies according to species and ecotypes, ecological conditions and the growth phase. The comparative studies on the chemical composition of different species and ecotype, are lacking and as a result only general information can be listed.

The values of dry matter of 5 to 7 percent may be taken as a fair estimate. The ash content generally measures between 10 and 20 percent of dry matter.

The carbon content is fairly constant at about 43 percent. The nitrogen percent varies considerably from 2 to 7 percent on dry weight basis. In natural conditions values near 3.5% are frequent. Both the varieties of *A. pinnata* are generally poor in nitrogen than the other species. The carbon/ nitrogen ratio which determines the state of decomposition at the time of incorporation in soil depends on the nitrogen percent and thus varies a great deal from 6.1 to 21.5 with possible average C/N ratio 12: .

The soluble carbohydrates represent about 3.5 percent, starch 4 to 10 per cent and lipids 3 to 6 percent of the dry matter. The cellulose and hemicellulose content usually varies between one specie to another. Young populations generally contain around 5 percent of lignin where as at the time of maturity (Phase IV), the content may exceed 30 per cent.

IV. Effect of Environmental factors

Water: *Azolla* prefers to grow in a free floating state. It requires nearly saturated deep standing water. However, a 5-7 cm deep of continuously flowing water is best for its production. Its multiplication is drastically reduced if there is not enough water and soil is just moist. This is probably due to the peculiar structure of its stomata that do not control the transpiration, without water the plant dies within a few hours. It can also survive rooted in the soil if the soil is permanently humid but in that case, its productivity is lower, over population occurs indeed rapidly in such conditions, since the fronds are unable to fragment and disperse. *A. pinnata* generally is the least able to support this situations not because it is less efficient in holding water in but because its root system is less efficient in extracting water in those conditions. A water depth of at least 5-10 cm is recommended for its multiplications but higher water depth do not have any adverse effect until the availability of plant nutrients become limiting (Singh, 1989).

Temperature: A temperature of 25-30°c (Peters *et al.,* 1980) is considered optimum for most species but temperature response varies among different species. Some strains grow, though slowly at temperature below 10°C but can survive frost conditions for a few days. *A. filiculoides* and *A. rubra* are tolerant to low temperatures and cannot withstand temperature over 30°C whereas A. *caroliniana, A. microphylla, A. nilotica* and some varieties of *A. pinnata* are tolerant to high temperature (Tung and Watanabe, 1983; Watanabe and Berja, 1983). High temperature also encourages the building up of insects and pest population.

It should be noted that *Azolla* culture in China since 1960 has been made possible by introduction of exotic species with different temperature and also development of methods to protect *Azolla* against extreme winter or summer temperatures.

Light: High light intensity adversely affects the growth of *Azolla*. Half of the full sunlight is normally good for the tiny fern. The growth rate of *Azolla* increases with light intensity up to u Einstein. M-2.S1. which corresponds to about 50 klux or 50% of maximum natural light intensity. The growth limiting effect on *Azolla* is more pronounced when it is grown alongwith rice, because rice plants cause varying degree of shading to *Azolla* depending on the canopy and growth stage (Singh and Singh, 1997). The light saturation for nitrogenase activity and photosynthesis under laboratory is reported between 5 and 20 klux but the growth under field conditions is maximum at 50 klux (Singh and Singh, 1990).

There also appears to be synergistic interaction between tolerance to high temperature and light intensity. At higher light intensities the optimum temperature for growth of *Azolla* moves to higher side. The extreme temperature results in reddish brown colouration of the fern but this change has not been reported to affect the nitrogen fixation. In rice fields, *Azolla* gets enough light from transplanting to panicle initiation stage or about the first 50-55 days of transplanting thereafter, the full canopy of rice crop shades the *Azolla* resulting in retardation in its growth.

Soil pH: *Azolla* grows satisfactorily in a pH range from 3.5 to 10. Indirect effects may cause unavailability of certain essential elements and at pH levels the growth of microsymbiont (*Cynobacteria*) may be reduced resulting in poor growth of *Azolla*. The average soil pH between 4.0 to 8.0 for successful growth of *Azolla* was reported but the optimum pH appears to be between 5 to7 (Peters *et al.*, 1980; Singh, 1979). Highly acidic and alkaline soils are not suited for multiplication of *Azolla*.

Salinity: The information on tolerance of *Azolla* to salinity is inconclusive but electrical conductivity beyond 3 mhos., cm-1 is probably always unfavourable to growth of *Azolla anabaena* which is more sensitive to salinity than *Azolla* and this results in production of *Anabaena*-free *Azolla*, inefficient in nitrogen fixation. Saline soils usually having pH above 7 are deficient in phosphorus, Zinc and copper which is not conducive for the growth of *Azolla*.

Relative humidity: The best relative humidity for *Azolla* growth is 85-90%. Low relative humidity below 60% dries up the *azolla* fronds while high relative humidity makes the plant susceptible to diseases caused by fungal infection.

Mineral Nutrients

Nitrogen: *Azolla* is capable of fixing atmospheric nitrogen even if the field is fertilized with mineral nitrogen and inhibiting effect of NH+, NO3 on nitrogenase is not reported. It appears that within *Azolla*-anabaena symbiosis, the nitrogenase seems to be protected in one way or other depending on the species and the ecotypes.

Application of urea or ammonium sulphate improves the yield of both the rice crop and A. *pinnata* (Manna and Singh 1988, Singh *et al.,* 1992). These findings suggest that the combined use of nitrogenous fertilizer and *Azolla* increased the yield of rice.

Phosphorus: Phosphorus is a key element and its deficiency results in poor growth, red colouration, curling of root hairs and reduced N content. A deficiency in phosphorus (<0.22% of the dry matter) is accompanied by reduction of nitrogenase activity and nitrogen content in *Azolla. Azolla* can accumulate phosphorus in the range of 1 to 2% of the dry matter depending on the P availability. The average phosphorus content of the soils in which *Azolla* grew satisfactorily was 20-25 ppm, however, it varied depending on the nature of species (Kushari and Watanabe, 1991). There are conflicting reports regarding the levels of P necessary for sustained growth. In laboratory studies, a concentration of 0.06 ppm is reported as being sufficient whereas in field conditions 0.3 to 1 ppm is required. Among the *Azolla* spp. examined, A. *pinnata* appears to have lowest P requirement. Sah *et al.* (1989) reported from studies under conditions of continuous growth medium that minimum P concentration was about 15 ppm. In addition, their findings have shown that the internal tissue - P necessary for sustained growth of *Azolla* is about 0.1% of the dry weight. The research study involving different strains of A. *pinnata,* A. *mexicana* and *A. microphylla* showed that A. *pinnata* was more productive as measured by total biomass than any of the other species. However, when P concentrations were higher (>6 ppm.) A. *microphylla* (IRRI4018) and to a lesser extent *A. mexicana* were more productive than *A. pinnata* species (Kushari and Watanabe, 1991). Kushari and Watanabe (1992) reported that *A. pinnata* var. *pinnata* (IRRI 7001) grew best at low phosphorus levels (0.02 and 0.03 ppm), however, at 0.06 ppm, *A. microphylla* (IRRI 4018) outperformed all the three species tested. Based on these studies, it appears that *A. pinnata* is better performer under field conditions even with low P content.

Potassium: Positive effects of potassic fertilizer on the growth of *Azolla* and also a trend of luxury consumption of the element by this fern has been observed.

The cases of deficiency in iron and molybdenum as well as cases of iron toxicity on the growth of *Azolla* have been also observed.

Insect pests and diseases: Many insects especially larvae of *pyralidea* (Lepidoptera) and *Chironomidae (Diptera)* feed on *Azolla* especially in tropics and at high atmospheric humidity. Gasteropods (snails) have been reported as pests which feed on both roots and leaves of *Azolla* but less damaging as compared to other insects. Insecticides commonly recommended to control rice insects, are equally effective against major insects of *Azolla*. The use of insecticides increases the growth and nitrogen fixation by fem. Insecticides such as carbofuran, increases the growth and nitrogen fixation by fern. Insecticides such as carbofuran, methomyl, monocrotophos, and diazinon can be sprayed for control of insect which should not cause environmental problems. If the *Azolla* is to be recycled through fish pond or is to be fed to livestock or poultry the toxic and non-biodegradable pesticides should not be used. The selection of *Azolla* species which are resistant to insect pest is essential to avoid use of pesticides. The biological control and use of biopesticides such as neem extract or other techniques should be employed. Some natural products such as of Neem *(Azadirachta indica)* origin or *Albizza lebbek* (Sirish) are reported to be effective against insects feeding on *Azolla* (Reynaud and Franche, 1986, Kushari and Teheruzzaman, 1991). Predators usually feed on insect larvae and if farmers are aware of these predators being present in their fields, they need not use pesticides and other means of artificial control indiscriminately.

The high temperature, humid environment and the population densities along with the damage to fern by insects induce secondary infections by fungi such as *Rhizoctonia* and *Sclerotium* and some bacteria. This usually occurs when the culture becomes old, or its mat is left unharvested for a longer period particularly in hot summer. Fungal infections can be controlled by use of fungicides. Algal growth should be removed manually so that growth of *Azolla* is not hampered.

Precautions to Ward off Insect Pests

- The use of healthy *Azolla* inoculum should be, done. Farmers should discard *Azolla* plants or portions that show signs of insect infestations as these may harbour eggs and larvae.
- The use of *Azolla* species or varieties that are observed to be resistant to known pests and those better adapted to environmental conditions in a locality, should be popularised.
- Avoid overcrowding the *Azolla* mat. As soon as the *Azolla* cover becomes full, half of it should be harvested to prevent the sudden increase in *Azolla* pests due to increase in their food resource.

- Grow a mixed culture of *Azolla* species or varieties in the same pond or paddy field. This makes the *Azolla* crop less prone to insect infestations than a culture of single *Azolla* species or varieties.
- Besides the cultural management, it is advised to the farmer to keep a watch on other animals, that are natural enemies of *Azolla* pests such as insect parasites that kill their host either in their egg, larval or pupal stages.
- Predators usually feed on larvae. Some beetles are also predators of *Azolla* pests.
- The use of chemical pesticides should be the last resort to control *Azolla* pests. When use of pesticides becomes necessary as the last resort, farmers should choose and apply them judiciously, preferring pesticides with short residual toxicity and persistence.

As much as possible, they should refrain from spraying *Azolla* to be used as animal feed or as food for man.

Different Methods for propagation and use: *Azolla* is to be kept in vegetative form in order to make it available whenever needed for use. It is almost impossible to obtain spores for the conservation of the material in dry state. Moreover, even when this is possible, the time required to produce sufficient biomass from spores is usually too long.

Nurseries: Standing water can normally be used for maintenance of an *Azolla* population. *Azolla* develops in the paddy field as soon as they are flooded, either from spores of previous seasons or vegetative parts which have survived in soil. The spontaneous presence of *Azolla* all the year round is doubtful and the vegetative forms have to be kept active in nurseries especially designed for that purpose. The maintenance of healthy *Azolla* culture should be made the responsibility of village communities. The farmers should get the effective healthy inoculant at appropriate time to multiply them.

A few hundred square meters having depth of water not exceeding 5 to 10 cm is desirable but may require frequent adjustments of water level depending on the climatic conditions such as rainfall etc. Slight shading of *Azolla* beds to avoid scorching sun under certain climatic conditions may be desirable. The use of creepers is recommended for protection from wind, if necessary.

Maintenance

Weed removal: In the growth plase the biomass production of *Azolla* is maximum at 1kg. m2. This prevents the growth of other plants such as *Lemna, Salvinia, Eichhornia, Spirodela* or other alga. Overcrowiding of *Azolla* plants should be avoided because it reduces the productivity. Excess biomass should be removed depending on the daily need or at least once a week.

Fertilization: Some biomass are regularly harvested, thus the nutrient removal is taking place. Hence the plant nutrients except nitrogen should be applied. Fertility of nursery can be maintained by recycling of crop wastes, animal residues, household ash and waste material. Split phosphate fertilization is required depending on the harvest and yield of biomass.

V. *Azolla* as Green Manure in Rice

Azolla is quickly decomposed being rich, in nitrogen content and a low C/N ratio organic material. It is used in areas where water is available before transplanting. Fresh *Azolla* is inoculated at the rate of 1 to 2 tonnes/ ha in well prepared and leveled field and the water level is maintained at 5-10 cm depth. Single superphosphate (25-50kg/ha) in split doses and insecticide (2-3kg/ha) are also applied for better growth of *Azolla*. The duration of monoculture to form a thick mat over the surface of water takes about 10-20 days. Water is drained from the field and *Azolla* growth is ploughed and incorporated into the soil. Depending on the resources, incorporation is done by hand, by rotary or conical weeders by ploughing or harrowing

The extent of preliminary draining is an important factor for a successful incorporation. If rice transplantation follows the *Azolla* monoculture which may be carried out in such way that about 20 per cent of the *Azolla* will float again upon flooding and form a new carpet between the rice plants while the bulk 80 per cent of its biomass shall remain underground and will be mineralized into available form of plant nutrients. One crop of *Azolla* normally gives 10-15 tonnes green manure.

Advantages of Monoculture

In tropical soils, the fertilizing effect of the monoculture is apparent during the first few weeks of growth and development of rice. It is generally comparable to 30 units of nitrogen. While functioning as green manure, *Azolla* monoculture also inhibits the growth of weeds.

Dual Cropping: Growing *Azolla* simultaneously with rice seedlings after transplanting is called dual cropping/culture (Plate 44). About a week after transplanting rice seedling, the field is inoculated with fresh, *Azolla* at the rate of 0.5-1.0 t/ha. *A microphylla* inoculated at the rate of 200 kg/ha could multiply faster having 15-25 tonnes of biomass within 2-3 weeks period. Lower loads of inoculation shall take longer duration to establish in rice fields. *Azolla* technology is very efficient in terms of nitrogen fixation and biomass production during wet season due to better environmental conditions prevailing during the second season of rice in India particularly cloudy days coupled with low temperatures (18-25±1°C) favours its multiplication.

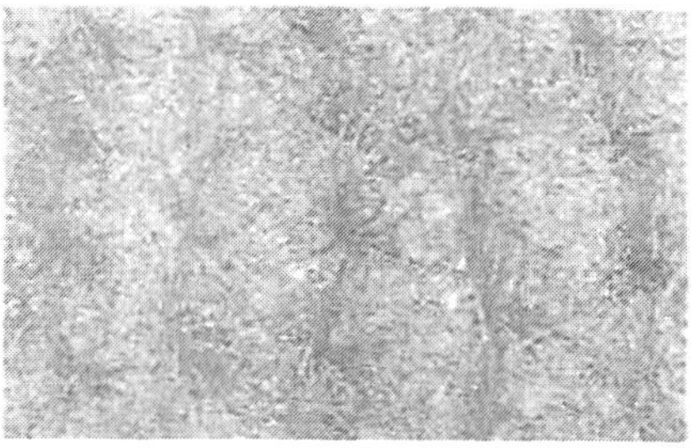

Plate 42: Rice-*Azolla* intercrop. The simultaneous presence of three *Azolla* ecotypes increase the probability of a high produc- tivity.

Intercropped *Azolla* is not fertilized but if soil is poor in phosphorus, it can be applied at the rate of 20 kg/ha including the recommended dose of phosphorus for the rice crop. Insecticides such as furadan at 2.5 kg/ha are applied to control insect pests. *Azolla* forms a thick mat within 15-20 days and it starts decaying. Fields is drained off and *Azolla* is incorporated in soil. It is possible to raise two dual crops during a single rice crop period. The growth of *Azolla* is significantly influenced by rate and time of inoculant, method of planting, density of rice, application of fertilizers and pesticides (Singh, 1989; Singh and Singh, 1990). On an average one crop of *Azolla* grown as green manure provides 30 kg N/ha and response of rice is equivalent to the application of 30 kg/N/ha as chemical nitrogen. *Azolla* green manuring plus one dual cropping gives about 60 kg N/ha and *Azolla* green manuring plus two dual cropping gives about 90kg N/ha.

Azolla as Biofertilizer in Rice Crop

Azolla has been used as a bio-fertilizer for rice crop in many countries and regions such as China, Vietnam, Philippines, Thailand, Korea, India, Bangladesh, Sri Lanka, Nepal, Burma, Indonesia, Brazil and West Africa but its full potential has not been harnessed in agriculture. The symbiotic association draws the energy from photosynthesis carried out by fern to fix atmospheric nitrogen which is of the order of 150-200 kg ha^{-1}. *Azolla* in rice system contributes about 40-60kg N ha^{-1} per crop (Kannaiyan 1989). The advantage of using *Azolla* as bio-fertilizer for rice crop is because of its quick decomposition and mineralization of its N in wet soil and rendered available to rice crop (Kannaiyan, 1990). The growth rate of *Azolla* is very rapid so much so that it doubles it weights within 2-3 days. Due to fast relative growth rate the biomass production is heavy in a short time which can be used as biofertilizer for rice. *Azolla* has the potential of maintaining an exponential growth rate under optimum conditions. The growth of *Azolla* is initially slow both in the growth medium and in rice fields followed by fast growth (Kannaiyan,

1993). Phosphorus fertilization is necessary for its growth and N2 fixation by *Azolla* (Kannaiyan *et al.,* 1981 and Singh and Singh, 1989). *Azolla* grown in phosphorus deficient solution had a reddish brown discolouration that spread from the centre of the frond to the tip of the body with reduction in size of the frond. Singh (1977) recommended 4-8 kg P2 05/ha per week for large-scale multiplication of *A. pinnata* under field conditions. Kannaiyan *et al.* (1982) have demonstrated the effectiveness of split application of phosphorus for *Azolla* growth and multiplication. Animal dung @ 1 to 1.5t ha^{-1} or cattle slurry 2000-3000 L ha^{-1} can be used as an alternate of superphosphate (Singh *et al.*, 1993).

VI. Effect on Crop Yield

Srinivasan (1980) reported that *Azolla* as green manure could reduce the nitrogen input by 25-30% in lowland rice. *Azolla* incorporation as green manure increased the yield of rice by 9-38% (Singh) 1997), 35.8% (Kannaiyan, *et al.*, 1981), 44% (Bartharkur and Talukdar, 1983). Subramani and Kannaiyan (1987) reported that application of gypsum at 5 t/ha with 60 kg N/ha as urea and *Azolla* inoculation as dual crop recorded the highest grain yield of rice. Jayanthi *et al.* (1994) have reported that incorporation of 10 t/ha fresh *Azolla* biomass along with 50 kg N/ha resulted in increased N uptake, available nitrogen in soil and rice yield. Gopalaswamy *et al.* (1994) have investigated the potential of *Azolla* hybrids and reported that inoculation of *Azolla* hybrids significantly improved the soil organic carbon status besides increasing the rice yield.

In wetland rice ecosystem the productivity of rice could be increased by raising the efficiency of applied nitrogen either by biological means or by chemical fertilization. Ladha *et al.* (1988) have established that the inoculation of *Azolla* coupled with urea super granule (USG) application have recorded significantly higher grain yield of rice than PU (prilled urea) application. Thangaraju and kannaiyan (1990) reported that *A. microphylla* and *A. filiculoides* coupled with USG application significantly increased the grain yield of rice. Kannaiyan (1992) reported increased rice yield by the combined application of USG and *A. microphylla.* Shanmugasundaram and Kannaiyan (1994) have reported the inoculation of *Azolla* hybrid RS- SK-TNAUD[1] with PU as well as USG recorded higher grain yield and N uptake in rice. The nitrogen fixing green manure plants viz. *Azolla* and *Sesbania rostrata* could be effectively utilized by incorporation with USG to increase the rice production in lowland rice ecosystem (Kumar *et al.* 1995). The nitrogen use efficiency in lowland rice system could be increased substantially by adopting the integrated nutrient management system using nitrogen fixing green manures or *Azolla* biofertilizer.

Azolla inoculation considerably increased the soil microbial as well as enzyme activity (Kumar, 1994 and Kannaiyan, 1994). The incorporation of 22.5-37.5t fresh *Azolla* ha^{-1} before planting increased the rice yield by 9.6-13.0 per cent (Zhang Zhuang *et al.*, 1987). Incorporation of 5 t *Azolla* ha^{-1} with 100 percent recommended nitrogen level yielded 6.7 t ha^{-1} which is more or less equal to the incorporation of 5t *Azolla* ha^{-1} with 75 per cent of the recommended dose of nitrogen (Patil and Mandhare 1991). Alexander *et al.* (1992) have observed maximum rice yield increase with an output ratio recovery of 2.07 with 7.5 tonnes *Azolla* ha^{-1} incorporation along with 90 kg N ha^{-1}. An increase in the grain yield of rice by the application of *Azolla* as green manure, dual crop or both with rice was equivalent to 30-40 kg N ha^{-1} and this has been well documented by several investigators in various countries such as China (Liu Chung Chu, 1987) the Philippines (Mabbayad, 1987; Watanabe, 1987), India (Singh 1997; Kannaiyan, 1981), Thailand (Loudhapasitipom and Kanareugsa, 1987) Sri Lanka (Kulasooriya *et al.*, 1987) USA (Talley *et al.*, 1977) and West Africa (Diara *et al.*, 1987). Inoculation of *Azolla* hubrids significantly improved the soil organic carbon besides increasing the rice yield (Gopalaswamy, *et al.*, 1994). The efficiency of *Azolla* may be improved physiologically by using the growth regulator gibberellic acid GA, (Gopalaswamy and Kanniyan, 1996a). *Azolla* hybrid AH-C, was found to be saline tolerant (Gopalaswamy and Kanniyan, 1996b). Rice yield was significantly increased with *Azolla* hybrids in combination with 75 kg N ha^{-1} Urea Super Granule (Gopalaswamy and Kannaiyan, 1996c).

In rice, *Azolla* can be a component of Integrated plant nutrient supply (IPNS) system with chemical fertilizes and biological inputs. The integrated use of *Azolla* and low levels of chemical fertilizes gave higher grain yields (Singh et *al.*, 1991). Application of FYM and urea super granules cause less reduction in the growth of *Azolla* as compared to ammonium sulphate. The inhibitory effect of nitrogenous fertilizes on *Azolla* can be reduced by their deep placement in furrows which also improves fertilizer use efficiency.

Future Prospects

Azolla has a great potential as nitrogen fixing green manure. It is easy to grow and incorporate in soil under rice in flooded conditions. However, It has yet to become popular with rice farmers due to some constrains such as environmental and technological aspects. The most important aspect is the availability of appropriate fresh inoculum, its multiplication and transportation. Its maintenance and preservation throughout the year is of great importance since it survives for a short period under dry condition. A suitable technology for production of *Azolla* sporocarps, their proper storage for effective utilization in Indian conditions has to be developed as soon as possible. Temperature sensitivity, incidence of pests and diseases and phosphorus requirement can be solved by selecting appropriate strains which will be temperature tolerant, resistant to insects and diseases with low P requirement. Some progress in this area has been reported by Indian workers. It is necessary to set up *Azolla* production centres in villages and at block levels in the country for timely supply of fresh and efficient inoculum to rice farmers.

References

Alexander, D, Sadanandan N. and Karunakaran, K. 1992. In New trends in Biotechnology. N.S. Subba Rao, C. Balagopalan and S.V. Ramakrishnan (Eds.) Oxford and IBH Publ. Co. New Delhi. pp. 255- 262.

Baratharkar, H.B. and Talukdar 1983. Intl. Rice Research Newsletter 8 (2):20.

Becking, J.H.1979. Environmental Requirements of *Azolla* for use in tropical rice production. In: Nitrogen and rice : IRRI, Manila Pp 354-374.

Diara, H.F., Van Brandt, H., Diop, A.M. and Van Hove C. 1987 Proc. Workshop on *Azolla* use. Fuzhou Fujian China. IRRI, Manila Pp.354- 347.

Gopalaswamy G. and Kannaiyan, S. 1996a. Inoculation of *Azolla* hybrids for rice. Paper in X southern Reg. Conf. on Microbial Inoculants held at Poondi, Thanjavur 10-11 Dec. 1996.

Gopalaswamy, G., and Kannaiyan, S. 1996b. Abst., 37th Ann. Conf. Assoc. Microbiologist, India p. 132.

Gopalaswamy, G., Anthoni Rya S., Kumpuchettey, N. and Abdulkareen, A. 1994 34th Ann. Conf. Assoc. Microbiologist. India. PAU, Ludhiana. P. 161.

Jayanthi, M., Rajagopal, V. and Shantaram, M.V. 1994. Natl. Seminar on *Azolla* and Algal Biofertilizers in rice. Tamil Nadu Agricultural Univ., Coimbatore, Tamil Nadu p.23.

Kannaiyan, S. 1981. *Azolla* biofertilizer for rice. Transfer Training Seminar, IRRI, Manila, Pp. 1-11.

Kannaiyan, S. 1989. *Azolla* biofertilizer for rice. In: Proc. Natl. Seminar on Biotech. Mol. Biol., Kottayam, Kerala P. 15.

Kannaiyan, S., 1990. Proc. Curr. Trends in Biotechnology. Cochin Univ. Science and Technology, Cochin, p. 16-27.

Kannaiyan, S. 1992. Tech Bull. Tamil Nadu Agric. Univ., Univ., Coimbatore Pp. 1-56.

Kannaiyan, S. 1993. Proc. Indian Natl. Sci. Acad. 59: 21-31.

Kannaiyan, S. 1994 Natl. Seminar on *Azolla* and Algal biofertilizers for rice. Tamil Nadu Agric. Univ. Coimbtore, Pp.25-26.

Kannaiyan, S., Thangaraju, M and Oblisami, G. 1981. Tamil Nadu Agric. Univ. Newsletter. 8: 2.

Kikuchi, M., Watanabe, I and Haws, L.D. 1984. In :Organic Matter and Rice. IRRI, Manila, Pp 569-592.

Kulasooriya, S.A. Hirmburegama, W.K. and Abeysekara, S.W. 1987. Proc. Workshop on *Azolla* use. Fuzhou Fujian China. Intl. Rice Res. Inst. Los Banos, Manila Pp. 131-140.

Kushari, D.P. AND Tehruzzaman, Q. 1991. In: Biological Nitrogen Fixation Associated with rice Production (eds., S.K. Dutta and C. Sloger) oxford and IBH Publishing Co., New Delhi Pp. 109-118.

Kushari, D.P. and Watanabe, I. 1991. Soil Sci. Plant Nutr. 37:271- 282.

Kushari, D.P. and Watanabe, I. 1992. Soil Sci. Plant Nutr. 38:65-73.

Kumar, K. 1994. Studies on the development of *Azolla* mutants and their nitrogen fixing potential, microbial decomposition and biofertilizer value for rice. Ph.D. Thesis. Tamil nadu Agri. Univ. Coimbatore.

Kumar, K., Thangaraju, M and Kannaiyan S. 1995 In: Rice Management Biotechnology (Ed., S. Kannaiyan) Assoc. Publ. Co. New Delhi. Pp.100-101.

Ladha, J;K., Watanabe, I and Saono, S. 1988. In: Green Manure in Rice Farming. Inter. Rice Res. Inst. Los Banos, Philippines. Pp.165- 183.

Liu Chung Chu 1987. Proc. Workshop on *Azolla* use. Fuzhou, Fujian, China, Int. Rice Res. Institute, Los Banos, Manila Pp.67-76.

Loudhapaasitiporn, L. and Kanareugsa, C. 1987. *Azolla* use in Thailand. In: proc. on the workshop on *Azolla* use. Fuzhou Fujian, China Inter, Rice Res. Inst. Los Banos, Manila Pp. 119-122.

Mabbayad, B.B.1987. The *Azolla* Programme in Philippines Proc. Workshop on *Azolla* use. Fuzhou, Fujian China. Inter Rice Res. Institute Manila Pp.101-108.

Manna, A.B. and Singh, P.K. 1988. Plant and Soil 107: 171.

Patil, P.L. and Mandhare, V.K. 1991. In Abstract: 31st Ann. Conf. Assoc. Micobiologist, India, Tamil Nadu, Agric. Univ. Coimbatore P. 135.

Peters, G.A. Toja Jr, R.E. Evans W.R. Christ, D.K. Mayne, B.C. and Poole, R.E. 1980. Plant Cell Environ. 3: 261-269.

Reynaud, P.A. and Franche, C, 1986. *Azolla* pinnata var. Africana from Molecular Biology to use as Green manure, Dakar, Senegal, p. 15.

Sah, R.N,.Goyal, S.S. and Rains, S.W. 1989. Crop Science 29:1033- 1037.

Shanmugasundram, R. and Kannaiyan, S. 1994. Studies on the development of *Azolla* and algal biofertilizer for rice. Tamil Nadu Agric. Univ. Coimbatore (Abst). P. 17.

Singh A.L. and Singh, P.K.1998. Nitrogen Fixation in Indian Rice Fields (*Azolla* and Blue-green Algae) Agrobotanical Publisher, Bikaner, India. P. 236.

Singh, P.K.1977. Riso 26: 125-137.

Singh, P.K. 1979. In: Nitrogen and rice. IRRI. Manila, P. 407-418.

Singh, P.K.1989. Appl. Agric. Res. 4:149-161.

Singh. P.K. 1997. Int. Rice Res Newsletter 2(7): 7.

Singh. P.K. and Singh, D.P.1990. *Azolla* and rice cultivation. In Biofertilizers (ed) L.L. Bhandari, S. N. Saxena and K.K. Vyas Pp. 39-45, Scientific Publishers Jodhpur.

Singh, P.K. and Singh, D.P. 1997. *Azolla*-Anabaena Symbiosis. In Biotechnological Approaches in Soil Microorganisms for Sustainable crop Production (ed) K.R. Dadarwal Scientific Publishers, Jodhpur, Pp. 93-107.

Singh, P.K. Singh D.P., Manna, A. B. Singh, R.P. and Bisoyi, R. N. 1991. In Biological Nitrogen Fixaton Associated with Rice Production (eds.) S.K. Datta and C. Slogge. Oxford and IBH Calcutta Pp. 95- 107.

Singh, P.K. Singh, D. P. and Singh, R.P. 1992. Biochem. Physiol. Pflanzen 188:121-127.

Singh, P.K., Singh, D.P. and Satpathy, K.B. 1993. In: Biogas Slurry utilization. Consortium on Rural Technology, New Delhi Pp.135

Srinivasan, G. 1980. Inte. Rice Res. Newsletter 6: 25.

Subramani, S. and Kannaiyan, S. 1987. Trop. Agric. 65: 226-228.

Talley, S.N.Talley, B.J. and Rains, D.W. 1977. Nitrogen fixation by *Azolla* in rice fields. In: Genetic Engineering for Nitrogen Fixation (ed) Alexander Hollaender. Plenum Publishing Company, New York, Pp. 250-282.

Thangaraju, M., and Kannaiyan, S. 1989. 29th Annual Conf. Assoc. Mircrobiologist, India. HAU, Hisar, Abstr.P.87.

Tung, H.F. and Watanabe, I. 1983. New Phytol. 93: 423-443. Watanable, 1.1987.Inte. Rice Res. Institute, Philippines Pp. 197- 208.

Watanable, I and Berja N.S.1983. Aquat. Bot. 15:175-182. Watanable, I., Berja, N. S. and Rosari Del D.C. 1980. Soil Sci Plant Nutr. 26:301-307.

Zhang Zhuang-Ta, Ke Yu-Si, De-Quan. L., Bing-Yuan D. and Liu Xi- lian 1987. Proc. Workshop on *Azolla* use. Fuzhou, Fujian, China, Inte. Rice Res. Inst. Philippines Pp. 141-145.

15

Asymbiotic Nitrogen Fixation *Azotobacter* and Others

There are several bacteria which fix atmospheric nitrogen and do not have a symbiotic partner such as a plant. There are conclusive evidences regarding nitrogen gains due to free living bacteria (Dart and Day, 1975) The diazotrophic bacteria have been classified based on the mode of nutrition and oxygen requirement (Alexander, 1977, Knowles, 1977). Free living nitrogen fixing bacteria fix nitrogen in soil and their population in rhizosphere of plants are significantly higher than in non-rhizosphere soils

They can be categorized into obligate aerobic, micro-aerophilic, facultative or anaerobic bacteria (Table 108).

Table 108: List of nitrogen fixing free living bacteria

Category	
Aerobic	*Azotobacter, Azotococcus, Beijerinckia*
Micro-aerophilic	*Azospirillum, Aquaspirillum, Mycobacterium Herbaspirillum, Corynebacterium, Thiobacillus*
Facultative	*Bacillus, Klebsiella, Enterobacter, Escherichia, Pseudomonas, Rhodopseudomonas, Rhodospirillum*
Anaerobic	*Clostridum, Desulfovibrio, Desulfotomaculum*

Although the economic and ecological significance of many of these bacteria have yet to be established, the most promising *diazotrophs* are *Azotobacter, Azospirillum, Beijerinckia* and *Clostridium*. *Azotobacter* and Beijerinckia are some of the most preponderant, non-symbiotic nitrogen fixing bacteria. Studies on these organisms in Indian soils have been done earlier by Rangaswamy and Sadasivam (1964) Sen and Barooah (1964) and Jagtap and Bhide (1971). Among the free living bacteria, *Azotobacter* is the most intensively investigated. Several species viz., *A. chroococcum*, *A. vinelandii, A. macrocytogenes, A. agilis* and *A. insignis* were reported but *A. chroococcum* is the most common specie encountered in arable soils.

I. *Azotobacter*

The efficiency of nitrogen fixation is directly dependent on the amount of carbohydate utilised by the organism

Firstly it was isolated and described by Beijerinck (1901). It occurs both in soil and rhizosphere but not in rhizoplane (Mishushtin, 1970). Its distribution and abundance in soil mainly depends on soil pH and organic matter content. The population of *Azotobacter* in temperate soil was studied by Darzneik (1961) which rarely exceeds 103 to 104 in soil. In Indian soils, the population of A. *chroococcum* generally also varies between 103 to 104 /g soil (Gaur *et al.*, 1975, Gaur and Misra 1978).

Taxonomy

The morphological and biochemical characteristics of different species of *Azotobacter* have been reported in Bergey's Manual of Determinative Bacteriology (Bachanan and Gibson, 1974).

Azotobacter group comprises several genera such as *Azotobacter, Azomonas, Azotococcus, Beijerenckia* and *Derxia.*

Habitat

A. chroococcum, A. Beijerinckii, A. vinelandii are found both soil and water. *Azomonas agilis* and *A. insigins* are found in water only where as A. *pasapali* is found in soil only. Except *A. agilis* and *A. insignis,* cyst formation is present in other species. They are motile except in *A. beijerinckii.* The pigment formation in *A. beijerinckii* and *A.* chrococcum takes place in old cultures where as in A. *vinelandii,* it occurs in young cultures. In other species, the pigment formation does not take place. The characteristics and colour of pigments are variable. The pigments of *A. chroococcum* are dark brown to black and water insoluble. The yellowish to pale brown pigments of A. *beijerinckii* do not diffuse in water whereas yellowish-green pigments of *A. vinelandii* diffuse into water.

Azotobacter species are incapable in utilizing starch. Most the species except *A. agilis* and *A. insignis* utilize mannitol or rhamnose benzoate.

They are mostly similar in morphological and physiological characters and are highly aerobic. They are large granular bacteria, ovoid or rod shaped and sometimes motile which form spreading glutinous colonies on nitrogen free agar media. *Azotobacters* except nif mutants artificially prepared in laboratory fix atmospheric nitrogen in ordinary air efficiently. However, *Derxia gummosa* has problem in initiating growth in ordinary air. It can be initiated by either inoculating the plates at low oxygen tension (10% O2) or by adding a small amount of ammonium salts to the agar medium. *Azotobacter chroococcum* is widely distributed in many cultivated soils than the other species which are more suited to aquatic environments.

Characteristics of *A. Chroococcum*

The bacteria is aerobic and can be isolated on nitrogen-free medium from neutral to alkaline soils. Its growth is highly pleomorphic, In young cultures, it is short rod shape but assumes spherical shape in old cultures, length and breadth varying from 4 to 6.5μm and 3.7 to 6.0pm respectively changing from short rods to coccal or oval form in old cells. The cells are heavily capsulated, non-spore forming and varying in degree of motility. The characteristic growth of the colonies with characteristic pigments form the basis of distinguishing strain characters (Plate 43). All strains show utilization of citrate, acidification of litmus milk and negative M.R. and V. P. tests.

It not only fixes atmospheric nitrogen and provides to plant but also produces a variety of growth promoting substances such as indole acetic acid (IAA), gilbberellins, Vitamin B and antibiotic and anti-fungal substances. Another important characteristic of *Azotobacter* activity in the rhizosphere of the crop is the excretion of ammonia, thus benefiting the plant. Such strains were better competitors with higher survival in soils compared to non-ammonia excreting strains (Satvir Singh and Lakshminarayana, 1982).

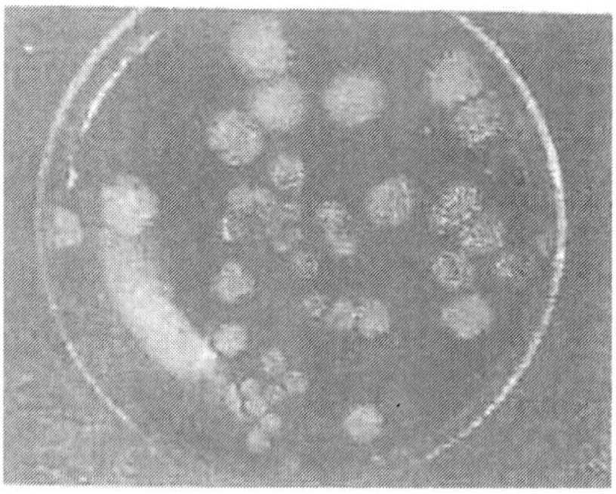

Plate 43: Colonies of *Azotobacter chroococcum*

Factors Influencing *A. Chroococcum* Population in Soils

The soil type such as fertility, moisture, temperature, pH and organic.

matter content influences the growth and multiplication of *Azotobacter* in soils. Rhizosphere is considered to be congenial habitat for their growth where they are found in higher numbers than in non-rhizosphere soils.

The specificity of *Azotobacter* to plant has been controversial till Dobereiner (1972) introduced a new species of *A. paspali*.

Nitrogen Fixation Range

Individual strains vary in nitrogen fixation ranging from traces to 25mg N/g of sucrose utilised. Both high and low nitrogen fixation has been reported by several workers (Loptina, 1949; Kluyver and Becking, 1955; Sundara Rao and Iswaran, 1959; James, 1970; Mishustin, 1970; Abd-el- Maleek, 1971; Patil, 1980; Gaur and Rai, 1982)

Effect of Organic Matter and Humic Substances on *Azotobacter*

Azotobacter can utilize various carbohydrates, organic acids, Polyhydric alcohol, glycerol and mannitol. Gaur and coworkers reported that humic and fulvic acids influence the growth of *Azotobacter* and also nitrogen fixation (Gaur and Mathur, 1964, Gaur *et al.,* 1968-69, Bhardwaj and Gaur, 1970) Further the inoculation of *Azotobocter* in humic acid amended soil increased the crop productivity significantly (Gaur and Bhardwaj, 1971, Mathur and Gaur, 1977).

Root exudates influence both the growth and nitrogen fixation capacity of *Azotobacter* in the rhizosphere. Organic acids and sugar in the exudates serve as carbon substrates while amino acids are inhibitory (Vancura and Macura 1961, Zinoveva, 1962).

The favourable pH, high phosphorus content and organic matter level are congenial for better results with *Azotobacter*.

Their numbers were increased with addition of straw, manure and compost containg cellulose fractions (Nepomiluev and Shishov, 1962; Erofeev and Vastov 1964; Kundu and Gaur 1980). Gaur and Rai (1982) investigated the effect of paddy straw, partially degraded paddy straw or FYM on nitrogen fixation by *Azotobacter*. FYM was found a better carbon source for N, fixation by *Azotobacter* while partially decomposed paddy straw was found better in case of *Azospirillum.*

Synthesis of Growth Promoting Substances

The cultures of A. *chroococcum* synthesize considerable quantities of biologically active growth promoting substances. Foremost among these are the vitamins of "B" group such as nicotinic and pantothenic acids, biotin and heteroauxins and gibberellins. It also produces cytokinin-like- substance.

A low N_2-fixing M-4 strain of *A.chroococcum* had been found to have broad spectrum utility to enhance growth and yield of crops such as maize, sorghum, cotton and potato (Shende and Apte, 1982). Thus it was clear from observation that high nitrogen fixation can not be a sole criterion for selection of A. *chroococcum.* The strain M-6 did not produce IAA but little of gibberellic acid, fixed very little of nitrogen and yet it promoted growth of maize and other crops.

While probing the effect of inoculation by *Azotobacter* on the growth of the plant, it becomes clear that manifestation of such effect begins spontaneously even during the germination of inoculated seeds. The improvement in germination of seeds of cotton, rice, maize, wheat, soybean and bengal gram was confirmed from time to time (Banerjee, 1970; Shende *et al.*, 1977). It can be attributed to the production of growth promoting substances. Relevance of the improved seed germination is of the paramount importance as the inoculation with *Azotoacter* can ensure the maintenance of proper plant stand of field crops per unit area. The stimulatory effect of *Azotobacter* on tomato ad wheat plants was reported by different workers (Jackson *et al.*, 1964, Rovira, 1965, Brown and Burlingham, 1968).

III. Strategy for Selection of *Azotobacter* Strains

It is felt that nitrogen fixing ability of a strain can not be a sole criterion to select it for efficiency because a few low nitrogen fixing strains were also found excellent in field performance than the high nitrogen fixing strains. Therefore, nutritional, growth promoting and therapeutic capabilities of different strains of *A. chroococcum* should be considered.

The strains should be homologous (crop-wise selection) forming large size colonies with brown to black pigments with at least a moderate capacity to fix nitrogen. They should be able to improve seed germination and should establish in the rhizosphere of crop plants and finally they should augment crop productivity under field conditions.

Isolation, Purification and Culturing

Isolation: Representative soil samples are collected and used for preparing serial dilutions. Ten grams sample is suspended in 90ml of sterilized water and serial dilution upto 1:1000 should be prepared.

Pipette out 1 ml of 1:10 to 1: 1000 dilutions of soil suspension and pour on cool Jensen's nitrogen free agar medium in petri plates and incubate at 28°C (±2°C) for 24-48 hours. Streak the individual colony on Jensen's agar medium. Carry out purification and confirmation test by growing the isolates on silica gel plates.

The isolated colony is then selected and rechecked for uniform growth by streak method. The colonies of *Azotobacter chroococcum* vary in size from 0.5 to 1.5 cm in diameter and become brown to black due to pigmentation after 48 hours.

Preparation of Silica Gel for Purification

A 20 per cent solution of water soluble crystalline sodium sillicate is prepared by slow heating. This alkaline solution is cooled and filtered. A known volume of sodium sillicate solution and equal volume of 33 per cent HCL is poured into a conical flask. The mixture thus prepared in small lots, is then poured into sterilized petriplates (12-15cc in 9 cm diameter petriplates) and left undisturbed overnight for gelation to take place which gives a transparent look. These plates are then washed in running tap- water for 48 hours, to get rid off excess acid and chloride ions. The pH of the draining water is tested by methyl red indicator. The gel made free of chlorides by washing it in running water and is tested by $AgNO_3$ solution. The plates are then kept inverted to drain off the water.

The sterilization of plates is done by exposing them to ultra-violet light of 2437 A° wave length at a distance of 70 cm for $^1/_2$ hour in the inoculation chamber. Two ml of sterilized nutrient mixture (Jensen's broth) are poured on the sterilized gel plates and allowed to evaporate by exposure of the plates to dust-free air in the plating room. Afterwards these plates are exposed to UV light for one hour. These plates are ready now to receive *Azotobacter* culture suspension of convenient dilution for purification. Streaking method can also be followed.

Maintenance and Culturing

Azotobacter isolates are maintained on nitrogen free media and periodically sub-cultured. Lypholization of culture helps in preservation of different isolates. The pure and efficient cultures are subjected to vigorous testing in pot and field conditions. If the culture proves beneficial for crops it is then released for mass production of the Inoculant.

Production of *Azotobacter* Inoculants

In India presently carrier based *Azotobacter* inoculants are being produced by Research Institutes, Agricultural Universities and Commercial/ Industrial factories. The carriers such as lignite, charcoal-soil, or pressmud plus charcoal are being used for mass production of inoculants for supply to farmers.

Large scale production of *Azotobacter* culture is done by growing it in large flasks on rotary shakers, or in batch fermentors depending on the scale of production. The media used for cultivation of *Azotobacter* is given in chapter 21.

The powdered carrier is neutralized with Ca CO, and autoclaved. The broth is mixed with the sterilised moist carrier. The mixture is cured in trays for3-5 days, packed in polythene bags and sealed. The cultures can be used for seed or soil inoculation. The procedure for inoculation of seeds are usually supplied with the packet.

IV. Standards for Quality Control of *Azotobacter* Inoculant (Indian Standard (IS-8268: 2001)

1. *Azotobacter* Inoculant (Al) shall be carrier based, the colour depending on carrier.
2. Al shall contain minimum of 10^7 viable cells of *Aztobacter chroococcum* per gram carrier material within 15 days of manufacture and 106 within 15 days before the expiry date marked on the packet: Al colonies are gummy, raised without striations, viscous and often sticky. The pigmentation varies from very light brown to black.
3. Al shall have a maximum of six months expiry date from the date of manufacture.
4. The pH of the carrier medium shall be between 6.5 and 7.5.
5. The maximum amount of nitrogen fixed by *Azotobacter chroococum* shall be not less than 10 mg per gram of sucrose utilised.
6. The carrier material shall be either peat, lignite, peat soil, humus or similar material neutralized by calcium carbonate and sterilized. It shall be in the powder form capable of passing through 10^6-micron IS sieve.
7. Al shall have no contamination with other microorganisms at 10^5 dilution.
8. Al shall be packed in 50 to75 micron low density polythene bags or in any other suitable containers.
9. Each packet shall be marked legibly with pertinent information such as name of the inoculant, manufacturer (name and address), name of the carrier, date of manufacturer, date of expiry, net mass, crops (S) for which intended and storage instructions.
10. Al shall be stored by the manufacturer in a cool place, away from direct heat, at a temperature preferably 15°C and in no case exceeding 30°C.

V. Crop Responses

Mishustin and Shilnikova (1969) summarised the results of the field trials conducted in U.S.S.R. on different crops such as wheat, oats, barely, maize, sugarbeet, cabbage and potato. These results showed that the yield of these crops were increased by 7 to 12 per cent due to *Azotobacter* inoculant (Mishustin and Navmova, 1962).

Several field experiments were conducted in India on different crops such as wheat (Sundra Rao *et* al., 1963 Mehrotra and Lehri 1971), rice (Mehrotra and Lehri, 1971) vegetables such as onion (Joi and Shinde 1976) and brinjal, tomato and cabbage (Mehrotra and Lehri 1971, Lehri and Mehrotra, 1972).

The beneficial effect of *Azotobacter* on crop-plants were also studied by other workers (Dobereiner and Campbelo, 1971a, b, Becking 1971). It is generally agreed that there is certain increase in the yield of crops (Krassilnikov, 1945; Shende, 1972; Badgire and Bindu, 1976; Oblisami *et al.*, 1976; Mishustin and Cherepkov, 1979) which is possible with the establishment of *Azotobacter* in the rhizosphere of various crop-plants like wheat, paddy and maize. Manib *et al.*, (1979a) reported that bacterization of barley by *Azotobacter* increased the yield as well as soil nitrogen through nitrogen fixation. N-balance in soil showed higher gains in inoculated treatments over uninoculated (control).

Manib *et al.* (1979b) grew, barley, wheat, maize, broadbean, fenugreek, lentil and cotton in low nitrogen soils with and without inoculation with a local strain of A. *chroococcum*. Increase in *Azotobacter* population both in the nutrient solution and rhizosplane was maximum for the cereals and minimum for fenugreek and lentil. Plant dry weight was increased by 5 to 12 per cent as a result of inoculation and N-content of plants and nutrient solution was also increased. But Morris *et al.* (1978) did not find any beneficial effect of *Azotobacter* inoculation on com, grain sorghum, cotton, cucumbers, oats and wheat. They attributed the local conditions of soil fertility and environment as main constrains for non -beneficial effect of *Azotobacter* strains.

Azotobacter inoculant increased the growth of rice seedlings (Purushothamon *et al.*, 1971) and paddy yields over NPK applied at the rate of 120, 75 and 25 kg ha^{-1} respectively (Wani *et al.*, 1976). *Azotobacter* inoculant in a field experiment increased the crop yield substantially where paddy straw was incorporated at 5 tonnes ha^{-1}. It could save 25 per cent of fertiliser nitrogen (Kasiranja *et al.*, 1976). Combined application of organic manure and A. *chroococcum* proved better for growth and yield of paddy (Rangaranjan and Muthukrishnan 1976). Rice varieties responded to root inoculation of paddy seedlings with *A. chroococcum* (Rao *et al.*, 1977). The relative effect of A. *chroococcum* and its interactions with nitrogen and phosphorus on the yield of paddy crop were studied by Kundu and Gaur (1984). The results showed significant response of paddy to the inoculant and interaction with nitrogen up to 80kg Nha^{-1}, but beyond 40 kg N ha the response was less marked.

Cotton crop seems to respond to *Azotobacter* inoculant in soil. The yield of seed cotton was increased by 50 per cent after inoculation of *Azotobacter* over no inoculation and no fertiliser treatment. *Azotobacter* inoculation together with 40kg N ha^{-1} was superior to 80kg N ha^{-1} (Kurdikeri and Koraddi 1977). The effect of cotton seed inoculation with A. *chroococcum* was studied in a field trial at three levels of fertility (Kundu and Gaur 1980a). *Azotobacter* inoculation increased the yield of cotton significantly when nitrogen and phosphorus (P_2O) were applied at 40 and 50 kg ha^{-1}, respectively. The combined inoculation of phosphorus solubilising microorganism and *A. chroococcum* proved better in augmenting the yield of cotton.

Agarwal *et al.* (1977) obtained higher yields of sugarcane with *Azotobacter* alongwith phosphate than of control although the culture alone was also effective.

Maize yields were augmented by *Azotobacter* inoculation but nitrogen and *Azotobacter* interaction was not significant (Reddy *et al.* 1977). On the other hand, after a 2 years trial on maize conducted in Kulu valley it was concluded that there was significant interaction between nitrogenous fertiliser and *Azotobacter* (Singh *et al.*, 1977). The maize yield with 60 kg N ha^{-1} plus inoculation was greater than with 60 kg N ha1 alone, Halemani and Kudasmannavar (1979) obtained higher yields of maize on seed treatment with *Azotobacter* with or without 75 kg N ha^{-1}, but found no effect of inoculation with 150 kg N ha^{-1}.

Sanoria and Rao (1975) investigated the effect of bacterisation of sorghum and wheat seed with *A. chroococcum* and noted both positive and negative responses. Likewise, Dhillon *et al.* (1980) reported non- significant responses of wheat to *Azotobacter* trials conducted at Ludhiana.

Kundu and Gaur (1980b) reported that wheat responded to *Azotobacter* inoculant and the combined inoculation with phospho-bacteria, viz. *Pseudomonas striata* and *Bacillus polymyxa*, in both sterilized and unsterilized Delhi alluvial soils; as a result the grain yield, nitrogen and phosphate uptake were increased. The combined inoculation of *Azotobacter* with *B. polymyxa* gave the maximum response.

Iswaran *et al.* (1979) studied the effect of foliar spraying of A. *chroococcum* isolated from water hyacinth on the yield of wheat and rice in plots in Delhi soil and reported higher yields. Nair and Tauro (1979) reported an increase in the yield of wheat grain due to inoculant with Pseudomonas *azotogensis* together with 80 kg N ha^{-1}. Poi and Kabi (1979) observed increased yields of wheat grain due to *Azotobacter* inoculant but the *Azotobacter* strains seemed to be crop-specific.

Inoculation of rice with *Beijerinckia* alone was not effective in increasing the grain yield but bacterisation in combination with 40 kg N ha^{-1} recorded significant increase in yield comparable to that obtained from the application of 80kg N ha1 (Balasundram and Sen, 1971). Inoculation of paddy seedlings with *Beijerincka indica* increased the paddy yields, which was higher than with 50 kg N ha^{-1}. There was no significant interaction between the nitrogen application and the bio-fertiliser. Increased nitrogen fixation and tillering caused greater paddy yields (Jagtap and Bhide 1977).

In soil inoculation tests at various levels of N-fertilization, Badve *et al.* (1976) reported the highest *Azotobacter* population (4.45 x 10^3/g of soil) and highest amount of nitrogen fixed (0.84 g soil) in a treatment receiving 40 kg N/ha, *plus Azotobacter* inoculation. Brown *et* al. (1979) also reported the increase in yield due to *Azotobacter* inoculation at different levels of nitrogen application. Kundu and Gaur (1980b) reported that application of 2 percent FYM stimulated *Azotobacter* population thus resulting in more uptake of nutrients by wheat. Stimulation of *Azotobacter* population by phosphobacteria with the release of hormones has also been reported by Barea *et al.* (1978).

Bhide and Purandare (1979) reported the occurrence of *Azotobacter* in the cortical cells of roots of *Cynodon dactylon* thus paving the way to establish loose symbiosis between *Azotobacter* and the crop.

Field trials were conducted on wheat, maize, sorghum, cotton, potato and mustard plants by Shinde and associates (Shende *et al.*, 1977; Shende and Apte, 1982). The yield increases were varying ranging from 6.7 to 71.7% with failures at some locations (Table 109). The failures were probably due to reason that the same culture was used for different crops. Recent trials indicate that homologous cultures should be used for specific crops and even sometimes for the specific varieties of a crop.

Narula *et al.* (1980) isolated efficient ammonia excreting nitrogen fixing strains of *A. chroococcum* from local soils. A soil isolate -103 with high ARA, marginal ammonia excretion and high hydrogenease activity increased the yield of wheat and barley crops (Sindhu *et al.*, 1994).

Table 109: Effect of *Azotobacter chroococcum* on different crops

Crop	Location	Control (qha^{-1})	*Azotobacter* (qha^{-1})	% increase in yield	CO at 5%
Maize	IARI	7.8	13.4	71.7	4.80
Sorghum	Pali Dharwar	12.8	14.0	9.3	1.22
		23.6	32.6	38.1	10.56
Cotton	Indore Khand-wa Surat	366	401	9.5	104
		559	768	20.6	165
		1254	1339	6.7	241

Many other reports showed marked increases in growth and grain yields of cereals such as wheat, pearl millet, maize, barley, sorghum as well as cotton and mustard crops when they were inoculated with *Azotobacter*, (Jadhav *et al.*,

1991; Lakshminarayana, 1993; Narula and Yadav, 1989, Reddi and Reddy, 1981, Sindhu *et al.*, 1994). Effect of *Azotobacter* on Plantation and Vegetable Crops Oblisami *et al.* (1985) studied the effect of *Azotobacter* inoculant and growth regulators on the growth of cashewnut plants in pots either alone or in combination with GA (180 ppm). *Azotobacter* alone caused better growth of root system and compared with growth regulators. Shoot length was also increased with *Azotobacter* treatment alone. The population of *Azotobacter* was maximum in rhizosphere in treatment with *Azotobacter* alone or *Azotobacter* + GA.

Siddique *et al.* (1993) reported the effect of *Azotobacter* inoculation after one month in mulberry cultivation. The vegetative growth including plant height, number of leaves and their size were increased significantly with *A. chroococcum.*

Improvement in germination, chlorophyll content and dry matter in vegetables like carrot, radish and chilies were improved due to *Azotobacter* inoculation (Khullar and Chahal, 1977, Rao and Sharma, 1981). The yield of green chilies was improved by 10.8 percent and was at par with 60kg N/ ha (Shivappa shetty *et al.*, 1976). Potato whose nutrient requirements are quite high responded to *Azotobacter* treatment. *Azotobacter* strains P-2 and M-4 Increased the yield of potato tubers by 6.4 and 12.2 percent respectively (Shende *et al.*, 1986). Verma and Shinde (1993) reported that vegetable crops in general and potato, onion and brinjal in particular responded well to *Azotobacter* treatment.

Combined Inoculations with *Rhizobium*

Nodulation and yield of soybean was improved due to combined inoculation of seeds with peat based culture of *R. Japonicum*, *Azotobacter chroococcum* and *Beijerinckia indica* (Apte and Iswaran, 1974). Beneficial effects of combined inoculation of *Rhizobium* and *Azotobacter* in improving the nodulation and grain yield of Chickpea (Rawat and Sanoria, 1977; Tosh and Sanoria, 1978), soybean (Sharma and Rao, 1978, Jauhri *et al.*, 1979) and green gram (Jauhri *et al.*, 1979) have been recorded, Gaur and Rai (Unpublished) studied the associative effect of *Azotobacter* with *Rhizobium* on the yield of mungbean (*Vigna radiata*) under field condition (Table 110). The maximum yield was recorded in treatment incorporating 20 kg N/ha followed in order by treatment receiving 10 kg N/ha and combined inoculation and treatment incorporation 10 kg N/ha with *Rhizobium* inoculation. But non-significant difference was found among these treatments. Likewise nitrogen uptake by the crop was also influenced. *Rhizobium* or *Azotobacter* inoculation alone contributed about 10kg. N/ha. However, negative effects of combined inoculation of *Azotobacter* and *Rhizobium* on nodulation and yield of chickpea was reported (Paul and Verma, 1999).

Table 110: Associative effect of *Rhizobium* and *Azotobacter* on the yield of mungbean (*Vigna radiata*) Grain

Treatments	Grain yield (q/ha)	Straw yield (q/ha)	Husk yield (q/ha)	Total N Uptake (kg/ha)
Control	4.21	6.67	1.56	26.66
10 kgN	6.13	9.67	2.64	40.89
10 kg N+*Rhizobium*	7.83	11.98	3.60	53.65
10 kg N+*Azotobacter*	6.36	9.75	2.83	42.44
10 kg N+*Rhizobium* + *Azotobacter*	8.25	12.33	3.75	55.20
20 kgN	8.55	12.49	3.85	54.75
C.D. at 5%	1.04	1.41	0.62	6.89

Other Nitrogenous Biofertilizers

Some Plant-bacterial Interactions

Although a number of different heterotrophic N, fixing bacteria have been isolated from the roots and rhizosphere of various gramineae species but only a few examples of predominant bacteria in the root or rhizosphere have been shown. *Paspalum notatum -Azotobacter paspali* is one of the examples of such specific interaction (Dobereiner, 1970; Dobereiner and Day, 1975).

Larson and Neal (1971) found that infection of roots by a nitrogen fixing *Bacillus species* was far more successful in case of certain root rot- resistant lines of spring wheat than other cultivars. Bacteria of the genus *Campylobacter* seem to be the organisms associated with *Spartina alterniflora* for most of their nitrogenase activity (Partiquin *et al.*, 1981). Watanabe and Barraquio (1979) reported evidence of association of *Achromobacter* with roots of low land rice. Moreover, the occurrence of *Bacillus* spp. and various members of the Enterobacteraceae have been shown to be of numerical importance in roots and rhizosphere of sugarcane (Ruschel, 1981).

Proteus vulgaris/Klebsiella planticola/Bacillus subtilis

Few bacteria were isolated from rhizosphere soil, which promoted the growth of sunflower. The organisms were characterised and identified as *Proteus vulgaris, Klebsiella planticola* and *Bacillu subtilis*.

Plant growth promoting potential of these rhizobacteria was evaluated by inoculating sunflower CV-OSH 1001 with carrier based inoculant singly and in combination (Malik, Gaur and Kavimandan, 1995). Sunflower varieties grown in pots without any addition of chemical fertilizer responded to seed inoculation. Increase in diameter of flower-head, plant height and seed yield was recorded. Mixed inoculation in this case was found to be more effective in enhancing growth of sunflower. The increase in seed yield of sunflower cv. OSH 1001, ranged from 49-62% over uninoculated plants grown in field. Significant improvements observed due to bacterisation with nitrogen fixing (strain A, B) and P-solubilizing bacteria (strain C), in absence of any added fertilizer, indicated the possibility of evolving these ecofriently inputs for the marginal farmers which could be alternatively used as plant nutrients.

Another variety of Sunflower cv. DSH1 was inoculated with *P. vulgaris, B. subtilis* and *Klebiella* sp. and grown in field condition. No inorganic fertilizer were applied. Increase in seed yield of sunflower varied from 34% due to inoculation with *B. subtilis* to about 56% due to bacterization with *P. vulgaris* over control plot (Table 111). Increases in plant height, diameter and weight of flower heads were also observed due to bacterisation (Malik *et al.*, 1996)

In a similarly conducted pot trial, Ganga, a cultivar of maize and Pusa 31, a cultivar of cotton were grown with seed inoculation with the 3 above-mentioned cultures, singly and in combination. Gain yield of maize increased more than 100% due to bacterisation as compared to uninoculated control (Table 111). Co-inoculation with *P. vulgaris* and *B. subtilis,* and *K. planticola* and *B. subtitlis* were relatively less effective in increasing grain yield of maize as compared to single inoculations.

Bacterial inoculation resulted in increased plant height, number of bolls/plant and fibre and seed yield of cotton crop. Fibre yield was enhanced by 76 to 174% due to inoculation over uninoculated control where as seed yield was augmented by 80 to 129%.

The results obtained with these experiments suggest a distinct possibility of increasing yield of sunflower, maize and cotton by using suitable bacterial inoculants.

Table 111: Effect of bacterial inoculation on yield of crops

Treatment	Sunflower	Maize	Cotton	
			Seed	Fibre
Uninoculated control	100	100	100	100
Proteus vulgaris	156	259	229	253
Klebsiella planticola	152	238	186	195
Bacillus subtilis	134	238	180	176
P.vulgaris + B. subtilis	-	209	229	274
K. Planticola+ B. Subtilis	-	230	204	229

Relative average yield/plant

Singh and Panwar (1997) investigated the effect of different nitrogen fixing bacteria *(Proteus vulgaris, Klebsiella planticola, Azospirillum brasilense)* and phosphorus solubilizing bacteria *(Bacillus subtilis)* on wheat C.V. HD 2428 under field conditions. The inoculated plants showed larger leaf area, higher chlorophyll content, nitrate reductase activity, total biomass production and grain yield as compared to uninoculated control. The grain and straw analysis showed that the P content was considerably higher with *Bacillus subtilis* whereas N content was highest in plants inoculated with Azospirilum *brasilense*. The grain yield was higher as a result of more effective ears, grain number/ ear and seed size in the inoculated plants over uninoculated control (Table 112).

Table 112: Yield components as affected by nitrogen fixing and phosphorus solubilising bacteria in wheat.

Treatments	Ear No. perm^2	Ear weight (g/m^2)	1000 seed weight (9)	Grain No. percent	Grain weight (g/m^2)
Control	470	760.5	39.0	37.8	562
Azospirillum	565	858.0	42.4	39.0	630
Bacillus subtilis	581	875.5	43.0	42.2	643
Klebsiella planticola	485	810.0	41.4	38.4	620
Proteus vulgaris	505	855.5	42.1	39.7	630
CD AT 5%	84	80.1	2.4	NS	56

The influence of *Kurthia* sp,. *Bacillus sphericus*, *Proteus vulgaris* and *Klebsiella planticola* alone and mixed inoculants was examined on the growth and seed yield of rapeseed cv Toria-9 in pot culture and compared with application of inorganic nitrogen and phosphorus (Malik and Kavimandan 1999). Bacteria inoculated rapeseed registered greater plant height, plant biomass and seed yield as compared to control and many of the bacterial treatments were comparable to the yield obtained with the use of chemical fertilizers. Mixed inoculants comprising of *Klebsiella planticola* and *Bacillus subtilis* and *Proteus vulgaris* and *B. subtilis* were the most effective in enhancing the productivity of rapeseed. This is in confirmity with current views on the role of microbial consortia for enhancing plant nutrition and crop growth.

Acetobacter diazotrophicus is a saccharophilic bacterium and is associated with plants such as sugarcane, sweet potato and sweet sorghum. These bacteria can be multiplied in nitrogen free malate medium and mixed with carrier material and used for inoculating sugarcane crop.

The roots, stems, leaves and trash of Brazilian sugarcane varieties contain substantial numbers (upto 10^7 g-1 fresh weight) of diazotrophic bacteria such as *Glucon Acetobacter diazotrophicus* (Yamada *et al.*, 1997, 1998; Franke *et al.* 1999). However, the specific microoranisms responsible has yet to be determined (James 2000).

There is evidence to show that *Acetobacter diazotrophicus* is a major contributor to endophyte biological nitrogen fixation, which is based on N balance studies showing as high as 150kg N/ha/yr in sugarcane (Dobereiner *et al.,* 1993).

Azoarcus is another N2 fixing endophyte found in roots of kallar grass *(Leptochloa fusca)* which yields 20-40 tonnes of hay/ha/yr without any addition of chemical fertilizer in saline-sodic and alkaline soils having low fertility (Sandhu and Malik, 1975).

Recent studies have demonstrated the endophytic colonization of rice roots by other nitrogen fixing bacteria *Alcaligens faecalis* (Zhou and you 1988, you and song 1994). Hurek *et al.* (1994) showed that Azocarcus BH 72 has ability to invade and colonize rice and able to fix nitrogen. *Herbaspirillum* have also been isolated form stems and leaves of rice (Baldani *et al.*, 1992). According to Dobereiner *et al.* (1995) and Boddey *et al.* (2000) sugarcane plants contain substantial numbers of *Herbaspirillum* spp.

References

Abd-el-malek, y.1971. Pl. Soil (sp.Vol.) 423-442.

Agarwal, M.L., Shishodia, O.P.S. Khan, Z.A. and Dayal R.1977. Indian Sugar Crops. 4: 66.

Alexander, M. 1977. Introduction to Soil Microbiology, John Wiley and Sons, N.Y.

Apte, R. and Iswaran V. 1974. Proc. Natn. Sci Acad. 40B: 482-485. Bachanan, R.E. and Gibbons, N.E. 1974. Bergey's Manual of Determinative Bacteriology. 8th ed. The William and Wilkins Co. Baltimore, P 1268.

Badgire, D.R. and Bindu, K.J. 1976. Madras Agric. J. 63: 603-605.

Badve, D., Konde, B.K.and More,B.B.1976. Food farming and Agriculture 8:23-24.

Balasundram, V.R. and Sen, A. 1971. Ind J. Agric. Sci 41: 700-704.

Baldani, VLD, James, E, Baldani JI and Dobereiner, J. 1992. An Acad. Bras. 64: 431.

Banerjee, S.1976. Morphological and physiological characteristics of Aztobacter isolated from soils of Nainital Tarai. M. Sc. Thesis. Pant University of Agriculture and Tech. Pant Nagar (Nainital) India.

Barea, J.M. Ocampo, J.A.Azcon, R., Olivares, I. and Montoya 1978. Ecol. Bull. (Stockholm) No. 26: 325-330.

Becking, J. H. 1971. Proc. Res. Coor. Meeting. Sofia Intern. Atomic Energy Agency, Vienna, pp 189-222.

Beijerinck, M.W. 1901. Zentralblatt fur Bakteriology, Abt II, 7:561.

Bhardwaj,K.R and Gaur, AC. 1970, Folia Microbiol. 15: 364-367

Bhide, V.P.and purandre, A.G. 1979. Curr. Sci. 48: 913-914.

Boddey, R.M., desilva, L.G., Reis, V.M., Alves, B.J.R. and Urquiaga, S. 2000. Assessment of bacterial nitrogen fixation in grass spp. In: Prokaryotic Nitrogen Fixation: a model system for analysis of a biological process (ed.) Triplet, E.W. Horizon Scientific Press, U.K. pp 705-726.

Darznik, Y.O. 1961, Microbiologiya 30: 1042-1044.

Dart, P.J. and Day, J.M. 1972. Non-symbiotic Nitrogen Fixation in soil In: Soil Microbiology (ed: N. Walker) 225-252. Butterworth, Lond.

Dhillon, G.S., Kler, D.S. and Chahal, V.P.S. 1980. Ind. J. Agron. 25 : 533.

Dobereiner, J 1970. Zentralblatt Fur Bakteriology 124: 224 0-230.

Dobereiner, J.and Campbelo A.B.1971 a. Soil Biol. Biochem. 6: 107-112.

Dobereiner, J.and Compbelo A.B. 1971b Pl.Soil (Sp. issue) 457- 470.

Dobereiner, J. Reis, V.M. Paula M. A. and Olivares, F. 1993. In: New Horizons in Nitrogen Fixation. (Eds. Palacios, R., Mora, J. and Newton, W.E.) Kluver Academic Publishers. The Netherland, p61.

Dobereiner, J. Baldani, V.L.D. and Reis, V.M. 1995 In: Azospirillum VI and related microorganisms (eds.) Fendrik, M. delgallo *et al.* Heidelberg: Springer-Verlag, Berlin, pp 3-14.

Doreiner, J. and Day. J.M. 1974. Symposium on Nitrogen Fixation by Free-living Microorganisms (Ed. Stewart, W.O.P) Cambridge University press, Cambridge, U.K. Pp. 39-56.

Erofeev, N.S. and Vastov, G.S. 1964. Irestiya Akad. Nauk, USSR, Ser. Biol. 5: 668.

Franke, I.H. Fegan, M., Hayward, G: Lenord E, Stackebrandt, E. and Sly, L.I. 1999. Int. J. Sys. Bacteriol 49: 1681-1693.

Gaur; A.C. and Bhardwaj, K.K.K. 1971. Pl. Soil 35: 613-621.

Gaur A.C. and Misra, K.C. 1978. Zbl. Bakt. Abt 11, 33: 357-361 Gaur, A.C. and Mathur, R.S. 1964. Sci Cult. 32:319.

Gaur A.C. Mathur, R.S. and Kavimandan, S.K. 1968-69 Agrochemica 13:81-84.

Gaur, A.C. Sadasivam, K.V. Kavimandan, S.K. Mathur, R.S. and Vimal, O.P. 1975 Sci and Cult. 41: 136-138.

Gaur A.C. and Rai, S.N. 1982. Paper Presented in Natl. Symp. On Facets of Indian Chemistry in 2000 AD. University of Allahadad Symposium Handbook pp. 28-29.

Halemani, H.L. and Kudannavar, B.T. 1979.Curr. Res. 8:172.

Hofer, A.W. 1944. J. Bacteriol 48: 497-701.

Hurek, T., Rheinhold- Hurek B., Von Montagu, M and Kollenber, E 1994. J. bacteriol. 176: 1913.

Iswaran, V., Bhatnagar, R.S. Jauhri, K.S. and Sen, A. 1979. Ind. J. Microbiol. 19: 83-84.

Jackson, R.M. Brown, M.E. and Burlingham. S.K. 1964. Nature, London 203:851-852.

Jagtap, V.D. and Bhide, V.P. 1977 J. Maharashtra Agric. University. 2:243.

Jagtap, V.D. and Bhide, V.p. 1977Malatma Phule Agric University, 2:61-74.

Jadhav, V. Shaikh, A.A. and Harinarayana, G. 1991: Indian J. Agric, Sic. 61: 268-71.

James, E. 1974. Strain Variation in *Azotobacter* chroococcum. M.Sc. Thesis, P.G. School, IARI, New Delhi.

James, E.K. 2000. Field Crops Res., 65: 197-209

Jauhri, K.S. Bhantnagar, R.S. and Iswaran, V. 1979. Pl. Soil 15 : 105-108.

Jensen, V. 1940 Proc. Linn. Soc. New South Wales 65: 543-556. Joi, M.B. and Shinde, P.A.1976. Jour Maharashtra Agricultural University 1: 161-162.

Khullar, S. and Chahl, V.P.S. 1977. Ind J. Microbiol, 17:45. Kluyver, A,V. and Becking, J.H. 1955. Drax- Sauomatai Tieteaket Toimitusk.Ser A. 11:367.

Knowles, R. 1977. In: A treatise on Dinitrogen Fixation, Sec IV (Eds, R.W.F. Hardy and A. H. Gibson) 33-83. John Wiley and Sons, N.Y. Krasilnikov, N.A. 1945. Microbiological Principles of Bacterial Fertilizers Izd. AN. USSR, Academy of Sciences of USSR, Moscow.

Kundu, B. S. and Gaur, A. C. 1984. Pl. Soil 79: 227-234.

Kundu, B.S. and Gaur, A.C. 1980a Ind J. Microbiol. 20:225-229.

Kundu, B.S. and Gaur, A.C. 1980b. Pl. Soil 57: 223-230.

Kundu, B.S. and Gaur, A.C. 1980c Curr. Sci. 48: 159.

Kurdikori. S. and Koradde, V. R. 1977.Curr. Res. 6:33.

Lakshminarayan, K. 1993. Proc. Indian Natl. Sci. Acd. 59 B: 303- 308.

Larson, R. I and Neal, J. L. 1971. Bull Ecol. Stockholm 26: 331- 341.

Lautrop, H. 1984. Bergey's Manual Systematic Bacteriology. Vol 1. (Eds. N. R. Krieg and J. H. Holt.) Williams and Wilkins Co. Baltimore, USA, p 491.

Lehri, L. K. and Mehrotra, C. L. 1972. Indian J. Agric. Res. 9 (3): 301-304.

Loptina, G. V. 1949. Proc. All Union Scientific Res. Inst. of Agri. Microbiology, Selhozgic, 108.

Malik, R. D. S. and Kavimandan, S. K. and Tilak, K. V. B. R. 1996, Indian Fmg. (Feb. Issue) 15.

Malik, R. D. S., Gaur, A. C. and Kavimandan, 1995. Indian J. Exptl. Biol. 33:311-312.

Mathur, R. S. and Gaur, A. C. 1977.. Mikrobiologiya 46: 149-154.

Mehrotra, C. L. and Lehri, L. K. 1971. J. Indian Soc. Soil Sci. 19: 243-248.

Misustin, E. N. 1970. Pl. Soil 32: 545-554.

Mishustin, E. N. and Naumora, A. N. 1962. Mikrobiologiya 31: 543- 555.

Mishustin, E. N. and Cherepkov, M.I. 1979. Izyestiya. Akad. Nauk, USSR. Biolochskaya 5: 656-678.

Mishustin, E. N. and Shilnikova, V. K. 1969. In Soil Biol. Rev. Res. UNESCO, PARIS pp. 72-124.

Monib, M., Abdel-el-Melek, Y., Hosney, I. and Fayex, M. 1979a. Zbl. Bakt. Abt II 134: 140-148.

Monib, M., Abdel-el-Malek, Y, Hosney, I and Fayex, M. 1979b. Zbl. Bkt. Abt II, 134: 149-155.

Morris, H. D., Gidden, J, and Brown, A. R. 1978. Georgia Agric. Res. 19: 1-9.

Mulder, E. G. 1939. Over der bietekens vonkoper voor de groci von planten eu Microogranisms. Thesis Wageningen, Netherland.

Narula, N. and Yadava, K. S. 1989. In: Bilogical Nitrogen Fixation Research Status in India. (Eds. Dadarwal, K. S. and Yadav, K. S.). The Society for Pl. Physiol & Biochemistry, New Delhi pp 87-124. Narula, N., Lakshminaryana, K. and Tauro P. 1980. Biotechnol. Bioeng. 23: 467-470.

Nepomiluev, V. F. and Shishkov, L. L. 1962. Microbiologiya 31 : 294-300.

Oblisami, G., Natrajan, T. and Balaraman, K. 1976. Madras Agric. J. 63: 590-594.

Oblisami, G., Pappiah, C. M. and Shamugavelu, K. G. 1985. Acta Hort. No. 108:40-95.

Pandey, A. and Kumar, S. 1989. J. Scien. Indus, Res. 48: 134-144

Patil, P. L. and Medhane, N. S. 1974. Pl. Soil; 40: 221-223.

Patil, R. B. 1967. Biochem. Biophy. Acta 136: 1-5.

Patriquin, D. G., Boyl, C. D., Livingstone, D. C., and Mc-Clung, C. R. 1981. Associative Nitrogen Fixation (Eds, P.B. Vose and A. P. Ruschal) CRC Press, Palm Beach, Florida, USA.

Paul, S. and Verma, O. P. 1999. Indian J. Microbiol 39: 249-251.

Pois, S. C. and Kabi, M. C. 1979. Ind. J. Agric. Sci. 49: 478. Purushotamon, D., Kasirajan, C. and Rangaswami, G. 1976. Curr. Sci. 45: 807.

Rangaswamy, G. and Sadasivam, K. V. 1964. J. Indian Soc. Soil Sci. 12: 43-49.

Rangrajan, M. and Muthukrishnana, P. 1976. Madras Agric. J. 63: 611.

Rao, A. V. and Sharma, R. L. 1981. Indian J. Microbiol. 21: 25-27.

Rao, L. J., Venkatachari, A., Rao, W. V. B. S. and Reddy, K. R. 1977. Curr. Sci. 46:50.

Rawat, A. K. and Sanoria, C. L. 1977. Curr. Sci. 46: 373.

Reddy, M. R. and Reddy, G. B. 1981. Indian J. Agron. 26: 408-412. Rovira, A. D. 1965. Effect of *Azotobacter* on wheat. In: Plant - Microbe Relationships (Eds Vancura V and Macura J.) Czechoslovakia Acad. Sci., Prague. pp 193-200.

Ruschel, A. P. 1981. Biological Nitrogen Fixation in Sugarcane, In: Intl. Workshop on Biological Nitrogen Fixation Technology for Tropical Agri. March, 1981, Cale. Columbia.

Sandhu, G. R. and Malik, K. A. 1975. Nucleus (Karachi) 12:35.

Sanoria, C. L. and Rao, W.V.B.S. 1975. Indian J. Agric. Sci. 45:224.

Sasson, A. and Daste, P. 1963. C. R. Acad. Sci. 257: 3516-3518.

Satyavir Singh and Lakshminarayana, K. 1982. Pl. Soil 69: 79-84.

Sen, A. and Barooh, P P. 1964. Application of *Azotobacter* and Beijerinckia in Indian soils Bull. Natl. inst. Soc. India. 26: 227-237.

Sharma, R. L. and Rao, A. V. 1978. Fertil Technol. 15:50.

Shende, S. T. 1972. Natn. Acad. Sci. India. Allahabad (Sp. Vol.) 191-192.

Shende, S. T. and Apte, R. G. 1982. *Azotobacter* Inoculation as remunerative input for agricultural crops. Proc. Natl. Symposium on Biological Nitrogen Fixation, New Delhi, India pp. 532-543.

Shende, S. T., Apte, R. G. and Singh, T. 1977. Indian J. Genetic Pl. Breeding 35: 314.

Shende, S. T., Verma, O. P. and Bisht, N. S. 1986. Seed and Farms, 12:41-43.

Shivappashetty, K. S., Khan, A. N. A., Subramanyan, M. B. and Patil, R. B. 1976. Madras Agri. J. 63: 606-610.

Siddique, S. A., Anamika, Nakul and Shukla, N. P. 1993. Indian Forester 199: 559-563.

Sindhu, S. S., Lakshminarayana, K. and Singh, D. 1994. Ind. J.

Expt. Biol. 32: 423-426.

Singh, C.M., Sood, B. R. and Modgal, S. C. 1977. Food Farming and Agri. 8:1.

Singh, O.and Panwar, J. D. S. 1997. Indian J. Plant Physiol, 21: 211-213.

Sundara Rao, W. V. B. Mann, H. S., Pal, N. B. and Mathur, S. P. 1963. Indian J. Agric. Sci. 33: 279-290.

Sundara, Rao, W. V. B. and Iswaran, V. 1959. J. Indian Soc. Soil Sci. 7: 91-95.

Tosh, B. N. and Sanoria, C.L. 1978. Fertil. Technolg. 15: 35.

Vancura, V., and Macura, J. 1961. Folia Microbiol 6: 250-259.

Verma, O. P. and Shende, S. T. 1993. Biofertilizer Newsletter 1 :6- 10.

Wani, S. P, Shinde and Konde, B. K. 1976. Curr. Res. 5 (12): 209- 210.

Watnabe, I. and Barraquios, W. L. 1979. Nature. 227: 565-566.

Yamada, Y.K. Hoshino, K. and Ishikawa, T. 1997. Bio.Sci. Biotech. Biochem. 61: 1244-1251.

Yamada, Y. Hoshino, K. and Ishikawa, T. 1998. Int. J. Syst. Bacterial 48:327-328

You, C. B. and Song, W. 1994. In: Biological Nitrogen Fixation (Eds: Yanfu, N., Kennedy, I. R. and Tingwei, C.) Qindaga Ocean University Press, Qindaga, China 7.

Zhou, F. Y. and You, C. B. 1988. Scientific Agricultura Sinica 21: 7.

Zinoveva, Kh. G. 1962. *Azotobacter* in Agriculture, Kieve.

16

Azospirillum as Biofertilizer

The accumulation of nitrogen under non-leguminous and cereal cropping systems seem to be generally universal at least in the tropics. Free-living heterotrophic nitrogen fixing bacteria are very widely distributed taxonomically, geographically and ecologically. But only a few genera and species are known to possess the potential for high amount of nitrogen fixation. *Azospirillum* has shown high nitrogen fixation attributes. The association of *Spirillum lipoperum (Azospirillum brasilense*) and *Digitaria decumbens* was reported by Dobereiner and Day (1974).

I. Spirillum

The description of members of genus *Spirillum* was available as early as 1670s by Leeuwenhoek (Dobell 1960 and Muller 1773). But the genus *Spirillurn* was created in 1832 by distinguishing it from phototrophic spirilla which were placed in a separate genus *Thiospirillum* (Ehrenberg, 1838). The subsequent developments of the taxonomy of the chemo- heterotrophic spirilla which generated knowledge of their characteristics, were extensively reviewed by Giesberger (1936) and Williams (1959 a, b).

An important observation from Giesberger's use of defined media for nutritional studies of *Spirilla* was their apparent preference for calcium salts of organic acids rather than the sodium salts. Although he considered that calcium was required for proper balance of the media, further studies (Hylemon et *al.*, 1973) indicated that fresh water spirilla have a low tolerance to sodium ions rather than a requirement for calcium ions.

A firm base was provided by Giesberger's report for the study and classification of the genus, which was poorly described although widely distributed in nature. *Spirillum* species were Gram negative and catalase positive and used to possess biopolar tufts of flagella. The organisms were aerobic and chemo-heterotrophic, which oxidized rather than fermented various organic substances, were able to grow in simple defined media and contained 'volutin granules. Martinez (1963) conclusively proved that volutin granules are, in fact, ß-Polyhdroxy-butarate (PHB) rather than poly- phosphate.

A comprehensive study of the genus *Spirillum* was made by Williams and Rottenberg (1957). As a result, considerable amount of new information appeared such as several species of *Spirillum* seemed to possess only a single flagellum at each pole when flagelia stains were examined. However, subsequent electron microscopy revealed that tufts of flagella were, in fact, present in these species. The individual flagellum has a tendency to aggregate into single fassicles that appeared as a single flagellum by ordinary staining (Williams, 1960; Williams and Chapman 1961). They also noted that all of their strains were catalase positive, though Clayton and Beston (1955) found a negative reaction for an organism called *Spirillum maricuniense.*

Isolation: Beijerinck (1922) observed in enrichment culture of *Azotobacter* chroococcum, the organism initially named *Azotobacter spirillum* was isolated in 1925. Although the spirillum fixed nitrogen in mixed culture, but not in pure cultures. In general, the cells were rod shaped or slightly curved, but on dilute bouillon, agar, they were definitely spiral. The cells possessed lipid droplets and polybeta-hydroxy butyrate (PHB) granules *A. brasilense* can accumulate upto 70% of its dry weight as PHB and fat droplets (Okon and Itzigsohn 1992). Calcium malate or lactate supported the growth but Beijerinck (1925) did not state whether sugar could be used by pure cultures. He considered S. *lipoferum* to be a connecting link between *Azotobacter* and *Spirillum* even suggested for including *Azotobacter* in the family *Spirillaceae.* Later studies by Schroder (1932) also failed to demonstrate nitrogen fixation in pure culture. Peptones and ammonium salts were necessary for growth and variety of sugar and organic acids could be used as carbon sources. Enrichment culture techniques using semi-solid malate medium is used to isolate the organism where it forms subsurface pellicle. A wide array of colonies on the petri-plates indicates that there may be different bacterial forms capable of growing in absence of combined nitrogen (Dart and Wani 1982). Several aerobic *Spirilla* that apparently could fix nitrogen were isolated by Russian investigators during 1956 to 1965 (Dalton, 1974) but confirmation by the use of $^{15}N_2$ was not done.

Rediscovery: *Spirillum lipoferum* was forgotten until 1963, when Becking isolated Vibrio or *Spirillum* that assimilated 15N2. The organisms could not be grown in a completely nitrogen-free medium, however, required 0.01 to 0.05 percent yeast extract. In broth cultures, the organism became localized below, rather than at the surface of the medium. In general, the cells were only slightly curved, but in the peptone media they were spirals. Many intracellular PHB granules were present, and the cells possessed a single polar flagellum.

II. Taxonomy

The composition of deoxyribonucleic acid (DNA) bases of 39 strains of *Spirillum* were determined by Hylemon *et al.* (1973). An uniform methodology provided a basis for comparison of the strains. The results together with previous serological data and with previous results for two strains of *Spirillum* volutans indicated that the present genus *Spirillum* should be divided into three genera, with the original name being restricted to obligatory micro-aerophilic fresh water forms having a large cell diameter and a DNA base composition of 36 to 38 mol percent guanine plus cytosine (G+C). This genus contained only one species, *Spirillum volutans*. The generic name *Aquaspirillum* was proposed for the aerobic, fresh water forms having a DNA base composition of 49 to 65 mol% G+C. Thirteen species are included in this genus The third generic name was proposed for marine forms which did not attack carbohydrates and which possess a DNA base composition of 42 to 48 mol% G+C.

It was not clear whether *S. lipoferum* is closely related to *Aquaspirillum*, Vibrio or to members of *Azotobacter*iaceae i.e., Azomonas, *Azotobacter*, *Beijerinckia* or *Derxía* (Krieg, 1976). By then the type of flagellation was not conclusively determined, the type of motility exhibited by strains was indicative of polar flagellation which was more typical of vibrios than spirilla. Further more, Dobereiner and Day (1974) had indicated that their strains were catalase positive, indole negative and unable to acidify sugar media. These characteristics were typical of the genus *Aquaspirillum*. The possession of nitrogenase activity of *S. lipoferum* does not appear to preclude its inclusion in the genus *Aquaspirillum* since *A. peregrinum* can reduce acetylene when cultured by the methods of Dobereiner and Day (Strength *et al.*, 1976). The possibility that *S. lipoferum* might be closely related to polar flagellated members of *Azotobacteriaceae*, as for example, *Azomonas insignis* and *Azomonas* microcytogenes (Baillie et *al.*, 1962) should also be considered.

Tarrand *et al.* (1978) conclusively established that *S. lipoferum* does not belong either to Vibrio, *Pseudomonas*, *Aquaspirillum*, *Spirillum* or to *Derxia gummosa*. The organisms were included in the new genus, *Azospirillum* which retains the term spirillum in new generic name to minimize confusion.

The new genus *Azospirillum* was further divided into two distinct groups. A. *brasilense* (Group I) and *A. lipoferum* (Group II). The Group II was distinguished from group I by its ability to use glucose as a sole carbon source for growth in nitrogen free medium, by their production of an acid reaction in a peptone based glucose medium, and due to its requirement of biotin and by their formation of wide, longer-S-shaped or helical cells in semi-solid nitrogen free malate medium, Strains belonging to group I and Group II were named as

A. *brasilense* sp. Nov. and *A. lipoferum* (Beijerinck) comb. Nov. respectively. Morphological, physiological and nutritional characteristics of different strains of *Spirillum* and *S. lipoferum* were studies by *Hylemon et al.* (1973), Okon *et al.* (1976 a, b) and Tarrand *et al.* (1978). But immunofluoresence and immunodifusion studies have also indicated antigenic difference (DePolli *et al.* 1980). *A. brasilense* contains two subgroup nir+ and nir- based on ability to denitrify, nitrate (Neyra *et al.*, 1977) Both groups possess assimilatory and dissimilatory nitrate reductase but *A. brasilense* nir- does dissimilate nitrate (Magalhaes *et al.,* 1978).

All strains of Hylemon *et al.* (1973) gave catalase positive test but negative test for catalase was also noted by Tarrand *et al.* (1978), with their five strains of *Azospirillum lipoferum.* The catalase reaction was divided by Tarrand *et al.* (1978) into three categories i.e. strong, weak and negative.

Ali *Spirillum* strains were negative for indole and HS production, penctinase and cellulase activity, hydrolysis of starch and gelatin, methyl red test and Voges Proskauer test. Some of the strains turned litmus milk alkaline (Giesberger, 1936; Williams and Rottenberg, 1957, Hylemon *et al.*, 1973, Krieg, 1976, Tarrand *et al.,* 1978). Sizes of all colonies ranged from 0.5 to 1.5 mm in diameter except for six colonies which formed pin- point. All colonies were white, circular and convex with smooth edges, except for two strains, which formed white colonies with fimbriated edges. The fimbriated colonies were formed partially below surface (Hylemon *et al.,* 1973). Okon *et al.* (1976 a) noted that colonies of *S. lipoferum* tend to develop a light pink pigment on nutrient or potato agar. However, certain strains formed very deep pink colonies on MPSS agar. The reference strain sp. 7 in which no such deep pink colonies have been found exhibited nearly 100 percent homology with deep pink strains isolated from *Cynodon lactylon* (Eskew *et al.*, 1977). The pigments may be probably due to b-orc- type cytochrome.

III. Occurrence and Geographic Distribution

Initial observation of Dobereiner and Day (1974) gave the indication that *Azospirillum* is confined to the root system of those tropical grasses where a C_4 (Hatch and Slack) pathway is operative. However, *Azospirillum* has been isolated from soils and roots of a variety of plants both from temperate and tropical regions (Dobereiner *et al.*, 1976, Okon *et al.* 1977, Von Bulow and Dobreiner, 1975).

Besides C_4 plants, several C_3 plants including weeds were investigated and an abundant distribution of the organisms in the roots was recorded. Closely associated with *Azospirillum* are some coccoid bacteria and protozoans. By repeated enrichment in the semisolid medium pure culture can be obtained.

Dobereiner et *al.* (1976) reported that less than 10 percent of soil and roots from temperate regions contain *Azospirillum* where as more than 50 percent of tropical samples were positive. Alluvial soils were more favourable for them than eroded hilly soils. Although pH of the soil had a strong influence on the distribution of *Azospirillum,* sporadic occurrences could be demonstrated in soils having pH as low as 4.

Azospirillum was also isolated from tropical grasses and xerophytes growing in arid zones of India (Rao and Ventateswarlu 1982 a,b). Most of these grasses belong to C_4 type and xerophytes (CAM type) have a high level of water use efficiency (Sprent, 1985). Most isolates of *Azospirillum brasilense* were tolerant to chloride and sulphates but sensitive to bicarbonates (Rao and Venkateswarlu, 1985).

A new nitrogen fixing bacterium, *Herbaspirillum*, taxonomically related to *Azospirillum* was isolated from Brazil (Baldani *et al.*, 1986). It was found to be associated with grasses.

A new species *Azospirillum amazonense* was reported during 1983. It forms pink coloured colonies on potato agar (Kreig and Dobereiner, 1984) shows absence of lateral flagella and its ability to use sucrose as a sole carbon source for nitrogen fixation (Megalhaes *et al.,* 1983). Another specie, *Azospirillum seropedicae* was described (Baldani *et al.,* 1984). Their cells are smaller in diameter compared to other azospirilla, have 1-3 flagella at one or both the poles forming small smooth, green-centred colonies on N- free medium containing yeast extract (Kreig and Dobereiner, 1984).

IV. Nitrogen Fixation

Several aspects of nitrogen fixation and procedure for isolation and purification of *S. lipoferum* have been describe (Von Bulow and Dobereiner, 1975: Day and Dobereiner, 1976). The process of assimilatory and dissimilatory nitrate reduction in other bacteria is well known (Breed *et al.*, 1975; Burns and Hardy, 1975; Hewit, 1975). But Neyra *et al.*, (1977), for the first time reported that some of the strains of nitrogen fixing *S. lipoferum* were able to bring about denitrification. They further reported that denitrification and C_2H_2 reduction may occur simultaneously under certain conditions and in an unshaken semi-solid agar culture in air. Neyra and Burkum (1977) demonstrated that dissimilatory nitrate reductase occurs at membrane level. Better understanding of these transformations is necessary to assess the economic importance of these organisms.

S. lipoferum grows vigorously on malate, succinate, lactate or pyruvate, moderately on galactose or acetate and poorly on glucose and citrate (Okon *et al.* 1976 a, b). It reduces $^{15}N_2$ and actetylene, reduction rates decrease rapidly when pH of the culture rises above 7.8. The organisms are highly aerobic when grown in ammonia solutions. However, *S. lipoferum* reduces N_2 well only under microaerophilic condition. The optimum pressures of oxygen for acetylene reduction by a stagnant culture is 0.006 to 0.02 atm. depending upon the cell density. Shaking *S. lipoferum* with air temporarily inactivates its nitrogenase and reactivation is inhibited by chloramphenicol. The organism assimilated 20 to 24 mg of N/g of organic acid oxidized during growth. Saxena (1983) isolated *Azospirillum* spp. from roots and rhizosphere soil samples of sugarcane, wheat, maize, *jowar* and *bajra* crop plants. The nitrogen fixing ability of isolates showed significant variations (Table 113). The maximum quantity of nitrogen fixed by isolate 1 obtained from sugarcane root was 19.5 mg/g malate used and minimum of 10 mg N/g malate used by isolate no. 201 obtained from *Jowar* roots. The nitrogen fixing capacity of isolates from wheat crop (C_3 plants) was similar to C_4 plants such as sugarcane, *jowar*, and *bajra*.

Table 113: Nitrogen fixation by *Azospirillum* spp.

Name of the organism	**Isolate no.**	**Source**	**mgN/g malate**
Azospirillum brasilense	1	Sugarcane roots	19.50
Azospirillum lipoferum	2	Sugarcane roots	11,50
Azospirillum brasilense	5	Sugarcane roots	10.40
Azospirillum lipoferum	6	Sugar rhizosphere	12.50
Azospirillum brasilense	8	Sugar rhizosphere	10.50
Azospirillum brasilense	9	Sugar rhizosphere	10.08
Azospirillum brasilense	11	Wheat rhizosphere	15.10
Azospirillum lipoferum	12	Wheat rhizosohere	12.64
Azospirillum brasilense	15	Wheat rhizosphere	10.06
Azospirillum brasilense	18	Wheat roots	13.80
Azospirillum lipoferum	20	Wheat roots	19.00
Azospirillum lipoferum	102	Maize roots	13.50
Azospirillum brasilense	151	Maize rhizosphere	12.00
Azospirillum brasilense	154	Maize rhizosphere	19.00
Azospirillum brasilense	201	Jowar roots	10.00
Azospirillum brasilense	251	Jowar rhizosphere	19.00
Azospirillum lipoferum	254	Jowar rhizosphere	11.00
Azospirillum brasilense	304	Bajra roots	11.55
Azospirillum brasilense	354	Bajra rhizosphere	16.50
Azospirillum lipoferum	353	Bajra rhizosphere	11.50

The neutral pH was found to be the best for nitrogen fixation (nitrogenase activity). At pH 9, nitrogenase activity was marginal and at pH 5, it was negligible. (Table 114).

The soil maintained at one third of water holding capacity showed highest nitrogenase activity when inoculated with *A. lipoferum* (Table 115).

Table 114: Effect of pH on Nitrogenase activity of *Azospirillum lipoferum*

Treatments	**Incubation time (Hrs) (n moles/C_2H_4 hr/20 g soil)**				**Average (n moles/C2H4 hr/ 20 g soil)**
	1	**2**	**3**	**4**	
Soil (5.5 pH)	12.79	25.58	38.37	51.19	12.80
Soil (5.5 pH) + *A. Upoferum*	12.79	26.08	38.87	51.66	12.70
Soil(7pH)	14.93	29.86	43.79	58.72	14.93
Soil (7 pH) + *A. Upoferum*	546.13	1092.26	1638.39	2184.53	546.13
Soil(9pH)	12.79	26.08	37.50	52.00	12.70
Soil (9 pH) + *A. Upoferum*	21.33	42.66	63.99	85.33	21.33
C.D at 5%	-	-	-	-	4.45

Saxena V. and Gaur, A. C., IARI, New Delhi

Table 115: Effect of moisture level on Nitrogenase activity of Azospirillum lipoferum

Treatments	**Incubation time (Hrs) (n moles/C_2H_4 hr/20 g soil)**			**Average (n moles/C_2H_4 hr/20 g)**
	1	**2**	**3**	
Soil (Field capacity)	25.02	50.04	75.06	25.02
Soil (FC) + *A. Upoferum*	53.22	106.44	159.66	53.22
Soil (1/2 WHC)	18.77	36.44	55.21	18.77
Soil (1/2 WHC) + *A. Upoferum*	79.78	169.56	239.34	79.78
Soil (1/3 WHC)	19.91	40.01	59.73	19.91
Soil (1/3 WHQ+ *A. Upoferum*	711.11	1422.22	2133.33	711.11
C. D at 5%	4.29	4.16	5.59	4.45

Saxena, V. and Gaur, A. C., IARI, New Delhi

Herbicides such as Arelon, Tribunil and Dosanex at recommended field rates and half the field rate of application inhibited the nitrogen fixation by *Azospirillum* strains under cultural condition (Table 116). It would be safer to use biopesticides instead of chemical pesticides.

Table 116: Effect of different herbicides on nitrogen fixation by different *Azospirillum* strains (mg N fixed /g malate)

Treatment	Isolate Number	Rate of Application								
		Arelon			Tribunal			Dosanex		
		0	$^{1}/_{2}$X	X	0	$^{1}/_{2}$X	X	0	$^{1}/_{2}$X	X
A. Lipoferum	254	10.45	9.55	9.70	10.45	8.77	7.09	10.45	6.55	5.75
A. Lipoferum	12	12.64	10.84	6.72	12.64	10.45	7.84	12.64	6.55	5.30
A. brasilense	9	10.08	9.85	5:84	10.08	6.72	6.72	10.08	4.48	4.48
A. brasilense	304	10.61	9.61	9.05	10.61	7.84	5.60	10.61	6.85	4.25
A. brasilense	201	7.84	8.84	7.84	7.84	2.80	-	7.84	3.24	2.16
A. brasilense	154	14.56	12.84	9.84	14.56	5.60	-	14.56	8.72	691
CD at 5%			1.44			0.56			0.28	

X-recommended rate, 1/2 x - 50% of the recommended rate
Source: Saxena, V. and Gaur, A.C., IARI, New Delhi

Okon *et al.* (1976 b) studied the carbon and ammonia metabolism of *S. lipoferum*. They found intact cells and extracts from *S. lipoferum* rapidly oxidize malate, succinate, lactate and pyruvate. Glucose, galactose, fructose. acetate and citrate do not increase the rates of O_2 uptakes by cells above the endogenous rate. Cells grown on NH^{+}_{4} oxidize various substrates at about the same rates as did cells grown on N_2

Tarrand *et al.* (1978) also studied the ability of *Azospirillum* sp. to utilize different carbon sources either in N free semi-solid medium or by paper disc method. *A. brasilense* failed to grow in nitrogen free semi- solid medium with glucose used as sole carbon source but *A. lipoferum* was found to grow in the same medium. By paper disc method only alfa-ketoglutarate/ glucose supported the growth of *A. lipoferum* and *A. brasilense* failed to utilize alfa-ketoglutarate, mannitol, sorbitol and ribose. Nur *et al.* (1980) made comparative studies of their local strains with *A. brasilense* and noted some variations in their physiological and nutritional behaviour.

Plate 44: Electron micrograph of *Azospirillum brasilense* sp. (Isolate No. 1)

Rai and Gaur (1982) reported that 7 isolates belonged to A. brasilense (Plate 44) and 3 belonged to *A. lipoferum*. The nitrogen fixing capacity of these isolates varied from 9 mg to 18 mg per gram of malate used. Isolates having more denitrifying capacity fixed less nitrogen and tryptophane failed to serve as carbon source. D. fructose, sorbitol, sucrose, tyrosins and tryptophane failed to serve as carbon source. D. glucose was utilized by only A. *lipoferum* strains. All the strains except one grew well in ketoglutrate, L. alanine and L. glutumate. Lactate, pyruvate and succinate also supported the growth of *Azspirillum* except a few strains.

V. Azospirillum and Crop Productivity

As Azospirillum fixes appreciable amount of atmospheric nitrogen and remains in loose symbiosis with tropical plant roots, experiments were conducted to use *Azospirillum* as inoculant to supplement the nitrogen need of cereal and leguminous plants (Smith *et al.*, 1978; Vasil *et al.*, 1979; Barber *et al.*, 1979; Jagnow, 1979; Nur *et al.*, 1980 a,b: Umaligarcia *et al.*, 1980; Henis *et al.*, 1980; Berg *et al.*, 1979; Bradley, 1980; Cohen *et al.*, 1980; Baldani and Dobereiner 1980; Schank *et al.*, 1981).

Smith *et al.* (1978) were the first to use nitrogen fixing *Azospirillum* sp. as inoculant on a commercial scale to test whether these organisms supplement the nitrogen need of field crops or not. Trials conducted by these investigators in 1974 on *Panicum maximum* and Digitaria *decumbens* produced 80 percent and 61 percent more protein as compared to uninoculated controls. 1n 1975, inoculant produced significantly high protein with increase in dry matter yields in pearl millet and *P. maximum.* Using regression analaysis, projected yields of both pearl millets and guinea grass indicated that about 40 kg N/ha/year were substituted by inoculation. Though nitrogenous fertilizer was required to enhance the response to inoculation but higher rates of its application repressed the response. Acetylene reduction verified that N-fixation occurred but did not correlate with yield increase. Dewan and Subba Rao (1979) noted increase in the yield of root biomass of rice by a mixed inoculant of *A.* brasilense *and Azotobacter chroococcum.*

Nur *et al.*, (1980 b) conducted trials under green house condition on *Zea mays* and S. italica with *Azospirilum* isolated from *Cynodon dectylon* and found increase in dry matter yields and total nitrogen content. But Barber *et al.* (1976) in the trials conducted on millets *(Panicum glaucum)* at three levels of nitrogen fertilization with and without inoculation of nitrogen fixing *Azospirillum* cells did not find any increase in the total nitrogen content of the crop.

Henis and coworkers (1980) in Israel found increase, on the yield and nitrogen content of grasses by inoculation with *Azospirillum* sp. both under irrigated and non-irrigated conditions. Umaligarcia *et al.* (1980) found that young inoculated roots of pearl millets and guinea grass produced more mucilaginous sheath root hairs and lateral roots than did uninoculated sterile control.

Rai and Gaur (1982) tested one strain of *A. lipoferum* having high nitrogen fixing capacity with negative test for denitrification as inoculant to supplement the nitrogen need of wheat crop in field condition with different doses of N with and without the inoculant (Table 117). The result showed that the inoculant contributed about 40 Kg N/ha to the crop.

Table 117: Effect of *Azospirillum* on yield and N uptake by wheat

Treatment	Grain (Kg/ha)		Straw (Kg/ha)	
	Yield	N uptake	Yield	N uptake
Control	1260	24.80	1780	8.20
Inoculation	2070	42.70	2110	14.90
40kgN	2370	51.20	3240	18.70
40 kg N + inoculation	3110	65.50	3820	21.90
SOkgN	2960	61.00	3690	18.90
80 kg N + inoculation	4150	85.40	4890	26.50
C. D. at 5 %	*730*	—	680	—

Basal dressing: 60 kg P_2O_2/ha and 40 kg K_2O/ha

The results of the field trials indicated that *Azospirillum* inoculation was beneficial to crops like sorghum, pearl millet, finger millets etc. and the responses due to inoculation were comparable to chemical fertilizer applied to soil at 20-40 kg N/ha (Subba Rao *et al.*, 1982).

The benefit of Azospirillum inoculation in rice crop as nitrogen saver was reported by Balasubramanian and kumar (1987). The effect was more marked at lower level of chemical N and maximum increase in rice grain yield due to *Azospirillum* inoculation was found with 50% of the recommended nitrogen level in rainfed upland condition and with 75 percent of the recommended level of N fertilizer in lowland rice (Table 118).

Table 118: Effect of *Azospirillum* on low and upland Rice

N Level	Grain yield (Kg/ha)			
	Lowland		Upland	
	Untreated	Treated	Untreated	Treated
0%	2100	3100	2000	2200
50%	3300	4100	2300	2900
75%	3200	4400	2400	2900
100% C.D. at 5%	4200	5200	2900	3100
	200		400	

Gaur *et al.* (1989) conducted field experiments in sandy loam alluvial soil to assess the contribution of seed treatment with *Azospirillum* brasilense and phos-phobacteria *(Bacillus polymyxa, Pseudomonas striata)* alone or in combination with and without carbofuran granules applied to seed furrows to control dead hearts caused by shoot fly in sorghum as well as on the yield and uptake of nitrogen by the crop. The shootfly damage was quite severe in treatments where carbofuran was not applied. The percentage dead hearts was slightly lower when the seeds were inoculated with bacterial cultures (Table 119). A highly significant reduction in shoot fly damage was obtained with carbofuran alone or with bacterial inoculants (Plate 45). The efficiency of carbofuran in reducing the incidence of dead hearts was confirmed in the next field trial and the grain yield was significantly increased (Table 120). The rhizospheric population of *Azospirillum* in soil was increased due to the treatment at both the vegetative and post harvest stage of plants (Table 121).

Table 119: Effect of Carbofuran, *Azospirillum* and Phosphobacteria in control of disease caused by shoot fly in Sorghum.

Treatments	Dead hearts % due to shootfly
Control (untreated seeds)	43.9
Carbofuran	12.6
Azospirillum brasilense	30.8
Bacillus polymyxa	31.5
A. brasilense + *B. polymyxa*	39.9
A. brasilense + *B. polymyxa* + carbofuran	11.0
B. polymyxa + *carbofuran*	10.6
A. brasilense + *carbofuran*	11.0
N100 + P60	40.4
Carbofuran + N1 00 + P60	13.6
C. D. at 5%	2.3

Plate 45: Effect of carbofuran in checking of shootfly damage in sorghum

Table 120: Effect of bacterial inoculants and carbofuran on dead hearts, grain yield and N-uptake by sorghum crop and insecticide residues

Treatments	**Dead hearts% (20 days after germination)**	**Insecticide residue in soil on 20th day of germination (ppm)**	**Grain yield (kg/ha)**	**Nin rain (%)**	**N-uptake by grain (kg/ha)**
Control (untreated seeds only)	20.6 (26.4)		1670	1.57	26.2
Carbofuran	2.6 (7.6)	1.84	2900	1.47	42.6
Azospirillum brasilense	17.6 (24.6)		2800	1.70	46.0
A. brasilense+ carbofuran	2.6 (6.6)	0.60	3200	1.57	50.2
Psudomonas striata	17.2 (24.0)		2410	1.55	37.4
P. striata +carbofuran	2.7 (9.4)	2.86	2860	1.50	42.0
A brasilense +*P. striata*	12.4(19.8)		2430	1.57	38.3
A brasilense + *P. striata* carbofuran	1.7(7.4)	1.74	3130	1.57	49.1
N100+P60+carbofuran	3.5 (10.5)	1.00	3510	1.65	57.3
S.Em+	2.6		190		2.9
C.D. at 5%	12.8		550		11.3

The effect of *A. brasilense* was more marked than P. striata although the results were statistically at par. The combined inoculation of seeds with the culture did not show any additional gain in augmenting the grain yield and nitrogen content in vegetative parts were not augmented in the presence of the inoculants. The *Azospirillum population* in *rhizosphere* soil both at vegetative and harvest stage was more due to different treatments. The best effect was obtained due to *Azospirillum* inoculation (Table 121).

Table 121: Rhizospheric population of *Azospirillum* in selected treatment at vegetative and harvest stages of the sorghum crop

Treatment	Vegetative x10^7/g soil	Harvest x10^5/g soil
Control (Untreated seeds)	1.1	1.5
Carbofuran	13.6	5.4
Azospirillum brasilense	20.0	9.2
A. brasilense + carbofuran	24.3	9.2
P. striata + *A. brasilense*	15.6	8.0
P. striata + *A. brasilense* + carbofuran	2.42	1.7

In an evaluation of the world wide success of *Azotobacter* and Azospirillum, Wani (1990) concluded that statistically significant increases were obtained in approximately 60% of the trials conducted in USSR, Israel and India. The mean grain yield of pearl millets increased significantly due to *Azospirillum lipoferum inoculants.* Eleven experiments out of 24 experiments increased the yield significantly by about 18.7%. Ten experiments increased the yield by 9.3% but were not statistically significant. The remaining 3 experiments either showed no response or showed some reduction in yield (Wani and Lee, 1992).

Multilocational trials in India showed that seed inoculation with *A. brasilense* increased the mean grain yield of pearl millet at 6 and sorghum at 4 out of 9 location tested. The yield increase with pearl millet varied from 10 to 17% and with sorghum from 7 to 31% (Rao *et al.,* 1986).

In Israel, field experiments with *Azospirillum* were conducted with different cereals, varieties and different fertilization levels (Okon 1988). The summary of the *Azospirillum* trials conducted in large plots (200-1000 m2) and 4-6 replications are presented in table 122. In general, inoculation of C_4 plants such as maize, sorghum, *Panicum* and *Setaria* showed greater yield increases than inoculated spring wheat, a $C\text{-}_3$ plant. In most cases the effects of *Azospirillum* varied with the season, year and crop.

Table 122: Results of *Azospirillum* trials conduced in Israel

Crop	No. of Experiments	Increase in yields %		No. of experiments Showing	
		Range	Average	Significant increases	Increases above 5%
Com (Forage and Sweet com and grain corn)	18	0-38	16	7	11
Whpat (Forage and grain)	10	3-19	9	5	8
Sorghum (Forage and grain)	5	12-20	17	5	5
Setaria italics	2	38-47	43	—	2
Panicummiliaceum	1	13	13	1	1

The summer crops showed better performance because 75% of the experiments showed significant increase and 90% of the experiments showed increase above 5%. The optimum temperature for *Azospirillum* growth is at 32-35°C and it is possible that *Azospirillum* activity and biological nitrogen fixation were greater in summer, particularly in irrigated crops. During vegetative phase of wheat growth during winter the soil temperature in Israel ranges between 10 to 15°C, nevertheless inoculation of wheat with *Azospirillum* also showed significant increase in vegetative growth and grain yield but with lower increase than the summer crops.

Effect of *Azospirillum on Fruit Trees*

Jeeva *et al.* (1988) investigated the effect of *Azospirillum* on the growth and development of banana. The fertilizers in the form of urea, super-phosphate and muriate of potash were applied in two split doses in the third and fifth months after planting. *Azospirillum* culture was applied @ 2.5 kg/ha (in 3 splits) by mixing with FYM in the ratio of 1:20 on the 3rd, 60th, and 120th day after planting. The application of *Azospirillum* plus nitrogen fertilizer significantly improved the vegetative growth, production of leaves, bunch weight etc. *Azospirillum* with 75% of the recommended level of fertilizer had the same effect in increasing the plant growth and yield as that of 100% recommended level of fertilizer without inoculation.

Rao and Dass (1989) reported that soil inoculation with pure cell suspension of *Azospirillum* increased the plants height and dry weight in 6 months old budded plant of ber (*Zizyphus mauritiana*) and in root cuttings of pomegranate grown in pots. In both cases the increased growth was correlated with a marked increase in N uptake.

Singh and Sharma (1993) studied leaf nutrient composition of sweet orange as affected by combined use of bio and chemical fertilizers. Three grams of carrier based *Azospirillum* was applied to root at the time of transplantation. *Azospirillum* inoculation enhanced the N content in 4-7 month old mature leaves analyzed at 6 monthly intervals to almost same level as in full doses of N and P. According to Kumar *et al.* (1998) *Azospirillum* inoculated plants exhibited maximum of cashew seeding.

VI. Production of Growth Promoting Substance

Tien *et al.* (1979) found increased yield of pearl millet and attributed this increase due to IAA, GA and cytokinin-like substances. Indole acetic acid and indole lactic acids were formed from tryptophane by *A. brasilense.*

Umaligarcia *et al.,* (1980) found that young inoculated roots of pearl millets produced more mucilaginous sheath root hairs and lateral roots than control.

Okon (1982) found that inoculation with *A. brasilense* caused significant increase in crop yield. At early stages of growth, inoculation increased the numbers of roots and their branching. Root segments of inoculated corn and wheat showed significant increase (30-50 %) over control in the uptake rate of plant nutrients such as nitrogen, phosphorus and potassium.

Azospirillum brasilense (strains 201, 251, 8, 15, 304) and A. lipoferum (strains 20, 254, 12, 353, 102) were tested for the production of growth promoting substances in the culture medium (Saxena, 1983). *Azospirillum* strains were grown for 5 days at 37°C in 250 ml flasks containing 100 ml Okon's medium (Okon *et al.,* 1977). Bacterial cultures after their growth were centrifuged at 20000 rpm for 20 minutes. Each centrifugate was acidified to pH 3.0 with N-HCL. The acidified centrifugate was divided into 2 equal fractions A and B. Fraction A was extracted for auxin-like substances and fraction B for gibberellin-like substances.

Auxin-like substance: Fifty ml of supernatent fluid (fraction A) was shaken 3 times with ethyl acetate (10 ml) at 30 minutes interval, Combined ethyl acetate extract were evaporated to dryness in a current of air and residue dissolved in 3.0 ml methanol. Extracts were examined by paper partition chromatography using freshly prepared solvent mixture, isopropanol, ammonia solution (sp. Gr. 0.88): water (10:1:1 by volume).

Gibberellin-like substance: Fifty ml (fraction B) was shaken with 0.5 g activated charcoal for two hrs. The charcoal was separated by centrifugation, extracted with aqueous acetone (16 ml 95% *V/V* acetone in water). The acetone was evaporated in a current of air at room temperature and the moist residue shaken twice with 1.5 times its volume of ethyl acetate. The combined ethyl acetate extract was evaporated to dryness in cold air. The residue was dissolved in 3.0 ml methanol and examined by paper partition chromatography using freshly prepared solvent mixture as mentioned above.

Chromatography

0.1 ml of the extract was spotted in bands on narrow strips (45 cm X 6 cm) of Whatman no.1 chromatography paper. Authentic IAA, GA3 and control were also run. With freshly prepared solvent mixture (isopropanol: ammonia solution (sp gr 0.88):water, (10:1:1) spotted chromatograms were equilibrated in solvent-saturated air for 2 hours and then developed with solvent by descending chromatography. After the solvent phase reached to solvent front, marked at 30 cm from the base line, chromatograms were removed, marked with lead pencil to the level of maximum flow and then were air dried in dark. One set of strips of IAA and the other set for GA3 were dipped in 5% (V/V) conc. Sulphuric acid in methanol, dried in hot air and exposed for 15 minutes to U.V. radiation (wavelength 350 mu). IAA was identified by yellow fluorescence at Rf 0.3 and 0.5 and GA3 by green fluorescence at Rf 0.5 and 0.7. Chromatograms strips not treated with chromogenic reagents were dried for 7 days to remove solvents which were cut into 10 equal parts representing the sequences of Rf values 0.1 to 1.0 and were eluted separately for bioassays.

Bioassay for Auxins and Gibberellins

Auxin-like-substances: First internode test of Nitsch and Nitsch (1956) was followed to measure the auxin activity of the elutes. The procedure is given below:-

- Oat *(Avena sativa)* variety Kant seeds were soaked in tap water for 4 hours at room temperature.
- These were washed thoroughly with distilled water and planted on wet filter paper in aluminum trays.
- Trays were kept at an angle of 45°C to facilitate straight growth of internode. Trays were covered with black polythene paper and seeds were allowed to grow at 25°C for 4 days in complete darkness. Afterwards all manipulations were performed under green light.
- When the first internode reached about 2.5 cm in length and coleoptile of 0.5 cm, 4 mm sections were cut with a conventional cutter, 2 mm below the coleoptile node.
- The sections were washed in glass distilled water for one hour.
- Five sections were taken in a test tube containing 1 ml of the phosphate buffer (1.794 g K2 HPO_4, 1.019 g citric acid monohydrate and 20g sucrose dissolved in 1 litre).

- Chromatogram pieces of different Rf values were eluted separately in 3 ml distilled water. One ml of this elute was added to respective tube with 5 internode section.
- Tubes were allowed to rotate around a horizontal axis at about 1 rpm in the dark for about 24 hrs at 25°C.
- The length of sections were then measured using enlarger (X5).
- The percent increment over control in the length of internode sections was calculated.
- Likewise different concentrations of IAA ranging from 0.001ug to 20ug were also tested with the internode sections and a standard curve was drawn for determination of IAA in the culture supernatant.

Gibberellin-like-substances: Following the modified technique of Ogawa (1963) seeds of rice (Variety Indira, a mutant of Tainan-3) were tested for its second leaf sheath growth in response to chromatogram pieces prepared from different cultural extracts. The method was as follows:

- Rice (Oryza *sativa*) variety Indira seeds were kept in running water for 4 hours.
- These were allowed to sprout under immersed condition in a germination chamber at 25°C.
- Uniformly sprouted seeds were collected and 5 seeds were placed in each tube containing chromatogram pieces of each Rf and 4 ml distilled water.
- The tubes were kept under continuous illumination in a temperature controlled chamber at 26°C and light intensity of 160 foot candles.
- Two ml of double distilled water was added on the 4th day.
- The second leaf sheath had attained the maximum expansion in about 7 days. It was then measured and the data were compared with dosage response curve of different GA3 concentrations ranging from 0.001 to 20 ug.

The effect of elutes from chromatograms of each of the microbial extract were tested on the length of first internode sections of oats (Var. Kant) and the results showed the production of auxin-like substances and the nature and level of auxin produced varied among themselves (Table 123). None of the cultures showed auxin activity throughout the chromatogram strip. The maximum activity was observed between Rf 0.3 to 0.5 in all the cases.

Table 123: Auxin activity in the culture filtrate of *Azospirillum*

Culture	Isolate Number	Percentage increase in length of oat coleptile over control									
		Rf values									
		0.1	0.2	0.3	0.4	0.5	0.6	0.7	0.8	0.9	1.0
A. brasilense	201	7		16	20	20	16				
A. lipoferum	20		13	13	20	25			29		
A. lipoferum	254			20	29	26			15		
A. lipoferum	12			20	20	18	15				
A. brasilense	251		26	25	30	25	30				
A brasilense	8			30	31	26	30				
A. brasilense	15			17	12	16			19		
A brasilense	304			28	20	25	25				
A lipoferum	353			17	25	17	7				
A. lipofeium	102			32	17	16				5	

Source: Saxena V. and Gaur A. C. (IARI), New Delhi

The total activity of auxin was higher in *A. brasilense* (251) followed by *A. brasilense* (8), *A. lipoferum* (20) and *A. brasilense* (304). The other organisms occupied intermediate positions.

The substances at Rf 0.3 to 0.5 corresponded to the same position of authentic IAA and behaved similarly in the biossay test.

The maximum extension of 32 per cent in oat coleoptile was observed in culture extracts of *A. lipoferum* (str 102) at Rf 0.3 which corresponded to 2.5 μg IAA/ml of original culture. *A. brasilense* (strain 251) showed increase at Rf 0.3 which corresponded to 0.25 μg IAA/ml of the original of the culture.

A. *lipoferum* (20) showed 20 percent increase which was equivalent to 0.067 μg IAA/ml of original culture. A. *brasilense* (15) at Rf 0.4 showed 12 percent in growth producing 0.007 μg IAA/ml of the originai culture.

Other Rf values were also observed which could not be identified

Gibberellin-like-substance: The effects of elutes of each Rf zone from the chromatograms were tested on the growth of second leaf sheath of rice (variety Tainan 3 mutant Indira) and the per cent increase in length were recorded (Table 124).

Table 124: Gibberellin activity in the culture filtrate of *Azospirillum*

Culture	Isolate Number	Percentage increase in length of oat coleptile over control									
		Rf values									
		0.1	0.2	0.3	0.4	0.5	0.6	0.7	0.8	0.9	1.0
A. brasilense	201					10.2	20.5	33.5	25.5	15.0	10.0
A. lipofervm	20					11.5	25.5	30.7	21.0	18.5	9.5
A. lipoferum	254		15.5		10.5	25.8	30.9	28.7	26.5		5.8
A. lipoferum	12				17.5	20.5	26.8	29.7	30.8	10.5	9.7
A. brasilense	251				16.5	19.8	31.8	36.8	24.5		8.5
A. brasilense	8						29.7	35.8	27.8		12.6
A. brasilense	15						30.5	39.7	29.2	11.5	
A. brasilense	304						25.6	26.8	29.7		10.8
A. lipoferum	353					14.5	50.8	55.7	30.5	14.8	12.5
A. lipoferum	102						30.8	42.5	40.6	25.6	10.5

Source: Saxena V. and Gaur A. C. (IARI), New Delhi

The result showed that the percent increase in length of second leaf sheath over control was noted at 0.5-1.0 Rf values in most of the cultures. Some cultures showed the activity at Rf 0.4 also. Most of the cultures showed the highest activity at and around Rf 0.7.

Highest level of total gibberellins was observed in *A. lipoferum* (353) followed by *A. lipoferum* strains 102, 12 and 254. *A. lipoferum* strains were found superior to *A. brasilense* strains (201,251,8,15 and 304). The activity at Rf 0.5 to 0.7 corresponded to the same position as that of authentic GA3.

A. *lipoferum* (353) showed 55.7% increase in second leaf sheath of rice at Rf 0.7 followed by *A. lipoferum* (102) 42.5%, *A. brasilense* (15) 39.7% and *A. brasilense* (251) 36.8% increase which corresponded to 0.08, 0.05, 0.047 and 0.042 μg GA3/ml of original culture respectively.

The studies clearly showed that *Azospirillum* spp. besides nitrogen fixation are also good producers of plant growth promoting substances, such as IAA and GA3.

VII. Production Technology

Isolation of *Azospirillum*

The bacterium occurs in soil and outside and inside plant roots. Surface sterilization of roots by 70 percent alcohol and maintenance of micro-aerophilic conditions in the growth medium are essential to obtain pure culture of the bacteria. Enrichment techniques are necessary to obtain pure culture of the organism. Surface sterilized root pieces (about 0.5 cm) with 70 percent alcohol for 3-5 seconds are repeatedly washed with phosphate buffer of a pH 7.0 then they are planted in semi-solid, nitrogen free malate medium and incubated for 2 days at 35°C. The formation of white pellicles 2-4 mm below the upper surface of the medium shows the characteristic pattern of their growth.

The purification is essential by repeated transfer of the isolates into small screw caped tubes containing the specific medium. These cultures are serially diluted up to 10^{-7}. 0.1 ml from each dilution is inoculated into a tube containing the medium. The least dilution showing the best growth is supposed to be pure culture.

Characteristics

They are Gram negative, motile, vibroid and contain polyhydroxy butyrate granules. In the presence of ammonium salts, it grows aerobically and in microaerophilic condition, fixes atmosphereic nitrogen. These bacteria are heterotrophs.

Maintenance of Culture

Azospirillum cultures can be maintained on ammonium chloride agar medium (Okon *et al.*, 1977) and also on semi-solid malate medium in the tubes.

Carrier Based Biofertilizer

Bottles or flasks containing ammonium choloride in liquid medium can be used and incubated in a rotary shaker for 3 days at 35°C for mass production, Fermentors of different capacity can be used depending on the scale of production.

Carrier material for *Azospirillum*

A potential bacterial fertilizer should have a suitable carried material in order to carry it to farmers and his field, where it could multiply, maintain a sizeable inoculum and survive for considerable periods of time. Several carrier materials including peat, which is widely used as carrier for *Rhizobium* and *Azotobacter* were tested. Soil and farmyard manure in the ratio 1:1 sterilized for 3 hours consecutively for 3 days was found to be best suited as carrier for *Azospirillum (Lakshmi et al.*, 1977; Tilak *et al.*, 1979).

Enumeration of *Azospirillum* in Carrier Material

Once an organism is ready for field application, it is necessary to follow up its proliferation and survival in the carrier, as a quality control measure. A solid medium was devised by Okon *et al.*, (1976 a) by incorporating ammonium chloride in the malate medium thus enabling the bacterium to grow aerobically using ammoniacal nitrogen. This medium has been made more specific by incorporating one of the acidic dyes like bromothymol blue or bromocresol purple (Lakshmi *et al.*, 1977). Being a strong alkali producer the colonies, as they grow, change the color of the medium around then to the basic shade of the dye, blue or purple as the case may be. Moreover, the colonies themselves absorbed the dye which make counting easier, and to check the presence of contaminants.

Establishment of *Azospirillum* on Plant Roots

An organism to be used as an inoculant for crop improvement should be capable of quick and firm establishment in the root region. To establish the affinity of *Azospirillum* to crop plant roots, studies are carried out under aseptic conditions by growing plants on nitrogen free seedling agar and inoculating them with a 48 hours old culture of *Azospirillum*. The first sign of their adaptability to the root system is the proliferation and clustering of the bacteria around the root hairs (Lakshmi, *et* al., 1977) followed by entry through root hairs into the cortical layers often extending into the xylem.

References

Baillie, A, Hodgkiss, W. and Norris, J. R. 1962, J. Appl Bacteriol 25:116-119

Balasubramanian, A. and Kumar, K. 1987. Performance of Azospirillum biofertilizer in irrigated and rainfed upland rice IRRN 12:43.

Baldani, V. L. D. and Dobereiner, J. 1980. Soil Biol Biochem. 12:433-439.

Baldani. J. I. Baldani, V. L. D., Sevin, L. and Dobereiner, J. 1986. Characterisation of Herbaspirillum seropedicae gen. Nov. sp. Nov., a root associated nitrogen fixing bacterium. Int. J. Syst. Bacteriol 36(1):86-93.

Baldani, J. I., Baldani, V. L. D., Dobereiner, J. 1984. Third International symposium with non-legumes, Helsinki, Finland [Fide Falk, E. C., Johnston, J. L., Baldani, V. L. D., Dobereiner, J. and Kreign, N. R.. 1986. Deoxyribonucleic and ribonucleic acid homology studies of the genera Azospirillum and conglomeromonas. International Joumal Systematic Bacteriology 36(1):80-85]

Barber, L E., Tijepkema, Y. Russel, S. A. and Evans, H. J. 1976. Appl. Environ. Microbiol 32:108-113.

Becking, J. H. 1963. Antonie Von Leeuwenhoek 29:326

Berg, R. H., Vasil, V. and Vasil, I. K. 1979. Protoplasm 101:148-163

Bradley R., De oliveira, L. A., de Podesta Filho, J. A. and St. John, T. V. 1980. Agro-Ecosystem 6:249-266

Beijerinck, M. W. 1922. K. Ned Akad. Wet Versewet Versl. Gerwona. Vergad. Afd. Naturkd 30:431-438

Beijerinck, M. W. 1925. Zbl. Bakteriol. Parasitenkol Infektional Hyg. Abs. 263:353-359

Breed, R. S. Murray, E.G.D. and Smith, N. R.. 1957. Bergey's Manual of Determinative Bacteriology, 7th Edition, Williams and Wilkins, Baltimore, Md, USA.

Burns, R. C. and Hardy, R. W. F. 1975. Nitrogen Fixation in Bacteria and Higher Plants. Springer-Verlag, New York, Heidelberg-Berlin.

Clayton, H. R. and Beston, N. W. 1955. J. Gen. Microbiol. 12:519- 525.

Cohen, E. Okon, Y. Kigel, J and Nur. I. 1980. Plant Physiol. 66:746-49

Dalton, H. 1974.Crit. Rev. Microbiol 3:183-220.

Day, J. M. and Dobereiner, J. 1976. Soil Biol. Biochem. 8:45-50

Dart, P. J. and Wani, S. P. 1982. In: Non-symbiotic nitrogen fixation and organic matter in tropics. 12th International Congress. Soil Science, New Delhi pp 3-27.

De-polli, H. Bohlool, B. B. and Dobereiner, 1980 J. Arch. Microbiol 126:217-222

Dewan, G. I. and Subba Rao, N. S. 1979, Pl. Soil.53:295-302

Dobell, C. 1960. Antony Van Leeuwenhoek and his little animals. Dover Publications, New York.

Dobereiner, J. and Day, J. M. 1974. In: Proc. Inter. Symposium on N-Fixation interdisciplinary Discussion, Washington state University, Pullman, Washington, USA pp 518-529.

Dobereiner, J. Marriel, I. E. and Nery, M. 1976. Can. J. Microbiol. 22:1464-1473

Eskew, D. L. Focht, D. D. and Ting, I. P. 1977. Appl. Environ. Microbiol. 34:582-585.

Gaur, A. C. Srivastava, K. P., Saxena, Vandana and Agnihotri, N. P., 1989, Pesticide Research Jour. 1:63-66

Giesberger, 1936. PhD. Thesis Delft University, Delft, The Netherlands.

Henis, Y. Kopunik, Y. Sarig, S. Nur, I. and Okon, Y. 1980. J. Appi. Bact. 49. Abs VII-VIII.

Hewit, F. J. 1975, Annual Rev. Pl. Physiol. 26:73-100

Hylemon, P. B., Wells, J. S. Krieg, N. R. and Janna Sch., H. W. 1973. Internal. J. Syst. Bacteriol 23:340-380

Jagnow, G. 1979. Pflanzenerlir Bodenkd. 142: 399-410

Jeeva, S., Kaulse Karan M. and Shanmugavelu, K. G. 1988. South Indian Horticulture, 36 (1-2):1-4

Kreig, N. R. 1976. Bacteriol. Rev. 40:55-155

Kreig, N. R. and Dobereiner, J. 1984. The genus Azospirillum. In Bergey's Manual of Systematic Bacteriology. Vol. I (Eds. Kreig, N. R. and Holt, J. G.), Williams and Wilkins, Baltimore, USA pp 94- 104.

Kumar, D.P., Hegde, M., Bagyaraj, D.J. and Madhava Rao, A.R. 1998. The Cashew 12: 3-9

Lakshmi, V., Satyanaryana Rao, Vijayalakshmi, K., Lakshmi-Kumari, M. Tilak, KVBR and Subba Rao, N. S. 1977. Proc. Indian Acad. Sci 86(B):397-403

Magalhaes, L. M. S., Baldani, J. L., Souto, S. M., Kuykendall, J. R. and Dobereiner, J. 1983. Annals of Academy of Brazil Gen 55:417- 430.

Magalhaes, L. M. S., Neyra, C. A. and Dobereiner, J. 1978 Arch. Microbiol. 117:247-252

Martinez, R. J. 1963. Arch. Microbiol. 44:334-343

Muller, O. F. 1773. Succinate Historia 1:1-135

Neyra, C. A. and Van Berkam, P. 1977. Can J. Microbiol 23:306- 310.

Neyra, C. A., Dobereiner, J., Lalande, R. and Knowles, R. 1977 Can. J. Microbiol 23:300-305

Nur, I. Okon, Y. and Henis, Y. 1980 a Can. J. Microbiol, 26:306-310.

Nur, I. Okon, Y. and Henis, Y. 1980 b Can. J. Microbiol, 26:714-718.

Ogawa, Y. 1936. Plant and Cell Physiology 4:227-237

Okon, Y. Alberecht, S. L. and Burris, R. H. 1976 a. J. Bacteriol. 127:1248-1254

Okon, Y. Alberecht, S. L. and Burris, R. H. 1976 b. J. Bacteriol. 128:592-597.

Okon, Y. Alberecht, S. L. and Burris, R. H. 1977. Appl. Environ. Microbiol 33:85-88.

Okon, Y. 1988 in.: Biological Nitrogen Fixation: Recent Developments (Ed: Rao, NSS). Oxford and IBH CO Pvt. Ltd., New Delhi p75.

Okon, Y. and Itzigsohn, R. 1992. FEMS Microbiol. Rev. 103:131- 140

Rai, S. N. and Gaur, A. C. 1982 Pl.. Soil 69:233-238.

Rao, A. V. and Venkateswarlu, B. 1982 a. Ind. J. Expl. Biol. 20:316- 318.

Rao, A. V. and Vekateswarlu, B. 1982b Can. Microbiol. 28:778-782 Rao, A. V. and Vekateswarlu, B. 1985. Acta Microbiol (Hung) 32:222- 224

Rao, A. V. and Dass, H. C., 1989. Annals of Arid Zone 28(1-2):143- 147

Rao, NSS, Tilak, KVBR, Singh, C. S and Nair, PV 1986. In:Cereal Nitrogen Fixation (Ed. Wani S. P.) ICRISAT, Pantancheru, India, p. 69.

Saxena, Vandana 1983. Plant Growth Substances and nitrogen Fixation by Azobacter and Azospirillum species Ph.D. Thesis. Indian Agricultural Research Institute, New Delhi.

Schank, S. C. Weiwer, K. L. and Mac Rae, I. C. 1981. Appl. Environ, Microbiol. 41:342-345.

Singh, C. and Sharma, B. B. 1993. South Indian Horti. 41(3) 131-134

Smith, R. L. Schank, S. C. Bouton, J. H. and Quesenberry, K.H. 1978. Ecological Bull. No. 26:380-385.

Sprent, J. I. 1985.

In:Plant for Arid lands (Eds. Wickens, G. E. Goodin, J. R. and Field, D. V.) George Allen and Unwin, Lond Pp 215-229

Strength, W. J., Isani B., Linn, D. M., William, F. D. Vandermolen, G. E., Laughen, B. E. and Kreig, N. R. 1976 Bacteriol Rev. 40:55- 155

Subba Rao, N. S. 1982 Advances in Agril. Microbiology. (Ed. N. S. Subba Rao) Oxfords IBH Pub. Co. New Delhi. Pp219-242

Subba Rao, N. S., Tilak, K. V.B.R., Singh, C. S. and Gautam, R. C., 1982. Proc. Natl. Symp. Biol. Nitrogen Fixation. Dept. of Atomic Energy. IARI, New Delhi. Pp507-516.

Tarrand, I. J., Kreig, N. R. and Dobereiner, J. 1978. Can. J. Microbiol 24:967-980

Tien, T. M., Gaskin, M. H. and Hubell, D. H.. 1979. Appl. Environ. Microbiol 37:1012-1024.

Tilak, K.V.B.R., Lakshmi-Kumari, M. and Nautiyal, C. S. 1979. Curr. Sci. 48:412-413.

Umalegarcia, M. Hubbel, D. N. and Gassons, M. H. 1980. Appl. Environ. Microbiol. 39:219-226.

Vasil, V. Vasil, D. K., Zuberer, D. K. and Hubbel, D. H. 1979. Pflanzen physiol. 95:141-147.

Von Bulow, J. and Dobereiner, J. 1975. Proc. Natl. Acad. Sci. USA 72:2389-2393.

Wani, S. P. 1990. Indian J. Microbiol. 30:363.

Wani, S. P. and Lee, K. K.. 1992. In: Fertilizers, organic manures, Recyclable Wastes and Biofertilizers (Tandon, HLS ed.) Fertilizer Development Consultation organization, New Delhi, India. p.91

Williams, M. A. 1959a. J. Bacteriol. 78:374-377

Williams, M. A. 1959b. Inter. Bull. Bacteriol, Nomencl. Taxon, 9:193- 196

Williams, M. A. 1960. Inter. Bull Bacteriol. Nomencl. Taxon, 10:137- 151

Williams, M. A. and Chapman, G. B. 1961. J. Bacteriol. 81: 195-203

Williams, M. A. and Rottenberg, S. C. 1957. Intern. Bull. Bacteriol. Nomencl. Taxon. 7:49-111.

17

Blue Green Algae as Biofertilizer

Rice is grown in about 40 million hectares of land in India and nitrogen requirement for the crop is 60 lakh tonnes based on the present levels of consumption. Rice production has to be sustained and increased to provide staple diet to our increasing population. Blue green algae is a potentional biofertilizer in rice because of their capacity to meet both carbon and nitrogen requirement naturally from the air. They are microscopic algae which are collectively called phytoplankton produce basic organic matter forming the first link in food chain and alongwith replenish the oxygen content of atmosphere. Algae contain proteins, Vitamins, iodine, bromine and substances of antibiotic and stimulatory nature. The soil algae are divided into chlorophyta (green algae) cynophyta (blue green) bacillariophyta (diatoms) and xanthophyta (yellow-greens).

De (1939) suggested that fertility of rice fields is maintained due to the growth of blue-green algae. Different types of blue-green algae have been reported to colonize the rice fields. The cells such as heterocysts in *Anabaena*, Nostoc, *Tolypothrix* etc. which have capacity to fix atmospheric nitrogen under aerobic condition and unicellular and non- hetercoystous forms also fix atmospheric nitrogen under microaerophilic conditions.

Blue-green algae (Cynobacteria) are prokaryotic, photo-lithotrophic, microscopic plants generally forming non-fibrous slimy growth on the soil or water surface. They have similarities with bacteria except the oxygen evolving photosynthesis.

All India survey of rice field soil showed that out of 2213 soil samples examined only 33% were found to harbour useful nitrogen fixing blue - green algae with wide variations in their population.

Multiplication

They predominantly multiply vegetatively through fragmentation. Asexual reproduction is brought about by the formation of exospores, endospore and akinetes. None of the BGA have sexual reproduction and form motile stages, Food reserves in BGA is in the form of glycogen which does not react with iodine to give violet colour as in green algae.

I. Nitrogen Fixation

The relative contribution of cyanobacteria as percentage of total nitrogen fixed in rice field varies a great deal and is dependent on physico- chemical condition of soil, climate and biotic factors. Quantitative data of nitrogen fixed by BGA in different countries is presented in Table 125.

Table 125: Nitrogen fixed by BGA in different countries

Country	Nitrogen Fixed (kg/ha)	Reference
India	15-53	De and Biswas 1952, Singh 1961, Venkatraman 1979, Goyal 1993
Japan	123	Okuda and Yamaguchi,1 955
Mali	50-80	Traore *et al.*, 1978
Philippines	18-33	Watanabe and Lee 1975 Alimango and Yoshida, 1977
Senegal	0-30	Renaud and Roger, 1 978

De and Mandal (1956) reported nitrogen fixation by BGA in six rice growing soils of West Bengal, values ranging from 13.8 to 44.4 lb N acre1 in cropped areas. By fertilization with phosphate and molybdenum, value reached up to 70 lb acre^{-1} in six weeks. Watanabe *et al.* (1951) observed an addition of 20 lb nitrogen acre^{-1} by *Tolypothrix tenuis*.

The fixed nitrogen by BGA was partly released extracellulary during the growth and rest was mineralized after its death. In paddy fields the death of algal biomass occurs due to drying and wetting cycles during the crop growth or after the harvest of the crop. This helps the succeeding crop by providing nutrients and improves the soil fertility.

In submerged rice field soils, biological nitrogen fixation is essentially an algal process contributing about 30 kg/N/ha. Using ^{15}N, values ranging from 40-80 kg N/ha/year have been recorded at the International Rice Research Institute, Philippines in nitrogen fixing rice fields. No depression of nitrogenase activity occurs in the presence of NH, nitrogen below 40 ppm or in the presence of various pesticides used for rice. The nitrogen fixed and liberated by algae is taken up by the rice plants as demonstrated by: N15 techniques. Besides nitrogen, the algae also synthesize and liberate growth promoting substances like auxins and amino acids which help the growth of rice plants (Venkatraman and Neelkantan 1967, Kaushik 1998). The algae that are generally used for inoculation are species of *Aulosira, Tolypothrix,* Nostoc, Anabaena, Plectonema, Scytonema *etc.* as composite culture.

II. Effect of BGA on Rice Yield and Nitrogen Economy

Extensive field trials conducted so far have clearly demonstrated a positive effect of these algae on the growth and yield of rice plants under varied agroclimatic conditions. At reduced levels of nitrogenous fertilizers, algal supplementation gives the same crop yield as that with the full dose of fertilizers. It was observed that at very low level of fertilizer nitrogen with algal supplementation, the crop yield was comparable to that in the next higher level of nitrogen without any algal application thus compensating the effect of 25 kg/N/ha. The results reported by several workers show that in areas where commercial nitrogenous fertilizer are not used for various reasons, algal application can give the farmer the benefit of applying 20-30 kg with N/ha and where commercial nitrogenous fertilizers are used, the dose can be reduced by 30 percent by supplementing it with algae thus saving in chemical fertilizers (Table 126).

Table 126: Blue Green Algae as Biofertilizer Effect of algal application on the yield of Rice in the Absence and in presence of different levels of added chemical nitrogen fertilizer.

Place	**Vasudeva-nallur (ASD5)**	**Kandaya-nlluar (IR-20)**	**Aduthurai (IR-20H) +**	**Ambassan-mudra (IR-20H) +**	**Kanpur (Jaya)**	**Sobour (Br.34)**
Control			4780	4640		
Algae			5400	4952		
Superphosphate (112ykg/ha)	4825					
Superphosphate +Algae	5670					
P205(45kg/ha) + 45KgN/ha						3491
P205(45kg/ha) + 45KgN/ha+Algae						4064
90kgN/ha		4250				
90kgN/ha+Algae		4750				
120kgWha		4822				
25:25:25 NPK			5280	5474		
25:25:25:NPK+ Algae			6060	5631		
50:25:25 NPK			5380	5605		
120kgN/ha		4822			5699	
120kgN/ha+ Algae		5852			6321	

Source: Venkatraman G.S. (1987)

III. Effect of BGA Inocula on Soil Properties

Algae are conspicuous due to their contribution to soil structure and erosion control. The surface blooms reduce erosion losses by binding together of soil particles. The physical condition of soil such as the size of soil aggregates was significantly augmented due to growth of algae. (Roychoudhary *et al.*, 1979, Kaushik and Krishanamurti, 1981) resulting in improvement of hydraulic conductivity, infiltration rate and soil permeability. Algal inoculation maintained both total and organic nitrogen levels beyond the tillering stage of the crop (Chopra and Dubey, 1971, Singh and Bisoyi 1989).

A bulk of organic matter produced by algal growth remains in the soil and the nutrients are mineralized and made available to the succeeding crop (Sankarn, 1971). However, organic matter build up is not stable due to its rapid mineralization in tropical climate.

In salt affected soil algal inocula bring down the soil pH and percent exchangeable sodium and improve the soil physical properties (Kasuhik 1994). The potential of cynobacteria in reclamation of salt affected soils was emphasized by Kaushik (1989) and reported by him to be comparable to gypsum application.

Many of the herbicides at recommended rates of field application do not interfere with proliferation and nitrogen fixation by cynobacteria (Goyal 1998). Similarly the common insecticides and fungicides do not limit the algae growth (Venkatraman and Rajalakshmi, 1971, 1972, Sardeshpande and Goyal 1982, Goyal 1993). Many blue-greens can tolerance high level of pesticides but there exists a wide range of variation in their tolerance limit (Venkatraman and Rajlakashmi, 1971). *Tolypothrix tenuis* and *Aulosira fertilissima* tolerated high concentrations of 2, 4-D, dalapon and ceresan. However, A. *fertilissima* is sensitive to carbofuron, linuron, diuron and propazine. Some organophs insecticides have been found to promote BGA growth (Goyal 1998). The composite cultures approach is better than use of inocula having monocultures in this context.

Benefit of Algal Fertilization

It can be easily managed for its use as a biological input in rice cultivation. It adds about 30 kg N/ha/season. It improves efficiency of chemical fertilizers and is ecofriendly. It has an ameliorating effect on saline and saline - alkail soils. It improves physical condition of soil by adding organic matter and polysaccharides.

IV. Technology for Production of Algal Inoculum

The algal inoculum can be prepared by the farmers in their fields without any appreciable investment. Based on the natural ecology of these algae, a simple rural oriented open air tank culture method has been developed at Indian Agricultural Research Institute, New Delhi. The 'Starter Culture' for multiplication, a mixture of *Aulosira, Tolypothrix,* Nostoc, *Anabaena* and *Plectonema* is available and obtained in packets of 400 gms at nominal cost.

Multiplication of Algal Inoculum

Prepare shallow trays of galvanized iron sheet or cement ponds or polythene lined pits in an open space (say 6'*3*39") In areas with assured water supply and heavy soil, lined multiplication method can be used by carving out small plots bounded by earthen bunds. The pond can be of any size (Plate 46) but in any case the width should not be more than 1.5 meters. Place 3-4 kg soil for one square meter area of the pond and 100 gm of superphosphate. The soil should be sandy loam or loam and not heavy type. Mix appropriate quantity of river bed sand if the soil is heavy type. The soil should be taken from an open and clean place or ponds. Fill the trays with water to about 10 cm height and add 2 ml of malathion (50% Ec). Mix well and allow to stand for a day. After the soil settles down and the water in the tray becomes clear, sprinkle 100 gms of powered, 'Starter Culture' on the surface of Water. During summer, when the temperature is 35°C and above, the growth of algae in the ponds is rapid and in about 10-15 days a thick algal mat is formed floating on the surface of water (Plate 47). During this period intermittently water the ponds to maintain the water level at 10 cm. When a thick algal mat is formed, stop watering the ponds and allow the contents to dry. Do not drain out the water. When it is completely dry, the algal mat will separate out from the soil and form flakes. These flakes are collected and kept is sealed polythene bags. Fill the tray again with water add superphosphate and

Plate 46: Open air Algal cul- ture tank

Plate 47: Growth of blue-green algae in Algae ponds.

Malathion and repeat the process. In three harvests of algal flakes, the entire soil in the tray is exhausted. After three harvests, the process is to be restarted. In one harvest about 1 kg of algal flakes are obtained from 1 m2 area. The collected algal flakes can be used as 'starter culture' also.

Recommendations for Field Application

1. Broadcast 10-20 kg of algal flakes per hectare 3-4 days after transplantation of rice on standing water in the field. Keep a little standing water in the field for about 10 days after inoculation.
2. When nitrogenous fertilizers are used, reduce their dose by 1/3 and use algae inoculum to get the yield equivalent to the full recommended dose of the fertilizers.
3. Apply the algal inoculum for at least 3-4 consecutive seasons.
4. Recommended pest control measures and other crop management practices can be undertaken without any detrimental effect on the algal activity.
5. The sundried algal flakes (Plates 48) packed in polythene bags can be kept for three years without any loss in their viability.
6. The algal inoculum, if necessary can be mixed with chemical fertilizers and applied to the field. However, it should be used immediately after mixing.
7. During rainy season when the rate of evaporation is very low, the algal mat from the ponds can be removed and dried in covered place by spreading it into a thin layer on smooth surface covered with a thin layer of sand. Precautions

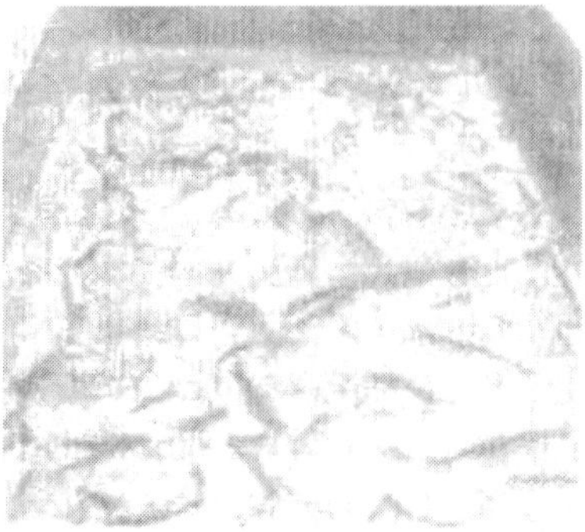

Plate 48: Sundried Algal flakes.

1. Do not allow the mosquitoes to breed in the ponds. If necessary, add malathion to prevent it.
2. Do not store the algal inoculum packets alongwith with agrochemical and fertilizers.
3. Ensure complete sun drying of algal flakes before packing.
4. Construct the ponds in such a way so that the length of the pond is against the normal wind directions.
5. The dried flakes, packed in polythene bags can be stored for three years without any loss in its viability.

Constrains in Adoption of BGA Technology

Although BGA technology is low cost and ecofriendly input for rice cultivation but it has not been adopted by the farmer in a significant way. In a recent study, the following constrains were highlighted (Nandan and Khairnar, 2003). There is lack of information and knowledge regarding benefits from this biofertilizer, including the marketing and availability aspects. The opportunity to visit field demonstrations which will directly involved the farmer for adoption of this technology is also lacking.

Similar findings were also reported by other investigators (Kamble and Shinde, 1998, Sharma *et al.*, 2000) The study has recommended, a need to organise 2-3 days training and demonstrations programmes for the farmers and estension workers for popularising this technology for rice culture.

References

Alimango, B.V. and Yoshida 1977. PI Soil 47: 239-244.
Chopra, T.S. and Dube, J.N. 1971 Pl, Soil 35: 453-462.
De, P.K. 1939. Proc. Royal Soc. Lond. B127:121.
De, P.K. and Biswas, NRD 1952. Indian J. Agri. Sci 22: 375-388.
De, P.K. and Mandal, L.N. 1956. Soil Sci 81: 453-458.

Goyal S.K. 1993. Proc. Indian Natn. Sci. Acad. B 59: 295.

Goyal S.K. 1998 In. Soil - Plant-Microbe Interaction in Relation to Integrated Nutrient Management (Ed.: B.D. Kaushik), Venus Publishers New Delhi pp 82-90.

Kamble, L.P. and Shinde, T.S. 1998. Maha J. Extn. 7: 219-21

Kaushik, B.D. 1989. Phykos. 28: 101-109.

Kaushik, B.D. 1994. In Natural Resources Management for Sustainable Agriculture and Environment (Ed. D.L. Deb.) Anghor Publishers (P)Ltd. Pp 403-416.

Kaushik, B.D. 1998. In Soil-Plant-Microbe Interactions in Relation to Integrated Nutrient Management (Ed. B.D. Kaushik) Venus Publisher, New Delhi pp 55-63.

Kaushik, B.D. and Krishanamurti. G.S.R. 1981. Phykos 20: 91-94.

Nandan, S.N. and Khairnar, S.P. 2003. Biofertilizer News letter II(2) 18-20.

Okuda, A and Yamguchi, M. 1955 Soil and Plant Food, 104.

Renaud P.A. and Roger, P.A.1978. Ecol. Bull Stockholm 26: 148- 157.

Roychodhary.P. Kaushik, B.D. Krishanamurti, G.S.R. and Venkatraman, G.S. 1979 Curr. Sci 48: 454-455.

Sankaran, A. 1971. Work done on blue green algae in relation to agriculture Bull. No 27, ICAR. New Delhi pp 28.

Sardeshpande, J.S. and Goyal, S.K. 1982. Proc. Natl. Symp BNF. Sharma, R., Singh, P.R. and Sharma, R.N. 2000. Maha. J. Extn. Edn. 19: 340-41

Singh, R.N. 1961. Role of Blue green Algae in Nitrogen Economy of Indian Agriculture, ICAR, New Delhi. pp 175.

Singh P.K. and Bisoyi, R.N.1989. Phykos 28: 181-195.

Traore, T.M., Royer P.A., Remaud, P.A. and Sasson, A. 1978. Cab ORSTOM Se Biol 13: 181-185.

Venkatraman, G.S. 1979. In Nitrogen and Rice, International Rice Research Institute, Manila, Philippines pp 311-321.

Venkatraman, G.S. 1987. In Organic Manures and Biofertilizers (Ed. A.C. Gaur) IARI New Delhi pp 12-17.

Venkatraman, G.S. and Neelkantan S.1967 J. Gen. App. Microbiol 13:53-61.

Venkatraman, G.S. and Rajalakshmi, B. 1971, Curri Sci 40: 143. Venkatraman, G.S. and Rajalakshmi, B. 1972, Ind. J. Agri. Sci 42: 119.

Watanabe, A.S. Nishigzaski, S and Konishi, C. 1951. Nature 108:748.

Watanabe, I.and Lee, K.K. 1975. In Biological Nitrogen Fixation in Farming system of the Tropics (Eds: A. Ayanaba and P.Dart) P 289- 305 John Wiley and Sons.

18

Phosphorus Solubilizing Microorganisms as Biofertilizer

Phosphorus is one of the essential plant nutrients, which is required in optimum amounts for proper growth of plants and soil biota. Being a constituent of ATP it is involved in various processes such as cell division, energy generation and transfer of genetic characteristics and regulation of metabolic pathways (Theodorou and Plaxton, 1993). Phosphorus constitutes about 0.2% plant dry weight (Schachtman *et al.,* 1998). The average soil contains 0.05% phosphorus but only one tenth of this is available to plants due to its poor solubility and chemical fixation in soil (Narayansamy *et al.*, 1981, Barber, 1984). About 98 percent of Indian soils have inadequate supply of available phosphorus (Ghosh and Hasan 1979, Hasan, 1994). A survey of Indian soils showed that out of 363 districts, only 2,2% were high, 51.5% medium and 46.3% were low in phosphorus content (Ghosh 1982).

The problem of P fertilization may become serious in coming years because of the fact that manufacture of phosphatic fertilizers requires the use of non-renewable resources such as high grade rock phosphate and sulphur which are becoming costlier. The situation is further aggravated by the fact that P is readily fixed in the soil and the average utilization efficiency of added fertilizer P by plants ranges from 15-25% only (Gaur 1982, 1990). The recent report showed that 42 per cent of soil samples fell in low category, 38 per cent in medium and 20 per cent in high category (Motsara, 2002). It has been observed that the vast alluvial tracts of the country are either low or medium in available P. Hence they need P application at recommended rates.

India has about 150 million tonnes of phosphatic rock deposits and most of them are low grade with impurities and not suited for superphosphate manufacture (Gaur 1985, 1990). Therefore, their utilization as direct fertilizer is of economic importance. Although rock phosphate is recommended in acidic soils, there are large areas of arable land world over which are non-acidic. In this context, the role of efficient rock phosphate dissolving microorganisms assumes greater importance for augmenting crop productivity.

Generally plants take up phosphorus as primary orthophosphate ion, H_2PO_4. The secondary orthophosphate, $H_2PO_4^{-2}$ is absorbed in smaller quantities of these two forms, plants prefer to take up monovalent $H_2PO_4^{-1}$ form (Furihata *et al.*, 1992). It helps in better growth of root system and is found in large quantities in plant seed and fruit. The deficiency of phosphorus reduces the size of plants and imparts it deep green colour.

I. Forms of Phosphorus and P Cycle

The inorganic unavailable forms are varisite, strengite, flourapatite, hydroxyapatite and tricalcium phosphate (Cosgroove, 1977). Phosphorus occurs in soil, manure, plant and microorganisms in various organic and inorganic combinations. Phosphate in soils' solution is generally negligible as compared with the unavailable forms.

The soil mineral P is normally present as orthophosphate ions complexed with Ca, Fe, Al, Mn and Si (Sanyal and De Datta 1991). In soils, insoluble forms of phosphorus are found as fluorapatite, hydroxyapatite or chlorapatite, as iron or aluminum phosphate or in combination with clay fraction. The calcium phosphates dominate in neutral to alkaline soils where as iron and aluminum phosphates occur in acid soils. In most alkaline soils, the activity of calcium is high and thus favours the formation of insoluble di and tricalcium phosphate. A part of the tricalcium may actually revert to hydroxyapatite.

Hydroxypatite is a constituent of bones. Bone meal contains 20 to 28 per cent total P_2O_5 and 50 per cent of this is citrate soluble. On alkaline soils, it is less effective as compared to soluble phosphatic fertilizer. Fluorapatite is a major constituent of rock phosphates. The total P_2O_5 contents of rock phosphates from different sources such as India, North Africa, USA, USSR etc., varies considerably.

The soil P cycle under undisturbed conditions such as forest or fallow soil is closed one and is sustainable. In farming system the P cycle is an open system because external sources of P have to be applied to soil to get good yields (Fig. 5). The most important aspects of the cycle are mineralization, solubilization and immobilization. The processes are of economic importance in P nutrition of plants.

The water-soluble phosphorus in rock phosphates is negligible. Although, rock phosphates are recommended in acidic soils but their use in neutral to alkaline soils may not be economical unless efficient rock phosphate solubilizing microorganisms are introduced in the rhizosphere of crops.

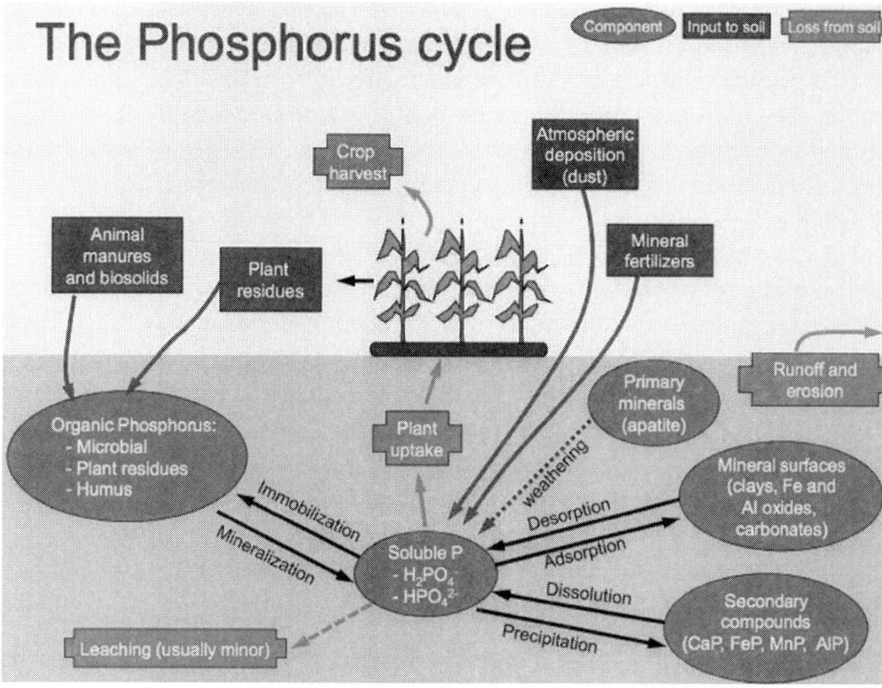

Fig. 5: Soil phosphorus cycle

Mineral Phosphate Solubilising Microorganisms (MPSM)

MPSM include different groups of microorganisms particularly bacteria, fungi and yeasts which have been reported to render inorganic phosphatic compounds soluble. (Gaur *et al.*, 1973, Gaur, 1990, Thakkar *et al.*, 1993, Narsian *et al.*, 1994, Narsian and Patel, 1995). Such bacteria and fungi grow in medium with insoluble phosphates such as tricalcium, ferric, aluminium and rock phosphates and bone meal. PSM besides assimilating phosphorus for their own requirement, release sufficient quantities in excess of their needs. The solubilization is not restricted to calcium salts only but salts of other cations as well are solubilized. The genera of bacteria and fungi such as *Pseudomonas*, *Bacillus*, *Aspergillus* and *Penicillium* have been reported to be active in the solubilization process (Gaur, 1990). The counts of PSB may range between 10^4-10^6/ g soil.

II. Occurrence and Isolation of Phosphate Dissolving Microorganisms

Many fungi and bacteria are potential solubilizers of bound phosphorus in soil (Goldstein, 1986, Gaur, 1990, Singh and Kapoor, 1992; Gupta *et al.*, 1998). Pikovskaya (1948) recommended a medium containing tricalcium phosphate as an insoluble phosphate source together with glucose, yeast extract and other salts. This medium is widely used for isolation, enumeration and maintenance of phosphate dissolving microorganisms, Addition of insoluble phosphate compounds like di-or tricalcium phosphate as fine powder in the medium has also been suggested by some workers (Johnston, 1951; Bunt and Rovira, 1955; Louw and Webley, 1958). Both the methods such as fresh precipitation of dicalcium phosphate or hydroxyapatite and incorporation of finally divided powder of insoluble phosphatic compounds in the medium are followed for isolation of phosphate dissolving microorganisms, enrichment culture techniques appears to be the most suitable. Bardiya and Gaur (1974) isolated very efficient and promising rock phosphate dissolving microorganisms by adding five per cent rock phosphate to soil for enrichment.

Katznelson and Bose (1959) observed that the capacity to solubilize phosphate by bacterial isolates depends upon the zone of their origin, those derived from the rhizosplane showed the highest activity, the rhizosphere intermediate and the control soil the least. They have shown that a majority of them come under the nutritional group requiring the unknown substances of yeast and soil extract for optimal growth.

Sperber (1957) reported that may of the bacterial isolates lost their ability to dissolve phosphate after a few sub-culturings on laboratory media. Louw and Webley (1959) reported that many isolates, which did not show any zone of solubilization on agar plates, released phosphate from rock phosphate when inoculated into liquid medium. Similar observation was made by Ahmad and Jha (1968).

The efficient phosphate solubilizing bacteria belong to the genera of *Bacillus* and *Psudomonas* (Gaur, 2002) although species of *Achromobacter, Micrococus, Sarcina, Enterobacter* and other have been reported to solubilize mineral phosphates (Table 127). These bacteria have been isolated from diverse sites such as rhizopshere of legumes, wheat, rice, arecanut palm, oat, jute, *dhaincha*, maize, chilli, desert soils of India, forest soils of Austria and Antartica lake, etc.

Table 127: Phosphate solubilizing bacteria and sites of isolation

Microoganism	**Site of Isolation**	**Author /Year**
Bacillus circulans, B. megaterium. B. Polymxa, B. Pulviferium pumilus, B. brews	Legume and Cereals, rhizosphere soils	Sundara Rao and Sinha 1963; Paul, 1966
B. megateriaum, Mkxococcus Sarcina	Broad beans and wheat rhizosphere soils	Taha *et al.*, 1969
Pseudomonas striata, P. rathonis	Rock phosphate deposit soil	Bardiya and Gaur, 1974
Bacillus, Pseudomonas and *Achromobacter*	Different crop rhizosphere soils	Barea *et al.*, 1976
Pseudomonas	Coconut and Cacao rhizosphore soils	Nair and Rao, 1977
P-solubilizing bacteria	*Zostera marina* rhizosphere soil	Craven and Hayasaka, 1982
Bacillus cereus and *Pseudomonas fluorescens*	Desert soil of Rajasthan, India	Venkateshwarlu *et al.*, 1984
Pseudomonas sp.	Wheat rhizosphere	Anas *et al.*, 1985
Pseudomonas aenjginosa, Bacillus megaterium	Jute and *Dhaincha* rhizosphore soils	Sattar and Gaur, 1985
PSB&PSF	Arecanut palm rhizosphere	Bopaiah, 1985
Pseudomonas fluorescens, *P. putida, P. syringae*	Antartica Lake	Rao *et al.*, 1989
Bacillus subtilis, B. cereus, B. megaterium, B. brews		Ali *et al.*, 1989
Baclus *circulans, B. sulitilis*	Oats and Pigeonpea rhizosphere soils	Gaind and Gaur, 1991
Pseudomonas aemginosa, P cepacia, P. fluorescens, P. putida	Wheat rhizosphere soil	Freitas and Germida, 1990
Pseudommas sp.	Forest soil (Austria)	IIlemer and Schinner, 1992
Bacillus licheniformis, B. mycoidas. B. megaterium	Paddy rhizosphere soil	Watanabe and Hayano, 1993
Bacillus megaterium, B. subtilis, B. *polymyxa, Serratia marcescens, Pseudomonas aeruginosa, Enterobacter aerogenes.*	Crops rhizophore soils	Thakkar *et al.*, 1993
Pseudomonas fluorescens	Rhizosphere/ Nonrhizosphere soils	Dave and Patel, 1999

Table 128: Phosphate solubilizing fungi/yeast and sites of isolation

Organism	Source	Author /Year
Aspergillus awamori, Penicillium digitatum	Legumes rhizosphere soil	Bardiya and Gaur, 1974
Aspergillus sp.		Banik and Dey, 1981
Aspergillus niger, Penicillium pinophyllum	Desert soil (Rajasthan, India)	Venkateshwarlu, et *al.*, 1984
Aspergillus sp, *Penicillium* sp. *Penicillium* sp.	Maize, Soybean and Chilli rhizosphere soils Forest soil (Austria)	Antarkanonda *et al.*,1990 Illemerand Schinner, 1992
Aspergillus aculeatus, A.niger	Rhizosphere soil	Narsian *et al.*, 1994
Schwanniomyces occidental	Legume rhizosphere soil	Bardiya and Gaur, 1974
Saccharomyces cerevisiae		Narsian and Patel, 1995, Abu Samaha and Patel, 2000

III. Solubilization of Insoluble Phosphates in Liquid Medium

Moreau (1959) observed that in a liquid medium containing tricalcium phosphate, the liberation of phosphorus was done mainly by bacteria and only by some fungi isolated from forest soils, most strains liberated 3 to 4 to 15 to 17 per cent of the total and some strains solubilized 20-40 per cent phosphorus within about 10 days. Myskow (1960) made an experimental study of microbial solubilization of phosphorus and found that 2 strains, of *Achromobacter* and *Pseudomonas* were efficient to dissolve calcium phosphate.

In vitro solubilization of insoluble phosphates by soil fungi isolated from forest tree seedbeds was shown by Agnihotri (1970). Of the 18 fungi tested, 14 solubilized tricalcium phosphate, 11 fluorapatite and 7 hydroxyapatite and the rest failed to solubilize them. *Aspergillus niger*, *A. flavus, Fusarium* oxysporum, *Scelerotium rolfsii*, *Cylindrocladium* sp. and *Penicillium* sp. showed significant solubilization of all the three phosphates tested. Damping off fungi viz., species of *Pythium* and *Rhizoctonia* failed to dissolve any of these phosphatic compounds.

Bajpai and sundra Rao (1971) selected four bacteria, *Bacillus megaterium* var. *phosphaticum* (USSR), *Bacillus megaterium* (11), *Bacillus circulans* (12) and *Escherichia freundii* (13) for solubilization of tricalcium phosphate in the liquid medium at different pH containing a known quantity of tricalcium phosphate and incubated for two weeks, The results showed the solubilzation of tricalcium phosphate at the wide range of pH. An Indian strain of *B. megaterium* solubilised good amounts of phosphate even at pH 9. The optimum pH for maximum solubilization of phosphate for each of the organisms was different. *Pseudomonas* spp. Isolated from Maharashtra soils could colubilize

13-58 per cent of tricalcium phosphate in a liquid medium. *P. putida* gave the maxiumum dissolution (Ostwal and Bhide 1972).

Bardiya and Gaur (1972) tested the solubilization of tricalcium phosphate by *Bacillus megateruim* var. *phosphaticum* (PB). *B. megaterium* (R3, 28), *B. polymyxa* (H_3, H_5, H_{10}), *B. circulans*, *Escherichia freundii* (13) and Czechoslovakian culture (Fosfo 24). *B. pulvifaciens* (R5) solubilized 76 mg P_2O_5 followed by *B. pulvifaciens* R1(57.0 mg), *B. megaterium* (PB) 36.5 mg, *B. pumilus* (31 mg), *B. megaterium* 28 (26 mg) *B. polymyxa*, H-5 (22 mg). The others solubilized varying amounts ranging between traces to small amounts.

Tricalcium phosphate was solubilised by PSM to a maximum extent followed by aluminium and iron phosphate (Banik and Day 1981). Bacteria were found most active followed by *Penicillium* and *Aspergillus*. *Streptomyces* were least effective. Similar, observation in respect to the activity of *streptomyces* against rock phosphate solubilization was made by Bardiya and Gaur (1974).

Ortuno *et al.* (1977-78) reported that solubilization of tricalcium phosphate started after 24 hours of incubation and reached a maximum on the 7th and 15th day with *Aspergillus niger* and *Pseudomonas fluorescens*, respectively.

The periodic solubilization of ferric and aluminium phosphate was investigated in liquid medium (Gaur and Gaind 1983). The results showed that the maximum solubilization of aluminium and iron phosphate by *Aspergillus* awamori was achieved on 10th and 11th day. The maximum amounts of P_2O_5 solubilised by P. *striata* from dicalcium, tricalcium, aluminium and iron phosphate were 30.8, 58.4, 4.24 and 26.6 per cent respectively.

Ahmad and Jha (1968) reported the solubilization of rock phosphate and hydroxyapatite by Gram-positive and Gram-negative rods and cocci- shaped bacteria and genera of fungi such as *Aspergillus*, *Penicillium* and *Rhizopus*. Gaur *et al.*, (1973) reported the presence of gram-positive and sporeforming bacteria and fungi. Among the isolates from rock phosphate, *Aspergillus carbonum* proved the best as they utilized three different types of rock phosphate as well as organic phosphates such as lecithin and calcium phytate. The cultures isolated from alluvial soil were identified as *Bacillus* sp. and *Aspergillus* spp. These cultures which were isolated as organic phosphate minerlizers were incapable of solubilizing rock phosphate. However, they solubilised tri-calcium phosphate to an appreciable extent. Phosphate solubilizing organisms especially adopted to high altitude from Himalayan soils were isolated. For example a fungal isolate, *Paecilomyces lilacinus* showed a higher degree of P-solubilizing activity at 10°C than 24°C (Pandey and Palni, 1998).

IV. Solubilization of Rock Phosphate

The earlier attempts on isolation of PSM were confined to solubilization of tri-calcium phosphate which were more easily solubilised than rock phosphate. The findings of Bardiya and Gaur (1972) showed that out of 14 tricalcium phosphate solubilizing bacteria tested, only one culture could solubilise 5 to 10 per cent of rock phosphate whereas others were not effective. Bardiya and Gaur (1974) isolated by enrichment technique, rock phosphate dissolving bacteria, fungi and yeasts from the rhizosphere of leguminous crop and soils samples obtained form rock phosphate deposits area. The summarized results showing range of solubilisation and pH change are presented in Tables 129 and 130.

Six isolates of yeast were obtained from rhizosphere soils of cowpea, soybean, black gram and groundnut. One of the isolates solubilised the least amout of 1.3 mg P_2O_5 whereas the other five cultures solubilised between 9 to 10 mg P_2O_5. The most effective culture was identified as *Schwanniomyces occidentalis*. Eleven isolates belonging to *Aspergilus* group solubilised from 3.1 to 26.6 mg P_2O_5. The pH of the medium became acidic and was found in the range of 3.3-5.5. *Aspergillus awamori,* isolated from the rhizosphere of soybean crop showed the maximum dissolution of rock phosphate. Six of *Penicillium* sp. showed varying degree of efficiency in rock phosphate solubilization.

Table 129: Efficiency of rhizosphere isolates in rock phosphate solubilization

Rhizosphere	Organism	No. of isolates	PH change	PA solubilised (mg/100 ml)
Control (uninoculated)	-	-	6.5	0.9
Cowpea, soybean, Black gram, groundnut	Yeast, *Asperglus* sp.	6	3.9-4.7	1.3-10.1
Green gram, soybean Black gram, groundnut	*Aspergillus* sp.	11	3.3-5.5	3.1-25.6
Green gram, soybean	*Penicillium* sp.	6	3.6-5.2	5.6-17.6
Cowpea, green gram	*Curvularia* sp.	3	4.5-5.7	4.1-11.0
Black gram	*Tricoderma* sp.	2	5.0-5.4	5.9-10.3

Table 130: Efficiency of isolates of rock phosphate deposit soils

Location	No. of isolate	Characterization	PH change	P_2O_5 solubilized (mg/100 ml)
Control	Uninoculated	-	6.2	1.31
Kanpura I	2	Actinomycetes	3.64.7	1.5-3.2
Matoon II				
Kanpura I	3	Yeasts	4.0-4.9	1.4-2.4
Matoon I & III				

Kanpura I & II	20	Gram (-) rods non- capsulated	2.8-3.2	9.5-15.2
Matoon I, II & III		non- sproulated motile bacteria		
Kanpura I	8	Gram (-) rods non- capsulated	3.2-3.5	5.3-9.0
Matoon I, II & III		non- sporulated non-motile bacteria		
Kanpura I	1	Gram (+) sporulated motile bacterium	5.9	1.5

Penicillium digitatum obtained from the rhizosphere of soybean crop showed maximum solubilization (17.6 mg P_2O_5). Three isolates of *Curvularia* sp. solubilised between 4.1 to 11.0 mg P_2O_5. Two isolates of *Trichoderma* sp. solubilised between 5.9 to 10.3 mg P_2O_5. The results indicated that rhizosphere of cowpea and soybean crops harboured efficient rock phosphate solubilizing fungi and yeasts.

The effective strains of bacteria were found in the soils samples of rock phosphate deposit area obtained from Udaipur. The soil samples from Kanpura and Matoon harboured the most efficient rock phophate dissolving bacteria as 20 isolates solubilising P_2O_5 in the range of 9.4 to 15.2 mg were obtained. The medium was found acidic with a pH range between 2.8 to 3.2. The isolates were Gram-negative, non-spore former, non-capsulating and motile rod shaped bacteria. They were identified as *Pseudomonas striata* and *P. rathonis*. Eight cultures of Gram-negative, rod shaped, non- capsulating and non-motile bacteria solubilizing comparatively less rock phosphate between 5.3 to 9.0 mg P_2O_5 were also isolated from the soils. They also produced less acidity.

The solubilization of 32p tagged hydroxyapatite by *Pseudomonas striata, Bacillus polymyxa* and *Aspergillus awamori* was studied in liquid medium (Arora and Gaur 1978). The extent of phosphorus solubilization was increased progressively upto 5th week of incubation, *A. awamori* was at the best followed by *P. striata* and *B. polymyxa.* At the sixth week of incubation, there was a substantial increase in the solubilization. This can be explained by the fact that during the process of phosphate solubilization, a part of the phosphate is assimilated by the concerned organisms which will be released after the autolysis of the aged cells.

The dissolution of different kinds of low grade powdered rock phosphate using efficient culture of bacteria such as *Pseudomonas striata, Bacillus polymyxa, B. megaterium, B. pulvifaciens, B. circulans* and *Citrobacter fraundii* and fungi viz., *Aspergillus awamori, Penicillium digitatum, Aspergillus niger* ANI and *A niger* AN30 were examined in liquid medium (Arora and Gaur 1979). P. *striata, B. megaterium* and *B. pulvifaciens, A. niger* and *P. digitatum* effectively solubilised tricalcium phosphate. Rock phosphates were also solubilised but relatively to a lesser extent than tricalcium phosphate. Among the fungi, *Aspergillus awamori* and *A. niger* while among bacteria, Pseudomonas striata and *Bacillus polymyxa* were found to be the best.

The phosphorus solubilizing ability of the selected efficient strains of fungi, *Aspergillus awamori* and *Penicillum digitatum* and bacteria, *Psudomonas striata* and *Bacillus polymyxa* were also confirmed on Pikoskaya's solid and liquid medium by other workers (Wani and Patil 1979, Wani *et al.*, 1979). The selected cultures exhibited zone of phosphorus solubilization on solid agar medium. The studies in liquid medium showed about 44-65 per cent solubilization of tricalcium phosphate, and about 14-28 per cent solubilization of rock phosphate. An overview of biotechnological techniques, mainly based on solubilization process involving agro-industrial residues, is presented (Vassilev and Vassilev 2003).

V. Factors Affecting Phosphate Solubilization

Carbon sources: Rose (1957) reported that glucose or xylose was the best source of energy for fungi in liquid media where as other workers reported that either yeast extract or soil extract was essential for proper growth of PSM under culture conditions (Katznelson and Bose 1959; Katznelson *et al.*, 1962). The solubilization of rock phosphate using different types of carbon sources was studied by Gaur (1990). Glucose, sucrose, galactose and arabinose were effective in that order. The yeast, *(Schwanniomyces occidentalis)* utilized a few carbon sources such as sucrose, glucose, arabinose, xylose and glactose in this order, *Penicillium digitatum* solubilised maximum P_2O_5 in presence of glucose followed by sucrose, mannitol, arabinose and xylose. The organism could use the carbon sources except galactose quite efficiently in relation to phosphate solubilization.

Gaur and Gaind (1983) reported that during 15th day of incubation in liquid medium maximum rock phosphate solubilization by *Aspergillus awamori* occurred with sucrose or mannitol as compared to glucose. It was observed that by increasing the concentration of glucose as carbon source from 1 to 3 per cent in broth the quantity of phosphate solubilised was enhanced. The results also indicated that greater acidity was produced when glucose was used at higher rate (3 and 2%) as compared to 1 per cent.

Nitrogen sources: The strains of *Pseudomonas striata* utilized different sources of nitrogen such as ammounium sulphate, aspragine, urea, ammonium nitrate, potassium nitrate, calcium nitrate except sodium nitrate. *Schwanniomyces occidentalis* utilized ammonium or amino nitrogen only. They could not utilize nitrate form of nitrogen. Ammonium sulphate was the best source both for the bacteria and the yeast. The fungi could solubilise a good amount of rock phosphate in presence of different nitrogen sources. The best sources for *Asepergillus awamori* was ammonium nitrate followed by aspragine. *Penicillium digitatum* was effective in presence of sodium nitrate followed by potassium nitrate.

Rock phosphate concentration: There was a significant improvement in the solubilization of rock phosphate with increasing concentration of P_2O_5 up to 100 mg added to 100 ml medium by bacteria or fungi and higher amounts than this did not show any further improvement (Fig. 6). Gaur and Sacher (1990) showed that lower the quantity or phosphate added to the medium, greater was the solubilization percentage by *Aspergillus awamori*.

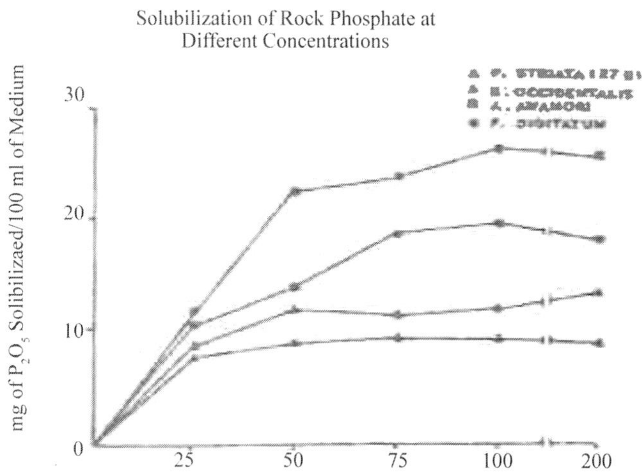

Fig. 6: Microbial solibilization of rock phosphate applied at varying rates

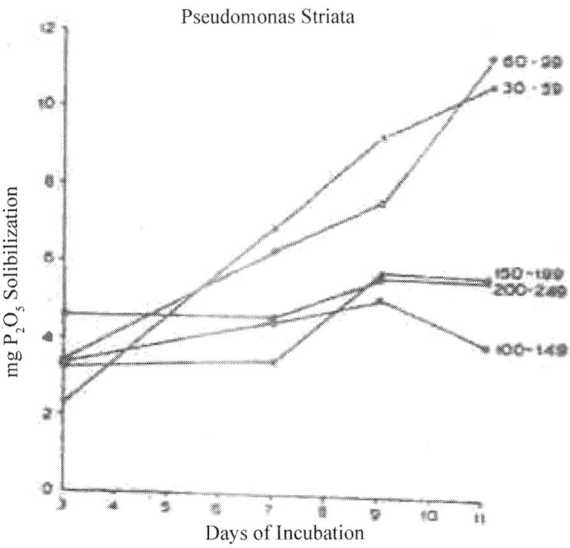

Fig. 7: Microbial solibilization of different mesh rock phos- phate by *Pseudomonas striata.*

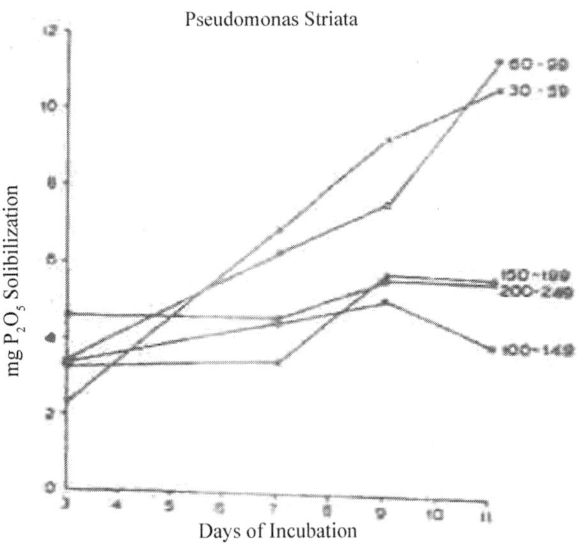

Fig. 8: Microbial solibilization of different mesh rock phos- phate by *Aspergills awamori.*

Particle size of rock phosphate: Gaur (1986) reported about the rates of solubilization of Mussorie rock phosphate of different mesh by *Pseudomonas* striata and Aspegillus awamori in the liquid medium (Figs. 7 & 8). The findings showed that the maximum phosphate was rendered soluble when rock phosphate particle size ranged between 30-59 and 60- 99 mesh. The efficiency of microbial solubilization was reduced in presence of finer rock phosphate particle of 150-250 mesh as compared to coarser particles. The results showed that grinding of rock phosphate beyond 60- 99 mesh is not advantageous for effective solubilization by fungi or bacteria.

pH: The results showed that the optimum pH for bacteria was around 6 and the activity of the fungi or the yeasts was the best in the pH range between 5 to 6. The fungal activity was greatly reduced at pH 7 to 8 (Fig.9).

Temperature: The optimum temperature for the best activity of Pseudomonas *striata* ranged between 20 to 30°C. The temperature range between 25-30°C was optimum (Fig. 9). For the fungi, 30°C was the optimum. The activity of these organisms was drastically reduced at higher temperature such as at 35°C and 40°C (Plate 49). There is a need to isolate active phosphate solubilising organisms which can grow at higher temperatures and at least tolerate upto 45°C.

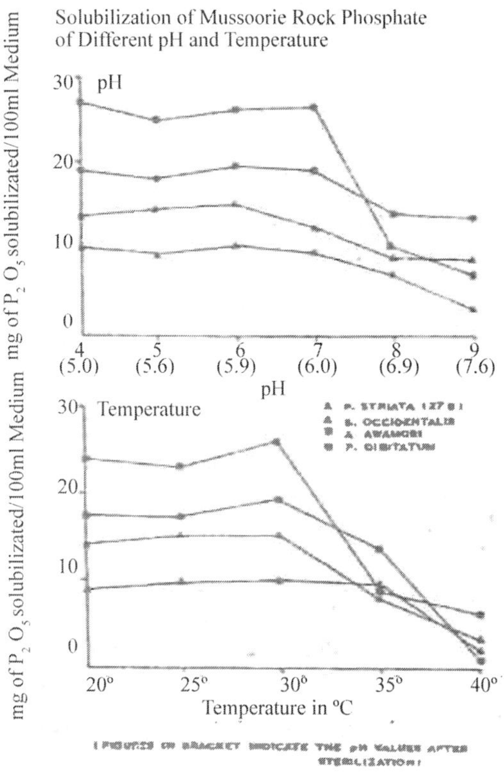

Fig. 9: Effect of pH and temperature on rate of rock phosphate solubilization by phosphate solubilizers.

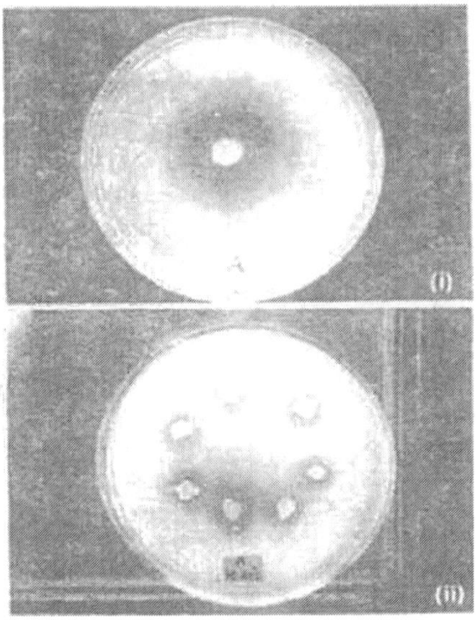

Plate 49: Solubilization of Tricalcium phosphate by *Pseudomonas striata* was reduced at 40°C as compared to 35°C. (i) 35°C (ii) 40°C

Aeration: A significant increase in rock phosphate solubilization was observed in shake culture as compared to static culture with 3 different strains of *Pseudomonas striata* and *Aspergillus awamori.*

Incubation period: Several workers have reported that under cultural conditions the phosphate solubilization increased progressively (Fig. 10) upto 10-15 days, (Gaur 1985 a, b).

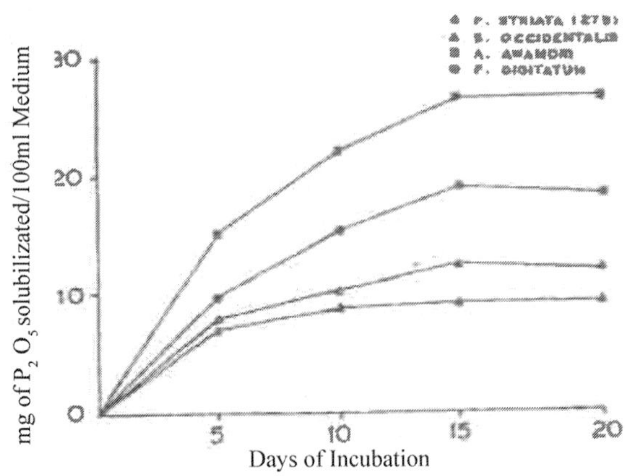

Fig. 10: Effect of incubation period on microbial rock phosphate solubilization

VI. Solubilisation or Rock Phosphate in Soils

Solubilisation of Mussorie rock phosphate in three different soils such as sandy loam alluvial soil of Delhi, a clay loam hilly soil of Dehradun and a red loam soil of Jablapur was examined. The soil varied in texture, organic matter status, pH and available P_2O_5 content. The red soil contained higher amount of available P_2O_5 as compared to alluvial and hilly soils containing about 19 ppm (Gaur 1990).

The results showed that inoculation of alluvial soil with PSM significantly augmented the available P_2O_5 content upto 45-60 days (Fig. 2 5 11). The fungi were found to be the P205 solubilized (ppm) Solubilization of rock phosphate in alluvial soil 90. Without farm yord manure to the alluvial soil poor in organic matter improved the activity of the phosphate solubilising microorganisms. The maximum solubilization was observed on 30th day and then the lowering trend was observed.

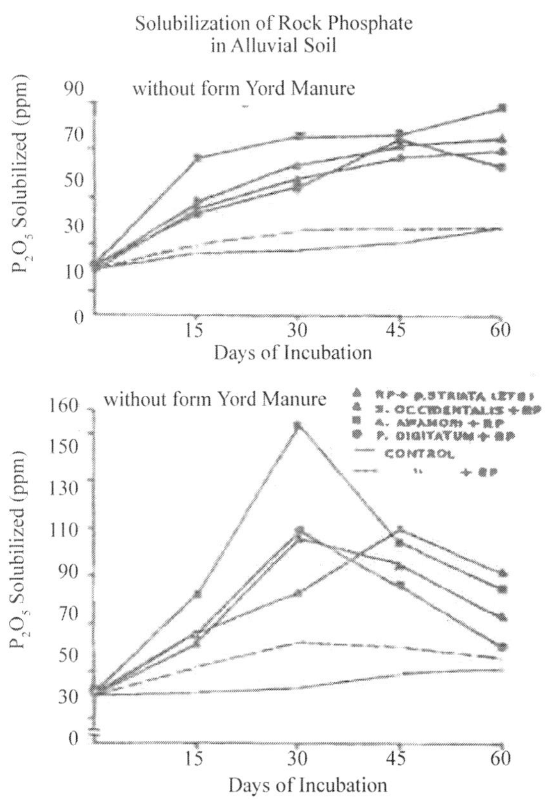

Fig. 11: Microbial solubilization results clearly showed that of rock phosphate in alluvial application of one per cent farm yard soil.

The maximum solubilization of Mussorie rock phosphate in a hilly clay loam soil occurred on 30th day of incubation with the fungi and *Pseudomonas striata* (Fig. 12). *Panicillium digitatum* performed very well in this soil and showed the maximum solubilisation. The amount of phosphate solubilised by microbial inoculant was about two to two and a half folds more than the control. However, with further incubation, phosphate availability of the soil was reduced. Similar results were observed in acidic red loam soil (Fig. 13). *Pseudomonas striata* showed the maximum solubilization on the 30th day of incubation. Although the soil was acidic and high in available P_2O_5 the inoculation by organisms was found beneficial.

Phosphate solubilizing microorganism do convert insoluble phosphates to available form in soils both under greenhouse and field conditions (Brown, 1974; Thakkar *et al.,* 1993, Tinker, 1980 Kucey *et al.,* 1989). The available phosphorus content of soils inoculated with phosphate solubilizing microorganisms have been found to increase in many studies (Gaur, 1990, Gaur *et al.,* 1973, Banik and Dey 1981). The bicarbonate soluble P content increased due to inoculation of soil with *Penicillium bilaji (*Kucey 1938). Similarly inoculation with *Bacillus lichenformis* in sandy soil significantly increased P in soil both in presence and absence of rock phosphate (Gupta *et al.*, 1993). It was reported that phosphate solubilizing capacity of terai and laterite soils of west Bengal (India) by PSM used was 4-5 time lesser for rock phosphate than Ca3(PO4)2. *Bacillus* sp. PBL, of laterite soil and *Aspegillus* sp. PFT1 of terai soil solubilised both $Ca_3(PO_4)_2$ and rock phosphate efficiently in culture media. However other *Bacillus* spp. (PBT1, PBT4 and PBL5). although solubilized $Ca_3(PO_4)_2$ but failed to solubilize rock phosphate (Kole and Hajra, 1997).

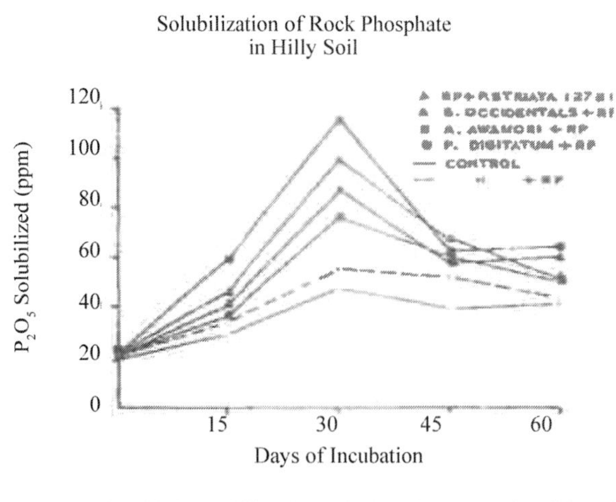

Fig. 12: Microbial solubilization of rock phosphate in hilly soil.

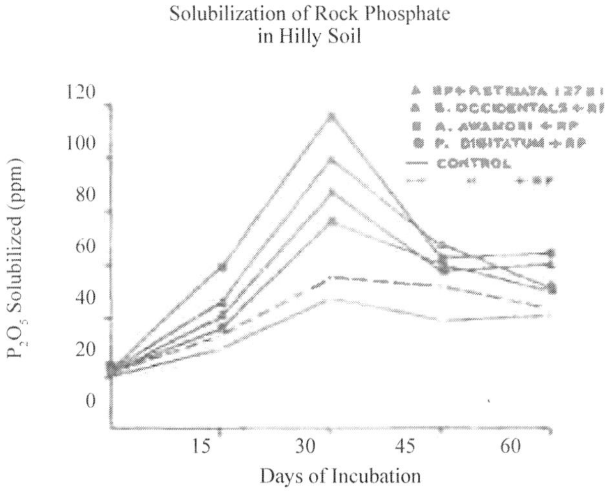

Fig. 13: Microbial solubilization of rock phosphate in red soil.

VII. Mechanisms of Rock Phosphate Solubilization

The production of organic acids by microorganisms is one of the important mechanisms but other products such as CO_2, H_2S and alkalinity production may also be one of the cause, A fall in pH during the growth of PSM in the culture medium has often been reported (Gerretsen, 1948, Sperber, 1958a, b, Swaby and Sperber 1958, Louw and Webley 1959, Chhonkar and Subba Rao 1967, Taha *et al.*, 1969, Gaur 1972, Bardiya and Gaur, 1972, 1974, Arora and Gaur 1979, Sattar and Gaur 1984). PSM produced monocarboxylic acid (acetic, formic), monocarboxylic hydroxy (lactic, gluconic, glycolic) monocarboxylic keto (2-keto gluconic) dicarboxylic (oxalic, succinic) dicarboxylic hydroxy acids (malic, maleic) and tricarboxylic (citric) in liquid media from simple carbobydrates (Sperber 1957, 1958, Swaby and Sperber 1958, Louw and Webley 1959 and Taha *et al.,* 1969). Tri and dicarboxylic acid are more effective than other types.

The production of organic acids by rock phosphate solubilising microorganism was investigated. Three strains of *Pseudomonas striata* produced seven acids such as malic, glyoxalic, succinic, fumaric, citric, tartaric, and alpha-ketoglutaric. In culture filtrate of *Schwanniomyces* occidentalis, 5 organic acids (succinic, fumaric, tartaric, citric and alpha- ketobutyric) were identified whereas *Asprgillus awamori* or *Penicillum digitatum* synthesized succinic, citric and tartaric acids only (Gaur 1990).

Their capability was further confirmed by electron-microscopic studies with an object to examine the morphological changes in the weathering of rock phosphate and tricalcium phosphates (Plate 50 a, b, c) by the action of phosphate solubilizing bacteria and the mechanism of the action involved in the process (Gaur *et al.*, 1979). Microphotographs showed that *Pseudomonas striata, Bacillus polymya,* B. *megaterium* and *B. pulvifaciens* were rod shaped, flagellated and had a size of 1.1 to 2.2 um (Plate 51). Drastic weathering of tricalcium and rock phosphate was observed due to action of bacteria,

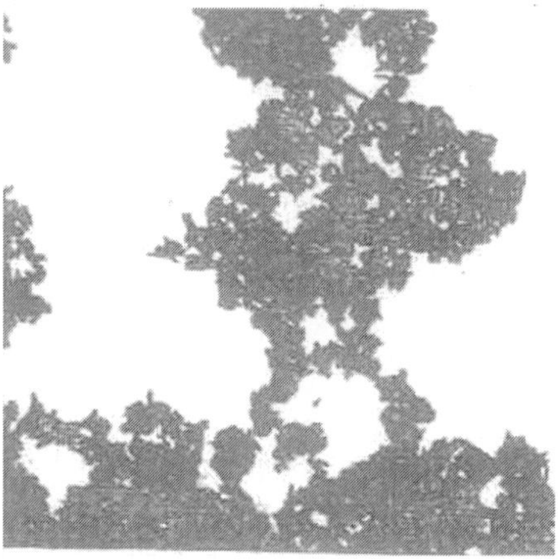

Plate 50a: Microphotograph of tricalcium phosphate

Plate 50b: Microphotograph of Udaipur rock phosphate.

Plate 50c: Microphotograph of Mussori rock phosphate.

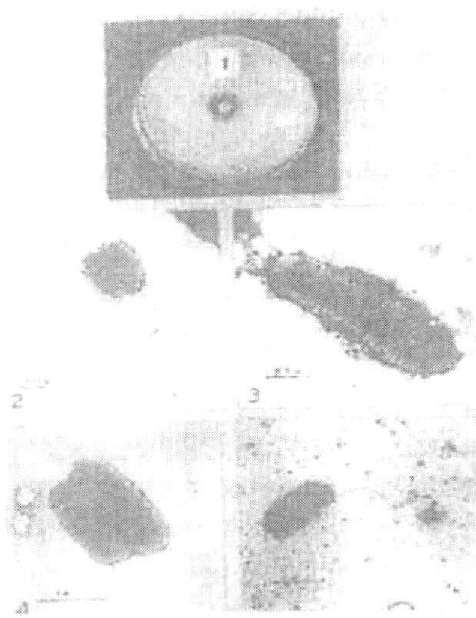

Plate 51: Microphotographs of phosphate solubilizing bacteria.

Plate 52a: Breakdown of tricalcium phosphate by *Pseudomonas striata.*

1. Colony with halo zone *(Pseudomonas striata)*
2. *Pseudomonas striata*
3. Bacillus *polymyxa*
4. B. *megaterium*
5. B. *pulvifaciens*

Plate 52b: Breakdown of rock phosphate by *Pseudomonas striata.*

P. striata in particular (Plate 54 a, b). The crystals were broken down into smaller fragments. The extent of solubilization and ultimate release of soluble phosphate also depends on the accessory minerals present in rock phosphates.

Pareek and Gaur (1973) studied the extent of solubilization of tricalcium phosphate and rock phosphate by organic acids, Aliphatic acids were found to be comparatively more effective in phosphate solubilization than phenolic acids and citric acid and fumaric acid showed the highest P solubilising ability (Table 131).

Table 131: Release of phosphorus from tricalcium phosphate and rock phosphate due to action of organic acids.

Acid	Tricalcium phosphate		Rock phosphate	
	pH	P released (ug/ml)	PH	P released (ug/ml)
Gortel	6.9	6.5	7.1	1.5
Citric	4.5	38.0	3.5	29.5
Fumaric	4.4	40.0	4.0	34.0
Salicylic	4.8	32.0	4.2	11.3
Vanillic	5.3	30.5	4.2	8.5
Protocatechuic	5.2	29.0	4.2	8.5
Benzoic	5.1	18.5	4.2	5.0
Phthalic	4.8	34.6	4.1	26.0
Pyrogallic	5.7	22.5	5.2	5.2

Chelating substance have also an important role in solubilization of insoluble phosphates. 2-Ketogluconic acid, a powerful chelator of calcium is produced by many aerobic bacteria and is very effective in solubilising insoluble phosphates such as hydroxyapatite, fluorapatite and aluminium phosphate (Duff *et al.*, 1963). Banik and Dey (1983) have also confirmed that microorganisms producing 2-ketogluconic acid were more efficient in -solubilization of insoluble phosphates.

Several type of organic acids which are produced as a result of microbial activity have been detected in soil (Chandrasekaran 1969, Gaur and Pareek 1974).

Proton and enzyme theory

Esterase like enzymes release phosphorus from organic compounds, PSM also produce phosphatase enzyme along with acids which cause the solubilisaiton of phosphorus in aquatic conditions (Al ghazali *et al.*, 1986). Phytase activity is known to be common in bacteria (Arthrobacter, *Proteus, Serratia, Pseudomonas* and *Bacillus*) and fungi *(Aspergillus, Penicillum)*. Bacteria, fungi and actinomycetes can use phospholipids (lecithin) and thus release the phosphate alongwith other products.

Proton extrusion: Solubilization without acid production is due to release of protons accompanying respiration or ammonium assimilation (Kucey 1983, Dighton and Boddy 1989, Parks *et al.* 1990). Insoluble phosphates in this manner are directly solubilised at the microbial cell surface.

Carbon dioxide production: The production of carbonic acid increases phosphate availability (Pierre, 1948, Waksman, 1952, Norman 1953 and Moreau 1959).

$Ca_3(PO_4)_2+CO_2+H_2O \rightarrow 2CaHPO_4+CaCO_3$

$Ca^3(PO_4)_2+2CO_2+H_2O \rightarrow 2Ca(H_2PO_4)_2+2CaCO_3$

Hydrogen sulphide: It is produced by fermentative microorganisms from sulphur containing amino acid as well as by an anaerobic sulphate respiring bacteria like *Desulfovibrio* and *Desulfotomaculum* (Doelle 1969). The H_2S reduces ferric phosphate to ferrous sulphide leading to release of H_2PO_4 ions. Increase in P_2O_5 availability in water logged soils amended with sulphur has been reported to be due to reduction of ferric phosphate resulting into solubilisation of iron and release of phosphorus (Swaby and Sperber 1958).

Inorganic acids: The nitrifying bacteria, *Nitrosomonas* and *Nitrobacter* which produce nitric acid from ammonium salt may be involved to a small extent. *Thiobacillus ferroooxidans* and *Ferrobacillus ferrooxidans* which oxidize iron and form sulphuric acid or convert ferric phosphate to ferric hydroxide ($Fe(OH)_3$) and release H_2PO_4 ions in solution. Thiobacilli which oxidize reduced sulphur compounds to sulphate, form sulphuric acid. These acids (nitric, sulphuric acid) bring down soil pH or may directly react with Fe, Al or Ca phosphates depending on soil pH and phosphorus is released by chemical reactions (Doelle, 1969).

Production of sideophores: These are the iron-chelating compounds produced by certain soil microorganisms. The role of siderophores in phosphorus solubilization appears to be important in acid soils where ferric phosphate is one of the major form of insoluble phosphates (Bossier *et al.*, 1988).

VIII. Development of Microphos Culture

By developing improved techniques for isolation of rock phosphate solubilising microorganism and systemic investigations, a few efficient bacteria such as *Pseudomonas striate, Bacillus polymyxa* and fungi *Aspergillus awamori* were selected for preparation of carrier based inoculant. (Fig. 14).

The survival/proliferation of the selected bacteria and fungi was studied in different carriers such as peat, soil, farmyard manure and dung cake powder. Sterilised peat powder proved to be the best carrier for survival and multiplication of the phosphate solubilising bacteria or fungi. The sporulating *Bacillus polymyxa* and *Aspergillus awamori* also survived well in soil (Gaur and Kundu 1981).

The shelf life of carrier based inoculant and the cheapness of material are the important criteria for the selection of a suitable carrier. Therefore, wood charcoal and soil mixture was studied as carrier for the preparation of Microphos culture (Gaur and Gaind 1984). The data showing numbers of phosphate solubilizing microorganisms in one gram carrier on wet basis at 4 monthly intervals are presented in table 132. *Pseudomonas striata* which is aerobic, non-spore former, Gram negative rod survive very well and their numbers proliferated to 82 x 10^7 in the second month as compared to 53 x 10^7 in the first month. However, decreasing trends were observed in the 3rd or 4th months. The population of aerobic, spore forming and Gram

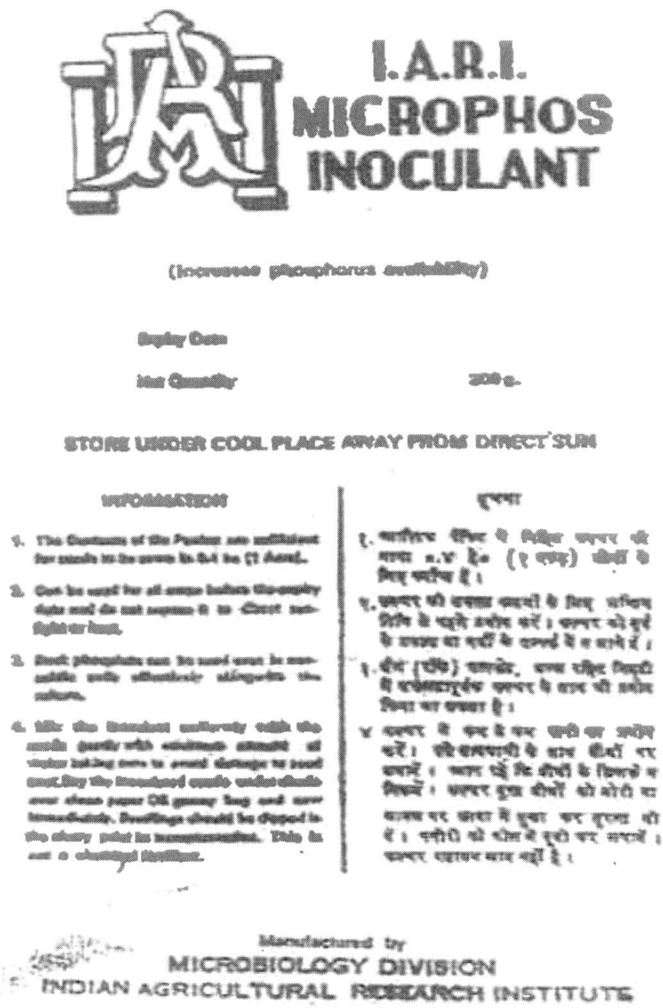

Fig. 14: IARI Microphos culture supplied in packets.

positive' rods, *Bacillus polymyxa* recorded as 38 x 10^7g carrier during first month, was found to decrease slowly, the lowest count 9 x 10^7/g was Inoted at the end of fourth month. *Aspergillus awamori* showed steady decrease from the highest count 6 x 10^7 /g to 3 x 10^7/g at the end of the fourth month. The results showed that wood charcoal and soil mixture is good for the survival and maintenance of these organisms for their supply to the farmers for field application. The culture's packets can be safely stored for three months around 30°C. However, PSM inoculants which can tolerate and grow at 35°-45°C, particularly shall be relevant for summer (Kharif) crops.

Table 132: Number of microorganisms (x10^7)/g carrier

Organisms	Incubation in months			
	1	2	3	4
Pseudomonas striata	53 (33.5)	62 (33.5)	46 (32.2)	20 (26.7)
Bacillus plymyxa	35(30.1)	27(30.1)	15(28.2)	9 (20.6)
Aspergillus awamori	5 (26.6)	6 (25.4)	4 (24.4)	3 (23.5)

Figures in parentheses are moisture percent of the carrier based culture.

IX. Microbial Inoculation and Crop Yields

Gerretsen (1948) reported the increased phosphorus uptake and yield of oat plant inoculated with pure cultures of phosphate dissolving microorganisms compared to uninoculated and this was supported by the other findings (Pikoskaya 1948, Dorosinsky and Lazarev, 1948).

In USSR the increase in yield due to use of *Bacillus megaterium* var. *phosphaticum* popularly known as phosphobacterin was of the order of 5 to 10 per cent in about one-third field trials. The beneficial effect of phosphobacterin is attributed to phosphate nutrition of plants, fungistatic activity in the root zone of higher plants and synthesis of growth promoting substances (Mishustin and Naumova 1962).

In India, field trials were conducted in different soil types, to evaluate the efficiency of seed inoculation with phosphabacterin. Sundara Rao (1965) reported beneficial effect on berseem (Egyptian clover). Significant increase in yield of maize was obtained with Fosfo-24 (Czechoslovakian culture) and on wheat in one out of three trials.

Taha *et al.*, (1969) reported about the effect of a Gram negative short rod bacteria which increased the yield and phosphorus uptake in barley crop.

Sharma and Singh (1971) observed that phosphobacterin alongwith bone meal application to soil increased the grain yield of rice and nitrogen and phosphorus content as compared to nitrogen in combination with bone meal treatments.

In addition to several studies, the effect of the Microphos cultures was tested on several crops such as wheat, paddy, Bengal gram, soybean, potato and cotton by Gaur and coworkers during 1972-95 (Gaur, 1990). The effects of phosphate solubilizing bacteria and fungi/yeast on the growth, yield and phosphorus uptake by difference crops with references are listed in table 133.

Table 133: Response of crops to phosphate solubilizing bacteria and fungi

Crop	Inoculant	Finding	Reference
Barley	Phosphobacterium	Increase in yield and P uptake	Krassilnikov. 1957
	Pseudomonas	Increase in yield	Myskow, 1961
	Pseudomonas and *Bacillus*	Increase in yield and P uptake	Taha *et al.*, 1969
Maize	*Phosphobacterium, Pseudomonas*	Biomass and yield increase	Dzhanaev, 1965, EdiPrimento *et al.*,1996
Oats	*Soil containing* PSM	Increase in yield and P uptake	Getter, 1969
Paddy	*Pseudomonas, Bacillus, spergittus* awamori	Increase in yield	Gaur, 1990, Chandra and Singh 1996
Wheat	*Phosphobacterium*	Increase in yield, protein and p uptake	Timoriu *et al.*, 1964
	PSM and VAM	Increase in biomass	Singh and Kapoor, 1999
	Pseudomonas and *Azotobacter*	Early emergence of seedlings, increase in yield and P uptake	Gaur. 1972, Kundu and Gaur, 1982
Cowpea	*Bacillus, Pseudomonas* and *Azotobacter*	Increase in biomass and nitrogen fixation	Gaur, 1990
Chickpea	Rhizobium + PSB or PSF.	Increase in yield and P uptake	Gaur, 1990 Zadi, Khan &Amil, 2003
Peas	Phosphobacteria Rhizobium + PSB	Increase in yield	Gaur 1990 Tyagi, Bhattacharya & Yadav 2004
Soybean	*Pseudomonas striata*	Increase in yield and P uptake	Gaur, 1979, 1985
Green gram	PSF	Increase in yield	Rao and Rao 1994
Blackgram	PSM	Increase in yield	Tomar, 1998
Groundnut	PSB	Increase in yield	Patel and Thakur, 1998
Mungbean	PSM. VAM	Increase in yield	Singh and Kapoor,1998
Cotton	*Azotobacter*, PSM	Increase in yield	Kundu & Gaur, 1980b.
Potatoes	*Pseudomonas, Bacillus*	Increase in yield	Shikina, 1961, Kundu and Gaur, 1980c
Tomatoes	Phoshobacterin	Increase in yield and sugar content	Zavyalov 1958

Table 134: Effect of phosphate solubilizing microorganisms on the grain yield and P uptake by wheat, rice, Soybean and chickpea

Treatments	Wheat			Rice			Soybean		Chickpea			
	Grain yield (kg/ha)	% increase over	P_2O_5 uptake (kg/ha)	Grain yield (kg/ha)	% increase over control	P_2O_5 uptake (kg/ha)	Grain yield (kg/ha)	% increase over control	P_2O_5 uptake (kg/ha)	Grain yield (kg/ha)	% increase over control	PA uptake (kg/ha
Control	3440	-	24.6	2490	-	6.7	1050	-	4.2	2370	-	5.7
P. striata	3730	8.4	29.6	2540	2.0	6.8	1786	70.1	6.5	2460	3.8	6.1
B. polymyxa	3660	6.3	27.8	2580	3.6	7.1	1730	64.8	5.6	2920	23.2	7.9
A. awamori	3620	5.2	27.2	2560	2.8	6,9	1724	64.2	6.1	2780	17.3	7.2
Rock phosphate @	3730	8.4	27.2	2720	9.2	7.4	1730	64.8	5.6	2560	8.0	7.2
60 kg P_2O_5/ha											•	
RP + *P. striata*	4350	26.5	34,9	2740	10.0	7.5	1892	80.2	6.9	2920	23.2	8.8
RP + *B. polymyxa*	4170	11.7	32,8	2820	13.3	7.8	1892	80.2	6.7	3140	32.5	10.7
RP+*A.awamori*	4170	21.2	32.2	2740	10.4	7.5	1814	72,8	6.4	3070	29.5	10.1
Superphosphate	4440	29.0	35.3	2870	15.3	8,3	1740	65.7	6.3	2850	20.3	9.1
CD at 5%	142			140			138			45		

Rock phosphate was applied @ 60 kg P_2O_5/ha for wheat, rice and chickpea and 200 kg P_2O_5 /ha for soybean. Similarly super-phosphate was applied @ 60 kg P_2O_5 /ha for wheat, rice and chickpea and 80 kg P_2O_5/ha for soybean.

The effect of phosphate solubilizing microorganisms such as *Pseudomonas striata*, *Bacillus polymyxa* and *Aspergillus amori* as inoculants was conducted under field conditions on wheat, rice, soybean and chickpea crops (Table 134). The results showed that yield of wheat, rice, soybean and chickpea was increased by 5.2 to 8.4%, 2 to 3.6%, 5.6 to 6.5% and 3.8 to 23.2% over control due to simple inoculation, by PSM. PSM solubilized the insoluble soil phosphate and made available to crops. The use of rock phosphate at basal dressing rate increased the yield of the crops appreciably. The use of cultures along with rock phosphate gave significant grain yield of all the four crops and the best combination of culture and rock phosphate was comparable to recommended application of superphosphate. The use of culture improved the level of phosphate uptake by crops due to use of these inoculants.

X. Phosphate Solubilizing Microorganisms (PSM) as Biofertilizer in Dry Land Agriculture

Nearly two-third of cultivated areas in India are rainfed. The major soil types in these regions are Alfisols and Vertisols. Besides fertility related constrains, soil moisture deficiency is a major limiting factor for the PSM biofertilizers. While nitrogen is universally deficient in most of the soil types, phosphorus is also of equal concern particularly in legume based cropping system and areas with double cropping system.

About 85% of course cereals, pulses, oilseeds, 65% of cotton and 50% of rice are grown in dry land agriculture. Vertisols in SAT regions have a major problem of P fixation with total P ranging from 150 to 1500 ppm while available P is as low as 3 to 5 ppm (Venkteswarlu, 2002).

Large number of field trials have been conducted under All India coordinated Research Project on Dry land Agriculture (AICRPDA) across different locations in the country on the performance of PSM biofertilizers with and without phosphatic sources such as superphosphate or rock phosphate. PSM biofertlizers developed by Gaur and coworkers (Gaur, 1990) have been tried both on legumes and cereals. PSM biofertilizers, namely *Pseudomonas striata, Aspergillus awamori, Bacillus polymyxa, B. circulans* and *Penicillium digitatum* were tried with different levels of P_2O_5 on crops such as pigeonpea, green gram, blackgram and soybean (Venkateswarlu, 2002). The residual effect of inoculation on succeeding crop was also examined. The PSM (MPS) population in soils under AICRPDA net work was also enumerated (Table 135). The population of PSM varied form 28000 to 63000 per gram dry soil.

The majority of soils have good PMS population. The continued application of fertilizer phosphorus reduced their numbers (Table 136). Significant increase in grain and straw yields of green gram and pigeon pea were recorded at some locations with the application of *Pseudomonas* striata and *Aspergillus awamori* alongwith 40 kg P_2O_5/ha (Table 137). Application of FYM at the rate of 10 tonnes per hectare always resulted in significant yield with PSM biofertilizers.

Table 135: Population of MPS in different locations under the AICRPDA network

Centre	Soil type	Organic carbon%	Available P (P_20_5 kg/ha)	Population X10^3 (g/dry soil)	Major crop/ cropping system
Anatapur	Alfisol	0.25-0.30	10-12	28	Groundnut
Akola	Vertisol	0.20-0.35	8-10	480	Sorghum/ Cotton
Hyderabad	AJfisol	0.50-0.75	8-12	360	Sorghum-castor
Bangalore	Alfisol	0.50-0.75	8-12	680	Ragi/ Groundnut
Jodhpur	Aridsols	0.15-0.22	12-18	590	Millet/pulses
Hisar	Aridsols	0.15-0.22	12-18	630	Millet/pulses
Dantiwada	Aridsols	0.25-0.30	20-35	78	Millets
Agra	Inceptisols	0.20-0.38	15-25	120	Pearlmillet
Hoshiarpur	Inceptisols	0.20-0.38	15-25	390	Maize
Varanasi	Inceptisols	0.10-0.25	15-35	160	Upland rice/ lentil
Rakh Dhiansar	Inceptisols	0.10-0.25	15-35	440	Maize
Bhubanewar	Oxisols	0.30-0.45	10-25	180	Upland rice
Ranchi	Oxisols	0.30-0.45	10-25	220	Upland rice
Rajkot	Vertic inceptisols	0.50-0.70	25-30	260	Groundnut
Jhansi	Vertic inceptisols	0.40-0.60	8-15	340	Sorghum
Kovilpatti	Vertic inceptisols	0.40-0.60	8-15	380	Cottan/ pearlmillet

CRIDA, unpublished

Table 136: Impact of continued P application on MPS population in vertisols at selected locations in India

Location	Cropping system	MRS population x 10^3/g dry soil	
		Plots with continued P application	Plots without P application
Akola	Cotton/ greengram	27	190
Indore	Soybean/ wheat	319	440
Solapur	Rabi sorghum	6.2	17
Hyderabad (Medak)	Pigeonpea	112	198

Source: Final Report of CRIDA-ICRISAT collaborative Project on P behaviour in vertisols

Table 137: Response of dryland crops to PSB at different locations under AICRPDA Network

Location	Soil type	Crop	Organism involved	Years of study	Responses (% increase in yield over control)
Hyderabad	Alfisd	Sorghum Castor	*Peudomonas striata* *Peudomonas striata*	2 1	8 NS
Hyderabad (Medak)	Verfeol	pigeonpea	*Peudomonas striata*	2	12
Indore	Vertisol	Soybean Sunflower	*Penicffiumdigitatum* *Asperojllusniger* *Pseudomonas striata*	2 2 2	NS 15 18*
Kovilpatti	Vertisol	Pearlmillet Greengram	*Bacillus subtillis* *Bacillus subtillis*	1 2	11 16*
Hisar	Aridisol	Pearlmillet	*Aspergillus niger*	2	12
Jodhpur	Aridisol	Mungbean Pearimtet	*Aspergillus awamori* *Pseudomonas florescence* *Aspergillus awamori*	2 2 2	21* 8 16
Ranchi	Oxisol	Upland rice	*Pseudomonas striata*	2	11

Source: Compiled from Annual Reports of CRIDA/AICRPDA

Extensive investigations on P behaviour in vertisol showed high P fixation capacity (upto 150 ppm) mainly due to calcium salts. Consistently increased availability in such soils was achieved by combined use of PSM biofertilizers (*Pseudomonas striata)* with organic manure applications (Venkateswarlu, 2002). Field studies are required on the integration of moist conservation practices and use of PSM biofertilizer under different cropping systems.

Biofertilizers in Chinese Agriculture

Biofertilizers for phosphate and potassium solubilization are more common in China than N-fixing inoculants. Both solid, liquid and granular forms are produced in China. *Bacillus megaterium* de Bary and *B. mucilaginosus* Krassikov are the main organisms used as mixed inoculant for potassium and phosphorus solubilization (Venkateswarlu, 2002).

XI. PSM in Alkaline and Alkali Soils

An alkaline soil has a pH greater than 7.0 and is included with calcareous soils (containing $CaCO_3$ and high amount of Ca^{**}) because there are many characteristics common in both the soils.

An alkali soil has a high exchangeable sodium percentage and may not contain excess soluble salts. Saline -alkali soils contain an excess of both soluble salts and exchangeable sodium. The conductivity of saturation extract of these soils is greater than 4 mmhos/cm at 25°C and the exchangeable sodium percentage is greater than 15 and pH between 8.5 and 10.0. These soils are frequently found in semi-arid and arid regions. Biological activity of these soils is generally low which can be reclaimed by amendments such as gypsum, pyrite and organic matter, followed by cropping with suitable crops. The reclamation methods are expensive and labour intensive.

Phosphorus in alkaline and calcareous soils occur chiefly in the form of hydroxyapatite, fluorapatite, chlorapatite, wagnerite and wavelite and in organic forms. Other compound of Fe and Al phosphates and other calcium and magnesium phosphates also occur.

In alkaline calcareous soils, the phosphate solubility increases above pH 8.5, is lowest when the pH is in the range of 7.5 to 8.5 and increases below pH 7.5. The release of P from the sorbed P is facilitated by root contact, dissolution by CO_2 produced by roots or by microbial activity. Soil microorganisms solubilize bound soil phosphates by secreting organic or inorganic acids or minerlizaton of organic phosphates.

The beneficial effect of PSM in increasing grain yield and P uptake by different crops has been reviewed (Gaur 1990, Patel, 2002). *Pseudomonas striata* was found more efficient than *Bacillus polymyxa* and *Aspergillus awamori.* Experiments conducted in India with PSM inoculation with or without addition of superphosphate, rock phosphate and pyrite have increased the yield of crops from marginal to significant levels. Wherever, insoluble sources of phosphates were used and inoculated with PSM, the available soil phosphorus was increased resulting, an improvement in P-uptake and crop yield. Pyrite is a promising input for increasing the productivity of alkaline soils by acting as an acidifying agent. In certain cases, the use of rock phosphates and PSM substituted the higher doses of phosphatic fertilizers thus recommending the direct application of rock phosphate combined with PSM inoculation.

XII. Role of PSM in Horticultural Crops

In commercial horticultural, superphosphate is applied indiscriminately for maximum fruit production alongwith nitrogenous fertilizers which has resulted in problems of soil salinity. The farmers have realized the use of biological inputs such as phosphate solubilizing bacteria (PSB) or fungi (PSF) to increase the availability of phosphorus to horticultural plants. *Bacillus polymyxa* alone and with arbusular mycorrhizae (VAM) has been successfully used in banana cultivation for sustainable production. Basal dose of super-phosphate could be completely saved besides advanced flowering due to inoculation with PSB. *Azospirillum,* nitrogen fixing bacteria could also to be inoculated with PSB/VAM to get the additional effect of nitrogen fixing bacteria (Sukhada, 1997). Similar results were obtained with *in vitro* grown banana. Even field application of PSB alone or in combination with *Glomus mosseae* and *G. fasciculatum* increased plant growth, nutrient uptake and yield in tissue cultured banana plants hardened with the same treatments. In pomegranate CV 'Ganesh' inoculation with PSM increased plant growth, uptake of P and fruit yield (Sukhada *et al.,* 2000). In grape cultivation, indiscriminate use of superphosphate in many orchards has caused salinity problems and hence PSB culture or 'Microphos' (IARI) is being used to meet the phosphorus needs of Grape fruit plants (Sukhada and Manmohan, 2002).

Jennifer (2001) reported the colonization of brinjal roots with *Bacillus polymyxa* (PSB). These bacteria occupied the endorhizosphere and moved into the stem through the intracellular spaces. They were found in root hairs, leaves and seeds. These bacteria produced growth hormones, gibberellins and zeatin ribosides which improved the growth of the plant. These bacteria on inoculation improved the yield of brinjal under field conditions. Growth promoting phosphate solubilizing *Bacillus* spp. were reported to occur in the rhizosphere of chilli and capsicum cultivars. The flower yield and phosphorus availability were increased by inoculation of Marigold *(Tagites erecta* L.) with PSB, *Bacillus* sp. (Chandrikapure *et al.*, 1999). Inocula of PSM are available in large quantities all over India as 'Microphos', 'Phosphobacteria', Superphos etc. (Sukhada and Manamohan, 2002). PSM has thus great potential in meeting the phosphorus demand of horticultural crops.

Role of Sulphur in Phosphate Solubilization

The process of manufacturing of phosphatic fertilizers involves the acidulation of high grade rock phosphates with mineral acids to convert insoluble P into a soluble form. Sulphuric acid has been the most the most commonly used acid for this purpose. Lipman *et al.*, (1916) advocated the use of rock phosphate and sulphur for increasing P availability in soil. This idea was floated on the premise that thiobacilli oxidize sulphur in the soil to sulphuric acid which, would then solubilize insoluble P. They incubated rock phosphate with soil and sulphur and found that there was increase in water soluble and ammonium citrate soluble P. In a green house study Neller (1956) found that addition of sulphur had a marked effect on P solubilization from rock phosphate and the availability of residual P from rock phosphate increased upto 13% when sulphur was added. A granular fertilizer prepared by pelleting rock phosphate with sulphur and inoculating with sulphur oxidizing bacteria has been prepared in New Zealand and Australia and such preparations are called "Biosuper" (Rajan, 1982, 1987) The term "Biosuper" was first used by Swaby (1975). Rajan and Edge (1980) have studied the pattern of dissolution of rock phosphates and biosuper to find out the extent of release of P during soil incubation. The oxidation of sulphur and dissolution of rock phosphate reached a maximum after 3-4 weeks after their addition. Biosuper was found to be a good P source for perennial ryegrass which increased the ryegrass yield by 13- 25% (Rajan and Edge, 1980, Rajan, 1987). Generally rock phosphate sulphur ratio of 5:1 has been used (Rajan and Gillingham 1986) for the preparation of Biosuper. Rajan (1983) evaluated the effect of sulphur content on the effectiveness of Biosuper and observed that ryegrass yield decreased slightly when ratio of rock phosphate to sulphur was widened to 6:1. Further widening of the ratio to 7:1 was not satisfactory as it reduced the yield drastically.

Sulphur is an expensive input and has to be imported in many developing countries. Moreover, the use of this treatment (rock phosphate + sulphur/ Pyrite + *Thiobacillus thiooxidans)* was not found superior to rock phosphate inoculated with phosphate dissolving fungi *(Aspergillus awamori)* where addition of either sulphur or pyrite was not required (Patil *et al.*, 1974-78). The result showed that phosphate solubilizing fungi increased the available phosphorus status of the soil, phosphorus uptake by the plant, its root development and dry matter yield of the crop was improved. The treatment involving rock phosphate + pyrite + *Thiobacillus thiooxidans* was only next in performance to the above treatment.

Effect of Pyrite in Rock Phosphate Solubilization

The microbial leaching of low grade ores to extract valuable metals like copper, zinc and uranium largely depends on the oxidizing action of *Thiobacillus ferrooxidans* and *T. thiooxidans* (Torma and Binhegy, 1984). Pyrite, a natural sulphur mineral is oxidized to produce sulphuric acid. Similar approach involving addition of pyrite to rock phosphate offers an attractive proposition for increasing the availability of P from rock phosphates.

Experiments conducted to determine if low grade pyrite could be used in combination with rock phosphate to improve the availability of P in neutral and alkaline soils has given contradictory results. Rastogi *et al.* (1976) observed mobilization of P from rock phosphate by pyrite in an alkaline soil but there was no effect in neutral and acidic soils. Khanna *et al.* (1979) observed that pyrite did not help in the solubilization of rock phosphate in a calcareous soil. On the other hand, Singh *et al.* (1979) have shown the utility of narrowing the ratio of rock phosphate to pyrite from 1:0.6 to 1:1 for improving the availability of P from rock phosphate.

In glasshouse experiments with Mussoorie rock phosphate with Maize (*Zea mays*) and berseem (*Trifolium alexandrinum)* as test crops in a neutral soil; showed that the addition of low grade pyrite brought about only a modest improvement in availability of phosphorus. Rock phosphate alone or rock phosphate with pyrite (1:5) were less efficient than superphosphate (Gupta and Mishra 1978).

Mishra *et al.* (1981) have reported that Mussorie rock phosphate mixed with pyrite (1:2.5) was a good source of P to potato in a neutral sandy loam soil and gave yields similar to single super-phosphate. Sharma *et al.* (1983) found that Mussorie rock phosphate was 40.5% as effective as super-phosphate in lentil (Lens *culinaris).* Its efficiency was improved to 50.4% when it was mixed with 25% by weight of pyrite.

XIII. PSM as Coinoculants with Nitrogen Fixing Bacteria

The combined use of PSM alongwith a nitrogen fixing bacteria such as *Rhizobium*, *Azotobacter* or *Azospirillum* depending on the cropping system can be used to meet both nitrogen and phosphorus requirement of the crop biologically. Synergestic interactions of certain specific PSM and nitrogen-fixing bacteria have been recorded. Ocampo *et al.*, (1975) observed that coinoculation of *Azotobacter* with PSB enhanced their population in the rhizosphere of lavender plants. Gaur and Kundu (1980a) observed three fold increase in *Azotobacter* population in wheat rhizosphere when it was inoculated with PSB compared to single inoculation. This also resulted in a significant increases in crop yields over single inoculations. The effect of coinoculation was clearly pronounced in the presence of farmyard manure applied to soil. The population of PSM was also augmented due to application of organic manure. Similar type of results were obtained with potatoes (Kundu and Gaur 1980a) rice (Kundu and Gaur, 1984) and cotton (Kundu and Gaur 1980b).

The coinoculation with *Rhizobium* and PSB has shown promise in improving the yield of legumes and phosphate uptake by the crops. Such benefits of coinoculation have been reported in soybean (Plate 55 a, b, c) and chickpea (Gaur 1979, 1985) and mungbean (Gaind and Gaur 1991, Gaur 1990). Algawadi and Gaur (1988a) reported increased nodulation, acetylene reductase activity, dry matter yield and N and P uptake in chickpea coinoculated with *Rhizobium* and PSB. The combined inoculation met not only the total P requirement of the crop through rock phosphate but also caused a saving of 10 kg N/ha. In sorghum, inoculation with PSB alongwith *Azospirillum brasilense* increased root associated AR activity, crop yield and P_2O_5 uptake over single inoculation with *A. brasilense* (Algawadi and Gaur, 1988b Gaur and Algawadi 1989).

Plate 53 a: Effect of coinoculant *Rhizobium* and PSB on soybean in superphosphate amended soil (60 kg P_2O_5/ha) Left: *Rhizobium*, Right: *Rhizobium* +PSB

In wheat an increase in dry matter production and P uptake from 10 to 27% and 15-34% respectively was noted when inoculated with *Penicillium bilaji* (Kucey, 1988). Nodulation, pod yield, oil content and P uptake in groundnut were improved significantly due to seed inoculation with PSM and there was saving of the fertilizer by the use of 75:25 ratio of MRP:SSP (Agasimani *et* al., 1994). A synergistic effect of PSB was observed with medium P rates of 30 Kg P_2O_5/ha in chickpea which was equal in response to 60 Kg P_2O_5 phosphatic fertilizer (Shinde and Saraf, 1994). Combined inoculation of *Rhizobium* and PSM resulted in a significant increase in dry matter content, grain yield, N and P uptake in soybean over uninoculated control (Gupta *et al.*, 1992).

Plate 53 b: Effect of coinoculant *Rhizobium* and PSB on soybean in rock phosphate amended soil (60 kg P_2O_5/ha) Left: *Rhizobium*, Right: *Rhizobium* +PSB

Plate 53 c: Effect of coinoculant *Rhizobium* and PSB on soybean crop in rock phosphate (30 kg P_2O_5 ha) amended soil. Left: *Rhizobium*, Right: *Rhizobium* + PSB

The studies with PSB and *Azotobacter* and their combination have shown encouraging results in the highly acidic soils of shillong where in, it has been observed that tuber seed treatment with the culture was better than soil application (Sharma *et al.,* 1998). The response of potato crop to PSB was more marked than that of *Azotobacter* (Table 138).

Table 138: Effect of PSB and *Azotobacter* on tuber yield in Shillong

Treatment	Tuber yield (Q/ha)	% increase over control
Control	120	-
Azotobacter (S)	135	12.5
Azotobacter (TT)	145	20.8
PSB (S)	154	28.3
PSB (TT)	162	35.0
Azotobacter + PSB (S)	176	46.6
Azotobacter + PSB (TT)	184	53.3
C. D. at 5%	25	-

PSB-Phosphate solubilising bacteria
S-Soil application, TT-Tuber treatment

Production of Growth Promoting Substances

Evidence continues to grow that microbial growth factors are present in the soil and the soil microflora includes many microorganisms for which growth factors are essential as they can not synthesize where as others are able to synthsize them. The quantitative data on the incidence of soil bacteria requiring a definite growth factor was reported by Lockhead and Thexton (1952). Out of 534 strains isolated from a field soil on a non- selective basis 41 (7.6 per cent) required vitamin B12 as essential nutrilite.

Lockhead (1957) examined the capacity of 316 culture for the synthesis of thiamin, biotin, riboflavin, Vitamin B12 and the terregens factor. It was observed that riboflavin was produced by a large number of isolates followed by thiamin and Vitamin B12.

Sankram (1960) reported that a large number of bacterial isolates from wheat, *berseem* and cowpea were able to synthesize vitamin B12.

Mishustin and Naumova (1962) reported that *azotobacter*in and phosphobacterin acted by suppressing fungi in the root zone and also produced biotic substances that stimulated plant growth.

Naumova *et al.,* (1962) reported that *Bacillus megaterium* var. *phosphaticum* and *B. fluorscens* accumulated in the medium biologically active compounds such as auxins, gibberellins, Vitamins etc., which can stimulate plant growth and inhibit the growth of fungi like *Fusarium* and *Alternaria.*

Brown (1972) reported that bacteria producing one or two growth promoting substances mostly belong to the genera of *Nocardia* and *Arthrobacter*. Species of *Flavobacterium* and *Brevibacterium* although not stimulated by roots, isolated from rhizosphere produced plant growth promoting substances.

Pseudomonas, *Flavobacterium* and *Brevibacterium* spp. isolated from root-free soil also produced growth hormones.

Barea *et al.*, (1976) reported that out of 50 PSB, twenty synthesized all the 3 type of plant hormones, 43 produced IAA, 29 formed gibberellins and 45 cultures produced cytokinin-like substances. Plant growth inhibitors were also detected in cultures of some isolates.

Azocon *et al.*, (1978) described the cell free supernatants and in some cases the whole bacterial cultures of phosphobacterin behaved as pure hormone in improving dry weight of plants and infection levels compared with mycorrhizal control plants.

Ana *et al.*, (1981) compared the relationship of vitamin production and ability to dissolve insoluble calcium phosphate by bacteria isolated from rhizoshpere, rhizoplane and control soil. Among isolates producing one or more vitamins, 76.4, 37.5 and 57.5 per cent of control, rhizosphere and rhizoplane isolates respectively solubilized phosphates. However, phosphate solublizers from rhizosphere and rhizoplane were found more active vitamin producers than solubilizers from control soil and non- solubilizers from any of the 3 regions. Production of vitamin B_{12}, riboflavin and niacin by rhizosphere isolates and of riboflavin by rhizoplane isolates was also found to be correlated with P-solubilizers but similar relationships were also observed for control isolates. Baya *et al.*, (1981) reported that a large percentage of PSB produced vitamin B12, riboflavin, niacin, pantothenate and biotin.

The tested phosphate dissolving bacteria, *Bacillus polpmyxa* (H5) *B. polymyxa* (P1) *B. pulvifaciens* (R5) and *Pseudomonas striata* (27) and fungi, *Aspergillus awamori* (F18), A. *niger* (An) and *Penicillium digitatum* (F20) produced auxins and gibberellins to an appreciable extent. The variation in nature and quantity of plant hormones was observed among the strains. The study suggested that beneficial effects of these organisms may also be due to the production of growth promoting substances besides phosphate solubilization (Sattar and Gaur, 1987).

XIV. Technology for Mass Production

Isolation of PSM: They can be isolated from rhizosphere, non-rhizosphere soils, rhizoplane and sample of rock phosphate, sediments, aquatic environments etc. The isolates and strains of PSM vary in their efficiency to convert insoluble phosphates into soluble form for uptake by the growing plants.

Techniques and Media

Pikovskaya (1948) suggested a medium containing tricalcium phosphate, yeast extract, glucose or sucrose and other salt containing essential nutrients for their growth. This medium has been widely used for isolation, maintenance and enumeration of PSM. Bardiya and Gaur (1974) suggested a strategy for isolation of efficient rock phosphate solubilizing organisms which show capability to dissolve different forms of insoluble phosphates (Gaur and Gaind, 1983) and organic phosphates (Gaur *et al.*, 1973).

Isolation of PSM from Soil

Standard methods are followed for preparation of serial dilutions of the soil suspension in sterile water blanks. The plates are incubated up to two weeks at 28 ± 2°C. The colonies showing halo zone (Plate 54 a, b) are picked up and transferred to slants. The compositions of Pikovskaya medium as well as modified Louw and Webley's medium (1958) are given in chapter 21.

Plate 54 a: Colonies of phosphate solubilizers showing halo zones

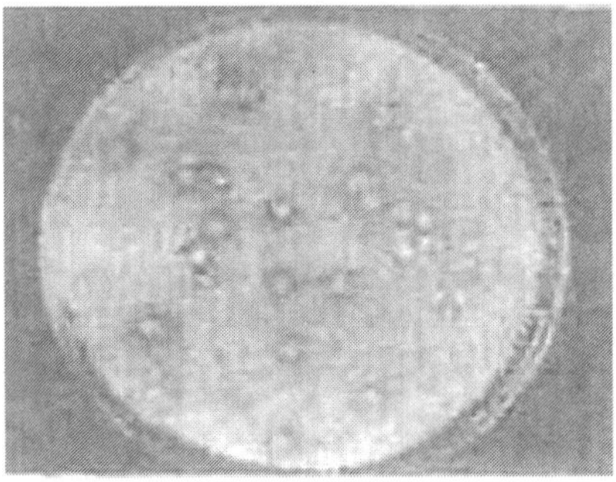

Plate 55 b: Colonies of phosphate solubilizers showing halo zones

Isolation by Enrichment Culture Technique

For isolation or rock phosphate solubilizing microorganisms, 5%, rock phosphate is added to a soil and incubated in moist conditions for about 2 weeks at 28 ± 2°C and then planted on the above medium. Enrichment can also done is liquid medium by successive transfers made at weekly intervals to enrich the medium with PSM. The cultures are then repeatedly plated to obtain pure culture of the microorganisms.

Screening for Efficient Strains for Solubilization of TCP on Solid Medium

The plates are prepared from the medium and pinpointed inoculations of strains of PSM are made on the plates under aseptic conditions. The plates are incubated for 4 to 5 days at 28 ± 2°C and observed for clearing zones around the colonies. The positive ones are supposed to be phosphate soulbilizers of inorganic or organic phosphates depending on the nature of the substrate used.

Finally microbial solubilization of different types of insoluble phosphates have to tested in liquid medium involving specific techniques and methods for phosphorus estimation and enumeration, production of siderophores, growth promoting substances etc., (Gaur 1990). The efficient strains thus selected should be tested in green house followed by field conditions to select the most competent and efficient strains for use as biofertilizer.

Preparation of PSM Inoculants

For production of PSM inoculants, a suitable and inexpensive carrier material has to be selected. Peat in not available in India and therefore alternate carrier such as wood charcoal and soil mixture after due studies was selected for the purpose (Gaur and Gaind, 1984). Two parts of ground charcoal (100 mesh) and one part of soil should be mixed alongwith 0.5 per cent KH_2PO_4 by weight and the moisture is maintained at 10 per cent. Trays containing the mixture (8 Kg/tray) are sterilized at 15 lb pressure for 3 hours alternately for 3 days. One ml of homogenised suspension of different cultures of *Pseudomonas striata,* Bacillus *polymyxa, Aspergillus awamori* or any other efficient selected culture are inoculated to the flasks containing 400 ml of sterilised above-mentioned broth and the culture are allowed to grow for 4-5 days. To 8 kg of the carrier 1200-1500 ml of specific broth culture are added and thoroughly mixed so as to maintain it at approximately 40 per cent moisture and pH nearly neutral. Two hundred grams of its inoculant are filled in polythene packets, incubated at room temperature at 26 ± 2°C in duplicate. The shelf life should be monitored at monthly intervals by counting the presence of PSM in the carrier by the techniques elaborated elsewhere (Gaur, 1990).

Directions for Use of PSM Inoculants

Dissolve 100 g of sugar or gur in 1/2 to 1 litre boiled and cooled water. Empty the contents of packet into the solution and slurry made. Pour the quantity of seeds intended to be sown in one acre or half a hectare in a suitable container and add the slurry. Continuously mix with a clean rod or hand so that the seeds are pelleted with culture properly. Spread the inoculated seeds on a clean cloth or gunny bag or newspaper in the shade to diy and avoid exposure to sunlight.

After drying, sow the seeds without delay. In case of transplantation crops like paddy or vegetables, the roots of the seedlings should be dipped in the culture suspension for 2-3 hours and then transplanted. While inoculating plantation trees, flower beds or roses, it should be mixed with good quality farmyard manure or compost and then should be applied in the pit near the roots of trees or applied to flowers near plants roots and mixed well so that it is not exposed to direct sun.

Prospects and Future Strategy

India will need 230-250 million tonnes of foodgrains by 2015. The food grain production has to be augmented keeping in view the fast depletion of resources of fossil fuels, increasing cost of chemical fertilizers and environmental pollution and costs. The conventional farming system is mainly dependent on regular applications of chemical fertilizers and pesticides resulting in loss of soil fertility and sustainability of agriculture. Positive role of microorganism in growth and nutrition of plants should be synthesized in the cropping system. The use of phosphate solubilizing biofertilizers has not only increased crop yields and phosphate nutrition but also can replace the use of phosphatic fertilizers by about 50 per cent.

The selection of PSM should be based also on other criteria such as synthesis of growth promoting substances, antagonism towards pathogens, compatible with beneficial microorganisms used in agriculture. The cultures should be developed which can perform under stress condition such as dryland agriculture, saline-alkali soils and at high temperatures during kharif season. The plant-microbe interaction studies are essential for improvement in selection of efficient phosphate solubilizing microorganisms for use as biofertilizer.

References

Abu samaha, A. and Patel, H. H. 2000. In: Ecology of Fungi (Eds. Bhat, D. J. and Raghu Kumar, S.) pp 82-86.

Agasimani, C.A., Mudlagiriyapha and Sreenivasa, M.N. 1994. Groundnut News 6:5.

Aginihotri, V. P. 1970. Can. J. Microbiol. 16: 877-880.

Ahmad, N. and Jha, K. K. 1968. J. Gen. Appl. Microbiol. 14: 89-95.

Alaghaozali, R., Mustafa, S. H., Mohammad, K. and Algzaw L. 1986. J. Bio. Sci. Res. 47: 157-172.

Algawadi, A. R. and Gaur A. C. 1988a. Pl. Soil 105: 241-246.

Algawadi, A. R. and Gaur. AC 1988b. Zentralbl. Microbiol. 143: 637-643.

Ana, M. Baya, Boethinling, R. S. and Ramos cormenzana, A. 1981. Soil Biol. Biochem. 13: 527-531.

Anas, I. Bossier, P. and Verstaete, W. 1985. Medeldelingen ande Facultelt Landbouww eterscappen Rijksuniversteit Gent 50:31- 40.

Antarikanonda, P. S., Sassanrakkit, S. Amaret, P. and Waiamai 1990. In: NRCT DOST-JSPS Joint Seminar on Biotechnology.

Arora, D. and Gaur A. C. 1978. Ind. J. Microbiol. 10:315.

Arora, D. and Gaur, A. C. 1979. Ind. J. Exptl. Biol. 17: 1258-1261.

Attamone, C. Norvell, W. A. Bjoerkman, T. and Hazman, G. E. 1999. Appl. Environ. Microbiol. 65: 2926-2933.

Azcon, G., De aguilar, C., and Barea, J. M. 1978. Can. J. Microbiol. 24: 520-524.

Bajpai, P. D. and Sundara Rao, W. V. B. 1971. Soil Sc. Plant Nutr. 17:41-43.

Banik, S. and Dey, B. K. 1981. Zentralblt. Bacteriol. 136: 478-486.

Banik, S. and Dey, B. K. 1983. Zentralblt. Microbiol 138:17-23.

Barber, S. A. 1984. Soil Nutrient Bioavailabity, Wiley and Sons, New York.

Bardiya, M. C. and Gaur, A.C. 1972. Indian J. Microbiol. 12: 269- 271.

Bardiya, M. C. and Gaur, A. C. 1974. Folia Microbiol. 19: 386-389.

Barea J., M., Navare, E. and Montoy, E. 1976. J. Appl. Bacteriol. 40 : 129-134.

Bopaiah, B. M. 1985. J. Plantation crops 13: 60-62.

Bossier, P., Hofte, M. and Verstrate, M. 1988. Adv. Microbiol., Eco. 10:385-413.

Brown, M. E. 1972. J. Appl. Bacteriol. 35: 443-451.

Brown, M. E. 1974. Ann. Rev. Phytopathol 12: 181-193.

Bunt, J. S. and Rovira A.D. 1955. J. Soil Sci. 6: 119-128.

Chandra, K. and Singh, J. 1996. Environ, Ecol. 14:39-41.

Chandrikapure, K. R., Sadwrte, K. T., Panchbai, D. M. and Shelke, B.D. 1999. Orissa Jour Horticulture 27:31-34.

Chandsekharan S. 1969 Pl. Soil 30: 299-304.

Cosgraoove, D. J. 1977. Adv. Microbial Ecol. 1:95-128.

Dave, Arti and Patel, H. H. 1999. Indian J. Microbiol. 39: 161-164.

Dighton. J. and Boddy, L. 1989. In: Nitrogen, Phosphorus and Sulphur utilization by fungi Cambridge University Press, Cambridge pp. 269-298.

Dorosinsky, L. M. and Lzarev, N. M. 1949. Agrobiologia 4: 39-45.

Doelle, H. W. 1969. In: Bacterial Metabolism. Academic. Press, London. P 59.

Duff, R. B., Webley, D. M. and Scott, R. P. 1963. Soil Sci. 95: 105- 114.

Dzhanaev, G. G. 1965. TudV Gorsk Sel Khoz Inst. 25:41-47.

Edi Premonto, M., Moawas, A. D. and Vlek PLG. 1996. Indones J. Crop Sci. 11: 13-23.

Forester, I. and Freier, K. 1988. Zentralbt, Microbiol. 143: 125-138.

Fretias, I. R. and Germinda, J.J. 1990. Can. J. Microbiol. 36: 265- 272.

Fruhata, T., Suzuki, M. and Sakuri, H. 1992. Pant Cell Physiol. 33 :1151-1157.

Gaind, S. and Gaur, A.C. 1990. Indian J. Microbiol. 30: 305-310.

Gaind, S. and Gaur, A. C. 1991. Pl. Soil 133: 141-149.

Gaur, A.C. 1972. In: Symposium on soil productinly. Proc. Nat. Acad. Sci. India (Spl. Volume) pp 259-268.

Gaur, A.C. 1979. Krishi Chanekya 1 : 7-10.

Gaur, A.C., Ostwal, K. P. and Mathur, R. S, 1980. Kheti, 32 :23-25.

Gaur, A.C.1982. A Practical Manual of Rural Composting FAO, pp. 102.

Gaur, A.C.1983. Biofertilizers and crop productivity. In: Advances in soil sciences (Ed. K. V. Paliwal) Books and Periodicals, New Delhi 1: 127-178.

Gaur, A.C.1985a. Phosphate solubilizing Microorganisms and Biofertlizer, Proc. Natl. Seminar on Development and use of Biofertlizers. Min. of Agric, New Delhi.

Gaur, A.C.1985b. Phosphate Solubilizing Microorganisms and their role in plant growth and crop yields. Proc. Nat. Symp. On Soil Biology, Hissar, pp 125-138.

Gaur, A.C.1986. Zentralbl. Microbiol. 14:103-105.

Gaur, A.C.1990. Phosphate Solubilizing Microorganisms as Biofertilizers. Omega Scientific Publishers, New Delhi pp176.

Gaur, A.C. 2002. Microbial phosphate solubilization-overview. Natl. Symp. on Mineral phosphate solubilization UAS, Dharwad, India.
Gaur, A.C. and Algawadi, A. R. 1989. Proc. Focal Theme Symposium Bot. Section, ISCA, Bangalore (1987): 35-46.
Gaur, A.C. and Gaind, S. 1983. Sci. Cult. 49: 110-112.
Gaur, A.C. and Gaind, S. 1984 Geobios. 11:227-229.
Gaur, A.C. and Gaind, S. 1992. In: Proc. Natl. seminar on Organic Farming J. N. K.V. V. Jabalpur and I.G.K.V.V., Raipur. Pp 134-142.
Gaur, A.C. and Ostwal, K. P. 1972. Indian J. Exptl. Biol. 10: 393- 394.
Gaur, A.C. and Sacher, S. 1980. Curr. Sci 49: 553-554.
Gaur, A.C. and Pareek, R. P. 1974. Indian Nat. Sci. Acad. 40B: 68- 76.
Gaur, A. C. and Kundu, B. S. 1980. Pl Soil 57: 223.
Gaur, A. C., Madan, M. and Ostwal, K. P. 1973. Ind. J. Exptl. Biol. 11:427-429.
Gaur, A.C., Arora, D. and Prakash, N. 1979. Folia Microbiol. 24: 314-317.
Geller, I. T. 1969. Izv. Akad. Nauk SSR Ser Biol. Part 1: 171-175.
Gerretsen, F.C. 1948. Pl. Soil 1: 51-81.
Ghosh A. B. and Hasan, R. 1979. Bull. Indian Soc. Soil Sci. 12:1- 8.
Ghosh A. B. 1982. F.A.I. Training Programme for Fertilizer Promotion Executives, New Delhi.
Goldstein, A. H. 1986. Am. J. Alter, Agric. 1: 51-57.
Gupta, R. Shanker, A. B., Saxena, R. K. and Kuhad, R. C. 1993. Folia Microbiol. 38:274-276.
Gupta, R. P., Bhardwaj, K. K. R. Marwah, B. C. and Tripathi, B. R. 1986. J. Ind. Soc. Soil Sci: 34: 498-504.
Gupta, R. P. and Mishra, B. 1978. Ind. J. Agri. Sci. 48: 239.
Gupta, S. B., Vyas, M. K. and Patil, S. K. 1992. J. Indian, Soc. Soil Sci. 40: 854-856.
Hasan, R. 1994. In: Phosphorus Research in India (Dev G. Ed.) Proc. Group Discussion held at IARI. PPICIP, Gurgaon, pp. 7-9.
Illmer, P. and Schinner F. 1992. Soil Biol. Biochem. 24: 389-395.
Jennifer 2000. Studies on the Beneficial Microbial Association in Brinjal Ph. D. Thesis. UAS. Banglore.
Johnston, H.W. 1951. Pl. Soil 32: 94-96.
Katznelson, H. and Bose, B. 1959. Can. J. Microbiol. 5: 79-85.
Katznelson, H., Peterson, E. A. and Rouatt, J. W. 1962. Can. J. Bot. 40: 1181-1186.
Khanna. S. S., Chaudhary, M.L. and Bathla, R. N. 1979. Bull. Indian. Soc. Soil Sci. 12: 545-549.
Kole, S. C. and Hajra, J. N. 1997. J. Interacad. 1(3): 198-205.
Karissilnikov, N. A. 1957. Mikrobiologiya, 26:659-672.
Kucey, R.M.N. 1983. Can. J. Soil. Sci. 63: 671-678.
Kucey, R.M.N.1988. Can J. Soil Sci. 68:261-266.
KKucey, R.M.N., Janzen, H.H. and Legget, M. E. 1989. Adv. Agron. 42:199-221.
Kundu, B. S. and Gaur, A.C. 1980a. Curr. Sci. 48: 159.
Kundu, B. S. and Gaur, A.C. 1980b. Pl. Soil 57: 223-230.
Kundu, B. S. and Gaur, A.C. 1980c. Ind. J. Microbiol. 20: 225-229.
Kundu, B. S. and Gaur, A.C. 1982. Curr. Sci. 51: 291-293.
Kundu, B. S. and Gaur, A.C. 1984. Pl. Soil. 76: 227-234.
Lipman, J. G., McLean, H. C. and Lint H.C. 1916. Soil Sci. 2: 499- 538.
Lckhead, A. G. and Thexton, R. H. 1952, J. Bacteriol. 63: 219-226.
Lockhead, A.G. 1957. Soil Sci. 48: 395-403.
Louw, H. A. and Webley, D. M. 1958. Nature, Lond, 182: 1317- 1318.
Louw, H. A., and Webley, D. M. 1959. J. Appl. Bact., 22: 227-223.

Mishra, B., Sharma, R. D. and Mishra, N. P. 1981. Potato Res. 24:183.
Mishustin, E. N. and Naumova, A.N. 1962. Microbiologiya 31: 543- 555.
Moreau, R. 1959. Comp. Rend. Akad. Sci. Paris 249: 1804-1806.
Myskow, W. 1960. Acta Microbiol. Polon 9: 367-383.
Myskow, W. 1961. Acta Microbiol. Polon. 10:395-401.
Nai, S. K and Rao N. S. S. 1977. J. Plantation Crops, 5: 67-70.
Narayanasamy G., Ghosh S. K. and Sarkar M. C. 1981. Fert News 26:3-9.
Narsian, V. and Patel, H. H. 1995. Indian J. Microbiol. 35:127-132.
Narsian, V. Thakkar, J. and Patel, H. H. 1994. Indian J. Microbiol. 34:113-118.
Narsian, V. and Patel, H. H. 2000. Soil Biol Biochem. 32: 559-565.
Naumova, A. N., Mishustin, E. N. and Marienko, V. M. 1962. Bull Acad. Sci. USSR 5 : 709-717.
Neller, J. R. 1956. Sol Sci. 82: 129.
Norman, A. G. 1953. Role of Microorganisms in nutrient availability in mineral nutrition of plants. Troug (ed.) Univ. Wisconson, Press, Madison, Wis.
Ocampo, J.A. Barea, J. M. and Montoya, E. 1975. Can. J. Microbiol. 21:1160-1165.
Ortuno, A., Hernansaez, A., Noguera, J., Morales, V. and Armero, T. (1977/78) Microbiologiya Espanola 31/31: 113-120.
Ostwal, K.P. and Bhide, V. P. 1972. Indian J. Exptl. Biol. 10: 153- 154.
Pandey, A. and Palni, L.M.S.. 1998. Microbes in Himalayan Soils : Biodiversity and Potential Applications. Jour. Scientific & Industrial Research 57:668-673
Pareek, R. P. and Gaur, A. C. 1993. Curr. Sc. 42: 278-279. Parks, E. J., Olsen, G. H., Brinckman, F. E. and Baldi, F. 1990. J. Indust. Microbiol. 5:183-190.
Patel, S. R. and Thakur, D. S. 1998. Indian J. Agri. Sci. 68: 119- 120.
Patil, R. B., Balasubramanian, A. and Rajashekharappa, B. J. 1974- 78. Terminal Report of Atomic Energy. Project-Role of Microorganisms in release of soil phosphorus to plant. UAS, Bangalore pp. 1-32.
Pierre, W. H. 1948. J. Amer. Soc., Agron. 40: 1-14.
Pikovskaya, R. I. 1948. Mikrobiologiya 17: 362-370.
Rajan, S.S.S. 1982. NZ. J. Agri. Res. 25: 355.
Rajan, S.S.S. 1983. Fert. Res. 4: 287-296.
Rajan, S.S.S. 1987. Fert. Res. 11:43-49.
Rajan, S.S.S. and Edge., E. A. 1980. N. Z.J. Agri. Res. 23: 451- 460.
Rajan, S.S.S. and Gillingham, A. G. 1986. N.Z.J. Exptl. Agri. 14: 313-318.
Rao, S. S., Saisree, N. S., Sheth, L. V., Reddy, G. S. N. and Bhargava, P. M. 1989. Appl. Environ. Microbiol. 55: 767-770.
Rao, V. V. and Rao, A. S. 1994. Indian J. Microbiol. 34 : 313-316.
Rastogi, R. C., Mishra, B. and Ghildyal, B. P. 1976. J. Indian Soc. Soil Sci., 24: 175-181.
Rose. R. E. 1957. N. Z. J. Sc. Technol. 38: 773-780. Schachtman, D. P, Reid, R. and Ayling, S. M. 1998. Plant Physiol. 166: 447-453.
Sankram, A. 1960. Rhizophere Effects on Soil Microorganisms. Ph. D. Thesis P. G. School, IARI, New Delhi.
Sanyal and De Datta, N. P. 1991.
Sattar, M. A. and Gaur, A. C. 1984. Proc. Intl. Congress of Soil Sci. (meeting) Dhaka, pp. 237-251.
Sattar, M. A. and Gaur, A. C. 1985. Bangladesh J. Microbiol. 2: 22- 28.
Sattar, M. A. and Gaur, A. C. 1987. Zbl. Microbiol. 142: 393-395.
Schwartz, S. M. and Martin, W. P. 1955. Soil Sci. Soc. Amer. Proc. 19:185-188.

Sharma, J. P. and Singh, M. 1971. Indian J. Agron. 16:15-18. Sharma S. N., Ray, S. B., Pandey, S. L. and Prasad, R. 1983. J. Agric. Sci. 101: 467.

Sharma, R. C., Sood, M. C. and Khan, M. A. 1998. Bull. Indian Inst. Soil Science. No. 2 pp. 59-72.

Shinde, v. S. and Saraf, C. S. 1994. Indian J. Pulse Res. 7:76-79.

Shikina, A. P. 1861. IZV. Akad Nauk Kazkh, SSR, Ser. Bot. Bochvobed 1: 13-14.

Singh, S. and Kapoor, K. K. 1992. Intl. J. Tropic Agric. 10: 209- 213.

Singh, S. and Kapoor, K. K. 1999. Biol Fert. Soil 28: 139-144.

Singh, D., Mannikar, N. D. and Srivas, N. C. 1979, J. Indian Soc. Soil Sci. 27: 334.

Soumaya, S. 1992. Fertilizer Research (Netherlands) 31: 263-267.

Sperber, J. 1957. Nature, 180:994-995.

Sperber, J. 1958a. Aust. J. Agri. Res. 9: 782-787.

Sperber, J. 1958b. Aust. J. Agric. Res. 9:778-781.

Sukhada Mohandas 1997. In: Soil organisms for sustainability (Eds. Rajagopal et. al.) ISSBE, UAS, Bangalore.

Sukhada, Mohandas and Manamohan, M. 2002. In: Natl. Symposium on Mineral Phosphate solubilization, UAS, Dharwad pp 15-17.

Sukhada, Mohandas, Prasanria Kumar, A. and Manamohan, M. 2000; Proc. Natl. Seminar on Hi-tech Horticulture, Bangalore June 26-28.

Sundara Rao, W. V. B. 1965. Bacterial Fertilizers. In Handbook of Manures and Fertilizers. ICAR Publications, New Delhi pp. 186.

Swaby, R. J. 1975. In: Sulphur in Australian Agriculture (ed. McLachlan, K. D.) Syaney University Press, Sydney pp. 213.

Swaby, R. J. and Sperber, J. I. 1958. In: Nutrition of Legumes, Academic Press, N. Y. pp 289-294.

Taha, S. M., Mahmoud, S. A. Z., EL-Damaty, A. A. and Abd-EL- Hafez., A. M. 1969. Pl. soil 31: 149-160.

Thakkar J., Narsian, V. and Patel, H. 1993. Indian. J. Exptl. Biol. 31 :743-746.

Theodoru, M. E. and Plaxton, W. C. 1993. Plant Physol. 101: 339- 344.

Timoriu, G., Rosu, g. and Ciobana, V. 1964. Inst. Agron. I Ionescu de la Brad Lasi Luc Stiint. 2: 67-72.

Torma, A. E. and Bahegy I. G. 1984. Trends Biotechnol. 2: 13-15.

Tinker, P. B. 1980. In: The Role of Phosphorus in Agriculture (Eds. Khaswneh, E.E. *et al.*) American Soc. Agron, Madison, pp. 617- 629.

Tomar, R. K. S. 1989. Indian J. Agric. Sci. 68: 81-83.

Tyagi, M.K. Bhattacharya, P. and Yadav, A.K. 2004. Biofertilizer News letter 12(2) 9-14

Vassilev, N. and Vassileva, M. 2003 Biotechnological Solubilization of rock phosphate on media containing agro-industrial wastes (Mini- review). Springer-Verlag.

Venkataeswarlu, B. 2002. In: Natl. Symposium on Mineral Phosphate Solubilization (Eds. Algawadi, A. R. *et al.*) UAS, Dharwad p. 6.

Vekateswarlu, B., Rao, A.V. and Raina, P. 1984. J. Indian Soc., Soil Sci. 32: 273-277.

Waksman, S. A. 1952. Soil Microbiology, John Wiley and Sons Inc. New York. Chapman and Hall Lt., London.

Wani, P. V. and Patil, L.K. 1979. Nat. Acad. Sci. Lett., 2: 360-362.

Wani, P V., More, B. B. and Patil, P. L. 1979. Ind. J. Microbiol. 19: 23-25.

Watanabe, K. and Hayano, K. 1993. Can. J. Microbiol. 39: 674- 680.

Zaidi, A ; Khan, M.S. and Amil, M. 2003 European Jour. Agronomy 19(1) 15-21.

19

Mycorrhizae as Bio-fertilizer

The term 'mycorrhizae' was coined more than a century ago (Frank 1885) describing that the roots of most plants are colonized by fungi and transformed into fungus roots organ. Mycorrhizae result from a mutualistic symbiosis between plant roots and certain fungi. Roots of most flowering plants live in mutual symbiosis with mycorrhizae which bio-tropically colonise the root cortex and extra-metricial mycelia help the plant to obtain plant nutrients from soil (Barea, 1991).

These fungi are ubiquitous in soil and are found in the roots of many Angiosperms, Gymnosperms, Pteridophytes and Thallophytes (Mosse *et al.*, 1981). The mycorrhizal fungi perform the function of root hairs. The fungus takes carbohydrates from the plant and in turn supplies the plants with nutrients, hormones and protects it from root pathogens. The dependency of fungal species on host plant ranges from obligate to facultative (Brundrett, 2002; Read and Perez-Moreno, 2003). The mycorrhizal plants have greater tolerance to toxic heavy metal, nigh soil temperature, soil salinity, unfavourable soil pH and to transplantation shocks.

They play an important role in increasing plant growth and nutrients uptake (Bagyaraj, 1992, VasanthaKrishna and Bagyaraj 1993). These fungi can be used to promote biotechnologically developed plant in soil (Varma and Schuepp 1994).

1. Types of Mycorrhizae

Frank (1885) distinguished two main types of mycorrhizae Viz., ectotrophic and endotrophic. According to recent information seven different types of mycorrhizae such as Ecto, Endo, Ectendo, Arbutoid, Monotropoid, Orchidaceous, and Ericoid are known (Harley and Smith 1983, Bhandari and Mukerji 1993).

Ectomycorrhiza

They form a sheath or a compact fungal mantle layer around roots and have inter-cellular hyphae in the cortex. They are commonly found on the roots of forest trees like pine, oak, beech and eucalyptus. These form a thick mantle (covering) outside the roots. Ectomycorrhiza is visible to the naked eye as fuzzy covering on the root. Inside the root they are present in intercellular spaces and forms a characteristic structure called "Hartig net". At times or as a stage in development, the "Hartig net" may become totally absent. This stage is known as 'perirhizic' mycorrhiza.

Basidomycetes belonging to Amanitataceae, Boletaceae, Cortinariaceae, Russiciaceae, Tricholomataceae, Rhizopogonaceeae and Sclerodermataceae are the fungal symbionts often encountered in these associations (Marx, 1972). Tree species belong to families Pinaceae, Salicaceae, Betulaceae, Fagaceae, Tibiaceae, Rosaceae, Leguminosae and Juglandaceae are the hosts (Meyer, 1973). Ectomycorrhizae benefit by improving plant growth of forest seedlings. The fungal partner of ectomycorrhiza can be grown on defined synthetic medium.

Ectendomycorrhiza

In this form, the sheath may be much reduced or even absent. The 'Hartig net" is well developed and the hyphae also penetrate into the host tissues.

Arbutoid Mycorrhiza

This form has a well developed 'Hartig net'. Sheath is usually much reduced *or* even absent. The fungal hyphae penetrate into the cells of the host in addition there is intracellular penetration. They form fungal coils.in the cells. These fungi belong to Basidiomycetes and the hosts belong to Ericales.

Monotropoid Mycorrhiza

The monotropoid mycorrhizal roots are found in Monotropaceae, a set of achlorphyllous plants. They have well developed fungal sheath enclosing the roots. They form 'Hartig net', the hypae enter the cells of the epidermis or cortex. Specialized underbranched haustoria peneterate the epidermal cells and go through a complicated development as the host plant grows and flowers. The fungus is often associated with the mycorrhizal roots of adjacent plants.

Orchidaceous Mycorrhiza

The host may be wholly or partially achlorophyllous for part of their life cycle. They are associated with Basidiomycetes that are cellulotytic and lignolytic. The fungus peneterates the host tissue and after a period of vegetative activity undergoes local or general disintergration. In nature orchid embryos normally become infected with the fungus. The fungus enters the protocorm (germinating seed) and stimulates the growth of protocorms mainly through carbon and phosphorus nutrition. Orchids are generally propagated through tissue culture. It is observed that although there is a lag in build up of mycorrhizal infection, the fungus spreads readily from plant within the glasshouse making them mycorrhizal. The fungus commonly associated with Orchids is *Tulasnella* spp. which can be cultured in laboratory media.

Ericoid Mycorrhiza

In many Ericaceae and related families, the hairlike roots are enmeshed in an extensive external weft of hyphae which also peneratrate the cortical cells of the rootlets. These fungal symbionts are members of Ascomycetes. The Basidiomycetes genus *Clavaria* has also been reported to form the symbiotic association but normally do not form the sheath.

Ericoid mycorrhizae are associated with health plants like blue berries, *Azaleas* spp., *Rhododendron* etc. The fungus involved is *Pezizella* ericae which can be cultured in the laboratory. Inoculation with the fungus has been reported to augment the yield of blue berries by about 15%. Commercial inoculum of fungus prepared in New Zealand is available in the trade name 'Mycoaid'.

II. Role of Ectomycorrhizae in Tree Species

Ectomycorrhizae rely on simpler carbohydrates for their nutrition since they are unable to use cellulose or lignified materials as carbon sources. Thus they depend on their hosts for sustenance. Many of them are unable to form fruit bodies when detached from their hosts (Laiho, 1970). It has also been observed that most species of Ectomycorrhizal fungi are not host specific. The ability of a single species of host to form mycorrhizae of more than one type is of wide occurrence under natural conditions. They may form either ectomycorrhiza or VA mycorrhiza. The species of Pinus may form ecto or ectendomycorrhiza.

Mycorrhizal fungi are perennial in nature and need a permanent nutrition from where they can translocate food to their growing tips. They use products of current photosynthesis. The non-specificity to particular hosts is of advantage to the fungus for its survival in that particular ecological site. It is quite common to find that the same plant host has a number of symbionts. Trappe (1962) reported that at least eight different species of fungi could form the mycorrhizal association with *Betula pendula* and *B. pubescens*. Giltrap (1979) extended the list to forty species of Basidiomycetes that form mycorrhizal association with *Betula*. Likewise, *Pinus sylvestris* and P. *storobus* were reported to be associated with more than dozen species of fungi belonging to several genera. However, *Alnus* is one such plant where only five out of 30 or more fungi present formed mycorrhizal association (Molina, 1979).

Molina and Trappe (1982) reported sporocarp specific host associations. This could be restricted to a particular species involving narrow host range, a single genus or a wider group (wide host range). This has been, therefore referred to as ecological specificity. Laccaria *laccata* and *Pisolithus tinctorius* are conifer-specific fungi whereas *Alpoa diplophoeus* is specific to *Alnus*. Rhizopogon fruits only with members of Pinaceace.

A large number of fungi form mycorrhizal associations in young plantations, but many fungi are eliminated as the plantations age. This may be caused due to shift in the root exudates, accumulation of litter and leachates from them, depletion of nutrients, progressive accumulation of phenolics and toxic compounds etc. These types of changes have far reaching influence, especially in replanting programmes when establishment of seedlings becomes problematic unless inoculated with appropriate symbionts.

Tree Responses to Ectomycorrhiza

Ashton (1956) reported that growth of *Eucalyptus regans* increased substantially in 17th month growth period when inoculated with *Meshophellia* sp. Pryor (1956) observed 88 per cent increase in the growth of 3 different species of *Eucalyptus* after 6 months of inoculation with *Scleroderma* sp. Laccaria *laccata* promoted growth of *Eucalyptus globulus* seedling by more than 200 per cent (Burges and Malajczuk, 1989). Increased nutrients uptake due to ectomycorrhiza by tree plants was reported (Lee, 1982). Total N and P content of *Pinus densiflora* seelings inoculated with *P. tinctorius* increased by 188 and 144 per cent respectively over control.

Positive response of Caribbean pine to inoculation with *Rhizopogon leuteolus* was reported by Ekwebelam (1979). Similar results were obtained by inoculation with *P. tinctorius* and *L. laccata* (Lee, 1982 and Valdes *et al.*, 1983). Quitas and Valder (1987) reported that 90 per cent of the *P. tinctorius* inoculated pine seedlings were myccorrhizal compared to 60 per cent in the uninoculated control. Volume and dry weight of the seedlings were augmented by 165 and 60 per cent respectively over control, Heinrich and Patrick (1985) obtained higher phosphorus uptake in *Eucalyptus* due to inoculation with ectomycorrhizal fungi.

The composite culture of ecto and endo-mycorrhizae was more beneficial than single type of infections presents in forests trees. Combined effect of G. mosseae and *Paxillus involutus* increased dry matter yields and P content of hybrid poplars more than the either fungus (Lopezaguillon and Garbaye, 1989). In *Eucalyptus*, VAM fungi were more prevalent on young seedlings where ectomycorrhizae took over as the plant aged (Chilvers et *al.*, 1987). However, both types of symbionts may be present simultaneously in the cortex of the same root. The VAM fungi colonized the inner part of the cortex where as ecto-mycorrhizae were found in the *Glomus* outer cell layer (Boudarga *et al.*, 1990). In *Eucalyptus* sp., *fasciculatum* and *P. tinctorius* form active and efficient mycorrhizae (Boudarga and Dexheimer, 1988).

Inhibition of Root Pathogens

Certain ecto-mycorrhizal fungi inhibited *Rhizoctonia solani, Pythium* sp. and *Fusarium oxysporum* (Sasek and Musilek, 1968). *Scleroderma aurantium* has also been shown to reduce the incidence of disease caused by *Pythium* and *Fusarium. Suillus bovinus* was effective against *R. solani* while *Amanita citrina, lactarius* helves and *Russulla fragilis* were effective against *P. debaryanum*. *Fomes annosus* was inhibited by antibiotics produced by several ectomycorrhizal fungi (Sasek and Musilek 1968). Water-soluble inhibitory compounds produced by *Boletus bovinus,* a fungi symbiont of *Picea abies* (Norway spruce) protected the seedlings from *Fomes annosus* (Hyppel 1968'ab). Marx (1969a) found that *Leucopaxillus cerealis* var. *piceina* inhibited *F. annosus* slightly while severely inhibited *Cylindrocladium scoparium,* many species of *Pythium*, *Rhizoctonia, Phytophthora* and *Scelerotium bataticola.* However, it was not effective against *Armillaria mellea, Fusarium oxysporum, Rhizoctonia crocorum* etc.

Production of Antibiotics and Its Role

The antibiotic produced by *Leucopaxillus cerealis* var. *piceina* was identified as diatretyne nitrile, polyacetylene (Marx, 1969 b). This could inhibit the growth of zoospores of *Phytopthora cinnamomi* at a low concentration of 50-70 ppb and could kill them at 2 ppm. *Laccara lacata* inhibited 16 species of *Pythium* and *Phytophthora.* Marx and Bryan (1969) reported that *Scleroderma,* a fungal symbiont of pine and pecan inhibited 5 species of *Phytophthora* and 4 *Pythium* species which caused damage to feeder roots. Park (1970) reported that *Lactarius* sp. a symbiont on barswood (*Tilia americana*) inhibited feeder root pathogens like *Pythium irregulare* and *R. solani*. *Pinus resinosa* seeds presoaked in the culture filtrate of *Lactarius* sp. has 93 per cent germination where as the control has only 7 per cent.

Marx (1970) observed that *Thelephora terrestris*, a widespread fungal symbion: on many trees inhibited *Pythium* spp. *Rhizopogan vinicolor,* a symbion: of doughlas-fir was reported to inhibit the growth of *Pythium* spp., *Fomes annosus* and *Poria weirii* (Zak, 1971) *Blotenus* sp. and *Suillus grevillei* reduced the mortality of pine seedlings caused by R. *solani* and *Fusarium oxysporum* (Kuo and Tang, 1981). Sylvia and Sinclair (1983) observed that a diffusible substance produced by *Laccaria laccata* inhibited mycelial growth and delayed spore germination of *Fusarium oxysporum*. Natarajan and Govindasamy (1990) observed that *Suillus brevipes* inhibited all the common root pathogens tested such as *Armillaria mellea, Cylindrocladium parvum, C. scoparium, Fusarium oxysporum*, *F. solani* and *Rhizoctonia solani.*

Mechanisms of Protection

Stack and Sinclair (1975) considered antibiotics and competition as inadequate explanation for root protection of Douglas-fir seedlings by *Laccara laccata* because the fungus did not inhibit *Fusarium oxysporum* in pure culture on media rich in nutrients. Park (1960) suggested that high levels of nutrients mask the antibiotic effect. Smiley and Cook (1973) and Elad *et* al. (1980) noted that low vigour of the pathogens may reduce its tolerance to antagonism.

Zak (1964) and Marx (1972) suggested a number of possible mechanisms by which mycorrhizal fungi might protect roots against soil pathogens. These include physical protection by mantle, antibiotic, competition of sugars from root exudates, stimulation of antigonistic microflora associated with the mantle and production of anti-microbial metabolites.

III. Endomycorrhizae (Arbuscular Mycorrhiza)

Arbuscular Mycorrhizae (AM) are geographically ubiquitous and occur in plants growing in different ecological regions (Hayman, 1982, Redecker *et al.*, 2000). Arbuscular mycorrhizae are placed in class Zygomycetes, order Glomales (Endogonodales) and family Endogonaceae. They are commonly associated with agricultural and horticultural crops, most shrubs, most tropical trees and some temperate tree species. (Thapar and Khan, 1973; Thomazini, 1974; Tinker 1982). Arbuscular mycorrhizae have been recorded in thousand genera of plants covering 200 families. There are at least three hundred thousand receptive plant hosts in the world (Kendrick and Berch 1985) and there are 120 species of arbuscular mycorrhizal fungi (Schenck and Perez 1987). However, there are a few families which are not involved in forming AM are Pinaceae, Betulaceae, Orchidaceae, Fumariceae, Commelinaceae, Urticaceae, Salicaceae, Fagaceae and Caesalpiniaceae. Lack of this association could be because of production of some toxic substances by these species that enhibit mycorrhizal colonization or lack of a production of signal substances that attract mycrrhizal association (Fontenla *et al.*, 1999).

Plate 55: Colonization by Asbuscular mycorrhiza in *Cajanus cajan Glomus mosseae* (Top) G. *constrictum* (Middle) *G. fasciculatum* (Bottom)

Arbuscular mycorrhizae are found to occur on the roots of all plants except on a few vascular plants such as Cruciferae, Chenopodiaceae, Polygonaceae, Cyperaceae Junaceae and Urticaceae which do not form (Smith and Gianianazzi-pearson 1988, Vierhielig *et al.,* 1996). This is the commonest type of mycorrhizae formed on the roots of an enormously wide variety of host plants by aseptate fungi belonging to Endogonaceae. These non-septate Zygomycetous fungi belong to genera of *Glomus, Gigaspora, Acaulospora, Entrophospora*, *Sclerocystis* and *Scutellospora*. There are over 200 species of VAM fungi reported so far. Obligate biotrophs do not grow on synthetic media and hence are classified according to morphological characteristics of the spores formed in the soil, and the relationship of the spore to the associated hyphal attachments.

The mycorrhizal condition is actually normal in most of the terrestrial plants and it greately enhances the uptake of plant nutrients from the soil (Peiffer and Bloss, 1988). It is estimated that over 2 billion hectares of degraded soils are available for revegatation. In such soils, plants usually cannot establish or grow normally without an appropriate mycosymbiont (Rhodes, 1980).

The ecological and economic interest in AM can be simply imagined from the fact that about four-fifths of all land plants including important food and horticulture crops form this type of mycorrhizae. Recent AM are quite similar to fungal formation observed in Rhynie fossil suggesting a plant -AM co-evolution (Newman, 1985).

Colonisation

AM colonization starts from hyphae arising from soil borne propagules. The hyphae penetrate the epidermal cells of young roots behind the meristemic region. The fungus grows throughout the cortex, but it does not invade the endodermis, stele or root meristem. Cells containing chloroplast are not infected. On reaching the cortex, hypae grow into cells by tree like dichotomous branching to give arbuscules. When colonization is well established, oval structures called vesicles (Plates 57) may form which have storage function (Barea, 1991). Vesicles are inter or intracellular, depending on the host species. Some species of the endophyte produce vesicles that remain relatively thin walled. In others, the vesicles become thick walled, and do not differ from chlamydospores in any important aspect and they may function as reproductive structures.

Mycorrhizal propagules can be severely influenced by damage to vegetation and soil resulting from natural disturbances or human intervention. These destructive processes include intense fires (Wicklow- Howard, 1989) soil erosion exposing subsurface soil (Day *et al.,* 1987, Habte *et al.,* 1988) by burrowing animal (Koid and Mooney, 1987) and volcanic activities (Allen, 1988). Human activities such as top soil removal/ disturbances and stock piling during mining (Stahl *et al.,* 1988) and clear- cut logging (Janos, 1987; Newman, 1985). Agricultural operations such as tillage (Evans and Miller, 1988) long fallow period (Thompson, 1987) soil compaction, growth of non-mycorrhizal crops, bioicide application (Hayman, 1986) and flooding of rice (Hall, 1987) can adversely affect mycorrhiza.

Reduction in mycorrhizal natural inocula following ecosystem disturbances is the main cause for the reduced symbiotic association. Introduction of mycorrhizal fungi as inocula will improve soil condition and its health (Danielson, 1985; Stahl *et al.,* 1988) Mycorrhiza play an important role in reclamation of mine soils (Wood, 1991).

Early studies in India have shown the occurrence of *Rhizophagus* in association with *Litchi chinensis* (Pandey and Misra, 1971) and of AM in Coconut (Lilly, 1975). Thapar and Khan (1985) during their survey in eight states at 21 locations isolated 15 species of AM belonging to various genera of which *Glomus macrocarpum* was the most widely prevalent species. Rao *et al.*, (1987) observed AM colonization in twenty aromatic plant species examined. The presence of AM on different tree species was recorded by different workers. (Jagpal and Mukerji, 1987; Sharma *et al.*, 1987 and Nair and Girija, 1988). Mohan Kumar *et al.* (1988) reported AM colonization in 53 plants species belonging to several families occurring along the coastline of Tamil Nadu. Rangarajan *et al.* (1988) during an extensive survey in Western Ghat region observed prevalence of AM association in plant communities existing at altitudes ranging from 300 to 2240 M above MSL. Raghupathy Mahadevan (1993) reported AM association in a large number of plants belonging to several families occurring in tropical plains of Tamil Nadu. Narayan and Bhat (1993) observed prevalence of AM colonization in 24 out of 25 forest tree species located in Coimbatore district. *Glomus* sp. was widely encountered in most of the tree species.

AM association has been reported from Indonesia (Suhardi and Tasiman, 1988). Chong (1987) reported presence of AM in forest and flowering ornamental trees from Malaysia. Redhead (1977) reported AM association in the moist low land rain forests and dry sahel regions of Nigeria, Hall (1975) and Hayman (1978) found AM association in New Zealand forests and the predominances of *Glomus* and *Acaulospora* spp. Hafeel and Guntilake (1988) found 95 per cent of tree species occurring in tropical forests of Sri Lanka to be endomycorrhizal. Chalermpongse (1987) found AM association in most tree species in the dry deciduous and semi evergreen dipterocarp ecosystem in Thailand.

Effect of Vitamins on Mycorrhizae

Ecto and orchidaceous mycorrhizae have been shown to be dependent on and stimulated by certain vitamins. Although plant roots may provide an adequate supply of vitamins for mycorrhizal development, rhizosphere/rhizoplane bacteria could also be a source of vitamins (Linderman, 1992) Thus inoculation of both mycorrhizal fungi and vitamin synthesizing organisms could result in further improvement of plant growth.

Singh and Kapoor (1999) reported that PSB (*Bacillus circulans*) and PSF (*Cladosporium herbarum)* and VA mycorrhizae *Glomus* sp. when inoculated on the seed and soil respectively improved the yield of wheat crop as well as N and P uptake. The increases were still higher when the soil was amended with Mussorie rock phosphate. Large population of PSB/ PSF were found in the rhizosphere of wheat crop when soils was amended with Mussorie rock phosphate and soil was inoculated with VAM, *Glomus* sp.

Specificity-Host Preference

It is well known that AM fungi are not host specific. However, the degree of VA mycorrhizal colonization and its effects can differ with different host-endophyte combinations. Plants like soybean and maize appear to be more mycorrhizal than other crops like potato or bean (Kruckelmann, 1975; Schenck and Kinlock, 1974, Strezemska, 1975). According to Woodhouse (1975) phosphorus deficiency in a host cell results in the release of a signal possible in the form of phosphates, that disrupts the action of protective host lectins. The fungi then invade the cells and the phosphate deficiency is corrected. Thereupon, lectins resume activity preventing pathogen entry. One theory is that phosphate deficiency includes membrane leakage which in turn favours fungal invasion (Graham *et al.*, 1981, Ratnayke *et al.*, 1978). Another suggestion is that when plant is under stress, it signals the fungus to invade and correct its deficiency (Gemma and Koske, 1988). This explains why most plant species are susceptible to VAM infection. However, Vesicular-arbuscular mycorrhizal fungi do not normally exhibit any species or strain specificity to a particular host plant, except, perhaps some host preferences. The same plant may be infected by more than two species at a given time (Molina, 1979; Graham *et al.*, 1982; Abbot and Robson, 1985). More than one ecto or endomycorrhizal species can coexist in one host. However, coexistence of ends and ectomycorrhiza is rare.

The extent to which a species or variety is mycorrhizal is determined by the background of the host, that is the kind of root exudates of the host plant or other barriers to AM infection. In tree crops very few reports are available on this aspects.

Since the early endophyte screening trials it has become obvious that VA mycorrhizal fungi differ greatly in their symbiotic effectiveness. Though a particular VA mycorrhizal fungus can infect and colonize many host plants, it has preferred host which exhibit maximum symbiotic response when colonized by a particular VA mycorrhizal fungus (Reena and Bagyaraj, 1990). The tailoring of plant-fungus combinations for maximum effect may lead to developing plants that will form association with certain specific VA mycorrhizal fungi (Gianinazzi *et al.*, 1989).

Another form of specificity is ecological specificity. Under controlled conditions, it may be possible to get mycorrhizal establishment in plants. However, when the same symbiont are introduced in the fields conditions, mycorrhizal infection may be retarded or prevented in certain areas while it may do better at other places. The ecology specificity does operate in nature which influences the prevalence of indigenous dominant VAM flora. Koske

(1987) suggested that selection of fungal flora is not so much dependent on nutrients, organic matter or other variable as it is on the climate.

IV. Screening and Selection of Efficient AM Fungi

Some of the commonly used spore types are *Glomus fasciculatum*, *G.* mossae, *G. etunicatum, G. tenue* and *Gigaspore margarita.* The fungi selected for inoculation into agricultural soils must be able to increase uptake of plant nutrients and persist in soil. The colonization, efficiency, survival etc. in the rhizoshere and in soil are important, Abbot and Robson (1982) have discussed this aspect for use in agriculture.

Govinda Rao *et al.* (1983) screened several fungi for best symbiotic response using host tested in pot, microplot and field trial. It is important to add same number of propagules of different fungi based on most probable number (Sieverding, 1991).

Role of VA-Mycorrhizal Fungi in Crop Growth

The AM fungal hypae proliferate both ouside and inside the root, with internal hyphae differentiating into arbuscules and vesicles (Plates 58). The arbuscules are formed only on the internal hypae in response to contact with host cortical cells, they are sites of nutrient exchange between the plant and the fungus. Vesicles which contain high concentrations of triglycerides at maturity are thought to serve as storage and survival organs. Thickwalled vesicles (Chlamydospores) may function as reproductive structures (Read 1997). The fungi being obligate biotrophs do not grow on synthetic media.

Benefits from AM symbiosis accrue because AM hyphae extend beyond the root hair and increase the absorptive surface area of the root. The following benefits are conferred by VAM on the host plants (Bagyaraj and Reddy, 2000):

Plate 56: Effect of inoculation on colonization by Arbuscular Mycorrhiza in the root system of *Cajanus cajan.* Showing vesicle and hyphae *Glomus mosseăe* (Top) *G. fasciculatum* (Middle) Uninoculated control (Bottom)

1. Enhance uptake of diffusion limited plant nutrients like phosphorus, zinc, copper etc. which are relatively immobile in the soil.
2. Augment water transport in plants, decrease transplant injury and promote establishment of plant in wastelands. Mycorrhizal plants recover better following moisture stress conditions, more water use efficiency and frequently have higher root-shoot ratio than non- mycorrhizal plants.
3. Reduce the vulnerability to disease caused by soil borne pathogens
4. Increase plant growth (biomass) by 20-30%.
5. Enable the plant to survive and grow better in problem soils such as coal spoils, sand dunes, saline alkaline soils, eroded-degraded sites and lands polluted by industrial wastes.
6. It has been recently reported that glomalin, an iron containing glycoprotein and hydrophobic is produced by hyphae of AM fungi plays an important role in soil structure. Glomalin concentration is highly correlated with the percentage of water stable aggregates in different soils (Wright and Upadhaya, 1998).

The role played by AM fungi in improving plant growth is of greater significance in tropical soils than temperate soils. This is mainly because the soils of tropics are low in soil fertility and are deficient both in N and P besides a few micronutrients. According to Ozanne (1980) less than 10% of total soil P enters the plant - animal cycle.

Experiments with ^{32}P labelled phosphorus conclusively proved that VA mycorrhizae can not solubilize unavailable inorganic sources but draw extra phosphorus only from the labile pool of soil solution by diffusion. The rate at which plant roots take phosphorus from soil solution in much faster than the rate in which phosphorus moves into soil solution by diffusion. This creates a phosphorus depletion zone around the root. The hyphae of the VA mycorrhizae travel much beyond the P depletion zone and scavenge a larger volume of soil and thus supply phosphorus to plants. The phosphorus taken up by the hyphae from soil solution passes through hyphae and gets deposited in arbuscules located in the cortex of host, thus short circuiting the nutrients to the host.

Mineral Nutrition

AM enhances plant growth by improving mineral nutrition (Barea and Azcon-Aguilar, 1983, Harley and Smith 1983; Hayman, 1982; Smith and Gianinazzi-Pearson, 1988). The available evidence suggests that VA mycorrhizae have significant effect on phosphorus uptake by plant. Greater soil exploration by mycorrhizal roots as a means of increasing phosphorus uptake is well known. In phosphate deficient soils immobile phosphate ions develop a phosphate depletion zone around the root. The hyphae reach beyond this zone and directly traslocate nutrients from the soil to root cortex (Hayman, 1983) in arbuscules where phosphorus is transferred to plant cell in exchange for carbon. Fungal hyphae are also known to absorb phosphorus from lower concentration compared to non-mycorrhizal roots. Tracer technique using 32, labelled phosphate in the experiments showed that AM hyphae obtain their extra phosphate from the labile pool rather than dissoluble phosphate (Raj *et al.*, 1981). Thus the increased uptake of phosphorus by plant occurs in three steps such as absorption of phosphate by hyphae from outside to internal cortical mycelia and finally the transfer of phosphate to cortical root cells, ready to be used by plants (Barea, 1991). Although genetypic and environmental factors influence symbiosis, a prolonged, stable and compatible relationship between plants and AM is based primarily upon the transfer of phosphorus from symbiont to the host that has a net phosphorus deficit (Barea, 1991). The increased plant growth inoculated with AM fungi is also due to greater availability of other plant elements like zinc, molybdenum, copper, potassium, sulphur, iron, manganese etc. Thus the efficiency of an endophyte will depend on the extent of mycelia spread in the soil and the number of connecting hyphae.

AM and Nitrogen Nutrition: The Efficiency

AM effectiveness has been found to decrease as result of fertilizer application in soil (Alexander and Fairley, 1983; Pareek *et al.,* 1990). However, AM improves growth, nodulation and nitrogen fixation by *Rhizobium* and *Frankia* (Barea and Azcon-Aguilar, 1983, Barea *et al.,* 1987; Hayman, 1986). AM inoculation also increased nodulation and nitrogen content and grain yield of pigeon pea. The effect was equivalent to *Rhizobium* inoculation. The effect was more pronounced with combined inoculation of *Rhizobium* and *G. fasciculatum* (Singh, 1996). AM increases nitrogen nutrition by different mechanisms: (1). Improvement in symbiotic nitrogen fixation by supplying phosphate and molybdenum to Azotrophs. (2) Direct uptake of combined nitrogen by AM fungi. (3) Facilitates N- transfer, a process by which a part of nitrogen fixed by nodulated plant benefits the non -N-fixing plant nearby (Newman 1985; Allen, 1988).

Plant Hormones, Water Deficits

Allen *et al.* (1982) showed that VA mycorrhiza directly affects the level of plant harmones such as cytokinins and gibberellin-like substances.

The beneficial effects of VAM fungi have been reported under drought and saline conditions (Nelson and Safir 1982; Peiffer and Bloss, 1988, Singh 1991). The important feature is that in drought conditions a mycorrhizal root has ability to tap additional water sources unavailable to non-mycorrhizal plant roots (Allen and Boosalis, 1983). Such effects are apparent under phosphate limitation as additional phosphate to non- mycorrhizal plants boosted their performance under drought and salinity (Nelson and Safir 1982). A commercial inoculum could be a substitute for chemical fertilizer (Skujins and Allen, 1986).

Protection Against Pathogens

Certain mycorrhizal associations benefit plant health by suppression of pathogens like *Phytophthora*, *Pythium*, *Rhizoctonia* (Duchesne *et al.,* 1987) and *Fusarium oxysporum* (Singh and Singh, 1995). The inoculation of VAM fungi also reduced the reproduction and development of root knot nematode, *Meloidogyne incognita* (Sharma *et al.*, 1995). The mechanism involved for VAM efficiency may be nutrient competition, a barrier effect caused by sheathing of root in fungal mantle or production of antibiotics effective against pathogens (Duchesne *et al.,* 1988; Sylvia 1983), Disease suppression may also result from activation of defence mechanism (Sylvia and Sinclair, 1983). Identification of these facts from plant host and VAM fungus responsible for disease suppressions may lead to genetic manipulation of VAM fungus and the plant host to maximize disease resistance potential of the host.

Other Effects

Analytical and physiological studies have shown that mycorrhizal plants have increased rates of respiration, photosynthesis and higher levels of sugar, amino acids, RNA etc. and larger and or more number of chloroplasts, mitochondria, xylem vessels, motor cells etc. (Hayman, 1983, Dyaladoss *et al.,* 1988). Changes in the root exudation and altered rhizosphere organisms beneficial to plant growth may result because of colonization of roots by VAM fungi (Gaur and Rana, 1990a,b; Sattar and Gaur, 1986).

Mycorrhizal colonization also augments the population of introduced inoculants like *Azotobacter, Azospirillum, Rhizobium* and PSM (Gaur and Rana, 1990 a, b; Sattar and Gaur 1989 a,b).

V. Interaction of AM and PSM

Symbiotic mircro-organisms such as AM and *Rhizobium* are known to improve plant growth and nutrition much more efficiently as compared to free-living micro-organisms (Brown 1974, Hayman 1975, Tinker and Sanders 1975, Gaur, 1987).

The mycorrhizal plants, particularly in low phosphate soils, are more efficient in phosphate uptake than non-mycorrhizal ones (Mosse, 1973) and in such soils satisfactory nodulation and nitrogen fixation are dependent on an adequate supply of phosphorus (Schreven 1958). Hence, there must be some synergistic interactions between *Rhizobium* and AM and such interactions have been described (Mosse *et al.,* 1976, Crush, 1974, Daft and El-Giahmi, 1974).

Plant inoculated with AM and PSB together were also able to mobilize significantly higher amounts of phosphorus than deficient soils (Azcon *et al.*, 1976, Barea *et al.,* 1975). PSB are known to produce substances like auxins, gibberellins and cytokinins (Barea *et al.* 1976, Sattar and Gaur 1987).

Azcon *et al.* (1978) tried to differentiate the role of PSB cells, rhizobial cells and their cell-fee supernatants on mycorrhizal infection and studied the influence of VAM and PSB on nodulation and nitrogen fixation by *Medicago sativa.* They observed positive interactions, between the PSB and VAM assayed on the formation of VA symbiosis.

The cell-free supernatant of PSB had a direct influence on the infection process, giving the highest percentage of VAM infection. Further, the cell- free supernatants of *Rhizobium* and PSB showed a positive influence on nodule formation by rhizobia and also exhibited a stimulatory effect on the growth of mycorrhizal plants. In turn, mycorrhizal plants were found to produce significantly more nodules than non-mycorrhizal plants. The inoculation of *Rhizobium* and PSB together gave highest yield and maximum N and P absorption by plants (Azcon *et al.,* 1978). These observations show definite synergistic interactions among *Rhizobium,* PSB and VAM.

Interaction of PSM and AM in Rhizosphere Soil of Wheat Crop

The findings with wheat crop indicated that microbial inoculation augmented their rhizosphere counts taken at 60 days of plant growth and at the harvest stage (Sattar and Gaur 1989a). The maximum count was recorded at the flowering stage. Phosphate solubilizing fungi (PSF) were more than the phosphate solubilizing bacteria (PSB). The PSF were less when the soil was inoculated with *Glomus fasciculatum* and PSB but the PSB were more with *G. fasciculatum* and less with PSF.

Phosphatic fertilizers and farmyard manure increased their numbers considerably. At the flowering stage, rock phosphate at 90 Kg P_2O_5/ha with FYM recorded maximum PSF, but at harvest stage the maximum was obtained with FYM only. Maximum PSB were observed with combined application of rock phosphate and superphosphate in FYM series at both the stages of plant growth.

Population of Phosphobacteria and Fungi in Rhizosphere Soil

A pronounced stimulation of PSM population was observed due to inoculants (Table 139 and 140). PSF population ranged from 5.13 x 10^3/g to 16.88 x 10^3/g rhizosphere soil at 60 days of plant growth and 2.13 x 103 /g to 10.13 x 103/g rhizosphere soil at harvest. The survival of PSF was greater in single inoculation and lower under mixed inoculation treatments. PSB number varied greatly as a result of seed inoculation. The numbers fluctuated from 7.75 x 10^3 to 60.88 x 10^3g soil at 60 days of plant growth and 4.63 x 10^3/g to 15.50 x 10^3/g soil at harvest. The PSB population was lowered in mixed but got stimulated with *G. fasciculatum*. PSF count ranged from 4.91 x 10^3/g to 10.09 x 10^3/g at 60 days growth and 1.64 x 10^3/g to 8.73 x 10^3g soil at harvest. The maximum numbers at 60 days growth were obtained with rock phosphate at 60 kg P_2O_5/ ha in the FYM series and at harvest the highest count was recorded with FYM treatment.

Table 139: Count of Phospho-microorganism at 60 days growth of wheat (x10³Vg rhizosphere soil)

Microbial inoculant	Phosphate sources and levels								
	0	$RP_{30}{}^{+}SP_{30}$	FYM	FYM+ RP_{60}	FYM+ RP_{30}	FYM+ SP_{30}	FYM+ Rp_{30} SP_{30}	FYM+ SP_{60}	Mean
Uninoculated	3	12	13	8	16	17	24	17	13.75
G. *fasciculatum* (GF)	7	14	18	20	20	13	29	22	17.88
A. awamori (F_{18})	8	18	28	16	14	24	42	26	22.01
A. niger (AN)	17	32	15	39	21	28	25	32	26.13
P. striata (27)	33	69	49	64	75	30	37	50	50.88
B.pulvifaciens (R_5)	25	53	32	64	50	32	38	30	40.50
GF+F_{18}	20	18	39	24	34	24	26	56	30.13
GF+AN	15	16	60	22	38	28	25	32	29.50
GF+27	81	63	78	70	59	40	69	68	66.01.
GF+R_5	62	58	58	59	66	53	50	51	56.13
GF+F_{18}+AN+27+R_5	14	43	14	26	40	21	20	23	25.13
Mean	25.9	36.0	36.7	37.5	39.4	28.2	35.0	37.0	

RP-Rock phosphate, SP-Superphosphate, FYM-Farm Yard Manure

Table 140: Count of Phospho-microorganism at harvest of wheat (x10^3/g rhizosphere soil)

Microbial inoculant	Phosphate sources and levels								
	0	RP_{30} $^+SP_{30}$	FYM	FYM+ RP_{60}	FYM+ RP_{30}	FYM+ SP_{30}	FYM+ RP_{30}+ SP_{30}	FYM+ SP_{60}	Mean
Uninoculated	8	11	13	16	8	3	11	9	9.88
G. fasciculatum (GF)	8	9	13	16	6	7	8	7	9.26
A. awamori (F_{18})	15	15	23	20	21	16	17	2	15.68
A niger (AN)	14	15	28	28	12	19	17	4	17.13
P. striata (27)	13	18	17	13	12	24	19	8	15.18
B. pulvifaciens (R_5)	16	20	27	13	23	15	24	9	18.38
GF+F_{18}	9	16	15	17	17	22	8	7	13.88
GF+AN	18	17	19	15	22	14	13	7	15.26
GF+27	19	21	33	14	16	21	14	11	18.63
GF+R_5	17	17	31	9	17	20	15	5	16.38
GF+F_{18}+AN+27+R_5	21	22	21	14	1C	13	18	6	15.63
Mean	14.36	16.54	21.55	15.91	14.91	15.81	14.91	6.82	

PSB numbers ranged from 21.00 to 32.00 x 10^3/g soil at 60 days growth and 5.18 to 12.82 x 10^3/g soil at harvest. The count was highest with rock phosphate at 30 kg P_2O_5ha at 60 days growth and with FYM alone at harvest. A depression or negative rhizosphere effect was noticed at a higher dose of superphosphate and at harvest stage.

Saric (1965) reported that the count of phosphate dissolvers in wheat rhizosphere increased continuously until maturity and then declined. Kundu and Gaur (1980) reported that the addition of 2 per cent FYM increased their population considerably.

Effect of AM, PSM and their Combinations on Lentil *(Lens Esculenta)*

Effect on nodulation: Both AM *(Glomus fasciculatum)* and PSB (*Pseudomonas striata* and *Bacillus pulvifaciens)* and PSF (*Aspergillus awamori* and *A. niger)* improved nodulation substantially (Sattar and Gaur, 1989b). Among the cultural treatments, *G. fasciculatum* produced the maximum number of mean nodules per pot (125.6) whereas the control has only 39 nodules, Single inoculation by all the PSM and AM showed higher numbers of nodules than their respective combination with *G. fasciculatum*. PSB were found to be slightly superior to PSF. The lowest number of nodules were recorded with mixed inoculum of all the cultures compared with other cultural treatments, although it was 14.7 percent higher than the control.

The different forms and levels of phosphatic fertilizers influenced nodulation considerably. Superphosphate at its lower rate (45 Kg P_2O_5) proved superior to its higher level (90 Kg P_2O_5) and both the levels of rock phosphate. The maximum mean number (85.6) was recorded with combined application of rock phosphate and super-phosphate at an equal rate of P_2O_5 in FYM amended soil, FYM alone produced 79.9 per cent more nodules than the control.

A several fold increase in nodule number was observed with *G. fasciculatum* with all the fertilizer forms and levels over control except at the highest rate (90 kg P_2O_5) of superphosphate. Among the phosphomicro-organisms, *P. striata* recorded the maximum number of nodules with most of the fertilizer treatments.

Delorenzini *et al.* (1979) reported the highest number of nodules in clovers with triple inoculation of PSB, *Rhizobium* and VAM followed by dual and single inoculations. Higher nodule numbers due to VAM were reported by several workers (Mosse *et al.,* 1976, Powel and Sithamparanathan 1977, Carlings *et al.*, 1978, Godse *et al.,* 1978, Smith *et al.,* 1979).

Increase in nodule numbers due to PSM and phosphatic fertilizer is not only because of its increased supply of phosphorus to higher plants but also to its direct effect on nodule bacteria. It stimulates the multiplication of rhizobia and was found conducive to the development of motile forms which are essentially required to migrate through the soil towards the root system.

Leghaemoglobin content: Leghaemoglobin content of lentil nodule was influenced with the addition of microbial inoculants and phosphatic fertilizers differently (Sattar and Gaur, 1986 b).

The maximum mean content of leghaemoglobin (37.88 mg/g nodule) was found with *G. fasciculatum*. *P. striata* alone and in combination with *G.*

fasciculatum produced 36.00 mg and 35.25 mg/g nodules which worked out to 66.0 mg and 56.7 per cent more than control respectively. *A. awamori* proved to be better than the rest of the inoculants.

A marked increase in leghaemoglobin content was observed with different phosphatic fertilizers. FYM and rock phosphate at higher level (RP90) was found to be superior to its lower level and both the levels of superphosphate, A maximum amount of leghaemoglobin was obtained when rock phosphate and superphosphate were applied together in FYM amended soil. FYM alone-enhanced leghaemoglobin content considerably.

With various fertilizer treatments, *G. fasciculatum* did better in most cases, whereas *P. striata* and *A. awamori* were found to be superior over other inoculants.

Jordan and Garrand (1951) showed the relationship between leghaemoglobin content and the growth of legumes. Increased leghaemoglobin content in soyabean due to FYM and poultry manure was observed by Dev and Tilak (1976). Shukla and Yadav (1982) reported that leghaemoblobin content increased with increasing phosphorus levels upto 50 ppm in chickpea and decreased at higher levels.

Dry matter yield at flowering: The results at 60 days of plant growth showed that dry matter yield was significantly augmented when PSM and VAM were used (Sattar and Gaur 1989b). *Pseudomonas* striata and *Glomus fasciculatum*, when used singly and in combination recorded higher yields than the rest of the inoculants. Mixed inoculation of all the cultures, though, gave greater yield but was not statistically superior to single inoculation.

Chemical phosphatic fertilizers increased dry matter production appreciably. Application of superphoshate at a lower level was more effective than application at a higher rate and rock phosphate applied at two rates. Maximum dry matter yield (5.5 g/pot) was obtained when rock phosphate and superphosphate were applied at equal rates of P_2O_5 in FYM amended soil. FYM alone augmented dry matter yield by 29.2 per cent over control. The yield was further augmented when *Pseudomonas striata* was used as seed inoculant and the soil was amended with combined application of rock phsphate, superphosphate and FYM. Delorenzini *et al.* (1979) reported increased plant growth of clover due to inoculation by PSB, VAM and *Rhizobium* with rock phosphate.

Grain and straw yield: PSF such as *Aspergillus awamori* and A. *niger* when used alone gave higher grain yield than PSB. The maximum increase of 48.2 per cent in grain yield and 30.3 per cent in straw yield was obtained when *Glomus fasciculatum* and *Pseudomonas striata* were used together as compared to control without any inoculation (Sattar and Gaur 1989 b). This was followed by *Aspergillus awamori* which increased the grain and straw yield by 35.2 and 29.6 per cent over the control respectively. *G. fasciculatum, A. niger* and *P. striata* when used as a single inoculum and A. *niger* combined *with* G. *fasciculatum* were statistically at par. Mixed inoculation involving different cultures did not show any desirable effect, rather the yield was decreased as compared to control, although the decrease was statistically non-significant.

The forms and levels of phosphatic fertilizer exhibited variable effects. The application of rock phosphate at 90 Kg. Kg P_2O_5/ha and superphosphate at 45 kg P_2O_5/ha gave similar yields and these two treatments were superior to lower level of rock phosphate and higher level of superphosphate. The maximum yield was obtained when rock phosphate and superphosphate were applied at 45 Kg P_2O_5/ha each in FYM amended soil. The higher rate of superphosphate application at 90 Kg P_2O_5/ha when combined with microbial inoculants showed lower yields. P. *striata,* when soil was amended with rock phosphate and superphosphate at equal rates in FYM series gave the highest grain (57.9 per cent) and straw yield (69.8 per cent) as compared to control.

Increase in yield due to use of phosphate solubilizing microorganisms have been reported in chickpea *(Cicer arietenum*) mungbean *(Vigna radiata)* wheat *(Triticum asestivum)* and rice *(Oryzae* sativa) by Gaur (1990). Increased yield due to *G. fasciculatum* has been reported in legumes by Crush (1974) and Kuo and Huang (1982). It is interesting to observe the beneficial interaction of *Pseudomonas* striata and VAM in augmenting the yield of lentil with different forms and levels of fertilizers.

Phosphorus uptake: The percentage of P_2O_5 in grain was increased from 0.532 to 0.789 and from 0.254 to 0.394 in straw by various treatments (Sattar and Gaur 1989 b). Among the single inoculations, *A. awamori* was found to be superior in P_2O_5 uptake both by grain (Table 141) and straw. In *G. fasciculaum*, *P. striata* proved to be the best and *B. pulvifaciens* second best in P_2O_5 uptake. Mixed culture treatment showed slightly better uptake than control.

The maximum P_2O_5 uptake was recorded with a combined application of rock phosphate and superphosphate in FYM series. Rock phosphate at 90 kg P_2O_5/ ha was superior to its lower level and both the levels of superphosphate. FYM showed a significant increase over control and increased the efficiency of phosphatic fertilizers. With microbial inoculation, further improvement was observed with all the forms and levels of phosphorus except with the higher level of superphosphate where P_2O_5 uptake was reduced.

Greater P_2O_5 uptake by different crops due to phosphate dissolving microbial inoculation has been reported by several workers. Similarly there are reports of higher P_2O_5 uptake due to *G. fasciculatum* and synergistic effects were observed with increased phosphorus uptake and dry matter production in finger millets with *Glomous fasciculatum* and *Bacillus circulans* (Gaur, 1990).

Table 141: Effect of inoculation and phosphatic fertilizers on phosphorus uptake by lentil grain (mg P_2O_5/pot)

Microbial inoculant	Phosphate sources and levels								
	0	RP_{45}+ Sp_{45}	FYM	FYM+ Rp_{90}	FYM+RP^{45}	FYM+ Sp_{45}	FYM+ RP_{45} + Sp_{45}	FYM+ Sp_{90}	Mean
Uninoculated	3.67	11.85	8.42	15.00	11.03	17.41	18.80	25.76	14.00
G. fasciculatum (GF)	7.35	16.53	13.17	23.46	17.62	22.12	25.51	19.95	18.21
P. striata (27)	8.22	15.58	11.41	23.41	18.40	19.29	25.73	15.70	17.21
GF+F_{18}	7.65	15.49	11.76	25.97	21.11	23.53	27.35	19.08	18.99
GF+AN	6.17	14.01	10.68	23.16	20.05	22.74	27.85	14.17	17.35
GF+27	13.66	18.71	18.41	28.37	27.59	27.47	33.97	19.22	23.43
GF+R_5	11.95	15.43	15.83	25.64	23.68	23.80	28.61	17.80	20.34
GF+F_{18}+AN+27+R_5	6.31	10.36	11.84	17.43	15.52	22.13	22.75	11.49	14.77
Mean	7.98	14.84	12.07	23.05	18.31	21.74	26.02	15.70	
	Inoculant			**Fertilizer**	**Inoculant x Fertilizer**				
S.Em. (±) C.D. (0.05)	0.53 1.48			0.45 1.26	1.50 4.18				

RP-Rock phosphate, SP-Superphosphate, FYM-Farm Yard Manure

Available soil P_2O_5: The available P_2O_5 content of soil was remarkably increased over control with microbial inoculants. *G. fasciculatum* alone and in combination with phospho-microorganism recorded maximum concentration of available P_2O_5 in soil (Sattar and Gaur 1984). A. *awamori* was superior to *A. niger* and P. *striata* to *B. pulvifaciens*. Rock phosphate at 90 Kg P_2O_5/ha was superior to its lower level and to both the levels of super-phosphate. The maximum amount was recorded with a combined application of rock phosphate and superphosphate in FYM amended soil. FYM per se increased available P_2O_5 of soil over control. With phosphatic fertilizers, *G. fasciculatum* alone and in combination with *P. striata* showed a higher amount of available soil P_2O_5 due to inoculation of phosphomicroorganisms.

The positive relationship between total number of phosphate solubilizing mycroorganisms, available P_2O_5 and leghaemoglobin content indicated that they play a viatal role in augmenting available soil phosphorus and synthesis of leghaemoglobin in the nodules.

Effect of AM, PSM and Combinations on Wheat Crop

Grain and Straw Yields: The response of wheat (Var. HD 2264) to VA mycorrhiza (*Glomus fasciculatum)* and PSB and their interactions in the presence and absence of phosphatic fertilizers was investigated in a greenhouse experiment (Sattar and Gaur, 1989a).

Microbial inoculation and fertilizers and their interaction significantly influenced grain (Table 142) and straw yield. An appreciable increase in grain yield over control was recorded with microbial inoculations. Among single inoculants the highest grain yield was obtained with *A. awamori* (17.2 g) followed by *P. striata* (17.0 g). The yields due to consortia of A. *niger* and *P. striata, B. pulvifaciens* and *G. fasciculatum* and PSF suppressed the straw and grain yields non-significantly as compared with control. On the other hand, combined inoculation with PSB and *G. fasciculatum* augmented straw and grain yields but not significantly over their respective control.

Table 142: Effect of inoculation and phosphatic fertilizer on grain yield of wheat (g/pot)

Microbial inoculant	Phosphate sources and levels								
	O	**RP_{45}+ SP_{45}**	**FYM**	**FYM+ RP_{90}**	**FYM+ RP_{45}**	**FYM+ SP_{45}**	**FYM+ RP_{45} + SP_{45}**	**FYM+ SP_{90}**	**Mean**
Uninoculated	6.75	13.03	10.32	14.22	13.25	16.63	18.13	20.08	14.05
G. *fasciculatum* (GF)	8.83	12.01	15.73	17.95	16.57	18.35	19.82	14.96	15.53
A. awamori (F_{18})	9.10	14.16	14.21	20.07	19.47	18.46	22.95	18.91	17.17
A niger (AN)	8.71	13.58	14.17	20.72	18.63	18.37	21.98	16.69	16.61
P. Striata (27)	9.89	13.79	13.92	19.11	17.92	20.75	21.87	18.85	17.02
B. pulvifaciens (R_5)	8.85	12.28	12.92	17.37	17.26	18.65	18.72	18.60	15.58
GF+F18	6.82	10.35	13.12	17.52	17.02	18.04	17.57	17.62	14.76
GF+AN	7.14	11.01	10.73	16.99	17.13	17.35	15.05	16.03	13.93
GF+27	9.84	14.77	15.77	19.27	19.39	18.21	21.11	16.92	16.91
GF+R_5	10.28	14.14	13.91	17.83	18.22	18.71	19.48	16.14	16.09
GF+F_{18}+AN+27+R_5	7.07	9.80	10.37	15.85	15.19	16.46	18.93	12.23	13.24
Mean	8.48	12.63	13.20	17.90	17.28	18.18	19.60	17.00	
	Inoculant			**Fertilizer**		**Inoculant x Fertilizer**			
S. Em. (±)	0.39			0.33		1.09			
C.D. (0.05)	1.07			0.92		3.04			

RP-Rock phosphate, SP-Superphosphate, FYM-Farm Yard Manure

There was a significant difference between two rates of rock phosphate but superphosphate at 30 Kg P_2O_5/ha gave significantly higher yield than 60 Kg P_2O_5/ha. The maximum grain and straw yields were recorded with rock phosphate and superphosphate applied together at equivalent rates in FYM amended soil. FYM also improved the efficiency of chemical fertilizers significantly.

Microbial inoculation in superphosphate series at a higher dose (60 Kg P_2O_5/ ha) decreased grain yield considerably. The highest grain yield was obtained with *A. awamori* when supplemented with rock phosphate and superphosphate at equal rates of P_2O_5 with FYM. However, the highest straw yield (32.66 g/pot) was recorded with *P. striata* plus *G. fasciculatum* in FYM and rock phosphate plus superphosphate treatment. *P. striata* showed better straw yields with most of the fertilizer combinations and *A. awamori* was effective only in a few cases.

Phosphorus uptake: Microbial inoculation and fertilizer, and their interactions have pronounced effects on P_2O_5 uptake by grain (Table 143) and straw yield. *P. striata* and alone together with *G. fasciculatum, yielded* higher uptake than single or composite inoculations (Sattar and Gaur, 1989 a). Among the phosphofungi, *A. awamori* proved to be superior. Mixed inoculation, though, gave higher uptake but was statistically at par with control.

Phosphorus uptake was significantly influenced by forms and levels of fertilizers. Superphosphate at lower level (30 Kg P_2O_5/ha) was found to be superior to its higher level and rock phosphate at the same level. Maximum uptake was recorded with the combined use of rock phosphate and superphosphate in FYM series. FYM also enhanced P_2O_5 uptake significantly. A depression in phosphorus uptake at a highest dose of superphosphate with cultural treatments was observed. On the whole, *P. striata* performed well with all the fertilizers forms and levels.

Table 143: Effect of inoculation and phosphatic fertilizer on phosphorus uptake by wheat grain (mg P_2O_5pot)

Microbial inoculant	**Phosphate sources and levels**								
	0	**RP_{45} + SP_{45}**	**FYM**	**FYM+ RP_{90}**	**FYM+RP_{45}**	**FYM+ SP_{45}**	**FYM+RP_{45} + SP_{45}**	**FYM+ SP_{90}**	**Mean**
Uninoculated	33.4	76.2	52.6	83.3	71.7	97.9	107.7	123.3	77.0
G. *fasciculatum* (GF)	49.1	75.7	89.7	114.7	99.2	116.0	136.0	85.6	96.8
A. awamori (F18)	49.9	90.0	79.5	123.4	114.1	117.6	149.6	116.5	105.1
A niger (AN)	46.4	84.6	77.1	126.4	107.5	115.4	141.4	100.5	99.9
P. striata (27)	55.5	88.4	80.6	120.6	105.9	132.0	148.3	116.5	106.0
B. *pulvifaciencs (R5)*	47.7	78.5	72.7	107.7	101.6	117.3	124.3	113.7	95.4
GF+F18	38.1	65.6	74.3	112.8	102.6	116.6	121.4	105.9	92.2
GF+AN	38.2	69.3	60.0	107.8	101.6	110.5	100.4	94.4	85.3
GF+27	56.2	95.0	91.8	125.8	124.1	125.2	147.6	106.8	109.1
GF+R5	57.5	90.4	79.2	114.9	115.3	119.1	133.2	99.8	101.1
GF+F18+AN+27+R5	38.3	59.8	57.6	97.5	91.3	103.6	123.1	69.9	80.1
Mean	46.4	79.4	74.1	112.3	103.2	115.6	130.3	103.0	
	Inoculant			**Fertilizer**	**Inoculant x Fertilizer**				
S. Em. (±)	2.4			2.0		6.8			
C.D. (0.05)	6.7			5.7		18.9			

RP-Rock phosphate, SP-Superphosphate, FYM-Farm Yard Manure

Available soil P_2O_5: Microbial inoculation and phosphatic fertilizers markedly influenced the available soil phosphorus (Sattar and Gaur, 1984). The maximum available P_2O_5 was obtained with *G. fasciculatum* (30.93 ppm) following by *P. striata* (28.28 ppm) and *B. pulvifaciens* (26.61 ppm).

A. *awamori* was superior to *A. niger* both in single and in combined inoculation with *G. fasciculatum*. Mixed inoculation showed lower values than all other cultural treatments excepts in the case of *A. niger*.

Addition of phosphatic fertilizer increased available P_2O_5 in soil differently. Rock phosphate at 60 Kg P_2O_5/ha was slightly better than at a lower dose and both the doses of superphosphate. The maximum available P_2O_5 (33.20 ppm) was recorded with rock phosphate and superphosphate when applied to soil at equal rates of P_2O_5 amended with FYM.

With different level of fertilizer treatment, *G. fasciculatum* did better than the other cultural treatments except at a higher dose of superphosphate where a suppression or negative result was observed in the presence of some of the inoculants.

VI. Mass Production of AM Fungi

The role of vesicular abruscular mycorrhiza in improving the growth and nutrition of plant has been established but the mass production for field application has been limited due to their inability to grow on artificial media. The most common technique for its limited production has been to grow symbiont with the host plant in pot culture (Mosse 1953). The availability of the inoculum of VA mycorrhizae for field applications is meager. Therefore, the propagation of axenic culture of these fungi may be the answer. However, *in vitro* cultivation under pure culture remains the most challenging task for the research workers (Johri and Mathur, 1989).

i) **Axenic culture of AM:** Surface sterilized clean spores (Hepper, 1985) are aseptically transferred to water agar and other media for suitable spore germination, Monoaxenic cultures of AM have been produced as whole plant cultures, excised root (root organ) culture or pure two member cultures (John *et al.,* 1981, Mosse, 1962). Infections were found to be the same as in case of whole plant systems. External mycelia grew vigourously but it was not possible to subculture from external mycelium. New methods are being developed for mass production of AM fungi which may be found in the right direction (Lui and Luo 1994, Smith 1994. Walker and Vestberg, 1994 and Zeze *et al.*, 1994).

ii) **Solution culture for inoculum production:** Macdonald (1981) used autoclavable hydroponic culture systems for the production of axenic

AM. Surface sterilized spores germinated on agar were utilized for infecting seedlings and incubated in greenhouse at 25°C till lateral roots appeared. The hydroponic culture was superior to root organ culture (Nelson and Safir 1982, and Newman 1985). The infected roots thus can be used for inoculation purposes.

Maintenance of AM in pot culture: They can be multiplied in the living roots of sorghum, maize, onion, rye and rhodes grass which are grown in sterilized soil. Rhodes grass (*Chloris gayana*) is considered the best host, perlite soil (1:1 vol.) mix to be the best substrate and use of peticides such as captan and furadan at half of the recommended rates checked the growth of contaminants without any adverse effect on VA mycorrhizae. Thus this technique to produce clean mycorrhizal inoculum can be used for developing "starter cultures" for supply to farmers. The farmers can further multiply it is in their own fields for inoculating crop plants. Plant growth period of 2-3 months gives a large crop of mycorrhizal spores to produce sizable amount of inoculum. The soil from the root zone as well as the roots are collected, ground to get uniform inoculum.

Nutrilink (AM Fungi), IARI: This biofertilizer, AM (Arbusular Mycorrhizae) is being produced on commercial scale by the Division of Microbiology, IARI, New Delhi. Presently the inoculant is mass multiplied in sterilized sand and soil mixture with a suitable host. The inoculum contains large number of fungal spores and infected root pieces. It can be used to improve the AM infection in plants and thus help in P nutrition of cereals, leguminous and vegetable crops besides horticultural and forest trees.

Method of application: AM should be applied in close proximity of the seed or plant. Depending on the cultural conditions, AM should be applied before sowing of the seeds especially with transplanted crops. Generally 3-5 Kg AM inoculant is recommended for one acre. The dose varies with the crops.

For transplanted crops such as chilli, tomato, brinjal, rice etc, it should be spread over the nursery bed first and then seeds are sown. Similar technique can be followed for transplanted seedlings and trees. This can also be used for forest and orchard plants. While preparing the nursery soil, it can be mixed with soil @ 20-50 g/plant. The same rate of inoculant can be used in case of established plants by mixing with surrounding soil of the tree plants (1-2 ft diameter) and also during interculture and application of organic manure etc. The cost of one packet containing 5 Kg is Rs.100, sufficient for one acre.

Inoculation methods: For successful mycorrhizal infection, the inoculum is placed in the root zone of actively growing plants which are not heavily fertilized. The quantity of infection by AM fungus depends on placement, inoculum potential, inoculum density and environmental factors. Various methods can be tried and tested for introducing VA mycorrhizal inoculum in field crops.

i) A reliable method of doing this is to place a single spore in a sterilized funnel filled with sand and to sow it with sorghum. The roots of the seedling gets colonized by the fungal spore and after two to three weeks the seedling is transferred to a pot. Pot cultures of AM fungi is maintained in the glass house to serve as the inoculum.

ii) Transplanted crops: Seedlings are grown in sterilized and unsterilized soil with efficient selected AM fungi in small nursery beds or other suitable containers which can be planted when mycorrhizal colonization has been established. This method has been used in important crops like chilli, finger millet, tomato, tobacco and wetland rice (Govinda Rao *et al.,* 1983, Sreemalu and Bagyaraj 1986 and Secilia and Bagyaraj 1994) horticultural crops such as citrus, mango, asters and marigold (Vinayak and Bagyaraj, 1990, Bagyaraj *et al.*, 1988) and forest tree species e.g. *Leucaena, Tamirindus indica,* Acacia *nilotica, Calliandra callothyrus* and *Casuarina equisetifolia* (Bagyaraj *et al.,* 1989, Reena and Bagyaraj 1990 a, b and Vasntha Krishana et *al.,* 1995).

iii) Directly sown plant

a) *Coating* of *seeds* with *AM inoculum:* Seeds are coated with *AM fungi* with an adhesive such as methyl cellulose. It is expected that the inoculum would stick to the seeds and subsequently infect the emerging radicals. It has proved satisfactory for large seeded plants such as citrus in field nurseries (Hattingh and Gerdemann, 1975).

b) *Mycorrhizal pellets:* The seeds are incorporated into the inoculum to form multi-seeded pellets. These pellets (about 1 cm in diameter) consist of soil inoculum from pot cultures stabilized with clay (Powel 1979, Hayman *et al.,* 1981) or peat inoculum.

c) *Fluid drilling:* Mixing of seeds and inoculum into an uniform suspension containing 4 % (w/v) methyl cellulose and then the slurry is applied in furrow. This has been reported to be successful under field conditions (Hayman *et al.,* 1981).

d) *Inoculation* in furrows: The inoculum is placed under or besides seeds sown in a furrow (Hayman *et al.,* 1981) which may be an effective method for inoculation of field crops. In field experiments, mycorrhizal inoculum was placed below the seeds. Very high rates of inoculum varying from 0.8 to 167 tonnes ha-1 with commonly used rates of 20 to 30 tonnes would be difficult to put in practice due to economic and practical problems.

e) *Precropping*: Population of AM propagules can be raised *in situ* by growing strongly mycorrhizal host plants and leaving the infected roots and the associates spores in the soil to infect the following crop. Judicious crop rotations might be practiced as means of manipulating raising of *in situ* AM propagules.

Constrains for commercialization: The yield of the crops would be augmented if the AM inocula become available to farmers. Several agencies including private companies have tried to develop commercial AM inoculants. Efficient strains have been identified (Duke *et al.,* 1986, Mahoney *et al.,* 1985). The culture methods have been developed and patented (Dixon and Buchesna, 1988; Jones and Huktchison, 1988). Several inoculant formulations have been worked out and techniques for applying inoculants to nursery and field soil have been formulated and tested (Wood, 1991). Despite these developments, AM has not been widely commercialized and used.

The market for AM inoculants exists for different agricultural and horticultural crops. The largest segments comprises of agronomic crops such as maize, soybean, wheat, cotton and others. Horticultural crops grown in fumigated soils and artificial soil mix provide another area for inoculation. At present their use is in agroforestory and reclamation of eroded soils and wastelands.

Mycorrhizal inoculants should be effective over a wide range of crops grown in diverse soils and agroclimatic conditions for their widespread and routine use. The cost of the inoculation should justify the profits and benefits accrued to the inoculated crops. Some of the constraints such as ability of the inocula to compete with native mycorrhiza, variations in plant genotype, phosphorus status of soils etc may limit their use.

References

Abbot, L. K. and Robson, A. D. 1982. Aust. J. Agric. Res 33:389- 408.

Abbott, L. K. and Robson, A. D. 1985. New Phytol. 99: 255-265.

Alexander, I.L. and Fairly, R.I. 1983. Pl. Soil 71:49-53.

Allen, M. F. 1988. Proc Roy. Soc. Edinb. 94: 63-71.

Allen, M. F., Moore, J. S. Jr and Christensen, M 1982, Can J. Bot 60: 468.

Allen, M. F. and Boosalis, M. C. 1983. New Phytol 93: 61-71.

Ashton D. H. 1956. Autoecology of Eucalyptus regnans. Ph. D. Thesis, University Melbourne, Autralia.

Azcon, R., Barea, J.M. and Hayman, D. S. 1976.Soil Biol. Biochem. 8:135-138.

Azcon. R., Azcon-G., De Aguilar, C. and Barea, J. M. 1978, New Phytol. 80: 359-364.

Bagyaraj, D. J. 1992. In: Methods in Microbiology Vol. 24 (Eds : Narris, J. R., Read, D. J. and Verma, A. K.) Academic Press, Lond. P 359.

Bagyaraj, D. J. Majunath A. and Govinda Rao, Y. S. 1988. J. Soil Biol. Ecol. 8: 98-103.

Bagyaraj, D. J., Byra Reddy, M. S. and Nalini, P. A. 1989. Forest Ecol. Manage 27: 791-801.

Bagyaraj, D. J. and Reddy, B. J. D. 2000. In: New Trends in Microbial Ecology (Eds: Bharat Rai and Dkhar, I.S.) I. S.N.E. Hill Univ. Pub., Shillong pp 108-117.

Barea, J. M. 1991. Adv Soil Sci. 15: 1-15.

Barea, J. M. and Azcon-Aguilar, C. 1983. Adv. Agron 3: 1-54.

Barea, J. M. Azcon-Aguilar, C. and Azcon, R. 1987. New Phytol 106: 717-725.

Barea, J. M., Azcon-Aguilar, C., Azcon, R. and Hayman, D. S. 1975. In: Endomycorrhiza (Eds. Sanders, F. E., Mosse, B. and Tinker, P. B.) Academic Press, London & New York pp 409-418.

Barea, J., Navare E. and Montoya, E. 1976. J. Appl. Bacteriol. 40: 129-134.

Bhandari, N. N. and E. Mukerji, K. G. 1993. The Haustorium. Research Studies Press Ltd. U. K. pp 235-246.

Boudarga, K. and Dexheimer, J. 1988. Bulletin de la societe' Botanique de France 135: 11-121.

Boudarga, K., Lapeyrie, F. and Dexheimer, J. 1990. New Phytol. 114:73-76.

Bowen, G. D. 1980. In: Tropical Mycorrhizal Research (ed. P. Mikola) Oxford University Press, London.

Brown, M. E. 1974. Ann. Rev. Phytopathol. 12: 181-198.

Brundrett, M. C. 2002. New Phytol. 154: 275-304.

Burges. T. and Malajczuk, N. 1989. Agric. Ecosystem, Environ 28:181-198.

Carlings, D. E, Richle, W. G. Brown, M. F. and Johnson 1978. Phytopath 68: 1590-1596.

Chalermopongse, A. 1987. Proc. 3rd-roundtable conference on Dipterocarps (ed. A. J. H. G. Kostermans), UNESCO.

Chang, K., PiHu, H.T. and Kao, P. C. 1986. Nitrogen Fixig Tree Research Reports 4: 40-41.

Chilvers, G. A., Lapeyrie, F. F. and Horan, D. P. 1987, New Phytol 107: 441-448.

Chong, I. 1987. Forest Research Report, Forest Dept. Sarawak No. FF 5. IV p 12.

Crush, J. R. 1974. New Phytol. 73: 743-749.

Daft, M. J. and E L- Giahmi, A. A. 1974. New phytol 73: 1139-1147.

Danielson, R. M. 1985. In: Soil Recalamation Practices (R. L. Date and D. A. Kelin, Eds) Marcel Dekkar, N. Y. pp. 173-201.

Day, L. D., Sylvania, D. M. and Collins, M. E. 1987. Soil Sci. Soc. Am. J. 51: 635-639.

Dayaladoss, D., Bagyaraj, D. J. Syamsunder, J. 1988. Indian J. Microbiol. 28: 276-280.

Delorenzini, C., Barea, J. M. and Oliverres 1979. Rev. Latinnoam. Microbiol 21: 129-134.

Dev, S. P. and Tilak, K. V. B. R. 1976. Indian J. Agric Sci. 48: 252- 256.

Dixon, R. K. and Buchesna, C. A. 1988. Pl. Soil 105: 265-271.

Duchesne, L. C., Peterson, R. L. and Ellis, B. E. 1987. Can. J. Bot. 66:558-562.

Duchesne, L. C., Peterson, R. L. and Ellis, B. E., 1988. New Phytol. 108: 471-476.

Duke, E. R. Johnson, C. R. and Koch, K. E. 1986. New Phytol. 104: 583-590.

Ekwebelam, S. A. 1979. Indian Forester 105: 750-757.

Elad, Y. Chet, I. and Ketan, J. 1980. Phytopathology 70: 119-121.

Evans, D. G. and Miller, M. H. 1988. New phytol. 110: 67-74.

Frank, A. B. 1885. Ber Dtsch. Bot. Ges., 3: 128-145.

Fontenla, S., Garcia-Romera, J. A. and Ocampo, J. A. 1999. Soil Biol. Biochem.31:1591-1597.

Gaur, A. C. 1987. Phosphate Solubilizing Biofertilizer in Crop productivity and their interaction with VA mycorrhiza Proc. Workshop Mycorrhiza Round Table, J.N.U., Delhi pp 505-529

Gaur, A. C. 1990. Phosphate Solubilizing Microorganisms as Biofertilizer. Omega Scientific Publishers, New Dehi pp. 178.

Gaur, A. C. and Rana, J. P.S. 1990a. Role of VA mycorrhiza and their interaction on growth and uptake of nutrients by wheat crop. Natl. conference on Mycorrhiza, HAU, Hissar, Feb. 1990.

Gaur, A. C. and Rana, J. P.S. 1990b. Effects of insecticide on wheat crop inoculated with PSB and VAM fungi. Natl Conference on Mycorrhiza-HAU, Hissar, Feb. 1990.

Gemma, J. N. and Koske, R. E. 1988. Trans. Br. Mycol. Soc. 91: 123-132.

Gianinazzi, S., Gianinazzi-Pearson, V. and Trouvelot, A. 1989. In: Biotechnology of fungi for Improving Plant Growth (eds: whipps, J. W. and Lumsden, R. D.) Cambridge Univ. Press, Cambridge pp 41- 54.

Gilmore, A. E. 1971. Experimental Studies on the establishment and stability of ectomycorrhizae. Ph. D. Thesis, Sheffield Univ., Sheffield.

Godse, D.B. Wani, S. P., Patel. R. B. and Bagyaraj, D. J. 1978. Curr, Sci. 47: 784-785.

Grahm, J. H, Lonard, R. T and Menge, J. A. 1981. Plant Physiol. 68 : 548-552.

Graham, J. HJ. Linderman, R. C. and Menge, J. A. 1982. New Phytol. 91: 183-189.

Govinda Rao, Y. S., Bagyaraj, D. J. and Rai, P. V. 1983. Zentralbl. Microbiol 139:415.

Habte, M. Fox, R. L. Aziz, R and EL-Swaify, S. A. 1988. Appl. Environ. Microbiol. 45: 945-950.

Hafeel, K. M. and Gunatilleke, IAVN. 1988. Proc. First Asian Conference on Mycorrhizae, Madras, India pp 37-45.

Hall, I. R. 1975. New Zealand. J. Bot. 15: 481-484.

Harley, J. L. and Smith, S. E. 1983. Mycorrhizal Symbiosis. Academic Press, N.Y. p 483.

Hattingh, M. J. and Gerdemann, J. W. 1975. Phytopathology 65: 1013-1016.

Hayman, D. S. 1978. New Zealand J. agric. Res. 21: 271-276.

Hayman, D. S. 1982. Phytopathology 72: 1119-1125.

Hayman, D.S. 1982. Practical Aspects of Vesicular Arbuscular Mycorrhiza. Advances in Agril. Microbilogy (Ed. N. S. Subba Rao). Pp 328-373.

Hayaman, D. S. 1983. Can. J. Bot. 61: 944-963.

Hayman, D. S. 1986. Mircens J. 2: 121-145.

Hayman, D. S. 1975. In: Soil Microbiology. A. Critical Review. (Ed. Walker, N) Butterworth, London and Boston, pp 67-92.

Hayman, D. S., Morris, E. J. and Page, R. T. 1981. Ann. Appl. Biol. 99:247-253.

Heinirch, P. A. and Patrick, J. W. 1985. Aust. J. Soil Sci. 23: 222-225.

Hepper, C. M. 1985. New Phytol. 101 :685-693.

Hyppel, A. 1968a. Stud Forest Suecica 64: 1-18.

Hyppel, A. 1968b. Stud Forest Suecica 66: 3-16.

Jagpal, R. and Mukerji, K. G. 1987. Distribution of VAM in Delhi ridge. In: Mycorrhiza roundtable (Eds: Verma, A K., Oka, A. K., Mukerji, K. G., Tilak K. V. B. R. and Jankraj) p 257.

Janos. D. P. 1987. In: Ecophysiology of Mycorrhizal Plants (Ed: Safir, G. R.) CRC Press, Boca Rotan, Florida pp. 107-134.

John, T. V. Hay. R. T. and Reid. C. P. P. 1981. New Phytol. 89: 81- 86.

Johri, B. N. and Mathew, J. 1989. In: Plant Microbe Interaction (Ed. Bilgrami, K. S.) N. P. H. Delhi pp. 269-303.

Jones, M. D. and Huktchison, T. C. 1988. New Phytol 108: 451-459.

Jordan, D. C. and Garrand, E. H. 1951. Can. J. Bot. 29:360-372.

Kendrick, B. and Berch, S. 1985. Comprehensive Biotechnol. 4: 109.

Koide, R. T. and Mooney, H. A. 1987. New Phytol., 107: 173-182.

Koske, R. E. 1987. Mycologia 79: 55-68.

Kuo, C G. and huang, R. S. 1982. Pl. Soil 64: 325-330.

Kuo, S. C. and Tang, F. 1981. Acta Bot Yunnanica 3: 359-366.
Kundu, B.S. and Gaur, A. C. 1980. Pl. soil 57: 223-230.
Kruchelmann, H. W. 1975. In: Endomycorrhizae (Eds. F. E. Sanders, B. Mosse and P. B. Tinker) pp, 511-525. Academic Press, London and New York.
Krishna, K. R. and Bagyaraj, D. J. 1983. Can. J. Bot. 61: 2349- 2351.
Laiho, O. 1970. Acta For. Fenn. 106: 1-65.
Lee, K. G. 1982. Korean J. Mycol. 10: 134-135.
Leong, J. 1986. Annual Review Phytopathology 24: 187-209.
Lilly, V. G. 1975. Curr. Sci. 44: 201-202.
Linderman, R. G. 1992. In: Mycorrhizae in sustainable agriculture (Eds.: Bethlenfalvay, G. J. and Linderman, R. G.) Amer. Soc. Agron. Madison, USA pp 45-70.
Lopezaguillon, R. and Garbaye, J. 1989. Agric. Ecosystems Environ. 29:263-266.
Lui, R. R. and Luo X. S. 1994. New Phytol. 128 89-92.
Macdonald. E. M. 1981, New Phytol 89: 87-93.
Mahoney, M. J., Chevone, B. I., Skelly, J. M. and Moore, L. D. 1985. Phytopathol. 76: 679-682.
Marx, D. H. 1969 a. Phytopathol. 59: 153-163.
Marx, D. H. 1969 b. Phytopathol. 59: 1128-1132.
Marx, D. H. 1970. Phytopathol. 60: 1472-1473.
Marx, D. H. 1972. Ann. Rev. Phytopathol. 10: 429-454.
Marx, D. H. and Bryan, W. E. 1969. Phytopathol. 59: 1128-1132.
Meyer, FH. 1973. In: Ectomycorrhizae their ecology and physiology (Eds.: G. C. Marks and T. T. Kozlowski) Academic Press, New York, pp. 75-105.
Mohan Kumar, V. I., Raghupati, S., Nirmala, C. B. and Mahadevan, A. 1988. Curr. Sci. 57: 367-368.
Molina. R. 1979. Can. J. Bot. 57: 1223-1225.
Molina. R. and Trappe, J. M. 1982. Forest Sci. 28: 423-457.
Mosse, B. 1953. Nature 171: 974.
Mosse, B. 1962, J. Gen. Microbiol 25: 209-250.
Mosse, B. 1973. Ann. Rev. Phytopathology 11: 171-176.
Mosse, B. 1990. In: Mycorrhizal Symbiosis and Plant Growth (Eds. : Bagyaraj, D. J. and Manjunath, A.) Proc. Second National Conference on Mycorrhizae, Bangalore, India p5-10.
Mosse, B., Powel, C. L. and Hayman, D. S. 1976. New Phytol 76: 331-342.
Mosse, B., Stribley, D. P. and Le Tacon, F. 1981. Adv. Microbial Eco. 5: 137-210.
Nair, S. K. and Girija, V. K. 1988. J. Plantation Crops. 16: 67-68.
Narayan Bhat, M. 1993. Acta Botanica Indica 25: 52-56.
Nataraj, K. and Govindaswamy, V. 1990. In: Proc National Conference mycorrhizae, HAU. Hissar, India (Eds.: Jalali, B.L. and Chand, H.) pp 98-99.
Nelson, C. S. and Safir, G. R. 1982. Planta 154: 407-413.
Newman, E. I. 1985. In: Nuclear Techniques to Study the Role of Mycorrhia in increasing Food Production (Ed: Danso, S. K.) Intern. Atomic Energy Agency, Vienna TecDoE-338 pp 63-68.
Ozanne, P. G. 1980: In: the Role of Phosphorus in Agriculture (Eds: Kasawneh. F. R., Sample E.C. and Kamprath, E. J.) Amer. Soc Agron. Madison, USA. Pp 559-589.
Pandey, S. and Mishra, A. P. 1971. Mycopatha and Mycol. Appl. 45: 337-354.
Park, J. Y. 1960. In: The ecology of soil fungi (Eds. D. Parkinson and J. S. Waid). Liverpool Univ. Press, Ligerpool p. 324.
Park, J. Y. 1970: Can. J. Microbiol 16: 798-800.
Pareek, R. P. Kumar, N and Chandra R. 1990. J. Ind. Soc. Soil 38: 186-188.

Peiffer, C. M. and Bloss, H. E. 1988. New Phytol. 108: 315-321.
Powel, C. L. 1979. New Phytol. 83: 81-85.
Powel, C. L. and Sithamparanthan., J. 1977 N. Z. J. Agri. Res. 20: 489-494.
Pryor, L. D. 1956. Nature 177:587-588.
Quintos, M. and Valder, M. 1987. Riv Latinoamerican de microbiologica 29: 189-192.
Raghupathy, S. and Mahdevan, A. 1993. Mycorrhiza 3: 123-126.
Raj, J., Bagyaraj, D J. and Manjuanth, A. 1981. Soil Biol Biochem. 13:105-108.
Rangrajan, M., Kandasamy, D., Oblisamy, G. and Narayanan, R. 1988. In: Proc. First Asian Conference on Mycorrhizae, Madras, India (Eds. Mahadevan, A., Raman, N. and Natrajan, K.) pp. 288- 290.
Rao, Y. S. G., Suresh, C. K. Suresh, N. S. and Bagyaraj, D.. 1987. Curr. Res. 16: 55-59.
Ratnayake, M., Leonard, R. T. and Monge, J. A. 1978. New Phytol. 81:543-552.
Read, D. J. 1997. Nature 388:517-518.
Read, D. J. and Perez-Moreno, J. 2003. New Phytol. 157: 457-492.
Reddell, P, Robson, A. D. and Bowen, G D. 1986. Aust., J. Bot. 34 : 435-444.
Redecker, D., Kodner, R. and Graham, L.E. 2000. Science, 289: 1920-1921.
Redhead, J. F. 1977. Trans. Br. Mycol. Soc. 69:275-280.
Reena, J. and Bagyaraj, D. J. 1990 a. World J. Microbial. Biotechnol. 6:59-63.
Reena, J. and Bagyaraj, D. J. 1990b. Arid Soil Res. Rehabil. 4: 261-268.
Rhodes, L H. 1980. Outlook agriculture 10: 275-281.
Saric, Z. 1965. Proc. Symp. Plant Microbes Relationships. Prague, p. 143-146.
Sasek, V. and Musilek 1968. Folia Microbiol. 13: 43-51.
Sattar, M. A. and Gaur, A. C. 1984. Proc. Inter. Symposium on soil test crop responses correlation studies. Inter Cong. Meeting of Soil Science, Dhaka pp 237-251.
Sattar, M.A. and Gaur, A. C 1986. Bangladesh J. Agric. Sci. 11:27- 30.
Sattar, M.A. and Gaur, A. C 1987. Zentralbl. Microbiol 142: 393- 395.
Sattar, M.A. and Gaur, A. C 1989 a. Bangaladesh J, Agric Sci. 14: 233-239.
Sattar, M.A. and Gaur, A. C 1989 b. Thai J. Agri. Sci. 22: 129-136. Schenck, N. C. and Kinloch, R. A. 1974. Plant Dis. Rept. 58: 169- 173.
Schenck, N. C. and Perez., Y. 1987. In: Manual of Identification of VA Mycorrhizal Fungi. Univ. of Florida Press, Gainsville.pp 1-245.
Schrevan, D. A. Van 1958. In: Nutrition of Legumes (Ed.: Hallswrolth, E G.) Butterworth, London pp 137-163.
Secilia, J. and Bagyaraj, D. J. 1987. Can. J. Microbiol 33: 1069- 1073.
Secilia, J. and Bagyaraj, D. J. 1994. Mycorrhiza 4: 265.
Sharma, S. K., Sharma, G. D. and Mishra, P. R. 1987. J. Indian Bot., Soc. 66: 226-268.
Sharma, M. P., Bhargava, S. Varma, M. K. and Adholeya, A. 1995. In Mycorrhizae : Biofertilizer for the future (Eds. A. Adholoya and S. Singh) TERI, New Delhi pp 115-118.
Shukla, U. C. and Yadav, O.P. 1982 Pl. Soil 65: 239-248.
Sieverding, E. 1991. Vesicular-Arbuscular Mycorrhizae Management in Tropical Agrosystems. Deutch Geselshafte for Technische Zusammen-arbeit, Eschborn.
Singh, C. S. 1996. Microbiol Res. 157: 87-92.
Singh, P. 1991. Field Crops Res. 28: 1-15.
Singh, R. S. and Singh, D. 1995. In: Mycorrhizae biofertilizer for future (Eds. Adhooleya, A. and Singh S.) TERI, New Delhi pp 119- 123.
Singh, S. and Kapoor, K. K. 1999. Biol. Fertl. Soil. 28: 139-144.
Skujins, J. and Allen, M. F. 1986. MIRCENS J. 2: 161-176.
Smiley. R. W. and Cook R. J. 1973. Phytopathology 63: 882-890.
Smith, J. E. 1994. Can. J. Microbiol. 40: 736-742.

Smith, S E. and Gianiazzi-pearson, V. 1988. Ann. Rev. Plant Physiol. & Plant Mol. Biol 39: 221-224.
Smith, S. E., Nicholas, D. J. D. and Smith, F. A. 1979. J. Plant Physiol. 6: 315-316.
Sreeramulu, K. R. and Bagyaraj, D. J. 1986. Pl. Soil 93: 299-302.
Stack, R. W. and Sinclair, W. A. 1975. Phytopathology 65: 468-472.
Stahl, P. D. Williams. S. E. and Christensen, M. 1988. Mycologia 76: 261-267.
Strzemska, J. 1975. In: Endomycorrhizas (eds. Sanders F. E., Moss, B. and Tinker, P. B.) Academic Press, London, pp 537-543.
Suhardi, J. and Tasiman, Y. F. 1988. In: Mycorrhizae for Green Asia (Eds. Mahadevan, A., Raman, A and Natrajan, K.), Madras, India pp 14-15.
Sylvia, D. M. 1983. Pl. Soil 71: 299-302.
Sylvia, D. M. and Sinclair, W. A. 1983. Phytopathol. 73: 384-390.
Thapar, H. S. and Khan, S. N. 1973. Proc. Indian Natl. Sci. Acad. 39: 687-694.
Thapar, H. S. and Khan, S. N. 1985. Indian J. Forestry 8: 5-7.
Thomazini, L. I. 1974. Pl. Soil 41: 707-711.
Thompson, J. P. 1987. Aust. J. Agric. Res. 38: 847-867.
Tinker, P. B. 1982. Mycorrhizas, the present position In. "Whiter Soil Research" Trans 12th Inter. Cong. Soil Science, New Delhi pp 150- 160.
Tinker, P. B. and Sanders, F. E. 1975. Soil Sci. 119: 363-368.
Trappe, J. M. 1962. Fungus associates of ectomycorrhizae. Bot Rev. 28: 538:606.
Valdes, M. Pina, F, and Grada, R. 1983. Boletine de la sociedad Mexicana de micologia No. 18: 65-70.
Varma, A. K. and Schuepp, Hans 1994. Mycorrhiza 5:29-37.
Vasanthakrishna, M. and Baryaraj, D. J. 1993. Arid Soil Research Rehabilitation 7:377-380.
Veirheilig, H, Isch, B. Raikhel, A.M., Wiemken, M. and Boller, T. 1996 Plant Soil. 183: 131-136.
Vinayak, K. and Bagyaraj, D. J. 1990. Biol. Agri Hort. 6: 305-311.
Walker, C. and Vestberg, M. 1994. Agri. Sci. Finland, 3: 223-240.
Wheeler, C. T. Miller, I.M. Naryanana, R. and Purushottam, 1991. soil Microorganism in Agro-forestry pp 142-166. In Biosphysical Research for Asian Agro-forestry (Eds. Avery, M. E., Cannell, M. G R.and Ong. C.) South Asia Books, USA.
Wicklow-Howard, M. 1989. Mycotaxon, 34: 253-257.
Wright, S. F. and Upadhayaya, A. 1998. Plant Soil 198: 97-107.
Wood, T. 1991. In: Hanbook of Applied Mycology. Vol. 4 (eds. Arora, D. K. Clander, R. P. and Mukerji, K. G.) Marcel-Dekker Inc. N. Y. pp. 823-871.
Wood house, H. W. 1975. In: Endomycorrhizas (Eds. Sanders, F. E., Mosse, B. and Tinker, P. B.) Academic Press, London and New York pp 209-240.
Zak, B. 1964, Ann Rev. Phytopathology 2: 377-392.
Zak, B. 1971. Can. J. Bot. 49: 1979-1084.
Zeze, A., Dulieu, H. and Gianiazzi-Pearson, V. 1994. Mycorrhiza 4: 251-254.

20

Production Technology for Biofertilizers

Introduction

Conventional agriculture is highly dependent on fossil fuel based mineral fertilizers to compensate for removal of plant nutrients by crops. Chemical fertilizers are not only in short supply but also are very costly. The preparation and use of organic manures are not being practiced in a scientific manner to harness the plant nutrients removed by plants, animals and humans.

The application of chemical fertilizers to soil alone and in an imbalanced way has resulted in deterioration of soil health and its quality. The use of chemical fertilizers has to be reduced to check the soil and water pollution and simultaneously the role of soil microorganisms has to be augmented by different ways such as greater use of biofertilizers. Moreover, the demand for organic food particularly both in developed and developing countries is on rise which opens up immense possibility to use organic manure and biofertilizers for production of foods grains, fruits, vegetables, spices, tea and coffee. Therefore, the production and the use of biofertilizers assumes greater significance and could be an excellent biological alternative to chemical fertilizers.

What are Biofertlizers

Microorganisms participate in various soil fertility processes such as decomposition of organic debris, mineralization of plant nutrients and in different biocylcles of carbon, nitrogen, phosphorus, potassium, sulphur, iron, manganese etc., in soil and nature. These microorganisms are important renewable and natural resources for producing biofertlizers.

Definition

Biofertilizers include biologically active, efficient and specific strains of bacteria, actinomycetes, algae, *azolla*, fungi or mycorrhizae which help to provide or increase the availability of plant nutrients through the processes of biological nitrogen fixation, phosphate solubilization or through production of growth promoting substances in the soil and plant environment. Biofertilizers are more aptly termed as Microbial/Bacterial or Fungal inoculants (Verma, 1993).

The carrier based biofertlizers containing symbiotic and asymbiotic nitrogen fixing bacteria, phosphate solubilizing miroorganisms and cellulolytic and lignolytic microbes are being employed in many countries to boost crop productivity. Biofertilizers are thus based on renewable energy sources which are cost effective and environment friendly.

I. Types of Biofertlizers

They can classified in the following categories.

1. *Nitrogen fixing microorganisms*: Symbiotic such as *Rhizobium, Bradyrhizobium*, *Frankia, Azolla.* Associative *Azospirillum* spp. and free living *Azotobacter*.
2. *Phospho-microorganisms*: Phosphate solubilizing bacteria (PSB) phosphate solubilizing fungi (PSF) Mobilizing/helper in phosphate uptake - Ectomycorrhizae and Endomycorrhizae.
3. *CompostInoculant*Cellulolyticorlignolyticfungiforrapiddecomposition of cellulose and lignin in organic wastes and for composting.
4. Plant growth promoting Rhizobacteria (PGPR) for increasing crop growth, and yield.
5. Microorganisms, which help to control, plant disease and insect pests.

Role and Prospects

Biofertilizers supplement major plant nutrients and improve the plant productivity for sustainable agriculture. They are important and cost effective inputs in agriculture, plantation and commercial crops. Microbial inoculants/ biofertilizers on their application multiply in rhizosphere soils and benefit the growing crops. If the soil conditions are favourable, the population of added microorganisms are built up in the rhizosphere of crop plants and frequent applications of microbial inoculants can be avoided. They are inexpensive and help in reducing the consumption of chemical fertilizers.

Nitrogen fixing biofertilizers provide atmospheric nitrogen directly to plants. Other biofertilizers render mineralization and solubilization of plant nutrients such as phosphorus, sulphur, iron, manganese etc. They improve the plant growth by producing growth-promoting substances. They also help in control of soil-borne disease. They improve soil properties and maintain good soil health. They improve the crop yields and uptake of plant nutrients. They are compatible with organic manures, fertilizers and agro-chemicals. Biofertilizers are non-polluting and eco-friendly. The cost of production of biofertilizers is low and so is the selling price. It has been estimated that one Kg of fertilizer nitrogen cost more than six rupees while biofertilizer nitrogen cost only twenty-five paise through *Rhizobium* and about fifty Paise via blue-green algae. On nutrient basis, one tonne of Rhizobium inoculant is equivalent to 100 tonnes of fertilizer nitrogen taking 50 Kg N fixed ha-1 by application of 500g *Rhizobium* culture. One tonne of blue-green algae considering 20 Kg N fixed ha-1 is equivalent to 2 tonnes of fertilizer nitrogen by application of 10 Kg BGA ha^{-1} (Verma and Bhattacharya, 1991).

II. Development of Biofertilizer Industry

The production of biofertilizer for inoculation of legumes was started as early as 1895 by Nobbe and Hiltner by extracts of crushed nodules and was named 'Nitragen' (Roughley, 1976). The initial preparations of legume inoculants were pure culture of *Rhizobium* growing on solid medium which were marketed in 8-10 ounces glass bottles. Rothamsted produced 2000 packets by 1929. Similar level of production was reported by Canada in 1905, Australia and Swedan in 1914 and India in 1956. The methods of preparing inoculants have since then undergone significant changes in the production technology and supply of inoculants. *Rhizobuim* inoculants are produced globally including both in developed and developing countries. Among the developed the most prominent countries are Australia, Canada, France, Italy, the Netherlands, New-Zealand, USSR, UK and USA. Among the developing countries are Argentina, Bangladesh, Brazil, Egypt, India, Indonesia, Kenya, Thailand and Venezuala.

A greater expansion of *Rhizobium* inoculant industry in USA occurred pluring 1929 to 1963 when production increased from 11 million bushel units per year to 25 million bushel units per year in 1963. In India, the use of *Rhizobium* inoculants was started with berseem crop by Joshi in 1920, followed by Madhok (1934). The spurt in demand for inoculant was initially noticed during 1964 with the introduction of soybean crop by J.N.KV., Jabalpur (M. P.) and GBPUAT, Pant Nagar (UP). Actual breakthrough in inoculant production occurred with the selection of peat as a carrier material. During 1948, agar based inoculants in Brazil were replaced with peat based commercial inoculants, known as "NODIST". This was subsequently introduced in U.K.

India in one of the largest producer and consumer of biofertilizers. The states of Gujrat, Madhya Pradesh, Tamil Nadu and Uttar Pradesh are the major biofertlizer producers of the country. The list of commercial biofertilizer units in India has been documented by Verma (1993) and Motsara et *al.* (1995). In India, there are different types of formulations of biofertilizers such as carrier based, granules, pellets and agar based. Out of these, only carrier based inoculants are most suitable and recommended by Bureau of Indian standards.

Earlier, in India, the production of biofertilizers was a small affair carried out by a few agricultural colleges, research institutes or state agricultural universities, However, during the last two decades, a good advancement in the production and use of biofertilizers has been made. Biofertilizers are slowly getting popular with farmers and the demand for biofertilizers other than *Rhizobium* inoculants has also increased. This has given a fresh impetus to industry to come forward and to take the commercial production of the biofertilizers in a big way. There are four categories of entrepreneurs in the biofertilizer industry in India such as state agricultural departments, Agricultural universities, institution agencies and private manufacturers. It was reported that there were 60 commercial biofertilizer production units in India with total production capacity of 2685 tonnes/year (Bhattacharya and Singh 1992). The production level was at about 1200 tonnes/year. Currently the installed capacity has gone well beyond 20,000 mt produced by more than 90 production units.

The Biotech Consortium Indian Ltd. (1992) has reported that state agricultural departments produce 54 per cent, agricultural universities 2 per cent, institutional agencies 28 per cent and private manufacturers 16 per cent of *Rhizobium* inoculants.

Dixit (2001) reported that requirement of different biofertilizer in tonnes per year would be as follows which show an enormous gap between the requirement and production levels.

Rhizobium 36,000, *Azotobacter* 162,000, *Azospirillum* 77,000, Blue-green algae 268,000 and phosphate solubilizing bacteria 276,000.

Need For Basic Facilities

- There should be modern microbiological laboratory with sufficient space, gas, electrical outlets, sinks and air lines, temperature control and storage facilities. The rooms should be maintained properly to make it contamination free.
- The basic equipments are laminar air flow cabinet, autoclave (small and large), BOD Incubator, dry-sterilising oven, rotary shaker, fermentor, mixer, sealer, balances and microscope and high grade chemicals and glasswares.
- Supply of gas, water and electricity without interruption should be ensured.
- The production of biofertilizers should be handled by technically qualified and experienced persons having basic training in Microbiology/ Bacteriology rather than by any person of other disciplines.

III. Production Technology

Biofertilizer production on commercial scale consists of the following steps:

1. Isolation and selection of efficient and competitive strains.
2. Media for broth culture.
3. Multiplication of mother culture in the appropriate broth.
4. Selection and preparation of carrier material.
5. Mass production and packing of the biofertilizer.
6. Storage and dispatch to farmers.

The above processes appear simple but they are quite complex in practice. Being living system, every step in the biofertilizer production is sophisticated and requires due care and precautions. The production and supply of poor quality inoculants will deprive the users of the benefits of these biofertilizers.

Selection of Efficient and Competitive Strains

This is one of the most important criteria for production of biofertilizers. Effective strains are isolated from natural sources depending on the type of biofertilizer. In case of legume inoculants, effective infective strains are isolated from healthy and pink nodule. More than 70 per cent of the area under pulses and oilseeds in the country falls in rainfed areas. *Rhizobiutn* inocuiant has to establish themselves in extreme condition of low moisture regimes in such soils and therefore, the plants growing vigorously in dryland conditions with good nodulation are selected for isolation of rhizobia. Other quality of the efficient strains besides nitrogen fixing capacity include their competition ability with native rhizobium strains, capacity to withstand extreme soil and weather conditions, better host infection ability, resistance to bacteriophage and ability for a long shelf life in carrier material and tolerance to pesticides, fertilizers and other agricultural chemicals.

Soil in the best medium for microorganisms to grow and multiply. A strain that is not able to fix nitrogen at a sufficient rate so as to meet the host's nutrient demand is either partially effective or totally ineffective. The selection of an appropriate strain for biofertilizer production is the most basic and important aspect.

Selection of Free Living and Associate Nitrogen Fixer

This is done based on the criteria of their ability to fix higher amounts of nitrogen in laboratory condition using specific growth medium, alongwith other characteristics like excretion of ammonia in the medium during active growth period, production of growth promoting substances and siderophores etc. The response of crops to the inocuiant and their establishment and survival in the rhizosphere and rhizoplane is also determined during the specific growth period of plants. Other characteristics like autotrophic growth on CO_2 and H_2, use of different sources of carbon, tolerance to wide pH and temperature range, ability to function as biopesticides and having phosphate solubilizing activity were also reported in some strains of *Azotobacter* (Shende *et al.*, 1973; Sundara Rao *et al.*, 1963). Similarly *Azospirillum* sp., may also impart, besides nitrogen fixation, some additive advantages. Selection of such strains is desired over the strains selected only for nitrogen gains.

Selection of Phosphate Solubilizing Microorganisms

Efficient and competitive phosphate solubilizing microorganisms are selected based on their ability to solubilize insoluble phosphate sources, eg., rock phosphate, tricalcium phosphate etc. Production of organic acids in the broth culture are determined as the organic acids are mainlyresponsible for phosphate soulubilizaiton (Gaur, 1990). Other characteristics such as thermotolerant, salt tolerant strains and those producing growth promoting substances may give additional benefits for plant growth and yield.

Multiplication of Mother Culture

Rhizobium: For large scale production of *Rhizobium*and *Bradyrhizobium* spp., yeast extract mannitol medium (Vincent, 1970) with and without glutamate is recommended. In some cases, for fast growing rhizobia, mannitol is substituted with sucrose. Studies done with other carbon sources such as molasses, commercial sugar and glycerol showed that most of the fast growing and a large number of slow growing strains could be grown well on these substrates but contaminants often develop on molasses based broth after incorporation in the carriers.

Azotobaeter: The medium having sucrose (Jensen, 1951) or mannitol (Subba Rao, 1977) are commonly used. These bacteria generally do not need any growth promoting substances in the medium. Addition of yeast extract as source of combined nitrogen may attract contaminants.

Azospirillurn: Okon's medium (Okon *et al.,* 1977) having macro and micronutrients with malate as carbon source and yeast extract as source for vitamins serves as an appropriate medium.

PSM: Phosphate solubilizing microorganisms are grown in Pikovskaya's medium having glucose as carbon source, ammonium sulphate as nitrogen source with yeast extract and tricalcium phosphate as insoluble phosphate sotirce (Gaur, 1990).

Maintenance of pH and temperature: For all the above organisms. The pH of the media is maintained around 7 and incubation temperature is around 30°C.

Inoculation: Mother culture from slants are transferred into small flasks of 250 to 500 ml capacity having 100-200 ml broth. After allowing the growth for 2-6 days on rotary or incubator shaker, the growth of inocula is used for further inoculation in fermentors or broth in flasks of 1-5 litre capacity with 5 to 10 per cent inoculum (v/v).

Simple glass jar fermentors or non-reactive metal fermentors of 5-10 litre capacity are commonly used for commercial production. The use of large fermentors require fool proof aseptic conditions to avoid

contaminants. The use of fermentors on a commercial scale is more economical than the shake flask method, the space and power requirement for the shaker is much greater than for fermentors. The size of fermentors can vary from a few litres to 100 or 200 litres depending on the working capacity. Different type of fermenters such as Niftal type LPG fermentors, glass fermentor and UPIL type fermentor are available with different modes of operation.

The contamination is hardly seen in *Azotobacter* multiplication as well as fast growing bacteria but in slow growing *Bradyrhizobium* strains which take about a week for growth to attain stationary phase, contamination with aerial spores are often observed.

IV. Quality Test of Broth

The broth quality before blending with the carrier should be tested for pH, Gram staining, agglutination with relevant serum and antibiotic resistance. The tests for these parameters have been described elsewhere.

The viable cell count reached after the required period of incubation in flasks or fermentors without contamination with other microorganisms is the most important test for assessing the quality of the broth. Normally, counts ranging from 109 to 1010ml-1 are preferred for a better quality broth. Such counts are attained within 3-5 days in case of *Azotobacter, Azospirillum*, fast growing rhizobia, *Pseudomonas* and *Bacillus*. In case of *Bradyrhizobium* strains, it take longer, about 6-7 days to reach such counts.

The viable counts are often determined by dilution plate method using specific media for each group. In case of rhizobia, the most probable number (MPN) in broth is determined by plant infection tests (Vincent, 1982). The agglutination test is very authentic to check the purity of broth of bacterial culture and this can be completed within 24 hours. The agglutination on microscopic slides, using antiserum prepared against the inoculant strains is often recommended to avoid addition of contaminants from the broth culture, if any developed. Broth cultures free from contaminants with viable counts around 5×10^9 to 10^{10} ml^{-1} are considered to be of high quality. In case of PSM, fungal mat is allowed to grow upto sporulation and counts upto 10^7 to 10^8 CFU m' are considered satisfactory for mixing into the carrier.

Inoculation Delivery System

Several delivery systems such as agar slope based liquid paraffin covered agar cultures (Gordon and Smith, 1947; Hartsell, 1953) lyophilized culture (Appleman and Sears, 1946) broth absorbed on porcelain (Norris, 1963) liquid culture in polypropylene bags (Van Schrevan *et al.*, 1953; Van Schrevan 1958) and carrier based (Feller 1918) are known.

Far mass scale production and for use by the farmers, the carrier based cultures packed in polythene bags and property sealed, is the common practice in almost all the biofertilizer producing countries. Peat is commonly used as carrier in Australia, France, and USA. (Iswaran *et al.*, 1969) Good quality peat is not available in India but peat-like organic material is available in Tamil Nadu, around 5.5 million tonnes. Hence, other alternative carriers such as lignite, wood charcoal, press mud, compost, clay and soil are used (Iswaran, 1974). As per the suitability, the peat is the best followed by lignite, charcoal, FYM, Soil, rice husk, coir dust, pressmud etc. Annual availability of lignite is around 3 million tonnes. Coal is available to the tune of 254 million tonnes in India. Peat, lignite and coal are non-renwable sources which have other economic uses. Biofertilizer units depend either on lignite obtained from Neyveli or Gobri or on coal from Rajasthan. Lignite can be procured from Singrauli (MP) Neyveli (Tamil Nadu) and Palana (Rajasthan).

The most important characteristics of a good carrier are its water holding capacity, particle size, be free from toxic substances, sterilisable and have readily adjustable pH to 6.5-7.0. The sticking characteristics on seeds and total surface area provided for absorption of microbial cells per unit weight of the material is equally relevant. Keeping in view these characteristics, lignite and charcoal are considered as suitable alternative to peat. Lignites have acidic pH require about 10% (w/w) addition of calcium carbonate to raise the pH to 6.0-7.0. Charcoal powder has pH near neutral to slightly alkaline (7.5-8) with water holding capacity (WHC) equivalent to peat, therefore it serves as a good carrier. The magnitude of carrier material required is so huge and increasing day by day and therefore, dependence only on non-renewables will escalate cost in future. There is a total carrier requirement of 3.4 lakh tonnes per annum as observed by Biotech consultation India Ltd. The use of rice husk ash, a cheap and renewable carrier for, production of bacterial inoculants was suggested by Athoni Raj *et al.* (2001).

Vermiculite

Vermiculite is a hydrated magnesium aluminium iron silicate and exfoliated at very high temperature (700-1,000°C). This processed vermiculite when supplemented with nutrient source and moisture under sterile condition can be used directly to multiply bacterial cultures to prepare bacterial inoculants. The multilaminate structure provided superior aeration and space for microbes. These properties make vermiculite attractive for production of inoculants for agricultural and horticultural plants (Weiss *et al.*, 1987).

Recently developed synthetic carrier (alginate and gel) have a potential for being developed into better carriers since they offer the advantage of constant quality and homogeneity. The future research should be focused on further development of a synthetic carrier.

Better growth of microbial inoculants are obtained in sterile carrier. Best method is Gamma irradiation for sterilization of carriers. While using autoclave, calcium carbonate mixed with lignite or charcoal-soil mixture is filled upto two third capacity of steel trays is sterilized at 15 PSI pressure for 3-4 hrs. continuously or one hour each for three successive days. Presterilization can be done of prepacked polypropylene sealed bags, keeping minimum moisture, inoculation broth is added by automatic syringes connected to fermentors or broth durms followed by mixing. In bulk sterilized carrier, broth is mixed in open and the required quantity is packed in polythene bags. The moisture level in the packed material is normally kept at one third of the WHC. After packing, the bags can be kept for curing at 30°C for 4-5 days. The quality control of the inoculants in the carrier is very important. A microbial load in the range of 10^8-$10^9 g^{-1}$ carrier is often desired to have so as to provide about 10^3-10^4 viable counts per seed but it should be around 10^5-10^6 when seeds are sown in adverse conditions.

Biofertilizer Packing Development of Mechanized System

A major problem encountered in conventional method is contamination of biofertilizers with unwanted microbes. The contamination affects both the quality and shelf life of the product. Although controlling contamination is easier during preparation of broth and carrier but serious problems are encountered during mixing, curing and packing of the inocula by using the physical methods followed presently.

A semi-automatic system has been designed and developed by MFL for packing microbial inoculants. It is claimed that the mechanized system for mixing and packing is advantageous over the conventional methods in checking contamination of the inocula (Ramasami, 2002). It is more suited for medium and large-scale production. Its feasibility should be examined by R&D workers.

V. Development of Liquid Biofertilizer

In India there are more than 100 commercial biofertilizer units (Motsara *et al.,* 1995) and most of them produce carrier based biofertilizer. As per existing ISI specifications, *Rhizobium*, *Azotobacter*, *Azospirillum* and *PSB* inoculants have been marketed as carrier-based (IS-8268:2001 and IS-9138:2002) *Azospirillum* (IS - 14806:2000) PSB (IS-14807-2000). FAO (1991) has also reported that most of the international producers supply carrier based inoculants. The choice of carriers should be based on its ability to support a large population of rhizobia and other bacteria over a long period of time. Inoculants based on sterile carrier are usually of higher quality than the non-sterile carrier (Somasegaran and Hoben, 1994) but the sterilization process is expensive. General experience has shown that most strains rarely exceed 10^9 cells/g even in sterilized carrier.

Due to many constrains and non-availability of good carrier materials, it is necessary to develop liquid biofertilizers. Liquid inoculants in the form of broth or suspension of cells growing on agar slants are used but it is not practised in marketing of biofertilizers (Nethery, 1991). First report on liquid inoculant came from Netherland (Van Scherevan *et al.,* 1953) but the result on survival of liquid inoculum on seed was not encouraging (Vincent, 1970). It was, however, observed that liquid inoculation was suited for large seeded grain legumes (Schiffman and Alpher 1968; Roughly, 1986).

The spray method of liquid *Rhizobium* inoculant directly onto the soil was also reported (Gault *et al.,* 1962). Inoculating rhizobia as liquid below the seed was found to be a suitable method for inoculating peanuts (Nambier *et al.,* 1984). During 1985-90, liquid inoculants in USA were reintroduced for commercial use in soybean crops.

Although the earlier experience with liquid biofertilizer is not satisfactory, but it has been observed that the liquid form is easy to use, does not require adhesive and no need for water supply (Nethery, 1991). The process of inoculation is quick and hassle free. In view of the above advantages, many commercial units have started production of liquid biofertilizer in the country and abroad (Table 144).

Table 144: Liquid biofertilizer producing companies

Name	Location	Product profile	Crops
Kumar Krishi Mitra Byproducts (I) Pvt. Ltd.	Pune, India	*Azotobacter, PSB*	Cereals and vegetables All crops
Cominco Fertilizer Ltd.	Saskatoon, Canada	*Rhizobium*	Lentil and Pea
MicrobioRhizogen Corporation	Saskatoon, Canada	*Rhizobium*	Lentil and Pea
Liphatech	Meyzieu, France	*Rhizobium*	Soybean
NifTAL	Hawaii, USA	*Rhizobium*	Soybean

New Technology

NIFTAL, USA is very much engaged in formulation of different types of liquid inoculants including glycerol media, yeast extract mannitol media with various additives like arabinose, polyvinyl pyrrolidone (PVP), Trehalose, Fe-EDTA, glycerol etc. and thus developed several liquid media like G-1, G-4, G-5, G-6 etc. and found good response with G-5 and G-6 especially for presence of PVP which perhaps is involved in increasing exopolysaccharides production which protects cells from desiccation and increased shelf life (Battacharya and Kumar, 2002). Apart from PVP, several gums have been used. The relative efficiency of field response to inoculation with G-5 medium liquid biofertilizer over control was significantly higher than carrier based inoculants (Hegde, 2002).

In case of production of liquid based *Azotobacter* and PSB some products in India are being produced based on cyst and spore formation. Encystation of *Azotobacter* can be induced in an encystation vessel containing 0.2% $CaCO_3$ and 0.3% butanol for 72 hrs (Inamdar *et al.*, 2000) and this liquid formulation has also been patented. In this liquid suspension maximum percentage of vegetative cells are transformed into cysts. Similarly, use of spore forming organisms with P-solubilizing capacity is important for liquid P-solubilizers.

There is no specified standards for liquid inoculum products by Bureau of Indian Standards (BIS). These products are gaining popularity both among producers and users because of reduced volume, cleanliness, cosmetic appearance etc (Crida 2003, Rupela 2004).

Quality and Constrains

All manufacturers of liquid inoculants claim that products have high titre value. Lipha Tech (France) claimed CFU of 17.21 x 10^{11} ml^{-1} in the liquid product (FAO, 1991). NifTAL has claimed 1 x 10^{10} cells ml-1 in G-6 liquid inoculant. NCL, Pune has reported TVC of 10^8 to 10^9 ml-1 in case of *Azotobacter* (liquid inoculant). However, many commercial preparations in USA were not found satisfactory.

There are sufficient data available in favour of carrier based inoculants but very few data are available on liquid inoculants which is a serious limitation. Most of the recommendations made in national and International conferences are in favour of carrier based inoculants rather than liquid ones. The use of liquid biofertlizer is still at R&D stage. It appears from the discussion that there is a need for more R&D studies to overcome the constrains.

VI. Methods for application of Biofertilizers

The carrier based inoculants of the diazotrophs and PSB/PSF can be applied by different methods such as seed treatment, root treatment, sets/tuber treatment and soil treatment (in case of standing crop). The seed, tuber and seedling inoculations are easy to carry out and cost less in field operations.

i) **Seed treatment:** 200g of carrier based culture is sufficient to inoculate 10-20 kg seeds of medium size required for 1 acre of land. If the seeds size are small the culture quantity may be reduced or increased if seeds are very large.

ii) **Seedling treatment:** This practice is applied for seedling treatment meant for transplantation. This method ensures a heavy load of microorganisms on the root system. A suspension of 1 to 2 kg of biofertilizer in 5-10 litres of water in shallow container is prepared. The roots of the seedlings are dipped in the slurry for 10-15 minutes followed by transplantation. This slurry is adequate for treatment of seedlings to be planted in 1 acre of land.

    Tuber/sets treatment: This method is applied to tuber and sets e.g. potato tuber and sugarcane sets. A slurry containing 2 to 4 Kg of biofertifeer in 40 to 80 litres of water is,prepared. The tubers or sets are dipped for 10-15 minutes before sowing in the field.

iii) **Soil treatment**: The soil treatment is practised in case of standing or plantation crops.

    Short-duration crops: Three to five kg of the biofertilizer is mixed with 60-100 kg soil or compost/ FYM which can be broadcasted in one acre of land at the time of sowing or last ploughing. The broadcasting may be done in the standing crops before irrigation.

    Long-duration crops: Four to six kg of the biofertiiizer is mixed in 80-100 kg of soil/ compost and broadcast this mixture in the field. It will be better to mix the inoculant in soil. In case of plantation crop, the mixture can be applied in the pit near the root system.

iv) **Top dressing of biofertilizer** Application of biofertilizers with organic manure (FYM) in the rhizosphere of the crop plants 15-20 days after plant emergence was reported to enhance fertilizer use efficiency and also that of the biofertilizer. The optimum temperature for the operation is around 30°C. The biofertilizers should be placed in the rhizosphere between 2 rows of the standing crop (15-20 days old) at a depth of 7 to 10 cm.

v) **Top dressing with Granular Biofertilizer:** The top dressing with granular biofertilizer is becoming popular. The granular biofertilizer can be prepared by different type of biofertilizers with solorized FYM (well decomposed FYM treated with direct sunlight during April-May for about a month). The mixed ratio of biofertilizer for different crops per hectare would be as follows:

For Leguminous crops: *Fthizobium,* 1.5 Kg; *Azotobacter,* O.SKg; PSB, 3Kg; Solarised FYM, 50 Kg.

For cereals/millets: *Azotobacter* or *Azospirillum,* 2 Kg; PSB, 3 Kg; Solarised FYM 50 Kg.

The mixture can be drilled into the standing crop (15-25 days old) between the rows.

Mixing of granular fertilizers with seeds : Granular biofertilizer can be mixed with seeds and drilled in soil during sowing. Besides drilling, broadcasting method can also be applied but for broadcasting in 1 acre of land, 100 Kg of solarised FYM is to be mixed with the specific doses of biofertilizers recommended above.

VII. Production Technology for Blue-Green Algae and *Azolla*

Soils kept under rice-rice cropping system continuously, normally have indigenous blue-greens and may not require fresh inoculations. But in cropping systems where rice is fallowed by another cereal or a legume crop, excessive tillage and less water may disturb the eco-system for the cyanobacteria. Cynobacterial inoculations are usually recommended as consortia for different species rather than monocultures. Also their cultivation do not require aseptic conditions. Therefore, the production can be carried out at farmers field from "mother culture". The method for preparation of starter or mother culture has been detailed in chapter 17.

The technology for cultivation of *Azolla* has been extensively discussed in the chapter 14.

VIII. Promotion of Biofertilizer

The state departments of agriculture, Indian Agricultural Research Institute, New Delhi, State Agricultural Universities and Ministry of Agriculture have significantly contributed for the promotion of different types of biofertilizers.

The ministry of Agriculture (GOI) launched a "National Project on the Development and Use of Biofertilizers" during 1983 to produce, distribute and promote the use of good quality biofertilizers. Under this project a National Biofertilizer Development Centre (NBDC) was established at Ghaziabad in January, 1988 and six Regional Biofertilizer Development Centres (RBDC) were started at Bangalore, Bhubaneshwar, Hisar, Imphal, Jabalpur and Nagpur along with 49 sub-centres which commenced their activities during 1991 -92. The technical and financial support was provided by FAO and UNDP.

The main objectives and activities of the project, was to ensure production, distribution and promotion of effective strains of *rhizobia*, blue-green algae, *Azotobacter, Azospirillum,* phosphate solubilizing microorganisms *(Bacillus* sp.., *Pseudomonas* sp. *Aspergilus, Penicillium* sp.) and cellulotytic cultures and to create an agency for their quality control, training programmes and dissemination of information about the biofertilizer

and their use in agriculture and related fields for the benefit of the farmers and the promotion of the programme are also important activities.

Under the project, it is envisaged to extend one time financial support upto Rs. 1,50,000 per BGA centre already established under administrative control of State Governments and Agricultural Universities to produce 10 mt soil based BGA per year per centre. There is also provision of one time financial support of Rs. 13 lakhs per unit for *Rhizobium* production. Twelve units belonging to State Government, agro-industries, fertilizer and seed companies have received such grants earlier (Verma and Bhattacharya, 1992).

A National Facility for Blue-green Algae collection and National Facility for *Rhizobium* germplasm collection at Indian Agricultural Research Institute, New Delhi and Mycorrhizal Collection at Tata Energy Research Institute, New Delhi were established by Department of Biotechnology, Govt. of India. These are important major developments that reflect our concern to harness biofertilizer in agricultural economy,

Promotion Activities of IFFCO/Use of Bio-fertilizer

IFFCO (1998) reported that a project on biofertliser was initiated in the state of West Bengal in 1995-96 with an objective to create awareness about the use of biofertilizer. The project was implemented in collaboration with Agricultural University and Cordet, Phulpur, Since 1995-96, IFFCO has supplied 45.72

MT of various types of biofertiliser in the project area (Table 145) and helped to create awareness regarding use of biofertilizers. In the initial two years, the material was supplied free of cost, however from 1997-98 onwards, material was supplied on 50% subsidy and an amount of Rs. *2*/- per packet was realized from the farmers. The results from the demonstrations of biofertilizers laid out in the project areas showed an increase in yield in the range of 1.2 to 8.7% due to application of *Azotobacter* over untreated plot in case of rice. Except in district Hooghly, crop response to the application of *Azotobacter* was not more than 3%. *Rhizobium* application increased the yield by 11.1 to 22.5% in case of blackgram and greengram crops. It was further observed that when *Azotobacter* + PSB were used, the yield increases in rice varied from 7.9 to 9.8% as against 1.2 to 8.7% in case of *Azotobacter* alone and 2.3 to 10.5% in PSB alone. It is apparent that use of *Rhizobium* for pulses and PSB for rice was quite beneficial in improving crop productivity.

Table 145: Distribution of biofertilizer packets (200g) in project areas of West Bengal

Bio-fertilizer (200 g packets)				
Year	***Azotobacter***	**PSB**	**Rhizobium**	**Total**
1995-96	14000	24400	12000	50400
1996-97	32000	44000	18000	94000
1997-98	12500	59700	12000	84200
Total	58500	128100	42000	228600

PSB-Phosphate solubilizing bacteria

National Productivity Awards

National Productivity Council, the premier organization spearheading the productivity movement in the country instituted National Productivity Awards for 14 subsectors of Agriculture in 1985-86 with the sponsorship of Ministry of Agriculture, Govt. in India. Realizing the benefits of biofertilizers to our Agricultural economy and with a view to encourage the manufacturers of these inoculants, the National Productivity award for biofertilizers producers were introduced in the same year, to promote productivity consciousness among them while recognizing the best performing units.

Promotional Activities and Productivity Efforts

With an objective of reducing the dependence on chemical fertilizers besides convincing and educating the farmers about the eco-friendly nature of biofertlizes, three units namely M/s T. Stanes & co. Ltd., Gujarat state Fertilizer Company Ltd. and NAFED have exhibited very high endeavour in this direction. Another three units are making feeble attempt towards biofertlizers promotion, rest of the units have shown average to high efforts for promotion of biofertilizers at grass root level. Again the units which have evinced very high inclination towards improving productivity levels by way of effecting quality improvement and achieving lower costs of production are M/s T. Stanes & Co. Ltd. Gujarat State Fertilizer Co. Ltd. and NAFED.

Biofertlizers have not yet become very popular in India inspite of their potential. The efforts to commercialize their large scale production have been intensified during the last 5 years. Both private and public sectors have shown interest in the production and marketing of biofertilizers. Many fertilizer producing units are making effort to produce and promote biofertilizers.

The capital subsidies provided by the Government has given impetus to produce and promote biofertilizer at the state level. At present there are 26 private companies engaged in the production of biofertilizers. The public sector has performed well in producing good quality inoculant at large scale. The private sector is also playing an increasing role in this endeavour.

The strengthening of agriculture extension activities and education of farmers/ users regarding the economy and benefits of the biofertilizers in cropping systems are required. The concept of organic farming and sustainable agriculture would lead to their greater usage in agriculture.

Guidance and Precautions in Using Biofertiliser

It is essential that biofertilizers to be used, should be of good quality (containing minimum 10^7/gm viable cells) and purchased from the reputed sources/manufacturer only.

The name of the crop should be mentioned on the packet and be sure to have the right type of biofertilizer before use.

The rhizobia are crop specific thus they should always be used in recommended pulses/ legume crops. *Azotobacter/Azospirillum* should be used for recommended non-legumes/ cereals/oil crops/ vegetables/ horticultural crops etc. Phosphate solubilising biofertiliser (PSB) can be used in all crops for supplying phosphorus. All biofertilisers have certain period for their use. Look for the expiry date on the packets and beyond that it should not be used on crop. Protect and keep away the biofertilizer from the direct sunlight and heat.

Store them before use in cool and shaded places at room temperature 25-28°C. In rural areas it is preferable to keep biofertilizer packets in earthen pots by putting the same in a pit of bigger size of the pot in shaded area upto neck, and this pit may be filled by sand around the pot which should remain saturated with fresh water. Put the biofertilizer packets and cover the mouth of pot with a lid/cloth.

Avoid direct contact of biofertiliser with pesticides/weedicides/ chemical fertilizer etc.

It is better to use biofertiliser just before sowing/planting of the crop in field because micro-organisms survive well in soil instead on seed/ seedling surface after treatment.

Avoid making seeds too wet or mixed in excess water while using biofertiliser and rough handling of seed treatment may also be avoided.

If seed has to be treated with any fungicide or insecticide or any toxic chemicals apply "FIR" formula for treatment of biofertilizers where "F" denotes first fungicide, the "I" which denotes insecticide and in last "R" denotes biofertiliser (*Rhizobium*/*Azotobacter*/PSB). Whenever, this is required follow the same sequence for seed treatment.

Once biofertiliser packets have been opened use its whole content and do not keep the opened packet of biofertiliser for future use.

If soil conditions are not favourable, then biofertiliser treated seeds should be coated with fine lime/gypsum @ 2-3 kg/ha in case of in acidic or alkali soil respectively and mix until seeds are evently pelleted.

Constrains and Future R&D

India is one of the largest producer and consumer of biofertilizer in the world, yet its quality is often viewed with Skepticism (Thompson, 1999, Bagyaraj, 2000, Bhattacharya, 2000). There are several constrains in the production and use of biofertilizers (Verma, 1992; Dadarwal *et al.*, 1997). They may be physical, chemical, biological, technological, infrastructural, financial, market related or concerned with human resource development. These are encountered at different stages of production and use of biofertiliser such as at levels of production, distribution, marketing, use of biofertiliser and field responses.

1. Technical Constrains

i) **Selection of strains:** The selection of an efficient and competitive strain is important aspect in biofertilizer production and quality control of the product. The genetic factors which control competence in nodulation by rhizobia and factors which control ecological establishment have not been explained yet at the molecular level. The genetic factors such as bacteriocin, siderophores, secondary metabolites, antibiotic production etc. have been reported in several beneficial strains. Future R&D is required to clone such factors into the strains capable of nitrogen fixation, phosphate solubilization and mineralization of other plants nutrients. Moreover, it is difficult to get effective strains which can grow in adverse soil condition. Attempts have been made to set up culture collection banks at IARI, New Delhi and under National Project on Biofertilizers at Ghaziabad. Some of the Agricultural Universities have also a good collection of agriculturally beneficial microorganisms.

ii) **Carrier:** The selection of carrier material for carrier based biofertilizer production is equally important. The carrier should be high in organic matter and WHC. It should be normal or have easily adjustable pH, no heat of wetting, near aseptic, be non-toxic, non-polluting and biodegradable, microbial growth supportive and amenable to nutrient supplements. The selected carrier should be available in adequate quantities at a reasonable price. Research on inexpensive and less bulky materials is of great relevance to minimize the storage problems and transportation cost.

    In India, biofertiliser production is generally done with non-sterile system which results in considerable contamination. Improved technology such as use of gamma irradiation of the carrier material, injection of liquid inoculum into polythene bags containing sterile carrier material etc. should be tested and used.

iii) **Lack of quality assurance:** The formulation of standards for quality control of inoculants is important criteria. The use of poor quality of inoculants often results in failure under field conditions and thus lack of confidence in the product. The Bureau of Indian Standards (BIS) has prepared ISI specification for *Rhizobium* and *Azotobacter* inoculants which was prepared in 1977 which has been revised Recently ISI specification for Azospirillum and PSB have also been prepared. Specifications for quality control of other biofertilizers are not available. The BIS has no network to check the quality of the biofertilizers samples. They send the samples to Research Institutes or State Agricultural Universities in order to label with ISI mark. Once a manufacturer gets an ISI mark they continue without routine checking. This results in substandard product reaching the farmer. There should be separate testing laboratories with adequate infrastructure and manpower to check the quality of the inoculants at the various stages of production, marketing and application. This can be done by strengthening the testing centers in Central Research Institutes and State Agricultural Universities competent to take up this work.

2. Constrains of Soil Fertility

Problem soils such as highly acidic and saline-alkali soils will adversely effect the population and activity of the inocula. In acidic soils after coating with the inoculum, pelleing of seeds with lime and in alkaline soils with gypsum is recommended which should be followed to tackle the problem. In soils low in available phosphorus (<8 ppm), application of phosphates with phosphate solubilizer inoculum will be essential for the proper functioning of nitrogen fixing or cellulolytic inocula.

3. Environmental and Weather Constrains

During summer season, in many parts of India temperature is very high (35 to 45°C) which results in deterioration of the quality of the inoculants during preparation and storage of inocula.

4. Biological Constrains

Soil flora and fauna can parasitize or devour used biofertilizer on seeds and soils. Rhizophages can destroy rhizobia in soil. Some protozoa like *Vorticella* and nematodes like *Meloidogyne* can act as predators on microbial inocula added to soil. *Azolla* is subjected to attack by pests. *Cyanobacteria* can be attacked by phages, myxobacteria and grazers such as snails and mosquitoe larvae.

5. Storage and Marketing

Poor quality of inoculants may be caused not only by poor standards but due to lack of proper storage and transportation infrastructure. The inoculants can be stored for six months or longer at 20°C but lose their effectiveness within few hours at 40°C or at higher temperature. The distribution and transportation from biofertilizer production units to farmers are subjected to high temperature particularly in Kharif Season resulting in loss of viability and low counts of the organisms. Necessary steps should be taken to improve storage and transportation facility to maintain the good quality of the agriculturally useful microbial inoculants.

The size of the packets vary from 75 to 500 g of inocula/packet. An uniform capacity of 200 g/packet may be recommended for proper designing of packing machine that can be manufactured and marketed all over the country.

6. Field Responses

The response of biofertilizer is not immediately observed as in the case of chemical fertilizers. The unfavoural soil conditions such as soil pH, high temperature, waterlogging, drought, deficiency of nutrients like P, cu, co, and Mo may hamper their effectiveness. The use of faulty methods of inoculation and build up of pesticide residues and nitrate levels in soil can reduce their proper establishment in soil and subsequently the desired influence on plants.

Some biofertilizer production units do not have well qualified technical and trained persons. Some organizations depend more on unskilled workers on contractual basis resulting in production of sub-standard biofertilizer. Intensive extension activities and demonstrations in farmers' field and training should be conducted to improve awareness among farmers regarding the availability of different types of biofertiiizers and their benefits in crops productivity.

Considering the constrains in the marketing of biofertiiizers, a financial support is being provided to fertilizer and seed producing companies with established marketing network by national project on biofertilizer established by the Ministry of Agriculture. The quality control standards should be strictly enforced to increase the acceptability and benefits of the microbial inoculants at a faster pace.

References

Athoni Raj S, Subramanian, P. and Singaravadivei, K. 2001. Biofortilizer Newsletter 9(1): 3-6.

Applernan, R. K. and Sears, K. S. 1946. J. Bacteriol. 52: 209.

Battacharya, P. and Singh, T. 1992. In: Proc. Nat. Seminar on organic Farming College of Agriculture, Indore pp. 166-178.

Battachaya, P. and Kumar R. 2002. Proc. Natl. Seminar on Development and use of Biofertiiizers. Biopesticides and Organic Manures, Bidhan Chandra Krishi Viswavidyalaya, Kalyani, West Bengal pp. 10-21.

CRIDA 2003. Liquid Rhizobium Inoculant for Improved Shelf Life. Annual Report 2002-03 GRIDA, Santosh Nagar, Hyderabad pp GO-61.

Dixit, S. K. 2001. Biofertiisers-A manual on Commercial Production Technology. Omega Scientific Publishers, New Delhi.

FAO 1991: Report on Expert Consultation on Legume Inoculant Production and Quality control. FAO, Rome, pp 148.

Feller, C. R. 1918. Soil Sci. 6:53.

Gaur, A. C, 1990. Phosphate Solubilizing Microorganisms as Biofertiizer. Omega Scientific Publishers, New Delhi. Pp 178.

Gault. R.,R. Chase, D. L. and Brockwefi. J. 1982. Aust. J. Expt. Agric. Anim. Husb. 22: 299-309.

Gordon, R.E. and Smith, N. R. 1947. J. Bacteriol. 53: 669.

Hartsell. S.E. 1953. Appl. Microbiol. 1: 36.

Hegde, S. V. 2002. Liquid biofertilizer in Indian agriculture. Fertilizer Newsletter 10(2): 17-22.

IFFCO. 1997-1998. Agricultural services and promotional activities. IFFCO, New Delhi p. 31-32.

Inamdar Sandeepa, Kanitkar, Ravi and Watve Milind, G. 2000. Curr. Sci. 78: 719-222.

Iswaran, V. 1974. In: A Manual for Proper Use of Inoculants and Pelleting for Legumes. Today and Tomorrow Printers, New Delhi.

Iswaran, V. Sundara Rao WVB, Magu S. P. and Jauhri, K. S. 1969.

Indian peat as a carrier for Rhizobium Curr. Sci. 38: 468-469.

Jensen, H. L. 1951. Proc. Soc. Appl. Bacteriol. 14: 89.

Joshi, N.V. 1920. Mem. Dep. Agric. India (Bact.) Ser 1: 246

Motsara, M. R. Bahttacharya, P. and Srivastava, B. 1995. Biofertilizer, Technology, Marketing and Uses. FDCO Publishers, New Delhi, pp 184.

Nambier, P. T. C., Dart, P. Srinivasa Rao, B and Ravi Shanker, H. N. 1984. Olegineux 39 :149-155.

Nethery Arthur A. 1991. Inoculant production with nonsterile carrier. In: Expert Consultation of Legume Inoculant Production and Quality Control, FAO, Rome, p 43-50.

NifTAL Report 1999. Liquid Inoculant Development Summary II.

Norris, D. O. 1963. Empire J. Expt. Agr. 31: 255.

Okon, Y, Albrcht, S.L. and Burris, R. H. 1977. Appl. Environ. Microbiol 33 :88.

Ramasami, S. 2002. Biofertilizer packing using mechanized system. Biofertiiizer Newsletter 10(1): 3-5.

Roughley, R.J. 1976. In: Symbiotic Nitrogen Fixation in Plants (Nutman, P S. ed) Cambridge Univ. Press, Cambridge, Lo-ridon. pp 125.

Roughley, R. J. 1986. In: Bionift Regional Symposium and Workshop Malysian Society of Soil Sciences. Pp. 147-154.

Rupela, O.P. 2004. Production and Quality Control of Microbial Inoculants: Some perspectives. Proc. Nat. Cong. on Quality Control of Biofertilizers (Eds. P. Bhattacharya and V. Dwivedi), NBDC, Ghaziabad Publ. pp 8-21.

Schiffmann, J. and Alper, Y. 1968. Expt. Agric 4: 219-226.

Shende, S. T., Arora, C. K. and Sen, A. 1973. Zbl. Bakt. Abt. II, 128: 668.

Somasegaran, P. and Hoben, H. J. 1994 Handbook for rhizobia: Methods in Legume - Rhizobium Symbiosis, Springer-Verlag, New York pp 450.

Subba Rao, N. S. 1977. In: Soil Microorganism and Plant Growth. Oxford & IBH Publ. Co New Delhi.

Sundara Rao, W.V. B., Mann, H. S. and Paul, N. B. 1963. Indian J. Agril. Sci. 33: 229.

Van Schrevan, D. A, Harmsen. G. W., Lindenbergh, D. J. and Otzen, D. 1953. Antonie Van Leeuwnhoek J. Microbiol Serol. 19: 300-308.

Van Schrevan, D. A. 1958. In: Nutrition of Legumes (Hallsworth, EG eds) Butterworh & Co. London. p 328.

Verma, L. N. 1993. Biofertlizer in Agriculture In: Organics in Soil health and Crop Production (Ed. PK. Thampan) Peekay Tree Crops Development Foundation, Cochin pp. 158-159.

Verma, L. N. and Bhattarcharya, P. 1991. Development and use of Biofertilizer in India. Fertilizer Scene -Annual, 1991 pp53-62. Verma, L. N. and Bhattacharya, P. 1992.

Production, Distribution and Promotion of Biofertilizers. In: Fertilizers Organic Manures, Recyclable Wastes and Biofertilizers (HLS Tandon, Ed.) FDCO, New Delhi.

Vincent, J.M. 1970. A Manual for the Practical Study of Root Nodule Bacteria, IBH Handbook pp. 116.

Vincent, J. M. 1982. In: Nitrogen Fixation in Legumes, Academic Press, London, 50.

Weiss, L.G., Bennet, M. L. and Paau, A. S. 1987. Appl. Environ.Microbiol. 53:2138-2140.

3. Extraction and Fractionation of Humic Substances

Regents: Sodium pyrophosphate and sodium hydroxide mixture: 44.6g $Na_4 P_2O_7 . 10H_2O$ and 4.0 g NaOH are dissolved in 1 litre of distilled water to give a solution 0.1 M with respect to each substance; the pH should be about 13.

Activated charcoal: Treat the commercial substance for 24 hours with 10% *v/v* HCI solution, wash out excess acid and grind to a paste.

Resins: OH Amberlite -IRA-400

H Amberlite -IR-120

The cation exchange resin is activated by 1 M HCL to the H^+ form and the anion exchange resin by 1 M NaOH to the OH- form. The resins are washed with distilled water to neutral reaction.

Procedure: 20.0g of air dry, 40 mesh compost is weighed into a 250 cm3 stoppered conical flask. 100-200 cm^3 of freshly prepared sodium pyrophosphate-sodium hydroxide mixture are added, preferably in the evening so that filtration can be done the next morning. Shake for 1 hour and stand for 18 hours. Filter or centrifuge at 10,000 rpm for 30 minutes. The residue is humin and the filtrate contains humic and fulvic acids.

Adjust the pH of the filtrate with conc. HCI to about 1 when humic acid begins to precipitate. Allow to stand for 12 hours and filter. A gel of humic acid remains on the filter paper while fulvic acid is collected as the soluble fraction.

Redissolve the humic acid in the pyrophosphate -hydroxide solution, re-precipitate and filter. Dissolve the residue in 0.1M NaOH solution and pass through a series of anion and cation exchange resins. This gives an extract of humic acids free from unbound mineral ions. Divide the solution into three parts, one for carbon determination, one for humic acid content by drying at 80°C to constant weight and one for freeze-drying and storage for further analysis. If a freeze drier is not available the gel of humic acid may be air dried by spreading it in a thin layer in a petri-dish, grinding to pass a 0.25 mm sieve and storing in lightly stoppered test tubes.

The filtrate from the separation of humic acid is passed over activated charcoal in a buchner funnel, the carbon will retain the fulvic acids. Wash several times with distilled water until the filtrate is slightly acidic. Remove the fulvic acids by washing the carbon with several small quantities of cold 0.2 M NaOH solution and make to standard volume. The solution is divided into three parts as described for humic acids.

Determination of organic carbon and total nitrogen in the acid fractions

The carbon content of acids may be detrmined by the Walkley-Black method (Hesse, 1971). The nitrogen content is measured by the micro or semi-micro Kjeldahl method (Hesse, 1971).

4. Determination of Total Nitrogen in Compost

The semi-micro Kjeldahl procedure with the salicylic acid modification to include all forms of nitrogen is followed and 1 g samples of compost are used; full details of the procedure can be found in Hesse (1971).

5. Determination of Total and Available Phosphorus in Compost

Total phosphorus is determined after digestion of the sample with acids using a 1 g sample (Hesse, 1971).

Available phosphorus is measured as the citric acid soluble fraction by the method of Dyer.

5 g of compost is placed in a stoppered flask with 50 cm3 of 1% citric acid solution. The flask is left for seven days with daily shaking. After filtering, the soluble phosphorus is determined colorimetrically as for total phosphorus.

II. Microbiological Analysis of Compost

The compost samples can be analyzed on a quantitative-qualitative basis for the following groups of organisms:

Bacteria, actinomycetes, fungi, *Azotobacter* and phosphate- solubilizing bacteria or fungi or phosphate solubilizing microorganisms which shall include all phosphate, solublizing bacteria, actinomycetes and fungi.

Apparatus: Autoclave, Incubators, Hot air oven, Inoculation Chamber or Laminar air flow cabinet, Sterile pipettes, Sterile water blanks (90ml), 250 ml conical flasks and 50 ml test tubes, Bacteriological filter, magnifying lens or colony counter.

Media Composition

The media used to culture various organisms are given below:-

1. Soil Extract Agar for Count of Total Bacteria and Actinomycetes (Allen, 1957)

Soil extract	100 ml (See preparation below)
Glucose	1.0 g
K_2HPO_5	0.5 g
Soil extract	100cm^3
Tap water	900 cm^3
Agar	15.0g
pH	6.8-7.0

Sterilize at 15 lbs psi pressure for 30 minutes.

Preparation: The soil extract is prepared by autoclaving 1 kg of garden soil with 1 litre of water for 30 minutes at 15 psi. A small amount of calcium carbonate is added and the soil suspension filtered through double filter paper until clear. The filtrate should be bottled and sterilised in 100 mi quantities.

2. Thornton's agar medium (Thornton 1922) for total bacteria Thornton's agar medium (Thorton, 1922)

Mannitol	1.0g	Ca CI_2	0.1 g
Asparagin	0.5 g	Na CI	0.1 g
K_2HPO_4	1.0g	Fe_2CI_3	0.002 g
KNO_3	0.5 g	Agar	15 g
$MgSO_4$	0.2 g	distilled water	1 litre

Dissolve the ingredients. Adjust to pH 7.4 with bromo-thymol-blue and sterilize at 15 psi pressure for 30 minutes.

3. Peptone-Dextrose Agar or Rose Bengal Agar Medium for Count of Fungi (Martin, 1950)

Peptone	5.0g
Dextrose	10.0g
KH_2P0_4	1.0g
$MgSO_4.7H_2O$	0.5g
Agar	20.0g
Distilled water	1 litre
Rose Bengal	33.3 mg
Streptomycin	30 mg mM

Streptomycin solution: The vial is opened aseptically and the contents (3g) added to 100 cm^3 of sterile distilled water and dissolved completely and sterilize through membrane filter. The streptomycin solution.can be stored for 4-6 months in a refrigerator.

Dissolve ingredients, adjust the pH to 5.5 using INHCI, then add Rose Bengal and 20 g agar. Sterilize the medium at 121°C at 15 psi pressure for 30 minutes. Cool the solution and add 1 ml of streptomycin sulphate solution (30 mg streptomycin/ml) and Shake well.

4. Medium for *Azotobacter* (Jensen, 1940)

Sucrose	20.0 g
$MgSO_4.7H_2O$	0.5 g
$FeSO_4$	0.001 g
K_2HPO_4	1.0g
NaCI	0.5 g
$NaMoO_4$	0.005 g
$CaCO_3$	2.0 g
Agar	15.0g
Distilled water	1 litre

Dissolve the ingredients, adjust the pH to 7 and sterlize at 15 lbs pressure for 30 minutes.

5. Medium for *Beijerinckia* (Becking, 1959)

Sucrose	20.0 g
K_2HPO_4	0.2 g
KH_2P0_4	0.8 g

$MgS0_4.7H_2O$	0.5 g
$FeCI_3$	0-1 g
Na_2MoO_4	0.05 g
$CaCI_2$	0.05 g
Agar	15.0g
Distilled water	1litre
PH	6.5

Dissolve the ingredients, adjust the pH at 6.5 and sterilized at 15 psi for 30 minutes.

6. Medium for Phosphorus Solubilizers (Pikovskaya, 1948 and Gaur 1990)

Glucose	10g
$Ca\ (PO_3)_8$	5g
$(NH_4)_2SO_4$	0.5 g
NaCl	0.2 g
$MgS0_4.7H_2O$	0.5 g
KCl	0.2 g
Yeast extract	0.5 g
$MnSO_4$	trace
$FeSO_4.7H_2O$	trace
Agar	15g 1
Distilled water	1litre
pH	7.0 ±0.2

Dissolve the ingredients except glucose and $Ca_3\ (PO_4)_2$ in distilled water, add agar, dissolve and filter. Add glucose and adjust the pH to 7 and sterilize at 10 lbs psi for 15 minutes.

For preparation of culture plates 10 ml of the stock suspension of $Ca_3\ (PO_4)_2$ is added asceptically to 90 ml sterilized medium and used.

7. Enumeration of Phosphate Solubilizing Organisms Using Soil Extract Agar (Bunt and Rovira, 1955 as modified by Louw and Webly, 1958)

Peptone	1.00 g
Yeast extract	1.00 g
Glucose	5.00 g
$(NH_4)_2SO_4$	0.50 g
K_2HPO_4	0.40 g
$MgCI_2$	0.10g

$CaCl_2$	0.10g
$MgS0_4.7H_2O$	0.05 g
$FeCI_3$	0.01 g
Soil extract	250cm^3
Tap water	750cm^3
Agar	15.0g
PH	6.6

Sterilize at lbs 10psi for twenty minutes.

A stock suspension of dicalcium phosphate 2.5% w/*v* in 0.5% w/v gum arabic is prepared by grinding together and adjusting to pH 8.4 before autoclaving.

For preparation of the plates, 1 cm 3 of this suspension is added to 9 cm3 of Bunt and Rovira's soil extract medium as modified by Louw and Webley. After thoroughly mixing by hand rolling, the medium dispersed insoluble compound is poured into petri dishes already containing the 1 cm3 aliquot of appropriate soil dilution. The plates are incubated at 25°C for 14 days and the total number of colonies and those showing halo zones are counted; the populations expressed as 10%/g of dry sample.

III. Procedure for Quantitative Determinations of Microorganisms

Compost to be analyzed is sieved through a 9 mesh sieve and about 5-10 g is placed in a tarred metal or glass container, weighed and dried over night at 105-110 °C. The dried sample is cooled in a desiccator and re-weighed. The moisture content is calculated.

10 g of the air dried sample is taken and added to 90 cm^3 of sterile distilled water in a 250 cm3 conical flask; mix thoroughly in a rotary shaker for 5 min.

Serial tenfold dilutions are prepared from the original dilution by pipetting 10 cm^3 of the solution while in motion into 90 cm^3 of sterile distilled water and continuing the series until the desired dilution is obtained. Serial dilutions for bacteria (general), 10^{-5} to 10^{-8}, Actinomycetes, 10^{-4} to 10^{-6}, Fungi 10^{-3} to 10^{-6}, *Azotobacter* and P-solubilizing organisms 10^{-1} to 10^{-4} are generally prepared. The plates are prepared by adding 1 cm^3 quantities of the desired dilution into clean, sterile petri dishes and pouring them with the respective medium. The plates are shaken clockwise and anti-clockwise to give a inform distribution and are allowed to set. After the plates are set, they are incubated at 30 ± 1°C. The plates are examined at intervals and counts of microbial population recorded. Examine the plates carefully, Select plates with 30-300 well distributed colonies and make total counts with the help of a magnifying lens or a colony counter. The number of colonies with a hand tally to avoid any mistake. Count the number of colonies more than 30 and less than 300 / plate. Calculate the viable cells per gram of compost or manure. The results are expressed as 10*/g dry sample.

Calculation

$$\text{Viable Count/g of oven dry organic manure} = \frac{A \times D}{W}$$

A = Average Count of 3 replicates

D = Dilution

W = Weight of oven dry manure.

IV. Rapid Tests for Compost Maturity

1. Plant Test

Only the plant test can give the final answer as to whether or not a compost is mature. However, most plants grow so slowly that the results are obtained too late. A feasible plant test is that using garden cress which gives results within about five days from planting (Spohn, 1969).

Procedure: An asbestos cement tray 38 x 28 x 6 cm is filled with the compost. 10 g of cress seeds are spread over the surface and the tray is covered with a glass plate until the seeds have germinated. The tray may be put on a slowly revolving, motor driven, table in order to ensure straight and even growth.

Every second day the tray should be weighed and any loss in moisture made up. After five days the cress is cut and weighed; a weight of 60-100 g is a good result. Unripe compost yields less or shows no germination. A further test is to taste the cress; unripe or putrid compost develops a bad taste in the cress. The cress test is rather rigid since only compost is used as a growth medium whereas normally compost would be mixed with soil.

2. Verticillium Test

The growth of this fungus is obtained within twenty four hours and the method can be followed in a well established laboratory (Chrometzka, cited by Spohn, 1969).

Procedure: 2 kg of peeled potatoes are boiled in one litre of water. 10 g of sugar and 25 g of agar agar (after 24 hours wetting) are added to the decoction which is then boiled for about 45 minutes to dissolve all the agar. If necessary adjust the pH to 4.8 with lactic acid.

Half fill test tubes with this culture medium and sterilize at 15 psi pressure at 121°C. Inoculate the slants with Verticillium culture using an inoculating needle and incubate at 22°C.

Growth test: Add 5 cm^3 of sterile water to the pure culture in the test tube aseptically, and shake. After a short while about a million spores will be present per cm^3 of fungal suspension. Pour the entire suspension (5 cm^3) onto a petri plate containing some of the culture medium. Place a 2 cm wide strip of compost across the plate and incubate at 22°C.

Evaluation: After twenty four hours fungal growth can be checked under a microscope. If a halo is observe around the compost sample, there is an indication of strong growth suppression and if the fungus grows into the compost and on its surface, ripeness of the compost is indicated. After three or four days the growth or inhibition of the fungus can be observed with the naked eye and with some practice, the degree of ripeness of the compost can be determined.

3. Starch-iodine Test (Lossin, 1970)

Three types of carbohydrate are found in compost; sugars, starch and cellulose. These constituents decrease in content with progressive decomposition of the material. Sugars are metabolized first followed by starch and then cellulose. Certain polysaccharaides (starches: amylose, amylopectin, glycogen and dextrans) form characteristically coloured complexes when combined with molecular iodine. Amylose, a straight chain starch, gives an intense blue-black colour; amylopectin, a moderately branched starch, forms from light blue to purple-red colours and glycogen, a highly branched starch, gives a reddish brown colour. Dextrans which are partial degradation products of starch, give reddish brown to yellow colours. The hydrolysis of starch may be followed by a gradual change in the iodine complex colour, blue black to light blue to purple red to red to yellow to colourless.

Reagents: lodine reagents prepared by dissolving 2 g Kl in 500 cm^3 of water and adding 0,.80 g l_2. Perchloric acid 36%.

Procedure: Select small (50-100g) samples of compost taken randomly to make a total composite sample of at least 1 kg. grind roughly and thoroughly mix. For the test take duplicate samples.

Place about 1 g of compost in a 100 cm^3 beaker, wet with a few drops of ethanol if dry, add 20 cm^3 of the perchloric acid and stir. Filter through a Whatman No. 90 paper, add 2 cm^3 of iodine reagent and stir. Place a few drops on a white tile and note the colour and precipitate. A finished compost gives a dark blue colour and a heavy precipitate. The colour sequence normally found as the age of the compost increases is blue-black-light blue-grey-green-yellow. A reddish colour may be found with partially degraded compost.

4. Sulphide Test (Lossin, 1969)

One spoonful of compost in a 400 cm^3 beaker is treated with 25 cm^3 of 18% hydrochloric acid. The beaker is covered with a lid on which strips of lead acetate paper are fixed., Different stages of black or brown colouration indicate sulphides originating from more or less putrifaction; ripe compost show no colouration of the strips.

It has been argued that hydrogen sulphide can be detected more sensitively by its smell than with lead acetate. Experiences show however, that the smell is sometime masked by other odours in compost and is difficult to detect. Iron sulphide is not toxic in soil but its presence indicates putrefaction in the compost which may have noxious effects; in a ripe compost all the sulphur is present as sulphate.

5. Nitrogen Test

When ammonium is present in the soil solution and is not yet oxidized to nitrate a growth depression of cress seedlings is caused. Total, organic, ammonium and nitrate nitrogen can be determined analytically but as this takes time and needs trained personnel a better procedure is that of Spohn (1978).

Reagents: Nessler's reagent (commercially available) naphthylamine, 0.2 g dissolved in 500 cm^3 of 20% acetic acid, sulphanilic acid, 3 g dissolved in 500 cm^3 of 20% acetic acid (210 cm^3 acetic acid 96% in one litre).

Griess reagent for nitrite, equal volumes of the naphthylamine and sulphanilic acid reagents mixed. It is recommended that these two reagents are kept separate until needed and in a refrigerator.

A. Cut Filter paper placed on glass plate

B. Covered with undivided filter paper

C. Ebonite ring placed and moistened sample pressed

D. Two halves moistened with clear soil solution

E. Tested for ammonia and nitrate

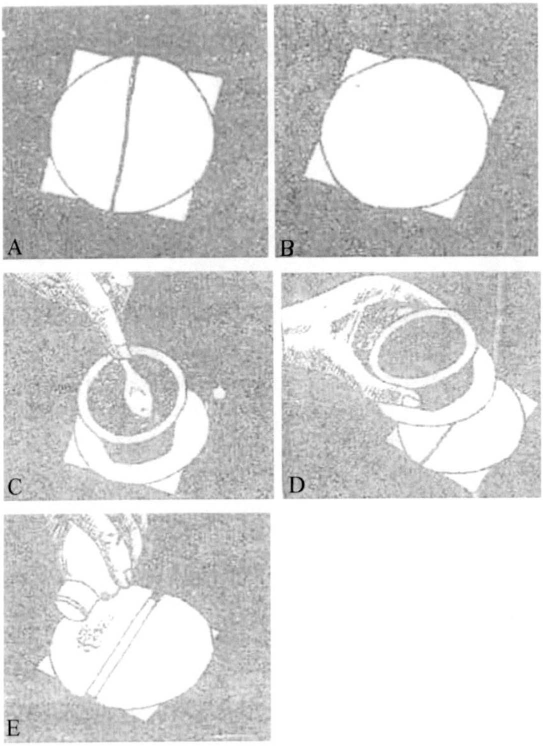

Fig. 15: The nitrogen test for compost maturity

Reducing powder, 95 g barium sulphate dried at 150°C mixed with 5 g zinc powder and 10 g manganese sulphate.

Procedure: A ring about 8 cm diameter, 4 cm high and made from plastic or ebonite is needed. The sample of compost is placed in a tray and moistened with distilled water; using a spoon the mixture is mixed until the surface appears slightly moist under pressure. A 9 cm filter paper is divided into two and placed on a glass or plastic plate (Fig. 15A). This paper is then covered with a second, undivided, filter paper (Fig. 15B). The ring is placed on the filter paper disc and filled with the moist compost sample, tamping with the spoon to moisten the upper paper (Fig. 15C). Then the ring and sample with upper paper disc are removed (Fig. 15D). This procedure results in the two half pieces of paper being moistened with clear extract of compost.

One of the pieces of filter is treated with 7 drops of Nessler's reagent (Fig. 15E). If it turns brown ammonia is present and indicates that the compost is not ready. The second piece of filter paper is lifted at one end and 7 drops each of the -naphthylamine and sulphnilic acid reagents placed under it. A red colour indicates the presence of nitrite. Some reducing powder is sprinkled on the surface of the paper if no red colour forms; if nitrate is present it will be reduced to nitrite and the red colour will form.

Well ripended compost has an abundance of nitrate and only traces of ammonium whereas unripe compost has only ammonium. If both compounds are present it may be that the sample includes ripe and unripe compost or that some mineral fertilizer has been added. If only traces of both are present, an intermediate stage of decomposition is indicated in which most of the nitrogen has been fixed in microbial cells.

This test as well as the nitrogen test gives quick results and can be made at the composting site.

V. Isolation of Cellulolytic Microorganisms

1. Isolation of Cellulotytic Culture from Natural Substrates

Cellulotytic culture can be isolated from natural substarates such as compost, farmyard manure, decomposing plant residues or soil.

Dilutions of the natural substrate of 1:10, 1:50, 1:100, 1:500 and 1:1000 are prepared. Place 1 cm^3 of each dilution to test tube containing 9- 10 cm^3 of Dubos medium.

Dubos Medium

$NaNO_3$	0.5 g
K_2HP0_4	1.0g
$MgSO_4$. $7H_2O$	0.5 g
KCI	0.5 g
$FeSO_4.7H_2O$	0.01 g
Distilled water	1 litre

A long strip of sterile filter paper is inserted in the test tube and fixed between the upper end and the cotton plug. The lower end of the strip just dips in the liquid. The tubes are incubated at 28±2°C (room temperature) and periodic observations for mottled areas or decomposition on the paper are made. As mottled or decomposed areas appear, the paper strip is removed and the affected areas macerated in a small quantity of sterile water. Plate these solutions on cellulose-peptone agar and Rose Bengal agar using the serial washing technique (Harley and Waid, 1955; Warcup, 1967) and incubate at 28°C.

Martin's Rose Bengal agar: Glucose 10.0 g, peptone 5.0 g, KH_2PO_4 1.0 g, $MgSO_4.7H_2O$ 0.5 g, Rose Bengal 33.3 mg, Streptomycin sulphate (10% solution) 3 cm^3, agar 15.0g and distilled water one litre.

Cellulose-peptone agar (Allen, 1953): Cellulose, 2.5 g, peptone 0.5 5g, K_2HPO_4 0.2 g, $MgSO_4$ 0.2 g, K_2CO_3 0.4 g, $CaCl_2$ 0.02 g, Fe_2 $(SO_4)_3$ 0.02 g, NaCI 0.02 g and one litre water.

When colonies of microorganisms develop, isolate them in pure culture for further testing.

2. Enrichment Technique

Prepare cellulose pellets from powdered cellulose adhered to glass beads. The pellets are then buried in sieved soil contained in a plastic bag or directly into a natural substrate material such as rice or wheat straw shredded into length of about 2-5 cm. Bring the moisture content to 50% of the water holding capacity of the material used and incubate at 28°C. After incubating for about a month retrieve the pelletes from the soil or straw and make initial isolations on cellulose peptone agar and Rose Bengal agar. Individual bacteria and fungal colonies can be isolated from the plates and screened for cellulolytic activity.

The isolation of fungi, bacteria and actinomycetes can also be done by enrichment technique using appropriate media.

Hans medium for bacteria: K_2HPO_4 0.5 g, KH_2PO_4 0.5 g, $(NH_4)_2$ SO_4 1.0 g, $MgSO_4.7H_2O$0.1 g, $CaCl_2$ 0.1 g, NaCl 6.0 g, yeast extract 0.1 g, cellulose (filter paper strips for enrichment) 10.0 g, water 1 litre, pH adjusted to 7.0.

Asparagine medium for fungi: $(NH_4)_2$ SO_4 0.5 g, L-asparagine 0.5 g, K_2HPO_4 1.0 g, KCI 0.5 g, $MgSO_4.7H_2O$ 0.2 g, $CaCl_2$ 0.1 g, yeast extract 0.5 g, cellulose (filter paper strips) 10.0g, distilled water 1 litre; pH 6.2.

Kenknight medium for actinomycetes: K_2HPO_4 1.0 g, $NaNO_3$ 0.1 g, KCI 0.1 g, $MgSO_4.7H_2O$ 0.1 g, cellulose (filter paper strips) 10.0g, distilled water 1 litre ; pH 7.0-7.2.

The inoculation from the appropriate samples are done in the respective media having immersed filter paper strips for each group of organisms. After ten days incubation at 30 ± 20°C, the filter paper and a few cm^3 of broth are transferred into freshly prepared and autoclaved broth. After repeated (4-5) such transfers, the medium will become sufficiently enriched for plating on an appropriate medium.

3. Isolation of Thermophilic Organisms

To isolate thermophilic organisms samples are collected from compost heaps during the thermophilic stage and incubated at temperatures ranging from 45° to 60°C for two or three days; they are then plated in the usual manner. Incubate the prepared plates at different temperatures between 40 and 55°C and observe the development of colonies. Isolate the organisms in pure culture and maintain at their respective optimum temperature.

4. Screening Organism for Decomposition of Cellulolytic Materials

The isolated organisms can be tested for their, cellulolytic activity to produce cellulose enzyme using pure cellulose, filter paper, wheat or rice straw as substrates. The media Czapek, Aspragagine and Mandel and Reese can be used for determining the extent of cellulase production by fungi. Other media are used for screening cellulolytic bacteria and actinomycetes.

Czapek medium: $NaNO_3$ 2.0 g, KCI 0.5 g, K_2HPO_4 1.0g, $MgSO_4.7H_2O$ 0.5 g, $FeSO_4.7H_2O$ 0.01 g, cellulose 10.0 g, distilled water 1 litre; pH 6.4- 7.0

Mandel and Reese medium (1957): Proteose peptone 1.0g, $(NH_4)_2SO_4$ 1.4 g, KH_2PO_4 2.0 g, urea 0.3 g, $MgSO_4.7H_2O$ 0.3 g, $CaCl_2$ 0.3 g, $MnSO_4.H_2O$ 1.6 mg, $ZnCl_2$ 1.7 mg, cellulose 10.0 g, distilled water 1 litre; pH 5.3.

Medium for bacteria (Dickermann and Staff, 1951): K_2HPO_4 0.8 g, KH_2PO_4 0.2 g, $MgSO_4.7H_2O$ 0.2 g, NaCI 0.2 g, Na NO_3 0.1 g, yeast extract 0.01 g, cellulose 20.0g, distilled water 1 dm^3.

Aliquots of 100 cm^3 of each medium are placed in 250 cm^3 conical flasks and autoclaved at 121°C for twenty minutes. They are then inoculated with selected organisms and incubated on rotary shakers at 30 ±1°C.

20 cm^3 samples are then transferred aseptically. The unused cellulose and mycelium are separated by centrifuging at 10000 rpm for twenty minutes. One percent merthiolate at a rate of 1 cm^3 per 100 cm^3 of extract, can be added to clear the light yellow supernatant liquid and the extra-cellular crude enzyme can be stored in a refrigerator without appreciable loss of enzyme activity. The activity can be measured by the release of reducing sugar per unit weight of the substrate.

The assay mixture consists of 2 cm^3 of enzyme solution in 18 cm^3 of 1% carboxy-methyl cellulose in McIlvaine's citrate-phosphate buffer (pH 6.0). The reaction mixture is incubated at 40°C for 60 minute and the reducing sugar as glucose are estimated using the method of Nelson (1944).

Procedure

Copper reagent A	Na_2SO_4 anhyd.	200g made to 1 litre with distilled
	Na_2CO_3 anhyd.	25 g water
	Na-K-tartrate	25 g
	$NaHCO_3$	20 g
Copper regent B	15 g $CuSO_4$. $5H_2O$ made to 100 cm^3 in distilled water and acidified with one or two drops of conc. H_2SO_4	
Arsenomolybdate reagent	$(NH_4)_2$ MoO_44H_2O 25 g dissolved in 450 cm^3 of distilled water and 21 cm^3 of conc. H_2SO_4 added. 3 g Na_2 $HASO_4$. $7H_2O$ dissolved in 25 cm^3 of water. Both solutions are mixed and placed in an incubator at 37°C for 48 hours.	
Standard glucose reagent	1.00 g pure, anhydrous glucose is dissolved in a solution of 0.2% benzoic acid and made to 100 cm^3. This solution gives 10 mg glucose cm^3. Aliquots of 0.25, 0.50, 0.75, 1.00, 1.25 and 1,50 cm^3 of the stock solution are diluted to 100 cm3 with 0.2% benzoic acid to give solutions having 25, 50, 75, 100, 125 and 150 ug cm^3 of glucose.	

To 1 cm^3 aliquots of each standard glucose solution, 2 cm^3 of alkaline copper reagent mixture (25 cm^3 of copper reagent A+ 1 cm^3 of reagent B mixed just before use) are added. Mix and boil the solutions for 20 minutes on a steam bath. Cool and add 2 cm^3 arsenomolybdate reagent; shake. Dilute to 25 cm^3 with water and read the blue colour at 540 nm. Prepare a standard curve by plotting optical density against concentration.

The reducing sugars of the experimental samples can then be determined by the same procedure and are an indicative of cellulase activity.

Garrett's method (1962) for testing cellulolytic activity: Weigh wads of six to ten oven-dried filter paper (Whatman No. 1/3, 5.5 cm) and place them in 250 cm^3 conical flasks. Add 15 cm^3 aliquots of mineral solution containing 5.0 g $NaNO_3$, 1.0 g K_2HPO_4, 0.5 g $MgSO_4.7H_2O$, 0.01 g Ca $(NO_3)_2$. $4H_2O$, 0.001 g $FeCl_3$, 0.5 g yeast extract to 1 litre in water; pH 6.5.

The flasks are sterilized at 15 psi pressure for 20 minutes. Then inoculate each flask with a 6 mm disc of the fungus to be tested, in triplicate. Set up controls using discs of blank agar as inocula.

Incubate the flasks at room temperature for 4 weeks (if thermophiles are being tested incubate under thermophilic conditions). Dry the contents of the flasks to constant weight at 105°C and calculate the mean loss in dry weight. This loss represents only that portion of the cellulose actually respired by the organisms and does not include the further portions converted to microbial mass.

VI. Media and Methods for Biofertlizers

1. Media for Rhizobium Culture

Yeast extract mannitol agar medium (YEMA) Vincent, 1970.

Mannitol	10.0g
K_2HPO_4	0.5 g
$MgSO_4 7H_2O$	0.2 g
NaCl	0.1 g
Yeast extract	1.0 g
Distilled water	1litre
Agar	15.0 g
pH	7 ± 0.2

Adjust the pH at 7+0.2 with N/10 HCI and N/10 NaOH sterilize at 15 lbs psi at 121°C pressure for 30 minutes

(For congo red, yeast extract mannitol agar medium add 2.5 ml of 1 per cent solution of congo red per litre of yeast extract mannitol broth i.e. before the addition of agar and sterilize).

Glucose peptone agar

Glucose	5.0 g
Peptone	10.0g
Agar	15.0g
Distilled water	1litre
Bromocresol purple (1.0% in ethanol)	10.0 ml
(Sterilize at 115° C for 20 minutes)	

Hofer's alkaline broth

Mannitol	10.0g
K_2HPO_4	0.5 g
$MgSO_4 7H_2O$	O 0.2 g

NaCl	0.1 g
$CaCO_3$	0.05 g
Yeast extract	1:0 g
Distilled water	1 litre

Adjust the pH at 11-13 and autoclave at 15 psi and 121°C for 30 minutes.

2. For studying root nodulation in legumes in aseptic condition, the following method can be followed Gibson's agar tube method

For studying root nodulation, the seedling nutrient agar of the following composition is used. The following nitrogen free commonly used media serve an example for preparation of seedling agar medium whose compositions are detailed below:

Jensen's medium (1942)		**Thornton's medium (1930)**	
$CaHPO_4$	1.0g	$Ca_3(PO_4)_3$	2.0 g
K_2HPO_4	0.2 g	K_2HP0_4	0.5 g
$MgSO_4 . 7H_2O$	0. 2 g	$MgSO_4 . 7H_2O$	0.2 g
NaCl	0.2g	NaCl	0.1 g
$FeCI_3$	0.1 g	$FePO_4$	1.0 g
Agar	15.0	$FeCl_3$	0.01 g
Trace lements	1 ml	Water	1 litre
Distilled Water	1 litre		

1. One litre of stock solution of trace elements is prepared with the following composition : Bo-50 mg, Mn-50 mg, Zn-5 mg, Mo -0.5 mg, Cu-2 mg. Add 1 ml of the stock solution to every 1 litre of either of the two media.
2. The pH of the media should be adjusted to 6.5 to 7.0 with N/10 Hclor N/10 NaOH. The media should be shaken well.
3. Autoclave at 15 psi and at 121°C for 30 minutes.

Sterilization of seeds before test: Seeds are rinsed with 55% ethanol. Then seeds are immersed for 3 minutes in 0.2% $Hg\ Cl_2$, acidified with 5 ml/l conc. HCI. After this treatment, seeds are washed with sterile water 5 times to get rid of the chemicals.

Germination of seeds: The seeds are germinated in folds of moist filter paper or preferably in plain agar (with water) in petri dishes. The petridishes are later inverted to provide seedlings with uniform straight roots.

The seedling is then carefully placed in the Gibson's tube so that the emerging roots just touch the surface of the agar slope. Add quarter strength sterile seedlings solution through ancther hole in the cap within 12 mm of the top of each tube.

The inoculum broth may be added through the hole made on the aluminum cap for watering and ventilation. The tubes are incubated at 30°C under artificial light (2,000 lux) No watering need to be done upto 2-3 days. After 4-5 days, quarter strength seedling solution should be added after 2-3 days intervals. After 7-10 days, nodulation in plant roots may be examined. After 4-5 weeks plant dry weight can be taken and increase in dry weight over control should be recorded which represents the nitrogen fixing efficiency of the inoculated *Rhizobium* strains.

3. Composition of Hydrogen uptake Medium

Part A:	Sodium gluconate -0.5 g, Sodium glutamate - 0.5 g, Yeast extract - 0.1 g, Distilled water -1 litre, Agar 15.0 g.
Part B:	Fe EDTA - 28.0 mg, H_3BO_3-3.0 mg, Zn SO_4. 7 H_2O - 2.0 mg, Mn SO_4 H_2O -10.0 mg, $NaMoO_4$ $2H_2O$ - 0.25 mg, $CuSO_4$ $5H_2O$-0.04 mg. $CoCl_2$. $6H_2O$ -0.025 mg, KI - 0.78 mg.
Part C:	KH_2PO_4 H_2O-0.2 g, NaH_2 PO_4H_2O-0.03 g,
Part D:	$CaCl_2$ $2H_2O$-150 mg $MgSO_{47}H_20$-250 mg.
Part E:	Sucrose 0.5 g, L-arabinose - 1.0 g, Inositol 100 mg, Thiamin * hydrochloride - 10 mg
Part F:	Nicotinic acid - 1.0 mg, Pyridoxan - HCI - 1.0 mg

Part B is added to part A (without agar) and mixed thoroughly. pH is adjusted to 7.0 and agar added to the medium (A+B) is then sterilized. Part C,D,E,F solutions are prepared by taking measured amount of distilled water from part A and by dissolving the ingredients mentioned, above.

Part C and D are autoclaved separately and then added (to A+B) at the time of pouring along with part E and F (sterilized using membrane filter). The plates containing the agar medium are incubated at 28°C.

The plates (containing the spots of the rhizobial cells on the above medium +TTC) are incubated at 28°C and the reduction of the TTC dye is observed upto 6 days.

Another modification of the TTC reduction method for the detection of hydrogen uptake system is the TTC reduction in the presence of "respiratory inhibitors".

In this method a thin suspension of cells of rhizobial strain in distilled water in spotted on the hydrogen uptake medium as described above and the plates are incubated till a visible growth is observed under a gas atmosphere of 10% H_2, 5% CO_2, 1% O_2 and rest N_2. To each plate 1ml of respiratory inhibitor (composition given below) solution is added and allowed *to* react for 5 min. Then 1 ml of 0.3% TTC (w/v) is added and spread so as to cover all the spots. The plates are again incubated under the above mentioned gas atmosphere and the reduction of the dye is checked after 30 minutes.

The composition of the respiratory inhibitors solution (mg per 100 ml) include: lodoacetate - 2975.33, Malonic acid - 1665.00, KH_2PO_4- 544.35, Hg Cl_2- 40.70. pH is adjusted to 5.5 with KOH.

4. Methylene Blue Reduction Method

Hydrogen uptake positive strains can also be detected by this method. Hydrogen uptake plates are incubated as described above with the difference that TTC is replaced with 2 ml of 10 nm methylene blue solution. The plates are incubated under the earlier mentioned gas atmosphere and the dye reduction is recorded.

Hydrogen uptake positive strains of *Rhizobium* reduce TTC to form pink coloured formzan without the presence of respiratory inhibitors. The reduction, should take place within 72 hours. Reduction in presence of respiratory inhibitor is more sure test than without the inhibition. Though methylene blue reduction is also not very reliable test for Hupt character yet reduction within 96 hours gives a good indication of the strain to be Hup^+.

5. Media for counts of Azospirillum (Okon *et al.*, 1977) Semi-solid malate medium

Sodium malate	5.0 g
KH_2PO_4	0.4 g
K_2HPO_4	0.1 g
$MgSO_4$	0.3 g
Na C12	0.1 g
Ca Cl_2	0.02 g
Fe CI_3	0.01 g
Na_2 MoO_4	0.002 g
Bromothymol blue (0.5% alcoholic solution)	1.0 g
Agar	1750 g
Distilled water	1litre
pH	6.8

References

Allen, O. N. 1957. Experiments in Soil Microbiology, 3rd ed. Burgess Publ. Co. Minneaplis, Minn.

Becking, J. A. 1959. Pl. Soil 11: 193-206.

Bunt, J. S. and Rovira, A. D. 1955. Microbiological studies of some sub-antartic Soils. J. Soil Sci. 6: 119.

Dickermann, J. M. and Staff, J. L. 1951. J. Bacteriol. 62: 133-134.

Garrett, S. D. 1962. Trans. British Mycol. Soc. 45: 115.

Gaur, A.C. 1990 Phosphate Solubilizing Microorganisms as Biofertilizer. Omega Scientific Publishers, New Delhi, p 148.

Harley, J. L. and Waid, J. S. 1955. Trans Brit. Mycol. 38: 104.

Hesse, P. R. 1971. A text book of soil chemical analysis John Murray, London. pp 528.

Jensen, H. L. 1940. Note on the biology of *Azotobacter*. Proc. Sòc. Appl. Bacteriol. 14: 89.

Jensen, H. L. 1942. Proc. Linn. Soc, N. S. W. 57: 98-108.

Lossin, R. D. 1970. Compost Sci. 11: 16.

Mandel, M. and Reese, R. T. 1957. J. Bacteriol. 37: 269-278.

Martin, J. P. 1950. Soil Sci. 69:215.

Nelson, N. 1944. J. Biol. Chem.. 153: 375.

Okon, Y., Albrecht, S. L. and Burris, R. H. 1977. Appl. Environ. Microbiol. 33:85-88.

Pikovskaya, R. I. 1948. Mikrobiologiya 17: 362-370.

Spohn, E. 1969. Compost Sci, 10:24.

Spohn, E. 1978. Compost Sci. Land utilization 19: 26

Thornton, H. G. 1922. Ann. Appl. Biol. 9: 241.

Thornton, H. G. 1930. Ann. Bot. 44: 385-392.

Vincent, J. M. 1970. A Manual for Practica! Study of Root Nodule Bacteria, Blackwell Scientific Publications, Oxford and Edinburgh.

Warcup, J. H. 1967. Fungi in soil. In soil Biology. Eds. Burges and Raw. Acad. Press. London p. 532.

Colour Plates

Chapter 3: Agricultural Residues, Tree Wastes Green Manure and Urban Wastes

Plate 6: Composting of city wastes

Plate 7: Semi-mechanized compost plant system

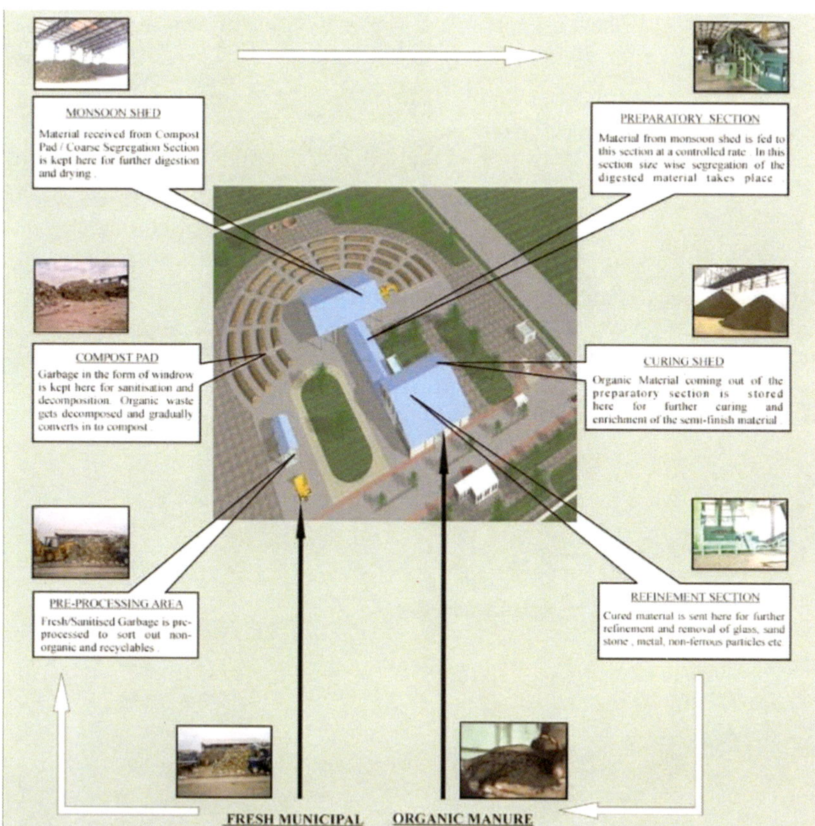

Plate 8: Compost prepared at semi-mechanized system

Chapter 10: Composting of Organic Wastes

Plate 15: Composting of high C/N ratio materials in pits

Plate 16: Rapid composting of farm wastes in cemented compost pits at IARI, New Delhi

Plate 17a: Effect of NK without phosphorus on rice

Chapter 18: Phosphorus Solubilizing Microorganisms as Biofertilizer

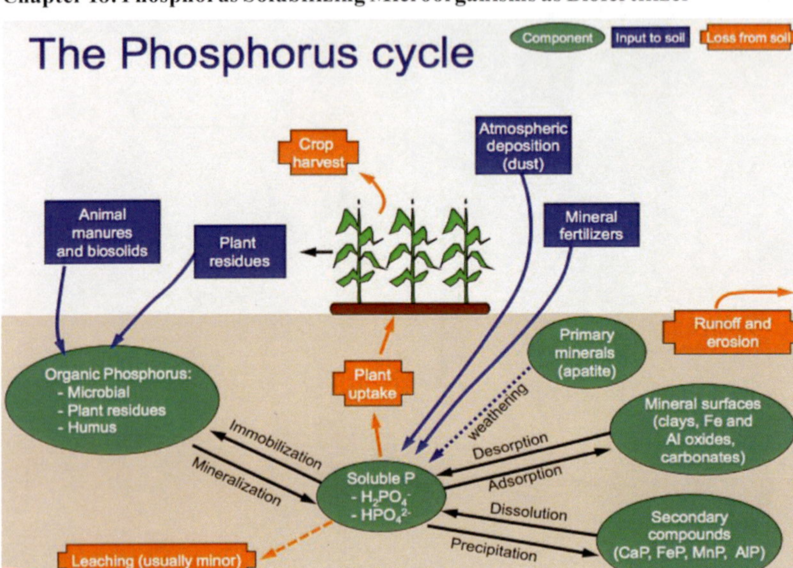

Fig. 5: Soil phosphorus cycle

Index

T

U

V

W